SYMMETRIC PROPERTIES OF REAL FUNCTIONS

PURE AND APPLIED MATHEMATICS

A Program of Monographs, Textbooks, and Lecture Notes

MONOGRAPHS AND TEXTBOOKS IN
PURE AND APPLIED MATHEMATICS

53. *C. Sadosky*, Interpolation of Operators and Singular Integrals (1979)
54. *J. Cronin*, Differential Equations (1980)
55. *C. W. Groetsch*, Elements of Applicable Functional Analysis (1980)
56. *I. Vaisman*, Foundations of Three-Dimensional Euclidean Geometry (1980)
57. *H. I. Freedan*, Deterministic Mathematical Models in Population Ecology (1980)
58. *S. B. Chae*, Lebesgue Integration (1980)
59. *C. S. Rees et al.*, Theory and Applications of Fourier Analysis (1981)
60. *L. Nachbin*, Introduction to Functional Analysis (R. M. Aron, trans.) (1981)
61. *G. Orzech and M. Orzech*, Plane Algebraic Curves (1981)
62. *R. Johnsonbaugh and W. E. Pfaffenberger*, Foundations of Mathematical Analysis (1981)
63. *W. L. Voxman and R. H. Goetschel*, Advanced Calculus (1981)
64. *L. J. Corwin and R. H. Szcarba*, Multivariable Calculus (1982)
65. *V. I. Istrăţescu*, Introduction to Linear Operator Theory (1981)
66. *R. D. Järvinen*, Finite and Infinite Dimensional Linear Spaces (1981)
67. *J. K. Beem and P. E. Ehrlich*, Global Lorentzian Geometry (1981)
68. *D. L. Armacost*, The Structure of Locally Compact Abelian Groups (1981)
69. *J. W. Brewer and M. K. Smith, eds.*, Emily Noether: A Tribute (1981)
70. *K. H. Kim*, Boolean Matrix Theory and Applications (1982)
71. *T. W. Wieting*, The Mathematical Theory of Chromatic Plane Ornaments (1982)
72. *D. B. Gauld*, Differential Topology (1982)
73. *R. L. Faber*, Foundations of Euclidean and Non-Euclidean Geometry (1983)
74. *M. Carmeli*, Statistical Theory and Random Matrices (1983)
75. *J. H. Carruth et al.*, The Theory of Topological Semigroups (1983)
76. *R. L. Faber*, Differential Geometry and Relativity Theory (1983)
77. *S. Barnett*, Polynomials and Linear Control Systems (1983)
78. *G. Karpilovsky*, Commutative Group Algebras (1983)
79. *F. Van Oystaeyen and A. Verschoren*, Relative Invariants of Rings (1983)
80. *I. Vaisman*, A First Course in Differential Geometry (1984)
81. *G. W. Swan*, Applications of Optimal Control Theory in Biomedicine (1984)
82. *T. Petrie and J. D. Randall*, Transformation Groups on Manifolds (1984)
83. *K. Goebel and S. Reich*, Uniform Convexity, Hyperbolic Geometry, and Nonexpansive Mappings (1984)
84. *T. Albu and C. Năstăsescu*, Relative Finiteness in Module Theory (1984)
85. *K. Hrbacek and T. Jech*, Introduction to Set Theory: Second Edition (1984)
86. *F. Van Oystaeyen and A. Verschoren*, Relative Invariants of Rings (1984)
87. *B. R. McDonald*, Linear Algebra Over Commutative Rings (1984)
88. *M. Namba*, Geometry of Projective Algebraic Curves (1984)
89. *G. F. Webb*, Theory of Nonlinear Age-Dependent Population Dynamics (1985)
90. *M. R. Bremner et al.*, Tables of Dominant Weight Multiplicities for Representations of Simple Lie Algebras (1985)
91. *A. E. Fekete*, Real Linear Algebra (1985)
92. *S. B. Chae*, Holomorphy and Calculus in Normed Spaces (1985)
93. *A. J. Jerri*, Introduction to Integral Equations with Applications (1985)
94. *G. Karpilovsky*, Projective Representations of Finite Groups (1985)
95. *L. Narici and E. Beckenstein*, Topological Vector Spaces (1985)
96. *J. Weeks*, The Shape of Space (1985)
97. *P. R. Gribik and K. O. Kortanek*, Extremal Methods of Operations Research (1985)
98. *J.-A. Chao and W. A. Woyczynski, eds.*, Probability Theory and Harmonic Analysis (1986)
99. *G. D. Crown et al.*, Abstract Algebra (1986)
100. *J. H. Carruth et al.*, The Theory of Topological Semigroups, Volume 2 (1986)
101. *R. S. Doran and V. A. Belfi*, Characterizations of C^*-Algebras (1986)
102. *M. W. Jeter*, Mathematical Programming (1986)
103. *M. Altman*, A Unified Theory of Nonlinear Operator and Evolution Equations with Applications (1986)
104. *A. Verschoren*, Relative Invariants of Sheaves (1987)
105. *R. A. Usmani*, Applied Linear Algebra (1987)
106. *P. Blass and J. Lang*, Zariski Surfaces and Differential Equations in Characteristic $p > 0$ (1987)
107. *J. A. Reneke et al.*, Structured Hereditary Systems (1987)

Additional Volumes in Preparation

SYMMETRIC PROPERTIES OF REAL FUNCTIONS

Brian S. Thomson

Simon Fraser University
British Columbia, Canada

CRC Press
Taylor & Francis Group
Boca Raton London New York

CRC Press is an imprint of the
Taylor & Francis Group, an **informa** business

CRC Press
Taylor & Francis Group
6000 Broken Sound Parkway NW, Suite 300
Boca Raton, FL 33487-2742

First issued in paperback 2019

© 1994 by Taylor & Francis Group, LLC
CRC Press is an imprint of Taylor & Francis Group, an Informa business

No claim to original U.S. Government works

ISBN-13: 978-0-8247-9230-5 (hbk)
ISBN-13: 978-0-367-40203-7 (pbk)

**Visit the Taylor & Francis Web site at
http://www.taylorandfrancis.com**

**and the CRC Press Web site at
http://www.crcpress.com**

Library of Congress Cataloging-in-Publication Data

Thomson, Brian S.
    Symmetric properties of real functions / Brian S. Thomson.
       p. cm. -- (Monographs and textbooks in pure and applied mathematics; 183)
    Includes bibliographical references and index.
    ISBN: 0-8247-9230-0 (acid-free)
    1. Symmetric functions.   I. Title.   II. Series.
QA212.T49   1994
515'.83--dc20

                                               94-14911
                                                  CIP

Preface

The title of this work is taken from the opening paragraph of an article by Mary Weiss [306]:

> "It is a familiar fact that symmetric properties of real functions play an important role in a number of problems. This is particularly true of problems in the theory of trigonometric series. The first symmetric derivative (Lebesgue's derivative), the second symmetric derivative (Schwarz's derivative) and their generalizations are familiar notions in the theory of trigonometric series, and the investigation of their properties is a legitimate topic in Real Variables."

By the symmetric properties of real functions we shall mean properties arising from the expression

$$f(x + t) = \frac{1}{2}[f(x + t) + f(x - t)] + \frac{1}{2}[f(x + t) - f(x - t)]$$

which defines the even and the odd parts of the function f at the point x. For example the continuity and differentiability properties of f can be analyzed in some cases by studying those properties separately in the odd and even parts. Thus we might be led to an investigation of "even continuity" and "odd continuity", and of "even differentiability" and "odd differentiability". This then focuses in on the expressions

$$f(x + t) - f(x - t) \quad \text{and} \quad f(x + t) + f(x - t) - 2f(x).$$

These are two aspects of a single entity; there are many parallels and no study of just one side is quite complete without the other.

It is well known by now that such ideas enjoy an important role in a variety of problems. In the study of trigonometric series in particular various symmetric derivatives come into play in an entirely natural way. The investigation of such notions has become a separate and recognizable topic of study in Real Variables. Larson [173], in an influential survey article, has called it "symmetric real analysis".

The history of this topic stretches back a century and a half and includes contributions from the most important figures in analysis. In every decade of this century one can find articles addressing such ideas. Perhaps the most fertile periods have been stimulated by three main influences. Above all Zygmund, who has impacted on so much of the analysis of this century, had long maintained an interest in the differentiability properties of real functions. He and his students have made major advances especially with regard to the symmetric structure of functions. The early years of *Fundamenta Mathematicae* saw some considerable interest in symmetric problems and dozens of papers and problems addressed such issues; many of the most famous analysts of the period asked or answered symmetric questions in its pages. More recently in my own time the *Real Analysis Exchange* once more introduced a host of readers to this symmetric literature and its unsolved problems; that the last two decades has produced so much new material in this topic is largely due to the energy and vitality of this journal.

In order to preserve the spirit of the early researches I have tried to present an account that pays attention to the history of the subject. Thus the early chapters present material that is not necessarily stated in the greatest generality and that is proved in mostly the original way. A reading of only the first two chapters affords then a classical view of the subject and its problems together with a perspective on how the ideas and techniques first evolved. Later material will tidy up the loose ends, prove the best versions of theorems and introduce simpler and more powerful techniques.

The background expected of the reader is just the usual assortment of real analysis concepts such as measure, category, density. The ideal reader has already become familiar with the arcana of the ordinary and approximate derivatives through such standard treatises as Saks [251] and Bruckner [18]. The present work is intended to extend many of these ideas to the symmetric setting where mostly different techniques and methods are required. While parts of the early chapters and selections from the later may serve to introduce the curious reader to this subject the main intent has been to provide a reference work for researchers in this area so that they may more easily find out what is known and where it has appeared. For this reason it is hoped that what is presented represents a more or less complete picture of the work done.

Chapter 1 begins the study by centering on the classical symmetric derivatives. The most important result presented here is due to Khintchine and is given as Theorem 1.23 with its original proof. Rather more elementary material appears too in order to fill out the necessary background to our study. In particular we learn most of the basic properties of the first and second symmetric derivatives in this chapter.

Chapter 2 develops the continuity properties of functions satisfying some kind of symmetric growth condition. The centerpiece of this chapter is a single important classical study of symmetric properties of real functions by Charzyński. This represents the first significant study of the continuity properties of functions that satisfy some kind of even or odd symmetry condition.

The rest of this chapter continues this theme in a number of its forms. Taken together with Chapter 1 this might be considered a brief introduction to the world of symmetric derivatives, symmetric continuity and like concepts. The reader who wishes to go on from here will find a more comprehensive treatment of these kind of ideas.

In Chapter 3 a variety of covering results are obtained that help to unify and clarify the arguments used in this subject.

Chapter 4 begins a systematic study of the regularity properties of functions that satisfy some kind of even symmetric condition. A large literature is surveyed mostly with proofs.

Chapter 5 is devoted to monotonicity and convexity theorems as they arise from the first or second symmetric derivative.

Chapter 6 does for the odd symmetry conditions what Chapter 4 has done for the even conditions. This is a rather complete survey of what is known about the regularity properties of functions that satisfy some kind of odd symmetric growth condition.

Chapter 7 returns to the derivative itself. We have defined a number of symmetric derivatives and this material attempts to survey the nature of the results so far known.

Chapter 8 introduces the variational theory associated with the first and second symmetric derivatives.

Chapter 9 concludes the story with an account of the various symmetric integrals. A primary motivation is the solution of the coefficient problem for trigonometric series.

We include an appendix as a location for material that is needed in the main body of the work but which would be cumbersome to include directly. It is hoped that for the most part the present material is sufficiently inclusive to save the reader numerous trips to the library for source material. Even so there have been plenty of places throughout where proofs have had to be omitted and the original sources should be consulted. Of course, for any serious student of this topic, we should insist that many of the original authors (notably Charzyński, Denjoy, Khintchine, Marcinkiewicz, Mazurkiewicz, Sierpinski and Zygmund) be read in any case!

The problems that are included at the end of the book are, apparently, at the time of writing open. Some of these have an impeccable and legitimate pedigree and will undoubtedly require new methods and insights. Many are just questions that appeared to be unanswered at the time of writing; they may succumb to standard methods.

Our subject has had a long history and many results are now known in their best form. There is much of this subject which can be enshrined now as "knowledge". Thus this monograph is not a report on a rapidly changing field; even so, however, a glance at the bibliography will show that research carries on in this area and it is clear that research will always continue even as the job of mining the results becomes tougher and more demanding. As far as I can tell it is a report on the current state of our knowledge in this area and should not obsolete itself except in small portions.

My main thanks are due to the legions of real analysts who have contributed to and deepened our understanding of this subject. More narrowly I would like to thank readers of preliminary versions for their suggestions for improvements, in particular Marshall Ash, Michael Evans, Chris Freiling, Paul Humke and Lee Larson all of whom have been major players in this area and whose comments have been invaluable. My thanks as well go to Hongjian Shi for a careful and detailed reading of several chapters. The manifold faults that remain are exclusively mine.

Brian S. Thomson

Contents

SYMMETRIC PROPERTIES OF REAL FUNCTIONS

1
The Derivative

1.1 Introduction

The first and second symmetric derivatives of a function f are defined by the expressions

$$\text{SD } f(x) = \lim_{t \to 0} \frac{f(x+t) - f(x-t)}{2t}.$$

$$\text{SD}_2 \, f(x) = \lim_{t \to 0} \frac{f(x+t) + f(x-t) - 2f(x)}{t^2}.$$

The first of these is, on occasion, called *Lebesgue's derivative* and the second *Riemann's derivative* or *Schwarz's derivative*. We begin our work with some classical studies of these derivatives. Although our concerns are broader than this, these studies provide us the opportunity to present many of the main themes and will lead us into the subject.

The first question that would obviously arise in a study of the symmetric derivatives is to ask how far properties of ordinary derivatives extend to symmetric derivatives. This is a natural starting point for our investigation and is the starting point taken by Khintchine [151] in 1927. This leads, by relatively familiar methods, to a number of elementary observations. We supplement this with several other important classical, but essentially elementary, results of Schwarz, Mazurkiewicz [197] and [198], and Auerbach [9]. The applications to trigonometric series that appear in Sections 1.4 and 1.7 remain among the central reasons why the subject has attracted attention.

The main result in Khintchine's fundamental paper, the first really technically interesting result in the subject, appears in Section 1.10. This asks the question: does the existence of the symmetric derivative of a function f on a set E say anything about the ordinary differentiability properties of f on that

1

set? The answer to this question will introduce us to some of the more subtle geometric arguments needed to study effectively this derivative in particular and symmetric properties of real functions in general.

1.2 Even and Odd Properties

1.2.1 The Even and Odd Parts of a Function

A function f is *even* if $f(t) = f(-t)$ everywhere; a function f is *odd* if $f(t) = -f(-t)$ everywhere. These conditions represent symmetries in the graph of the function, in the first case symmetry about the y–axis and in the second symmetry about the origin.

At any point x and for any function f we can study these symmetries in the ordinary difference

$$\Delta f(x, t) = f(x + t) - f(x).$$

This difference can be written as

$$f(x + t) - f(x) = \frac{f(x + t) - f(x - t)}{2} + \frac{f(x + t) + f(x - t) - 2f(x)}{2}.$$

We shall employ the notations

$$\Delta^1_s f(x, t) = f(x + t) - f(x - t)$$

and

$$\Delta^2_s f(x, t) = f(x + t) + f(x - t) - 2f(x)$$

occasionally to represent these odd and even increments. They are known as the *first symmetric difference* and *second symmetric difference* for the function f.

These expressions appear in a variety of studies. For example the connection of these ideas with trigonometric series is by now well known. For instance if

$$a_0/2 + \sum_{n=1}^{\infty} (a_n \cos nx + b_n \sin nx)$$

is the Fourier series of a function f then the convergence behavior or the summability behavior of the series at a point x or of the differentiated series (differentiated an even number of times) depends very much on properties of the function

$$t \to f(x + t) + f(x - t) - 2f(x)$$

near the point $t = 0$, that is to say on properties of the even difference. For a well known example, a theorem of Dini (see [313, Theorem 6.1, p. 52]) asserts that the Fourier series of a function f converges at a point x to the sum $f(x)$ provided that the integral

$$\int_0^\pi \frac{|f(x+t) + f(x-t) - 2f(x)|}{t}\, dt$$

is finite.

The odd difference of the function plays a dual role. For example the behavior of the conjugate Fourier series for f or of the differentiated series (differentiated an odd number of times) depends on the odd part of f at x, that is depends on properties of the function

$$t \rightarrow f(x+t) - f(x-t)$$

near the point $t = 0$. Again, for example, the result analogous to that of Dini cited above is that of Pringsheim (again see [313, Theorem 6.1, p. 52]) asserting that the conjugate Fourier series of a function f converges at a point x provided that the integral

$$\int_0^\pi \frac{|f(x+t) - f(x-t)|}{t}\, dt$$

is finite.

An accumulation of results of this kind has led to a study of such properties for their intrinsic interest. This offers some preliminary motivation for a study of even and odd properties of functions. In this chapter we begin by an investigation of the symmetric derivatives and follow where they lead. The even and odd structure of these two derivatives is, of course, obvious.

1.2.2 Even and Odd Properties of a Function

The ordinary difference can be written usefully as

$$\Delta f(x,t) = \tfrac{1}{2}\Delta^1_s f(x,t) + \tfrac{1}{2}\Delta^2_s f(x,t)$$

using these odd and even differences. It is natural in many contexts to examine the continuity and differentiability properties of f by studying those properties in these two parts. This follows a well known and useful tradition in analysis resulting in notions such as right and left hand continuity, upper and lower semicontinuity, right/left and upper/lower derivates (Dini derivatives) and so on. The goal is to split a property into smaller conditions allowing separate analysis. Thus we might be led to an investigation of "even continuity" and "odd continuity", and of "even differentiability" and "odd differentiability".

The continuity of a function f at a point x, that is the requirement that $\Delta f(x,t) \rightarrow 0$ as $t \rightarrow 0$, is equivalent to requiring both $\Delta^1_s f(x,t) \rightarrow 0$ and $\Delta^2_s f(x,t) \rightarrow 0$. This might split into two "semi-continuity" conditions. It seems natural to say that a function f is *even continuous* at x if $\Delta^2_s f(x,t) \rightarrow 0$ and is *odd continuous* at x if $\Delta^1_s f(x, t) \to 0$. The terminology, however, has not evolved in any such systematic manner. Continuity of the odd part of $\Delta f(x, t)$ is known as *symmetric continuity* at x while f is said to be *symmetric* at x if the even part is continuous. In Chapter 2 we study extensively such notions.

Similarly, differentiability of a function f at a point x is equivalent to the existence of both of the limits

$$\lim_{t \to 0} \frac{\Delta^1_s f(x, t)}{t} \quad \text{and} \quad \lim_{t \to 0} \frac{\Delta^2_s f(x, t)}{t}.$$

The derivative of the odd part is exactly the symmetric derivative SD $f(x)$ of the function f at the point x. On the other hand differentiability of the even part is easily checked to be equivalent to the requirement that $\Delta^2_s f(x, t) = o(t)$, i.e. that

$$\lim_{t \to 0} \frac{f(x + t) + f(x - t) - 2f(x)}{t} = 0,$$

and this condition is usually called the *smoothness* of the function f at the point x. (This notion was first considered by Riemann in his famous memoir on trigonometric series; the terminology is due to Zygmund.) The second symmetric derivative SD$_2$ $f(x)$ now can be viewed as a kind of second order even derivative.

In order to catalogue our many results in this work we shall use the following language. By an *odd property* of a function f we shall mean some property expressed in terms of the odd symmetric difference $\Delta^1_s f(x, t)$ and by an *even property* of a function f we shall mean some property expressed in terms of the even symmetric difference $\Delta^2_s f(x, t)$.

Note that a function is even if and only if its odd part vanishes. Thus the even symmetric structure of a function is expressed as an odd property. Similarly the odd symmetry of a function is expressed as an even property. Thus in general we can think of the study of the even/odd properties of a function as promoting the odd/even symmetric structure of the function.

1.2.3 Higher Order Symmetric Differences

The two symmetric differences $\Delta^1_s f(x, t)$ and $\Delta^2_s f(x, t)$ can be considered as the odd and even parts of $\Delta f(x, t)$. More commonly they are referred to as the first and second symmetric differences of f at x and our notation reflects this.

The terminology indicates the fact that the second difference can be viewed as an iteration of the first. Write

$$g(x) = \Delta^1_s f(x, h) = f(x + h) - f(x - h).$$

A direct and elementary computation shows that

$$\Delta^1_s g(x, h) = g(x + h) - g(x - h)$$
$$= \; = f(x + 2h) + f(x - 2h) - 2f(x) = \Delta^2_s f(x, 2h).$$

One can evidently continue in this fashion forming inductively a hierarchy of symmetric differences. A notation (cf. [313, Vol. II, p. 61]) that places all of these in the same context is to write instead

$$\Delta_s^n f(x, t) = \sum_{i=0}^{n} (-1)^i \frac{n!}{(n-i)!i!} f\left(x + it - \tfrac{1}{2}nt\right).$$

We mostly avoid using any such notation and we write out explicitly the even and odd expressions as needed.

In our study we confine our attention to what might be called the symmetric structure of a function, namely properties expressible in terms of the even and odd differences. There is, by now, a rich literature devoted to higher order symmetric derivatives and higher order smoothness conditions. Since these properties do not directly express properties of the odd and even parts of a function we can restrict the scope of our work by not reporting on this activity.

1.3 Elementary Considerations

The symmetric derivative has been defined as a kind of odd derivative. In any discussion of generalized derivatives, of which the symmetric derivative may be considered one of the simplest, it is natural to ask to what extent the usual rules of derivation apply. The first formal presentation of the symmetric derivative along these lines was given by Khintchine [151] in an early paper in Fundamenta Mathematicae. The elementary observations in this section are from that paper.

A simple example suffices to show that the properties of the symmetric derivative are very different from those of the ordinary derivative. For example the function $f(x) = |x|$ is everywhere symmetrically differentiable with

$$\text{SD}\, f(x) = \begin{cases} 1 & \text{if } x > 0 \\ -1 & \text{if } x < 0 \\ 0 & \text{if } x = 0 \end{cases}$$

and this dispenses with a number of the simpler properties that might be expected. The symmetric derivative can evidently exist when the ordinary derivative does not and the derivative itself need not have the intermediate value property (Darboux property). The characteristic function of any finite set has an everywhere vanishing symmetric derivative, but is not itself constant.

The elementary formula

$$\frac{f(x+h) - f(x-h)}{2h} = \frac{1}{2}\left\{ \frac{f(x+h) - f(x)}{h} + \frac{f(x-h) - f(x)}{-h} \right\}$$

illustrates the symmetric derivative as the average of the two one-sided derivatives at points where both exist. Khintchine [151] sketches an example to show that Rolle's theorem cannot hold: construct a continuous function f that vanishes at 0 and 1, has a positive derivative on $(0, \frac{1}{2})$, a negative derivative on $(\frac{1}{2}, 1)$, and such that $f'(\frac{1}{2}+0) = 0$ and $f'(\frac{1}{2}-0) = 1$. Here although $f(0) = f(1)$ and the symmetric derivative exists everywhere on the interval $(0, 1)$ it never vanishes. This example too shows that symmetric derivatives do not necessarily have the Darboux property.

We can summarize this discussion by the following elementary lemmas.

Lemma 1.1 *If $f'(x)$ exists then so too does $SDf(x)$ and they are equal.*

Lemma 1.2 *If the one-sided derivatives $f'_+(x)$ and $f'_-(x)$ exist then so too does $SDf(x)$ and*

$$SDf(x) = \frac{f'_+(x) + f'_-(x)}{2}.$$

Lemma 1.3 *There is a continuous function f that has a finite symmetric derivative everywhere but such that $SDf(x)$ does not have the Darboux property.*

A more interesting problem, which has subtle and curious answers, is to ask how far monotonicity theorems can be carried to this kind of derivative. If a function f has everywhere a positive symmetric derivative then is the function increasing or nearly so? We shall have occasion to return to this problem again but here we shall reproduce an elementary monotonicity theorem from [151, p. 216] with its original proof; by 1927 when Khintchine's paper appeared this theorem would probably have been known. The proof, it should be noted, is merely a minor modification of a well known proof of a monotonicity theorem for the Dini derivatives (*cf.* [251, pp. 203–204]) and the ideas go back at least to 1878. There is no special use of symmetric arguments here and the proof cannot be considered very informative for us. A full study of monotonicity theorems for the symmetric derivatives appears in Chapter 5.

Theorem 1.4 *A continuous function is necessarily increasing in any interval in which its symmetric derivative exists and is positive.*

Proof. Let us show that for every pair of points x_1 and x_2 in this interval $f(x_1) < f(x_2)$ if $x_1 < x_2$. If not then there are such points with $f(x_1) > f(x_2)$ or $f(x_1) = f(x_2)$. But the latter possibility reduces also to the former because a further point x_3 in between could be chosen with $f(x_3) \neq f(x_1)$. Let us suppose then that for some pair of points x_1 and x_2 with $x_1 < x_2$, $f(x_1) > f(x_2)$.

Choose any

$$f(x_1) > c > f(x_2)$$

and take the point

$$z = \inf\{x \in (x_1, x_2) : f(x) < c\}$$

which evidently lies on the level $f^{-1}(c)$. For some sequence of points $x_n \searrow z$, $f(x_n) < f(z) = c$. The symmetric points $2z - x_n$ on the other side of z must have $f(2z - x_n) \geq f(z) = c$ and so

$$f(x_n) - f(2z - x_n) \leq 0.$$

Hence SD $f(z)$ cannot be positive, and with this contradiction the theorem is established.

As a corollary we deduce that continuous functions with zero symmetric derivatives are constant. Hence functions with identical symmetric derivatives differ by a constant. This was probably first proved in Mazurkiewicz [197].

Corollary 1.5 *Two continuous functions having identical (finite) symmetric derivatives everywhere in an interval, must differ by a constant in that interval.*

Proof. If f and g have identical (finite) symmetric derivatives everywhere in an interval (a, b) then SD $(f(x) - g(x)) = 0$ there. For any $\epsilon > 0$ this means that $f(x) - g(x) + \epsilon x$ has a positive symmetric derivative and so, by the theorem, must be increasing; similarly $g(x) - f(x) + \epsilon x$ has a positive symmetric derivative and must also be increasing. As $\epsilon > 0$ is arbitrary $f - g$ must be constant as required.

This theorem can be looked at in another useful manner. As we have already observed, there is no Rolle's theorem available for the symmetric derivative and hence no mean value theorem. In its place we have the following assertion which we can view as a quasi-mean value theorem. For more theorems of this type see Chapter 5.

Corollary 1.6 *Let f be continuous in an interval $[a, b]$ and suppose that its symmetric derivative exists at each point in (a, b). Then there are points ξ_1 and ξ_2 in (a, b) such that*

$$SD f(\xi_1) \leq \frac{f(b) - f(a)}{b - a} \leq SD f(\xi_2).$$

Proof. Let $k = (f(b) - f(a))/(b - a)$ and suppose, contrary to the statement of the corollary that there is no point $a < \xi < b$ for which SD $f(\xi) \leq k$. Then the function $g(x) = f(x) - k(x - a)$ has a symmetric derivative that is positive in (a, b). By the theorem g is increasing and so $g(b) > g(a)$. This gives $f(b) > f(a) + k(b - a)$ which is a contradiction. It follows that SD $f(\xi) \leq k$ for at least one $\xi \in (a, b)$. In a similar manner it follows too that SD $f(\xi) \geq k$ for at least one $\xi \in (a, b)$.

1.4 Riemann's Theorems

Let us pass now to *even* derivatives. These notions first arose in Riemann's study of trigonometric series. In his investigation of series of the form

$$f(x) = a_0/2 + \sum_{n=1}^{\infty} (a_n \cos nx + b_n \sin nx), \tag{1.1}$$

where the coefficients converge to zero, he was led to a study of the function

$$R(x) = a_0 x^2/4 - \sum_{n=1}^{\infty} (a_n \cos nx + b_n \sin nx)/n^2$$

obtained by formally integrating the series (1.1) twice. This function R is now called the *Riemann function* for the series. The properties of the limits

$$\lim_{h \to 0+} \frac{R(x+h) + R(x-h) - 2R(x)}{h} \tag{1.2}$$

and

$$\lim_{h \to 0+} \frac{R(x+h) + R(x-h) - 2R(x)}{h^2} \tag{1.3}$$

constitute what is commonly called the Riemann theory of trigonometric series (cf. [313, Vol. I, Chapter 9]).

The mere existence of the limit in (1.2) forces that limit to be zero. A function with this property at a point x is said to be *smooth* at x. Riemann's second theorem (Theorem 1.10 below) says that the Riemann function is uniformly smooth.

The limit in (1.3) we have already introduced as the *second order symmetric derivative* and we use the terminology $SD_2 R(x)$ for this derivative. Any study of the first order symmetric derivative must lead soon to a confrontation with the second order symmetric derivative. The genesis of these derivatives in the expression for the even and odd parts of a function alone should suggest this.

This derivative also arises by integrating a first symmetric derivative. If $SD f(x_0) = c$ then

$$\frac{f(x_0 + t) - f(x_0 - t)}{2t} \to c \tag{1.4}$$

as $t \to 0$. If f is integrable and F is an indefinite integral of f then $SD_2 F(x_0) = c$ as some elementary computations show:

$$
\begin{aligned}
&F(x_0 + h) + F(x_0 - h) - 2F(x) \\
&= (F(x_0 + h) - F(x_0)) - (F(x_0) - F(x_0 - h)) \\
&= \int_0^h (f(x_0 + t) - f(x_0 - t))\, dt \\
&= ch^2 + \int_0^h ([f(x_0 + t) - f(x_0 - t)] - 2ct)\, dt.
\end{aligned}
$$

From this and (1.4) we can deduce

$$\frac{F(x_0 + h) + F(x_0 - h) - 2F(x)}{h^2} \to c.$$

Thus statements about the symmetric derivatives of integrable functions can be lifted up to the primitive function by using the second order symmetric derivative. One advantage here is that the primitive is generally a nicer function than the original function; it is absolutely continuous for example. Of course the method is limited to integrable functions but in earlier times this was not seen as much of a limitation.

As a first step in the study of the second symmetric derivative let us note that if $f''(x)$ exists then so too does the second symmetric derivative and both have the same value. Thus this latter derivative is a generalized version of the ordinary second order derivative. By an application of the mean-value theorem of the calculus, the fraction

$$\frac{f(x + h) + f(x - h) - 2f(x)}{h^2}$$

must have the value

$$\frac{f'(x + k) - f'(x - k)}{2k}$$

for some $0 < k < h$. Letting $h \to 0$, $k \to 0$ we obtain

$$\frac{f'(x + k) - f'(x - k)}{2k} \to f''(x)$$

which justifies our assertion.

We can summarize these observations as lemmas.

Lemma 1.7 *Let f be an integrable function with a symmetric derivative $SD f(x_0)$ at a point x_0. If F is an indefinite integral of f then $SD_2 F(x_0) = SD f(x_0)$.*

Lemma 1.8 *Let $F''(x_0)$ exist. Then $SD_2 F(x_0) = F''(x_0)$.*

Riemann's first theorem (Theorem 1.9 below) asserts that the second symmetric derivative of the Riemann function for a trigonometric series recovers the sum of the series at any point at which the series converges.

Theorem 1.9 (Riemann's first theorem) *Let R denote the Riemann function for the trigonometric series*

$$a_0/2 + \sum_{n=1}^{\infty} (a_n \cos nx + b_n \sin nx) \tag{1.5}$$

where a_n, $b_n \to 0$. If (1.5) converges to a finite sum s at a point x_0 then $SD_2 R(x_0) = s$.

Proof. Elementary trigonometric identities and the expression for R lead immediately to the fact that

$$\frac{R(x_0 + 2h) + R(x_0 - 2h) - 2R(x_0)}{4h^2}$$

$$= a_0/2 + \sum_{n=1}^{\infty} (a_n \cos nx_0 + b_n \sin nx_0) \left(\frac{\sin nh}{nh}\right)^2.$$

The rest of the proof then is just a matter of computing this limit as $h \to 0$. Details of this computation appear in [11, Vol. I, Theorem 2, pp. 190–191]. This may also be viewed as a regularity claim for a summability method as in [313, Vol. I, Theorem 2.4, pp. 319–320].

Theorem 1.10 (Riemann's second theorem) *Let R be the Riemann function for a trigonometric series*

$$a_0/2 + \sum_{n=1}^{\infty} (a_n \cos nx + b_n \sin nx) \tag{1.6}$$

with a_n, $b_n \to 0$. Then

$$\lim_{h \to 0+} \frac{R(x + h) + R(x - h) - 2R(x)}{h} = 0$$

uniformly in x.

Proof. This statement too may be viewed as a regularity claim for a summability method as in [313, Vol. I, Theorem 2.8, p. 320]. A more general assertion is proved too in [313, Vol. I, Theorem 2.11, pp. 320–321].

These two theorems illustrate the role of "even" derivatives in the study of trigonometric series. The odd derivatives play a role too. As a sample let us state a theorem due essentially to Fatou and closely related to the two theorems of Riemann.

Theorem 1.11 (Fatou) *Suppose that a_n and b_n are $o(1/n)$ or, more generally, that*

$$\sum_{k=1}^{n} k\sqrt{a_k^2 + b_k^2} = o(n).$$

Then, if $s_n(x)$ are the partial sums of a trigonometric series,

$$a_0/2 + \sum_{n=1}^{\infty} (a_n \cos nx + b_n \sin nx)$$

the series

$$L(x) = \frac{1}{2}a_0 x + \sum_{n=1}^{\infty} \left(a_n \sin nx - b_n \cos nx\right)/n$$

obtained by a single integration of the series converges uniformly and absolutely to a continuous, uniformly smooth function L. Moreover

$$\frac{L(x+h) - L(x-h)}{2h} - s_{N(h)}(x) \to 0 \qquad (1.7)$$

uniformly in x as $h \to 0$ where $N(h) = [1/h]$ (the integral part of $1/h$).

Proof. See [313, Vol. I, Theorem 2.16, p. 322].

We should point out that although there is a symmetric derivative (odd derivative) implicit in this assertion the function L is smooth (it has an even derivative) so that should SD $L(x)$ exist it must be simply $L'(x)$. Even so the expression in (1.7) makes it clear that it is symmetric differentiation rather than ordinary differentiation that dominates this kind of study.

1.5 Schwarz Theorem

One of the most interesting and immediate applications of Riemann's theory of trigonometric series is to the problem of uniqueness. The problem was posed by Heine: can a function have two different representations as the sum of convergent trigonometric series? By subtracting two such series we see that the question reduces to asking whether there can exist a trigonometric series

$$a_0/2 + \sum_{n=1}^{\infty} \left(a_n \cos nx + b_n \sin nx\right), \qquad (1.8)$$

converging everywhere to 0 but whose coefficients are not all zero. Heine had shown that for *uniform* convergence there was uniqueness but recognized that for pointwise convergence the problem was more delicate.

The problem was solved by Cantor [41, pp. 82–83] in a report from 1870. The solution stands as a landmark in the history of mathematics, not so much for the depth of the methods themselves but for the fact that it marked the beginning of a research program for Cantor that culminated in the development of modern set theory and the theory of transfinite ordinals.

Cantor's solution has three simple ingredients. He showed first that if the trigonometric series (1.8) converges everywhere to 0 then the coefficients a_n, $b_n \to 0$. Take then the Riemann function R for the series; by Riemann's first theorem (Theorem 1.9) SD$_2 R(x) = 0$ everywhere. In a letter, mailed to Cantor from Zurich, Schwarz supplied the second step in the proof: a continuous function F with SD$_2 F(x) = 0$ everywhere must be linear. Finally then one has a relation

$$R(x) = \alpha x + \beta = a_0 x^2 / 4 - \sum_{n=1}^{\infty} \left(a_n \cos nx + b_n \sin nx \right) / n^2.$$

Simple arguments show that this can hold only if all the coefficients vanish.

Our interests here are mostly in the property of the second order symmetric derivative expressed by this theorem of Schwarz which we now prove. The proof uses recognizably nineteenth century methods.

Theorem 1.12 (Schwarz) *Let F be continuous on the interval $[a, b]$ and suppose that $SD_2\, F(x) = 0$ at every point of (a, b). Then F is linear in this interval.*

Proof. Let $\epsilon > 0$ and define the functions

$$G(x) = F(x) - F(a) - \frac{F(b) - F(a)}{b - a}(x - a) + \epsilon(x - a)(x - b) \qquad (1.9)$$

and

$$H(x) = F(x) - F(a) - \frac{F(b) - F(a)}{b - a}(x - a) - \epsilon(x - a)(x - b).$$

We prove that $G(x) \le 0$ everywhere in $[a, b]$. If not then, because $G(a) = G(b) = 0$, there is a point $c \in (a, b)$ at which a positive maximum is attained. At such a point the derivative $SD_2\, G(c)$ cannot be positive and yet this would contradict the fact that $SD_2\, F(c) \ge 0$. An identical proof establishes that $H(x) \ge 0$ everywhere in $[a, b]$.

This gives

$$\left| F(x) - F(a) - \frac{F(b) - F(a)}{b - a}(x - a) \right| \le \epsilon(b - a)^2$$

for all $\epsilon > 0$. From this the linearity of F evidently must follow.

Later researchers extended this theorem to provide a more delicate analysis. This variant is apparently due to de la Vallée Poussin.

Theorem 1.13 *Let F be a continuous function on the interval $[a, b]$. Suppose that F satisfies the condition*

$$\limsup_{t \to 0} \left| \frac{F(x + t) + F(x - t) - 2F(x)}{t^2} \right| < +\infty \qquad (1.10)$$

at every point of (a, b) and suppose that $SD_2\, F(x) = 0$ at almost every point of (a, b). Then F is linear in this interval.

Proof. The proof uses a standard device from the early years of this century to push the argument used in the proof of Theorem 1.12 to handle the exceptional set of measure zero. We just insert the following addition into the proof of that theorem. Let E be the set of points x at which $\mathrm{SD}_2 F(x) = 0$ may fail. As E has measure zero there is a continuous increasing function $k(x)$ on $[a, b]$ whose derivative on E is infinite and such that $0 = k(a) < k(b) < \epsilon$. We may choose k so as to have a positive and finite or infinite derivative everywhere (see [18, p. 229]). Let $K(x) = \int_a^x k(x)\, dx$ and in the defining equation (1.9) for $G(x)$ add on the extra term $K(x)$.

Now

$$\mathrm{SD}_2 G(x) \geq \mathrm{SD}_2 F(x) + 2\epsilon + \mathrm{SD}_2 K(x)$$

at every point x. But $\mathrm{SD}_2 K(x)$ is everywhere positive and $\underline{\mathrm{SD}_2} K(x) = +\infty$ at every point x with $k'(x) = +\infty$. Thus we can conclude, as before, that $\underline{\mathrm{SD}_2} G(x) > 0$ at every point. The argument continues more or less unchanged. The same addendum is, of course, necessary to handle the function H and so the proof, in its modified form, works for this variant of Theorem 1.12.

There are further directions that this theorem of Schwarz can take us and that we will explore in greater depth in later chapters. Here we are content to state the results without proof. The first theorem asserts the Schwarz theorem in the context of a convexity result; if f and $-f$ are both convex then f is linear and so Theorem 1.12 above follows directly from this version. In Chapter 5 a complete account of convexity theorems can be found.

Theorem 1.14 *Let F be continuous on the interval $[a, b]$ and suppose that that $SD_2 F(x) \geq 0$ at every point of (a, b). Then F is convex in this interval.*

Yet another way of looking at the Schwarz theorem is due to de la Vallée Poussin. Note that Theorem 1.12 follows directly from this version too. These ideas will be explored and generalized in Chapter 9.

Theorem 1.15 *Let F be continuous on the interval $[a, b]$ and suppose that that $SD_2 F(x) = g(x)$ exists at every point of (a, b). If g is Lebesgue integrable on (a, b) then*

$$F(x) = \int_a^x \int_a^{t_1} g(t_2)\, dt_2\, dt_1 + c_0 + c_1 x$$

where c_0 and c_1 are constants.

1.6 Auerbach Theorem

We return now to the first order symmetric derivative, but armed with some of the tools obtained in our study of the second symmetric derivative. We know already something about a function with an everywhere vanishing symmetric derivative. Here is Auerbach's generalized version of this with its original 1926 proof. In Chapter 2 we will discover that a function that satisfies the condition (1.11) on an interval is continuous off a countable set and so the conclusion of the theorem can be considerably strengthened.

Theorem 1.16 (Auerbach) *Let f be a real valued measurable function on the interval $[a, b]$ that satisfies the condition*

$$\limsup_{t \to 0} \left| \frac{f(x+t) - f(x-t)}{2t} \right| < +\infty \qquad (1.11)$$

at every point of (a, b) and suppose that $SD\, f(x) = 0$ at almost every point of (a, b). Then f is almost everywhere constant in this interval.

Proof. If f is integrable then this theorem is almost a direct corollary of Theorem 1.13. Simply take F as an indefinite integral of f and then the conditions of the present theorem translate exactly into the conditions of that theorem. We see that

$$\limsup_{h \to 0} \left| \frac{F(x+h) + F(x-h) - 2F(x)}{h^2} \right|$$

$$= \limsup_{h \to 0} \left| \frac{1}{h^2} \int_0^h \frac{f(x+t) - f(x-t)}{2t} \, 2t \, dt \right|$$

$$\leq \limsup_{h \to 0} \left| \frac{f(x+h) - f(x-h)}{2h} \right|.$$

Thus the condition (1.11) for f is exactly the condition (1.10) for F. Also, since we assume that $SD\, f(x) = 0$ at almost every point of (a, b), it follows that $SD_2\, F(x) = 0$ at almost every point of (a, b). Theorem 1.13 now shows that F is linear and so $F' = f$ is almost everywhere constant.

Accordingly what we need to do is show how to extend the result from integrable functions to arbitrary measurable ones. Define the functions

$$f_n(x) = \frac{|f(x) + n| - |f(x) - n|}{2}.$$

These are measurable and bounded and so integrable. Moreover a simple computation shows that the symmetric difference quotient for the f_n is easily computed in terms of that for f:

$$\left| \frac{f_n(x+h) - f_n(x-h)}{2h} \right|$$

$$= \left| \frac{|f(x+h) + n| - |f(x-h) + n|}{4h} - \frac{|f(x+h) - n| - |f(x-h) - n|}{4h} \right|$$

$$\leq \left| \frac{|f(x+h) + n| - |f(x-h) + n|}{4h} \right| + \left| \frac{|f(x+h) - n| - |f(x-h) - n|}{4h} \right|$$

$$\leq \left| \frac{f(x+h) - f(h-h)}{2h} \right|.$$

Now from what we have already proved we know that each f_n is almost everywhere constant, say c_n, outside of a set N_n of measure zero. Since $f(x) = \lim_{n\to\infty} f_n(x)$ this means that $f(x) = \lim_{n\to\infty} c_n$ outside of $N = \bigcup_{n=1}^{\infty} N_n$. Thus f is constant off the measure zero set N and the proof is complete.

1.7 Borel Symmetric Derivative

Borel defined a variant on the ordinary derivative that he called a mean-value derivative (*derivée moyenne*); for an integrable function f the derivative is computed using the quotient

$$\frac{1}{h} \int_0^h \frac{f(x+t) - f(x)}{t}\, dt.$$

A symmetric version of this has some utility too and has been introduced by Khintchine [151, p. 220]. We present the definitions and prove the main observation of Khintchine, that this derivative is intimately linked to the second symmetric derivative. The proof is exactly the original one from [151, pp. 220–221]; a more detailed set of relations between these two derivation processes will appear in Section 7.5 much later.

The symmetric–Borel derivates of an integrable f at a point x are defined as

$$\overline{\text{SBD}}\, f(x) = \limsup_{h\to 0} \frac{1}{h} \left\{ \int_0^h \frac{f(x+t) - f(x-t)}{2t}\, dt \right\}$$

and

$$\underline{\text{SBD}}\, f(x) = \liminf_{h\to 0} \frac{1}{h} \left\{ \int_0^h \frac{f(x+t) - f(x-t)}{2t}\, dt \right\}.$$

If the two extreme derivates are equal and finite then f is said to have a symmetric Borel derivative at the point x and the common value is denoted as $\text{SBD}\, f(x)$. Here the integral may be taken as an improper Lebesgue integral, i.e. interpreted in the sense that f is integrable in a neighbourhood of the point x and $\int_0^h$ is taken to mean $\lim_{\epsilon\to 0+} \int_\epsilon^h$. (Later on the Denjoy–Perron integral will be used in this context.)

Theorem 1.17 (Khintchine) *Let f be integrable in a neighbourhood of a point x and let F be an indefinite integral of f. If the second symmetric derivative of F exists at a point x, then the symmetric Borel derivative of f exists there too and*

$$SD_2\, F(x) = SBD\, f(x).$$

Proof. The proof is obtained from the elementary differentiation identity

$$\frac{d}{dh}\left\{\frac{F(x+h)+F(x-h)-2F(x)}{h}\right\}$$
$$=\frac{f(x+h)-f(x-h)}{h}-\frac{F(x+h)+F(x-h)-2F(x)}{h^2}$$

which must hold almost everywhere in some interval $(0, \delta)$. We integrate this between the limits p and q for any $0 < p < q < \delta$ and divide by q to obtain

$$\frac{F(x+q)+F(x-q)-2F(x)}{q^2}-\frac{F(x+p)+F(x-p)-2F(x)}{pq}=$$
$$\frac{1}{q}\int_p^q\frac{f(x+t)-f(x-t)}{t}\,dt-\frac{1}{q}\int_p^q\frac{F(x+t)+F(x-t)-2F(x)}{t^2}\,dt.$$

The hypotheses of the theorem require that

$$\frac{F(x+h)+F(x-h)-2F(x)}{h^2} \to c$$

for some c as $h \to 0$. Now we let first $p \to 0$ and then $q \to 0$ in this identity. The two fractions on the left are easily handled and the second integral on the right requires only an application of the first mean value theorem for integrals. Thus we obtain

$$c = \lim_{q\to 0}\frac{1}{q}\int_p^q\frac{f(x+t)-f(x-t)}{t}\,dt - c$$

and hence

$$\lim_{q\to 0}\frac{1}{q}\int_p^q\frac{f(x+t)-f(x-t)}{2t}\,dt = c$$

exactly as required, completing the proof.

As an application of Theorem 1.17 we can produce a variant on the theorems of Section 1.4. The function

$$L(x) = a_0 x/2 + \sum_{n=1}^{\infty}\left(a_n \sin nx - b_n \cos nx\right)/n, \tag{1.12}$$

obtained by formally integrating the trigonometric series (1.1) once, is called the *Lebesgue function* for the series. If the coefficients a_n, $b_n \to 0$ then this series is known to converge almost everywhere to a Lebesgue integrable function. As a result of Theorem 1.17 the following is just a rewording of Riemann's first theorem (Theorem 1.9).

Theorem 1.18 *Let L be the Lebesgue function for the trigonometric series (1.5) with a_n, $b_n \to 0$. If (1.5) converges to a finite sum s at a point x_0, then $SBD\,L(x_0) = s$.*

1.8 Approximate Symmetric Derivative

To complete the picture suggested by Theorem 1.18 we point out that yet another symmetric derivative may be used in place of the Borel symmetric derivative in this application to convergent trigonometric series. This allows us to
introduce at an early stage a concept that will play an important role in the
sequel.

The symmetric derivative can be generalized by using an approximate limit
in place of the ordinary limit. This, in particular, allows for the differentiation
of a function defined merely almost everywhere (as was the case for the Borel
symmetric derivative). For any measurable function f defined almost everywhere
in a neighbourhood of a point x_0, we write

$$\mathrm{ASD}\, f(x_0) = \mathrm{app}\lim_{t \to 0} \frac{f(x_0 + t) - f(x_0 - t)}{2t},$$

employing an approximate limit, and refer to this as the *approximate symmetric
derivative* of f. When the ordinary symmetric derivative exists so too does the
approximate symmetric derivative, but not conversely of course. The study of
the approximate symmetric properties of the Lebesgue function of trigonometric
series was initiated by Rajchman and Zygmund; this material can be found
in [313, Chap. IX, §2]. The methods are very much rooted in the study of
trigonometric series and will not be reproduced here. The assumption on the
coefficients must be made only in order that the Lebesgue function exist almost
everywhere; as Zygmund remarks, the assumption

$$a_n, b_n = O(n^{\frac{1}{2}-\epsilon}) \ \ (\epsilon > 0)$$

would be enough.

Theorem 1.19 (Rajchman–Zygmund) *Let L be the Lebesgue function for
the trigonometric series (1.5) with a_n, $b_n \to 0$. If (1.5) converges to a finite
sum s at a point x_0, then $ASD\, L(x_0) = s$.*

Proof. This is proved in Zygmund [313, Vol. I, Theorem 2.22, p. 324].

This theorem seems to put us in position to offer a different proof of the
Cantor uniqueness theorem mentioned in Section 1.5. We wish to show that, if
a trigonometric series

$$a_0/2 + \sum_{n=1}^{\infty} (a_n \cos nx + b_n \sin nx)$$

converges everywhere to 0, then the coefficients a_n, $b_n = 0$. Take the Lebesgue
function L for the series; by the Rajchman–Zygmund theorem $\mathrm{ASD}\, L(x) = 0$

everywhere. It should follow that L is constant (almost everywhere constant that is) and it is straightforward then to show that the coefficients vanish as required. But, surprisingly, it is no easy matter to prove that a measurable function L for which ASD $L(x) = 0$ everywhere is equivalent to a constant function. The techniques for proving this were not developed until 1988. A full discussion will appear in Chapter 5.

1.9 Higher Order Symmetric Derivatives

The first and second symmetric derivatives can also be viewed as the initial steps in a hierarchy of symmetric derivatives. We shall not have too much to say in the sequel about these higher order derivatives but we can present the basic facts here.

The method of the elementary calculus whereby higher order derivatives are introduced merely by iterating the derivation process is not of much use in many parts of real analysis. The existence of $f''(x_0)$ requires the existence of $f'(x)$ for all x in a neighbourhood of x_0 and this is a severe restriction. The method proposed by Peano in 1891 avoids this by modeling higher order derivation on Taylor's Theorem.

Definition 1.20 Let $k \in \mathbb{N}$. A function f is said to have a k–th order Peano derivative at a point x_0 if

$$f(x_0 + h) = P(h) + o(h^k) \quad (\text{as } h \to 0)$$

where P is a polynomial of degree k or less. In this case we write

$$\mathrm{PD}^{(k)} f(x_0) = P^{(k)}(0).$$

Of course the first order Peano derivative is easily seen to be merely the ordinary derivative and $\mathrm{PD}^{(1)} f(x_0) = f'(x_0)$. For higher orders, $k = 2, 3 \ldots$, simple examples illustrate that the concept is much more general than any iterated scheme of differentiation. For a survey of material on this derivative see Evans and Weil [98] and the continuation of this survey in Weil [305].

It is entirely natural to develop symmetric versions of the Peano derivative. This was initiated by Ch. de la Vallée Poussin in 1908 who also established their role in certain problems arising from summability and differentiation questions in trigonometric series. We report on some of these ideas here; all this material may be found in [313, Vol. II, Ch. XI] which the reader should consult for proofs and further refinements.

Definition 1.21 Let $k \in \mathbb{N}$. A function f is said to have a k–th order symmetric Peano derivative at a point x_0 if

$$\frac{f(x_0 + h) + (-1)^k f(x_0 - h)}{2} = P(h) + o(h^k) \quad (\text{as } h \to 0) \tag{1.13}$$

where P is a polynomial of degree k or less. In this case we write

$$\text{SD}^{(k)} f(x_0) = P^{(k)}(0). \tag{1.14}$$

There are some immediate observations. One checks that the first two cases ($k = 1, 2$) give exactly the same derivatives as before, justifying our use of this notation in this case. The expression in (1.13) employs the even part of f for k even and the odd part of f for k odd. Thus these derivatives can be considered as arising from the symmetric structure of f: the even order derivatives represent even properties of the function and the odd order derivatives represent odd properties. The polynomial P appearing in (1.13) can contain only even or odd terms depending on whether k is even or odd. Thus, while the existence of the derivative $\text{SD}^{(k)} f(x_0)$ will require also the existence of $\text{SD}^{(k-2)} f(x_0)$, $\text{SD}^{(k-4)} f(x_0) \ldots$, there is no necessity for $\text{SD}^{(k-1)} f(x_0)$ to exist. Again this just reflects the even/odd nature of these derivatives.

The symmetric derivatives play an important role in the study of trigonometric series. We report on just one aspect. Note the similarity between the following theorem and Theorem 1.9, Theorem 1.18 and and Theorem 1.19.

Theorem 1.22 *Let* $k \in \mathbb{N}$, $\alpha \in \mathbb{R}$ *with* $k > \alpha + 1 > 0$. *Suppose that the trigonometric series (1.5) is summable* (C, α) *at* x_0 *to a finite sum* s_0. *Suppose that the series (1.5) integrated term by term* k *times converges*[1] *in a neighbourhood of* x_0 *to a sum* $f(x)$. *Then the* k*-th order symmetric derivative of* f *exists at* x_0 *and* $\text{SD}^{(k)} f(x_0) = s_0$.

Proof. This is proved in Zygmund [313, Vol. II, Theorem 2.1, p. 66].

In place of the definitions used here one might instead employ the higher order differences of Section 1.2.3. Thus it is reasonable to define

$$\text{SD}^{(k)} f(x_0) = \lim_{t \to 0} \frac{\Delta_s^k f(x_0, t)}{t^k} \tag{1.15}$$

as a k-th order symmetric derivative. In the literature these are called *Riemann's derivatives*. In fact if $\text{SD}^{(k)} f(x_0)$ exists in the sense of Definition 1.21 then it exists in the sense of (1.15). The cases $k = 1, 2$ are of course the same for both definitions.

We shall not pursue these ideas, however, as they do not represent symmetric properties in the narrow sense we use here.

---

[1]This is certainly the case if $|a_n| + |b_n| = o(n^\alpha)$.

1.10 Khintchine Theorem

Thus far our methods have been mostly of a classical nature and have not required an analysis that uses any subtle geometric features of the symmetric derivative. We turn now to what must be considered a fundamental property of symmetric derivates, one that has been a central theorem in the history of the subject and a focus for much generalization. This theorem shows that in the most frequent situations in analysis, those in which sets of measure zero may be ignored or play no role, the symmetric derivative and the ordinary derivative cannot be distinguished.

We reproduce the original proof of Khintchine from [151, p. 217] with commentary and some notational changes.

Theorem 1.23 (Khintchine) *Let f be a measurable function. Then f is differentiable at almost every point x at which*

$$\limsup_{h \to 0} \frac{f(x+h) - f(x-h)}{2h} < +\infty. \tag{1.16}$$

Proof. Let E denote the set of points x at which the inequality (1.16) holds. The first step in the proof invokes a standard device in derivation theory and one which we will repeat many times throughout this work. The relation (1.16) asserts a strictly local property; expressed in other words it asserts that at every point x in the set E there are positive numbers $\delta(x)$ and $M(x)$ so that

$$x \in E, 0 < h < \delta(x) \implies \frac{f(x+h) - f(x-h)}{2h} < M(x).$$

But such a local property does not allow for any argument about the existence of the derivative. Instead we would require such a statement to hold uniformly. The device we use is to partition the set E in such a way that this implication does indeed hold uniformly on each member of the partition.

Lemma 1.24 *If the limit*

$$\limsup_{h \to 0} \frac{f(x+h) - f(x-h)}{2h} < +\infty$$

holds at every point of a set E then there is a partition $\{E_n\}$ of the set E and a sequence of positive numbers $\{M_n\}$ so that

$$x \in E_n, 0 < h < 1/n \implies \frac{f(x+h) - f(x-h)}{2h} < M_n.$$

To prove Lemma 1.24 it is enough to define the sequence of sets

$$A_n = \{x \in E : \delta(x) < 1/n, \, M(x) < n\}$$

and then to define $M_n = n$, $E_1 = A_1$ and, for $n > 1$,

$$E_n = A_n \setminus \bigcup_{k<n} A_k.$$

The second step in Khintchine's proof is to obtain, from Lemma 1.24, estimates on the size of the Dini derivatives at points in the sets E_n. We use the notation $\underline{D}^+ f(x)$, $\underline{D}^- f(x)$, $\overline{D}^+ f(x)$ and $\overline{D}^- f(x)$ for the four Dini derivatives of a function f at a point x. The best we can hope for here is to work with density points of these sets, but since almost every point is a point of (exterior) density this will be enough for the conclusion of the theorem. Let us state this step as a separate lemma.

Lemma 1.25 *Suppose that f is measurable and, at every point x in a set E, that*

$$0 < h < \delta \implies \frac{f(x + h) - f(x - h)}{2h} < M.$$

Then at every point x_0 of exterior density of E the estimate $\overline{D} f(x_0) \leq M$ must hold.

To prove Lemma 1.25 we first simplify the computations by replacing $g(x) = f(x) - Mx$ so that the hypothesis now reads

$$x \in E, \, 0 < h < \delta \implies \frac{g(x + h) - g(x - h)}{2h} < 0$$

and the conclusion required is that at every point x_0 of exterior density of E we have $\overline{D} g(x_0) \leq 0$. It is enough to argue that $\overline{D}^+ g(x_0) \leq 0$ and then, by identical arguments, $\overline{D}^- g(x_0) \leq 0$ must follow too.

If x_0 is a point of density of E then we may choose a positive number $\delta_1 < \delta$ so that

$$|E \cap (x_0, x_0 + t)| > 3t/4 \tag{1.17}$$

if $0 < t < \delta_1$. The proof is completed now merely by showing that $g(x) \leq g(x_0)$ for all $x_0 < x < x_0 + \delta_1$ for then the inequality $\overline{D}^+ g(x) \leq 0$ must follow. This in turn may be proved by proving the impossibility of a point x_1 in this interval for which $g(x_0) < g(x_1)$. The remainder of the proof is devoted to obtaining a contradiction in this situation.

Define the two sets

$$A = \{x \in (x_0, x_1) : g(x_0) < g(x)\}$$

and

$$B = \{x \in (x_0, x_1) : g(x) < g(x_1)\}$$

Because $g(x_0) < g(x_1)$ these sets exhaust the interval (x_0, x_1) and because f, and hence g, is measurable both A and B are measurable so that one of these sets has measure exceeding $(x_1 - x_0)/2$.

If A has measure exceeding $(x_1 - x_0)/2$ then the set $A' = \frac{1}{2}(x_0 + A)$ is contained in $(x_0, (x_0+x_1)/2)$ and has measure at least $(x_1-x_0)/4$. This combined with (1.17) shows that the intersection $E \cap A' \cap (x_0, x_1)$ cannot be empty. Let z be a point in this intersection. Write $h = z - x_0$ and note that $0 < h < \delta$. Then $z - h = x_0$, $z \in E$ and $z + h \in A$ and we must have

$$g(z + h) - g(z - h) < 0$$

and

$$g(z - h) = g(x_0) < g(z + h)$$

which is the desired contradiction.

On the other hand if B has measure exceeding $(x_1 - x_0)/2$ then the set $B' = \frac{1}{2}(x_1 + B)$ is contained in $((x_0 + x_1)/2, x_1)$ and has measure at least $(x_1 - x_0)/4$. This combined again with (1.17) shows that the intersection $E \cap B' \cap (x_0, x_1)$ cannot be empty. If z is a point in this intersection then, with $h = x_1 - z$, $0 < h < \delta$, $z + h = x_1$, $z \in E$ and $z - h \in B$ we must have

$$g(z + h) - g(z - h) < 0$$

and

$$g(z + h) = g(x_1) > g(z - h)$$

which is again a contradiction. Thus in both cases we obtain a contradiction and Lemma 1.25 is proved.

Combining Lemmas 1.24 and 1.25 we can now prove Theorem 1.23. Together these facts and the fact that almost every point of any set is a point of exterior density of that set show that at almost every point $x \in E$ the derivate $\overline{D} f(x)$ is finite. But it is well known in the theory of the ordinary derivative that, at almost every such point, the function f must have a finite derivative (see [251, Theorems 7.2 and 10.1]). This completes the proof.

As an immediate corollary we conclude that, at least for measurable functions and in situations where sets of measure zero may be ignored, the symmetric derivative is not distinct from the ordinary derivative. In the statement of this corollary note that infinite symmetric derivatives would be allowed.

Corollary 1.26 (Khintchine) *If a measurable function f has a symmetric derivative at almost every point then f is almost everywhere differentiable in the ordinary sense.*

2

Continuity

2.1 Introduction

In Chapter 1 we have presented some classical studies of the symmetric derivatives, largely as an introduction to our broader topic. In this chapter we turn to a problem that arises from the same type of concern. We ask now for the continuity properties of a function that satisfies some kind of symmetric growth condition. While the intention of the chapter remains to introduce the reader to the subject we can present here nearly a complete picture of what is known; even so, later chapters will offer some refinements that require more technical apparatus than we are prepared to develop for the moment.

A natural starting point for our study is to ask for the continuity properties of functions whose even or odd parts are themselves continuous. The main classical result here is the theorem of Stein and Zygmund asserting that measurable functions are continuous at almost every point at which either of these parts is continuous.

As we move to stronger symmetric conditions the continuity properties become sharper. This leads us to the the study of Charzyński [42] that introduces new ideas and methods which have had a considerable impact on the development of our subject. The original problem addressed by Charzyński was posed by Steinhaus [277]: determine the continuity properties of a function f that satisfies the condition

$$\mathrm{SD}\, f(x) = \lim_{h \to 0} \frac{f(x+h) - f(x-h)}{2h} = 0$$

at every point x. Mazurkiewicz [197] had already demonstrated that for a measurable function f this condition could hold only if the discontinuity points of f

23

were nowhere dense, while Sierpiński [257] had shown that in this situation the discontinuity points of f were at most denumerable. Charzyński relaxed these conditions by removing the measurability hypothesis and asking instead that

$$\limsup_{h \to 0} \left| \frac{f(x+h) - f(x-h)}{2h} \right| < +\infty$$

at each point x. Originally he proved that this hypothesis was enough to show that the set of points of discontinuity was countable and nowhere dense. It was conjectured by Szpilrajn (Marczewski) that the set was in fact scattered (clairsemé)[1] and thus we are led to the final version of the theorem as it appeared in Charzyński's classical paper in *Fundamenta Mathematicae* in 1931. We present the original arguments in more or less the original form but with commentary. Like the Khintchine paper, which forms the centerpiece of Chapter 1, this stands as one of the most important and fundamental studies of symmetric properties and several new ideas and techniques are introduced here.

The remainder of the chapter is devoted to related concerns. In particular we address a problem that remains open: to determine the exact nature of the set of points of discontinuity of a symmetrically continuous function.

2.2 Even and Odd Continuity

Continuity of the odd part of a function is known as symmetric continuity; thus f is *symmetrically continuous* at a point x provided

$$\lim_{t \to 0} (f(x+t) - f(x-t)) = 0.$$

Certainly a point of ordinary continuity is a poin of symmetric continuity. The converse is of course false: the function $f(x) = x^{-2}$ is everywhere symmetrically continuous but discontinuous at $x = 0$. This example also shows that symmetric continuity has properties quite distinct from ordinary continuity: while the function $f(x) = x^{-2}$ is symmetrically continuous at $x = 0$ it is not bounded near that point nor is it defined at the point.

A symmetrically continuous function may evidently have many discontinuities. Indeed if C is any countable set and $x_1, x_2, \ldots$ is an enumeration of C then we may define $f(x) = 0$ for $x \notin C$ and $f(x_n) = 2^{-n}$; this function is everywhere symmetrically continuous and is discontinuous precisely at the points of C. One might ask, then, whether the symmetric continuity of a function imposes some regularity on the function and requires "most" points to be points of ordinary continuity. Our first result is the theorem of Stein and Zygmund giving the most important continuity property of symmetrically continuous functions.

---

[1]See Section A.1 in the Appendix.

Continuity of the even part of a function has evolved a different terminology: a function f is said to be *symmetric* at a point x provided

$$\lim_{t \to 0} (f(x+t) + f(x-t) - 2f(x)) = 0.$$

Thus a function f is symmetric at a point x if the even part of f at that point is continuous. While there is some analogy here between even continuity (symmetry) and odd continuity (symmetric continuity) simple examples distinguish the concepts. A symmetrically continuous function can have no jump discontinuities, so in particular a monotonic function that is symmetrically continuous is continuous. On the other hand any monotonic function f that satisfies at least the familiar condition

$$f(x) = \frac{f(x+0) + f(x-0)}{2}$$

at every point is symmetric. Thus, in particular, a symmetric function may have any countable set as a set of discontinuity points. The even version of the Stein and Zygmund theorem is nearly identical to the odd version.

2.2.1 Stein–Zygmund Theorem

The theorem of Stein and Zygmund that we now state and prove shows that, under a measurability assumption, symmetric continuity and ordinary continuity are equivalent up to a set of measure zero. (The measurability assumptions will be dropped to some extent in the investigations of Chapter 4 and Section 2.7.) There is an oversight in the original proof in [274, Lemma 9, p. 266], as observed in [12], but it is easily amended. This result was perhaps known long before 1964 when the paper appeared but Stein and Zygmund were unable to find an earlier published proof and to this date apparently no-one has found an earlier reference; the best is, perhaps, that of Mazurkiewicz [197] but he assumes the function is bounded and uses very classical methods. The proof here uses the same techniques that we have seen in the Khintchine proof (Theorem 1.23); as in that proof the key idea is to use a geometric argument at the density points of sets where some uniform symmetric continuity requirement is met.

The density computations needed here are explained in the Appendix (Section A.3).

Theorem 2.1 (Stein–Zygmund) *Let f be a measurable function that is symmetrically continuous at each point of a measurable set E. Then f is continuous at almost all points of E.*

Proof. Fix a positive number ϵ. The symmetric continuity of the function at the points of the set E assures us that there is a positive function δ on E so that

$$x \in E, \ 0 < |h| < \delta(x) \implies |f(x+h) - f(x-h)| < \epsilon.$$

By a device that we have seen before (see Lemma 1.24) we may construct a partition $\{E_n\}$ of E with the property that

$$x \in E_n, \ 0 < |h| < 1/n \implies |f(x+h) - f(x-h)| < \epsilon. \qquad (2.1)$$

We claim that at almost every point x in each E_n (and hence at almost every point of E) that

$$\limsup_{y \to x} |f(y) - f(x)| \leq 2\epsilon. \qquad (2.2)$$

We may assume that E is bounded. Then by our measurability assumptions on E and f we may choose, using Lusin's theorem, a compact set $P \subset E$ so that f is continuous on P and $|E \setminus P|$ is as small as we please. Choose $\delta < 1/n$ so that $|f(y) - f(x)| \leq \epsilon$ if $x, y \in P$ and $|y - x| < \delta$. Let x_0 denote any point that is simultaneously a point of density for both E_n and P; we may suppose in order to simplify the computations that $x_0 = 0$. Now then 0 is a point of density of the sets P, E_n and hence also of $2E_n$. By the density Lemma A.10 in the Appendix (Section A.3) there is a positive number $\kappa < \delta$ so that the intersection

$$(2E_n - x) \cap P \cap (x_0, x)$$

cannot be empty if $0 < x - x_0 < \kappa$. If y is any point in this intersection then, because $(x+y)/2 \in E_n$, the condition (2.1) gives that $|f(x) - f(y)| < \epsilon$. Because $y \in P$ and $|x_0 - y| < \delta$ we have too that $|f(x_0) - f(y)| < \epsilon$. From this we obtain that

$$|f(x) - f(x_0)| < 2\epsilon$$

if $x_0 < x < x_0 + \kappa$ and hence that

$$\limsup_{y \to x_0+} |f(y) - f(x_0)| \leq 2\epsilon$$

at every such point x_0. But $|E_n \setminus P| \leq |E \setminus P|$ and the latter can be made arbitrarily small; this means that this relation must hold at almost every point of E_n.

Of course the argument on the left is similar. We obtain then the relation

$$\limsup_{y \to x_0} |f(y) - f(x_0)| \leq 2\epsilon$$

at almost every point of the set E. Then, by taking a sequence of values $\epsilon_k \searrow 0$, we obtain a sequence of sets of measure zero containing all the discontinuity points of f in E.

From this theorem we derive an easy corollary.

Corollary 2.2 *Let f be a measurable function that is symmetrically continuous at every point. Then the set of points at which f has a discontinuity point in the ordinary sense is measure zero and first category.*

Proof. It is enough to recall that the set of points of discontinuity of any function is of type F_σ and that a measure zero F_σ must be first category.

2.2.2 Even Version of Stein–Zygmund Theorem

The even version of this theorem is also from Stein and Zygmund [274, Lemma 9, p. 266]. The continuity properties of symmetric functions were first studied in Mazurkiewicz [198]. His results were extended by Auerbach [9] and Neugebauer [213] who independently obtained the same theorem we now prove. We reproduce the original proof of Stein and Zygmund here. Note that the proof is almost identical to the proof of Theorem 2.1 and once again exploits the main idea of the Khintchine proof by arguing geometrically at points of density.

Theorem 2.3 (Stein–Zygmund) *Let f be a measurable function that is symmetric at each point x of a measurable set E. Then f is continuous at almost all points of E.*

Proof. Fix a positive number ϵ. The symmetry of the function at the points of the set E assures us that there is a positive function δ on E so that

$$x \in E, 0 < |h| < \delta(x) \implies |f(x+h) + f(x-h) - 2f(x)| < \epsilon.$$

By our usual device (see Lemma 1.24) there is a partition $\{E_n\}$ of E with the property that

$$x \in E_n, 0 < |h| < 1/n \implies |f(x+h) + f(x-h) - 2f(x)| < \epsilon. \qquad (2.3)$$

We claim that at almost every point x in E_n (and hence at almost every point of E) that

$$\limsup_{y \to x} |f(y) - f(x)| \le 4\epsilon. \qquad (2.4)$$

By our measurability assumptions on E and f we may choose, using Lusin's theorem, a closed set $P \subset E$ so that f is continuous on P and $|E \setminus P|$ is as small as we please. Choose $\delta < 1/n$ so that $|f(y) - f(x)| \le \epsilon$ if $x \in P$ and $|y - x| < \delta$. Let x_0 denote any point that is simultaneously a point of density for both E_n and P; we may suppose in order to simplify the computations that $x_0 = 0$. Now then 0 is a point of density of the sets P, E_n and hence also of $2E_n$ and $2P$. Employing again the density Lemma A.10 (remembering that P and $2P$ are measurable) we find a positive number $\kappa < \delta$ so that the intersection

$$(2E_n - x) \cap P \cap (2P - x) \cap (x_0, x)$$

cannot be empty if $0 < x - x_0 < \kappa$. If y is any point in this intersection then, because $(x + y)/2 \in E_n$, the condition (2.3) gives that

$$|f(x) + f(y) - 2f((x + y)/2)| < \epsilon.$$

Because $y \in P$ and $|x_0 - y| < \delta$ we have too that $|f(x_0) - f(y)| < \epsilon$; similarly, since $(x + y)/2 \in P$, $|f(x_0) - f((x + y)/2)| < \epsilon$.

From this and

$$|f(x) - f(x_0)| \leq |f(x) + f(y) - 2f((x + y)/2)|$$
$$+\ 2|f((x + y)/2) - f(x_0)| + |f(y) - f(x_0)|$$

we obtain that $|f(x_0) - f(x)| < 4\epsilon$ if $x_0 < x < x_0 + \kappa$ and hence that

$$\limsup_{y \to x_0+} |f(y) - f(x_0)| \leq 4\epsilon$$

at every such point x_0. Once again $|E_n \setminus P| \leq |E \setminus P|$ and the latter can be made arbitrarily small. Thus this relation must hold at almost every point of E_n.

Again the argument on the left is similar. From the relation

$$\limsup_{y \to x_0} |f(y) - f(x_0)| \leq 2\epsilon$$

at almost every point of the set E we can again obtain a sequence of sets of measure zero containing all the discontinuity points of f in E.

From this theorem we derive an easy corollary paralleling a similar statement, Corollary 2.2, for symmetrically continuous functions.

Corollary 2.4 *Let f be a measurable function that is symmetric at every point. Then the set of points at which f has a discontinuity point in the ordinary sense is measure zero and first category.*

2.2.3 A Reduction Theorem

The remarkable similarity that we have just seen between the continuity properties of functions given to satisfy even or odd symmetric growth conditions can be explained in part by the following theorem. Here we show that any such property proved for the odd condition implies immediately a corresponding version for the even condition. This theorem will allow us to reduce continuity proofs to a consideration of only the odd conditions. This is related to a similar, but narrower, idea in Marcus [188, Lemme 1, p. 202]. (A similar statement for "little o" $o(\phi(h))$ holds too.)

Theorem 2.5 *Let ϕ be a positive function on an interval $(0, \eta)$ and let f be a function possessing a dense set of points of continuity. If at every point x of a set E*

$$|f(x+h) + f(x-h) - 2f(x)| = O(\phi(h)) \tag{2.5}$$

or

$$|f(x+h) - f(x-h)| = O(\phi(h)) \tag{2.6}$$

as $h \to 0+$ then there is a measurable function g such that

$$|g(x+h) - g(x-h)| = O(\phi(h)) \tag{2.7}$$

as $h \to 0+$ at each $x \in E$ and f is continuous precisely at the points where g is continuous.

Proof. As usual we write

$$\omega_f(x) = \limsup_{h \to 0+} \{|f(x_1) - f(x_2)| : x_1, x_2 \in (x-h, x+h)\}$$

for the oscillation of the function f. For the function g in the statement of the theorem we have merely to take $g = \omega_f$ if this assumes only finite values or as

$$g(x) = \frac{\omega_f(x)}{\omega_f(x) + 1}$$

if there are infinite values. Not only is g measurable but it is upper semi-continuous (and bounded if we take the latter version).

The function f is continuous precisely when ω_f vanishes. If f has a dense set of points of continuity then ω_f vanishes on a dense set and so $\omega_f(x) = 0$ if x is a point of continuity of ω_f. In the other direction it is clear that if f is continuous at a point x then so too is ω_f. Finally g too has the same continuity points as f.

Here are the computations needed to check that the even symmetric condition (2.5) on f at a point will require that ω_f satisfies the parallel odd condition (2.7) there. (Similar computations will show that the odd symmetric condition (2.6) on f at a point will require that ω_f satisfies (2.7) there.) At any point $x \in E$ and for any some $C > 0$ we may choose $0 < \delta < \eta$ so that

$$|f(x+h) + f(x-h) - 2f(x)| \le C\phi(h) \tag{2.8}$$

if $0 < h < \delta$. If $0 < h_0 < \delta$ and t is any number $t < \omega_f(x + h_0)$ then we may choose sequences $\{x_n\}$ and $\{y_n\}$ converging to $x + h_0$ in such a way that $|f(x_n) - f(y_n)| > t$. Reflect these sequences about the point x by writing $x'_n = 2x - x_n$ and $y'_n = 2x - y_n$; these new sequences converge to $x - h_0$ and for sufficiently large n both of the inequalities

$$|f(x'_n) + f(x_n) - 2f(x)| \le C\phi(h)$$

and

$$|f(y_n') + f(y_n) - 2f(x)| \leq C\phi(h)$$

must hold because of the inequality (2.8). Thus we have

$$\begin{aligned}|f(x_n) - f(y_n)| &\leq |f(x_n) + f(x_n') - 2f(x)| \\ &+ |2f(x) - f(y_n) - f(y_n')| + |f(y_n') - f(x_n')| \\ &\leq |f(y_n') - f(x_n')| + 2C\phi(h).\end{aligned}$$

We can conclude from this that $\omega_f(x - h_0) \geq t - 2C\phi(h)$. This is true for every $t < \omega_f(x + h_0)$ and hence either $\omega_f(x + h_0)$ is infinite or

$$\omega_f(x + h_0) - \omega_f(x - h_0) \leq 2C\phi(h).$$

Note that if $\omega_f(x + h_0) = +\infty$ then necessarily $\omega_f(x - h_0) = +\infty$.
    Identical arguments show that

$$\omega_f(x - h_0) - \omega_f(x + h_0) < 2C\phi(h)$$

or else if $\omega_f(x - h_0) = +\infty$ then necessarily $\omega_f(x + h_0) = +\infty$.
    Hence we have proved that ω_f has the odd symmetric property required at x at least discounting infinite values. If ω_f is not finite then replacing it by g will supply a finite function with the property required in the statement of the theorem.

    A particular case of the theorem may be usefully expressed as a corollary. This shows that the continuity properties of symmetric or symmetrically continuous functions can be determined within a smaller class of functions (bounded, semicontinuous functions). Later we shall see that the measurability assumption in the case of symmetrically continuous functions may be dropped.

Corollary 2.6 *Let f be measurable and everywhere symmetrically continuous [symmetric]. Then there is an upper-semicontinuous function g that is everywhere symmetrically continuous with $0 \leq g \leq 1$ so that f is continuous precisely at the points $g^{-1}(0)$ where g is continuous.*

Proof. We already know that such functions have a dense set of points of continuity. Thus the theorem may be applied. We have only to take

$$g(x) = \frac{\omega_f(x)}{\omega_f(x) + 1}.$$

2.2.3.1 A Remark. The relation between the odd and even continuity properties of a function in Theorems 2.1 and 2.3 can be explained in another way. This shows that for measurable functions the a.e. continuity properties are less symmetric properties than inherent in any generalized continuity scheme. Thus the measurable case loses some of its interest for us as we have not really discovered *symmetric* properties of our functions; it remains as a useful first step in the analysis nonetheless.

The following theorem, from Stein and Zygmund [274, p. 249], is easily proved from the fact that measurable functions are approximately continuous almost everywhere. Then arguments similar to those used in Sections 2.2.1 and 2.2.2 above can be used. For nonmeasurable functions the analysis is much more intricate. Note that the situation $\alpha_1 = 1$, $\alpha_2 = -1$, $\alpha_3 = 0$, $\beta_1 = 1$, $\beta_2 = 1$ and $\beta_3 = -2$ includes the even situation in Theorem 2.3 and $\alpha_1 = 1$, $\alpha_2 = -1$, $\beta_1 = 1$, and $\beta_2 = -1$ includes the odd situation in Theorem 2.1.

Theorem 2.7 (Stein–Zygmund) *Let $\{\alpha_1, \alpha_2, \ldots \alpha_n\}$ be a finite sequence of distinct real numbers and let $\{\beta_1, \beta_2, \ldots \beta_n\}$ be another sequence for which $\sum_{i=1}^n \beta_i = 0$. Let f be measurable and suppose, at each point x in a measurable set E, that*

$$\lim_{t \to 0+} \sum_{i=1}^n \beta_i f(x + \alpha_i t) = 0.$$

Then f is almost everywhere continuous in E.

By the same methods one can also show the following.

Theorem 2.8 *Let $\{\alpha_1, \alpha_2, \ldots \alpha_n\}$ be a finite sequence of distinct real numbers and let $\{\beta_1, \beta_2, \ldots \beta_n\}$ be another sequence for which $\sum_{i=1}^n \beta_i = 0$. Let f be measurable and suppose, at each point x in a measurable set E, that*

$$\sum_{i=1}^n \beta_i f(x + \alpha_i t) = O(t)$$

as $t \to 0+$. Then f is bounded in a neighbourhood of almost every point in E.

For other continuity questions within the same framework see also Ash *et al.* [3]. This scheme can be extended to derivatives as well. For example Ash [4] and Humke and Laczkovich [134] study derivatives of the form

$$\lim_{t \to 0+} \frac{\sum_{i=1}^n \beta_i f(x + \alpha_i t)}{t}$$

which would generalize the symmetric derivative. But again the methods leave the realm of the symmetric considerations that dominate our study.

2.3 Charzyński Theorem

We are ready now to approach the important study of Charzyński [42]. We wish to characterize the set of continuity points of a function f that is given to satisfy everywhere the condition

$$f(x + h) - f(x - h) = O(h)$$

as $h \to 0$. The methods that we develop will show us incidentally how to handle functions that are everywhere symmetrically continuous but not given *a priori* to be measurable; this allows us to supplement the Stein–Zygmund theorem (Theorem 2.1).

A set of real numbers is scattered if it has the property that no nonempty subset of it may be formed that is without isolated points. For a further discussion of scattered sets see Section A.1 in the Appendix.

Theorem 2.9 (Charzyński) *Let the function f satisfy at every point x the condition*

$$\limsup_{h \to 0} \left| \frac{f(x + h) - f(x - h)}{2h} \right| < +\infty. \tag{2.9}$$

Then the set of points of discontinuity of f is scattered.

We shall reproduce the ideas of the proof in several stages. These ideas, albeit rather technical, have had a considerable influence on the direction of the subject and can be considered as cornerstones of the techniques developed over the years.

2.3.1 Charzyński's Lemma

The first lemma used in the proof is a simple geometric observation obtained by performing two rotations about continuity points of a function.

Lemma 2.10 *Let the function f satisfy the following conditions at the two points x_1 and x_2 where $h_0 = x_2 - x_1 > 0$:*

$$|f(x_1 + h) - f(x_1 - h)| < \epsilon$$

for all $0 < h < h_0$ and f is symmetrically continuous at the point x_2. Then there is a positive number $\delta < h_0$ such that for every $x_1 < x < x_1 + \delta$,

$$|f(x + h_0) - f(x - h_0)| < 2\epsilon.$$

Proof. The proof requires only the geometric argument that the translation $x - h_0 \to x + h_0$ can be obtained by two rotations, one about the point x_1 and

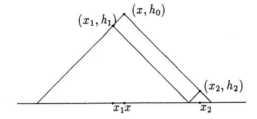

Fig. 2.1. Two rotations in Charzyński's Lemma.

one about the point x_2; of course in order to use the symmetric continuity at the point x_2 the first rotation must carry $x - h_0$ very close to x_2.

Here are the algebraic computations necessary to justify the notions; the figure (Figure 2.3.1) illustrates the two rotations in a geometrically obvious manner. In Chapter 4 we shall codify these notions as covering theorems. There is a $\kappa < h_0$ so that

$$|f(x_2 + h) - f(x_2 - h)| < \epsilon$$

for $0 < h < \kappa$. Now for any $x_1 < x < x_1 + \kappa$ we may choose h_1 and h_2 so that $x - h_0 = x_1 - h_1$, $x_2 + h_2 = x + h_0$, and then

$$h_2 = x - x_2 + h_0 = x - x_1 < \kappa$$

so that

$$|f(x_2 + h_2) - f(x_2 - h_2)| < \epsilon.$$

Together with

$$|f(x_1 + h_1) - f(x_1 - h_1)| < \epsilon$$

this gives

$$|f(x + h_0) - f(x - h_0)| < 2\epsilon$$

as required.

2.3.2 $D(f)$ is Nowhere Dense

For the next step of the proof let us produce the uniform partition that the inequality (2.9) induces (similar to the step used in the proof of Khintchine's Theorem.).

Thus we obtain the following partition of the line:

Lemma 2.11 *There are a sequence of disjoint sets $\{E_k\}$ covering the real line and a sequence of positive numbers $\{M_k\}$ such that*

$$x \in E_k, \, 0 < t < 1/k \Longrightarrow |f(x + t) - f(x - t)| < 2M_k t.$$

We shall apply 2.11 by appealing to the Baire category theorem to obtain that in any interval I there is a subinterval J in which some set E_k is dense. The argument now passes to intervals in which some set E_k is dense and the next lemma shows that in this kind of situation there is some control on the oscillation of the function f.

Lemma 2.12 *Suppose that for a set E there are positive numbers δ and M so that*

$$x \in E, \ 0 < t < \delta \Longrightarrow |f(x+t) - f(x-t)| < 2Mt.$$

Let $\epsilon > 0$ and suppose that E is dense in (a,b), that

$$b - a < \min\{\delta, \epsilon/M\}$$

and that $|f(a) - f(b)| > \epsilon$. Then every subinterval of (a,b) contains points z_1 and z_2 with $|f(z_1) - f(z_2)| > \epsilon/3$.

Proof. Let (c,d) be a subinterval of (a,b). Because E is dense in (a,b) there are points x_1 and x_2 in E so that $2x_1 - a$ and $2x_2 - b$ are contained in (c,d). The inequalities

$$|f(a) - f(b)| > \epsilon,$$

$$|f(2x_1 - a) - f(a)| < M(b-a) < \epsilon$$

and

$$|f(b) - f(2x_2 - b)| < M(b-a) < \epsilon$$

then provide

$$|f(2x_1 - a) - f(2x_2 - b)| > \epsilon/3$$

and this completes the proof of 2.12.

We are now in a position to prove that the set $D(f)$ of discontinuity points of f is nowhere dense. In order to obtain a contradiction let us suppose that this set is dense in some interval. Then the Baire category theorem and Lemma 2.11, together can be used to show that the following situation can be reached: there is an interval (a,b), a positive number ϵ and an index k so that E_k is dense in (a,b),

$$b - a < \frac{1}{2}\min\{1/k, \epsilon/M_k\},$$

and $|f(a) - f(b)| > 15\epsilon$. For any point $x \in E_k \cap (a,b)$ note that

$$|f(x+h) - f(x-h)| < 2M_k h < \epsilon \tag{2.10}$$

provided the points $x + h$ and $x - h$ lie inside the interval (a,b) since then $h < 1/k$ and $h < \epsilon/M_k$.

Now choose a point x_1 in $E_k \cap (\frac{1}{2}(a+b), b)$. We may use Lemma 2.12 to find points $z_1 < z_2$ from (x_1, b) with $z_2 - z_1 < \frac{1}{4}(b-a)$ and

$$|f(z_1) - f(z_2)| > 5\epsilon. \tag{2.11}$$

Set $h_0 = (z_2 - z_1)/2$. In this situation we can apply Lemma 2.10 (using x_1, $x_2 = x_1 + h_0$ and h_0 as here) to obtain a positive number δ so that

$$|f(x + h_0) - f(x - h_0)| < 2\epsilon \tag{2.12}$$

for all $x \in (x_1, x_1 + \delta)$. As E_k is dense in (a, b) we may choose a point $c \in (x_1, z_1) \cap E_k$ so that both of the points $2c - z_1$ and $2c - z_2$ lie in the interval (a, b) and so that

$$x_1 < 2c - \frac{1}{2}(z_1 + z_2) < x_1 + \delta. \tag{2.13}$$

As c is in E_k and all the points z_1, z_2, $2c - z_1$ and $2c - z_2$ are in (a, b) we can employ the inequality (2.10) to obtain

$$|f(z_1) - f(2c - z_1)| < \epsilon \tag{2.14}$$

and

$$|f(z_2) - f(2c - z_2)| < \epsilon. \tag{2.15}$$

By (2.13) the center of the interval $(2c - z_2, 2c - z_1)$ lies in the interval $(x_1, x_1 + \delta)$ and so from (2.12) we have

$$|f(2c - z_1) - f(2c - z_2)| < 2\epsilon. \tag{2.16}$$

The inequalities (2.14), (2.15) and (2.16) together show that $|f(z_1) - f(z_2)| < 4\epsilon$ and this is in contradiction to the inequality (2.11). This contradiction completes the proof.

Before passing to the next step in the overall argument let us observe that the full force of the assumption $f(x + t) - f(x - t) = O(t)$ in the assertion of Charzyński's theorem has yet to be applied. In fact the weaker assertion $f(x + t) - f(x - t) = O(t^\alpha)$ for $\alpha > 0$ would have worked to this point equally well. This will be used below to obtain a proof of Theorem 2.18.

2.3.3 $D(f)$ is Denumerable

We know from the preceding arguments that the the set of discontinuity points $D(f)$ is nowhere dense and we now proceed, using that fact, to prove that it is also denumerable.

Let $\omega_f(x)$ denote the oscillation of the function f at the point x. The set of pairs

$$T = \{(x, \omega_f(x)) : x \in \mathbb{R}\}$$

displays the set of discontinuity points conveniently; every point of T lying above the x–axis represents a point of discontinuity of f. Charzyński's argument considers the wedges

$$W(x_0, K) = \{(x, y) : x_0 < x \text{ and } K(x - x_0) < y\}$$

and his countability argument involves checking the sets $T \cap W(x, K)$; obviously if every such set is countable then there are only countably many discontinuity points as we wish to prove. He asks for us to construct at each point x_0 a point $z = z(x_0, K)$ to the right which is the supremum of the values z for which the set

$$\{(x, y) : x_0 < x < z \text{ and } y = \omega_f(x) > K(x - x_0)\}$$

is denumerable. If this supremum is finite at any point x_0 (in fact we need consider only $x \notin \overline{D(f)}$) then the theorem is contradicted; the strategy of the proof is to show that such a possibility can occur only if there is a point z_0 with

$$\limsup_{h \to 0} \left| \frac{f(z_0 + h) - f(z_0 - h)}{2h} \right| = +\infty.$$

Let us begin with a lemma that provides needed estimates on the symmetric growth of functions relative to points in T.

Lemma 2.13 *Let the function g satisfy the following condition at points $a < b$:*

$$\omega_g(b) > \omega_g(a) + K(b - a) \tag{2.17}$$

for some positive number K. Then there is a neighbourhood U of the point $(a + b)/2$ so that

$$\sup_{0 < h < b - a} \left| \frac{g(x + h) - g(x - h)}{2h} \right| > K/4. \tag{2.18}$$

at every point $x \in U$.

Proof. Choose a positive number δ so that

$$|g(a') - g(a'')| < \omega_g(a) + K(b - a)/8$$

if a' and a'' are in $(a - \delta, a + \delta)$. Let U be an open interval centered at $(a + b)/2$ with length smaller than both δ and $(b - a)/4$. Suppose that $x \in U$. We may choose points y' and y'' within $(b - a)/4$ and within $\delta/2$ of b so that

$$|g(y') - g(y'')| > \omega_g(b) - K(b - a)/8. \tag{2.19}$$

But then $2x - y'$ and $2x - y''$ are within δ of a so that

$$|g(2x - y') - g(2x - y'')| < \omega_g(a) + K(b-a)/8. \qquad (2.20)$$

Combining (2.17), (2.19) and (2.20) we have

$$|g(y') - g(2x - y')| + |g(y'') - g(2x - y'')|$$
$$\geq |g(y') - g(y'')| - |g(2x - y') - g(2x - y'')|$$
$$> \omega_g(b) - \omega_g(a) - K(b-a)/4 > 3K(b-a)/4.$$

From this inequality we easily derive the estimate (2.18) of the lemma.

Recall that at each point x_0 we define $z(x_0, K)$ to be the supremum of the values z for which the set

$$\{(x,y) : x_0 < x < z \text{ and } y = \omega_f(x) > K(x - x_0)\}$$

is denumerable. If this supremum is finite at any point x_0 then an obvious argument show that points $x \notin \overline{D(f)}$ can be chosen so that their corresponding point $z(x, 2K)$ (with a doubled slope) cannot exceed $z(x_0, K)$. Since $D(f)$ is nowhere dense we can obtain the following which we state as a lemma.

Lemma 2.14 *If* $z_1 = z(x_1, K) < +\infty$ *then for any* $\delta > 0$ *there is a point* $x_2 \notin \overline{D(f)}$ *so that*

$$z_1 - \delta < z_2 = z(x_2, 2K) \leq z_1$$

From Lemma 2.13 we obtain the following lemma.

Lemma 2.15 *If* $z = z(x, K) < +\infty$ *then then there is a closed neighbourhood* I *of the point* z *so that*

$$\sup_{0 < t < 1/K} \left| \frac{f(x+t) - f(x-t)}{2t} \right| > K/2. \qquad (2.21)$$

at every point $x \in I$.

Proof. By the definition of the point $z = z(x, K)$ we see that there are uncountably many $h > 0$ and close to zero so that $\omega_f(z+h) > 2K((z+h) - x)$ and for any $\epsilon > 0$ all but countably many $h > \epsilon$ satisfy

$$\omega_f(z - h) \leq 2K((z - h) - x).$$

Evidently then we may choose a number $0 < h < 1/2K$ so that

$$\omega_f(z + h) > \omega_f(z - h) + 4Kh.$$

From Lemma 2.13 we then, with $a = z - h$ and $b = z + h$, may obtain a closed interval centered at z and such that (2.21) holds as required.

Now we can give the closing arguments using these lemmas. If $D(f)$ is not denumerable then we may certainly select a point $x_1 \notin \overline{D(f)}$ so that $z(x_1, 1) < \infty$. Lemma 2.15 allows us to select a closed neighbourhood I_1 of $z(x_1, 1)$ so that

$$\sup_{0 < t < 1} \left| \frac{f(x + t) - f(x - t)}{2t} \right| > 1/2.$$

at every point $x \in I_1$.

Now by Lemma 2.14 we may select a point $x_2 \notin \overline{D(f)}$ so that $z(x_2, 2)$ lies in I_1; thus again we may select a closed neighbourhood I_2 of $z(x_2, 2)$ contained in I_1 so that

$$\sup_{0 < t < 1/2} \left| \frac{f(x + t) - f(x - t)}{2t} \right| > 1.$$

at every point $x \in I_2$.

Proceeding inductively, we arrive at a nested sequence of intervals $\{I_n\}$ so that

$$\sup_{0 < t < 2^{-n+1}} \left| \frac{f(x + t) - f(x - t)}{2t} \right| > 2^{n-2}.$$

at every point $x \in I_n$. The point x that belongs to all of the intervals then evidently has

$$\limsup_{t \to 0} \left| \frac{f(x + t) - f(x - t)}{2t} \right| = +\infty$$

which contradicts the fundamental hypothesis of our main theorem. This contradiction establishes that $D(f)$ is denumerable.

2.3.4 $D(f)$ is Scattered

We turn now to the final part of Charzyński's proof and establish that the set $D(f)$ is in fact scattered. We begin with a supplementary result; this constructs a countable set from the limits points of $D(f)$ that is used to obtain the scattered part of the proof.

Lemma 2.16 *Let the function f satisfy at every point x the condition*

$$\limsup_{h \to 0} \left| \frac{f(x + h) - f(x - h)}{2h} \right| < +\infty. \tag{2.22}$$

Then the set of points x at which

$$\limsup_{y \to x} \frac{\omega_f(y)}{|y - x|} = +\infty \tag{2.23}$$

is denumerable.

Proof. If E denotes the set of points at which (2.23) holds and $x \in E$ then there are positive numbers K and δ so that

$$\left| \frac{f(x+t) - f(x-t)}{2t} \right| < K \qquad (2.24)$$

if $0 < t < 2\delta$. Then if $y - x < \delta$ we must have

$$|\omega_f(2x - y) - \omega_f(y)| < 4K|y - x| \qquad (2.25)$$

for if not then, by Lemma 2.13, there is an $h < 2\delta$ with

$$\left| \frac{f(x+h) - f(x-h)}{2h} \right| \geq K$$

in violation of (2.24). Since

$$\limsup_{y \to x} \frac{\omega_f(y)}{|y - x|} = +\infty$$

we may choose a number y arbitrarily close to x so that $\omega_f(y) > 5K|y - x|$ and hence, in particular, from (2.25) we see that $\omega_f(2x - y)$ is positive.

Thus to each point $x \in E$ we may associate a pair of points

$$(2x - y, y) \in D(f) \times D(f).$$

This maps E one-one into a subset of the denumerable set $D(f) \times D(f)$ and so E must be denumerable as required.

Now let us complete the proof that $D(f)$ is scattered. We suppose in contradiction to what we wish to prove that $D(f)$ contains a set D_0 that is dense in itself. Let $\{e_1, e_2, e_3, \ldots\}$ be an enumeration of the set E from Lemma 2.16. We show that we may choose a nested sequence of intervals $\{[x_n - \epsilon_n, x_n + \epsilon_n]\}$ whose intersection must contain a single point e that necessarily belongs to E and yet each interval $[x_n - \epsilon_n, x_n + \epsilon_n]$ may be chosen to avoid the point e_n. This contradiction proves that there is no such set D_0 and so $D(f)$ must be scattered.

The construction is inductive. We start by choosing $x_1 \in D_0$, $x_1 \neq e_1$, and ϵ_1 smaller than $\min\{1, \omega_f(x_1), |x_1 - e_1|\}$ and then, for $n > 1$ we choose $x_n \in D_0$, $x_n \neq e_n$, $x_n - x_{n-1} < \epsilon_{n-1}$ and ϵ_n smaller than

$$\min\left\{ \frac{1}{n}, \frac{\omega_f(x_n)}{n}, |x_n - e_n|, x_n - x_{n-1} + \epsilon_{n-1}, x_{n-1} - x_n + \epsilon_{n-1} \right\}.$$

Since D_0 is dense in itself this is certainly possible. The sequence of intervals $\{[x_n - \epsilon_n, x_n + \epsilon_n]\}$ contains a single point x in their intersection and certainly x cannot be in E since it cannot agree with any member of the sequence $\{e_n\}$ that enumerates E. On the other hand $x_n \to x$ and

$$\frac{\omega_f(x_n)}{|x_n - x|} > \frac{\omega_f(x_n)}{\epsilon_n} > n$$

and so x must be in E. This contradiction establishes the theorem.

This proof that $D(f)$ is scattered has been reproduced from the original paper. For a more transparent proof of this last step see Lemma A.9 in the Appendix where a straightforward category argument supplies the scattered nature of the sets.

2.3.5 The Even Analogue of Charzyński's Theorem

The even analogue of the Charzyński Theorem merely replaces the odd condition in the statement of Theorem 2.9 with its even parallel. The addition of some regularity hypothesis, such as measurability, to the even analogues is always necessary; this will be explained in detail in Chapter 4. This is not necessary in the original Charzyński theorem. In fact the condition there already implies that the function f is measurable.

Theorem 2.17 *Let f be a measurable function that satisfies the condition*

$$\limsup_{h \to 0} \left| \frac{f(x + h) + f(x - h) - 2f(x)}{h} \right| < +\infty \qquad (2.26)$$

at every point x. Then f is continuous at every point with the exception only of a scattered set.

The literature on this theorem is rather less direct than that for the odd version. Auerbach [9] shows that for an integrable function and any $\alpha > 0$ the condition

$$\lim_{h \to 0} \left| \frac{f(x + h) + f(x - h) - 2f(x)}{h^\alpha} \right| = 0 \qquad (2.27)$$

holding at every point will require the set of discontinuity points to be measure zero and nowhere dense; we shall present a version of this as Theorem 2.18. Neugebauer [213] applies this to measurable functions with $\alpha = 1$ in (2.27) and obtains the same conclusion. In a later paper [212] he shows that the set is countable. Evans and Larson [96] carry this to the final version by showing that the set of discontinuities can be characterized as scattered but again asserted for condition (2.27) with $\alpha = 1$ rather than the slightly weaker (2.26). The statement and proof of Theorem 2.17 appears in [285].

For a proof we have here only to appeal to Theorem 2.5; if f satisfies condition (2.26) at every point x then, by the Stein–Zygmund theorem (Theorem 2.3) it has a dense set of points of continuity and so, by Theorem 2.5, there is a measurable function g with the same set of points of continuity and g satisfies the hypotheses of the original Charzyński theorem.

2.4 Mazurkiewicz–Auerbach Theorem

As an application of the methods of Charzyński we turn to some related results initiated by Mazurkiewicz [197] and Auerbach [9]. They determine the continuity properties of a function f that satisfies at every point x a condition of the form

$$|f(x+h) + f(x-h) - 2f(x)| = O(h^\alpha) \tag{2.28}$$

or

$$|f(x+h) - f(x-h)| = O(h^\alpha) \tag{2.29}$$

as $h \to 0+$. The methods here differ from theirs but it is appropriate to attribute this theorem to them.

Theorem 2.18 (Mazurkiewicz–Auerbach) *Suppose that f is a measurable function satisfying condition (2.28) [or (2.29)] for some $\alpha > 0$ at every point of an interval (a,b). Then the set of points of discontinuity of f is measure zero and nowhere dense in (a,b).*

Proof. Because of the Stein-Zygmund theorem we already know that the set of points where f is continuous has full measure in (a,b). Thus we can apply Theorem 2.5 to obtain a function g as there with

$$|g(x+h) - g(x-h)| = O(h^\alpha) \tag{2.30}$$

as $h \to 0+$ at every point of (a,b). The $\alpha > 0$ may depend on x. Recall that the set of points where g vanishes is precisely the set of points of continuity of g and f. In particular g vanishes on a dense set.

Now, using (2.30), and employing the methods of Section 2.3.2 with only minor modifications we see that the set of discontinuity points of g is nowhere dense.

2.5 Wolibner Theorem

Suppose that at every point x the function f has an everywhere vanishing symmetric derivative:

$$\text{SD } f(x) = \lim_{h \to 0} \frac{f(x+h) - f(x-h)}{2h} = 0.$$

From our experience in the elementary calculus we should expect that f should be constant or nearly so; of course the example of the characteristic function of any finite set shows that there is some limitation. We already know that, for continuous functions f, a vanishing symmetric derivative requires that f be constant (Corollary 1.5); we know too that a measurable function that has a zero symmetric derivative is almost everywhere constant (Theorem 1.16). Moreover, in view of Charzyński's theorem, such a function must be continuous except on a very small (scattered) set. Thus we are already very close to the statement of Wolibner's theorem (Corollary 2.20 below) which asserts that such a function (not assumed in advance to be measurable) is constant on its set of points of continuity.

The original theorem and method of Wolibner was used by Charzyński to obtain the more general theorem below. We ask instead what the condition

$$\limsup_{h \to 0} \left| \frac{f(x+h) - f(x-h)}{2h} \right| \leq M$$

imposes on the function f. If this were an ordinary derivative rather than a symmetric extreme derivate then f would be a Lipschitz function with Lipschitz constant not exceeding M:

$$|f(x) - f(y) \leq M|x - y|.$$

Guided by Wolibner's result and by his methods, Charzyński obtains the following theorem.

Theorem 2.19 (Charzyński) *Let the function f satisfy at every point x the condition*

$$\limsup_{h \to 0} \left| \frac{f(x+h) - f(x-h)}{2h} \right| \leq M.$$

Then there exists a function g satisfying a Lipschitz condition

$$|g(x) - g(y)| \leq M|x - y|$$

such that f and g agree on the set of points of continuity of f.

Proof. The proof is an elementary compactness argument very similar to Khintchine's proof for Theorem 1.4. If the theorem is false there is a number $M' > M$ and points x_1 and x_2 at which the function is continuous with nevertheless

$$f(x_2) - f(x_1) > M'|x_2 - x_1|.$$

We may suppose that $x_1 < x_2$ and define the function

$$g(x) = f(x) - M' \left(x - \frac{x_1 + x_2}{2} \right) - \frac{f(x_1) + f(x_2)}{2}.$$

Note that g is just f less a linear function so that they have the same points of continuity and, by a simple computation, $g(x_1) < 0 < g(x_2)$. Let z be the supremum of those numbers $x_1 < x < x_2$ with $g(t) < 0$ for all but denumerably many $x_1 \le t < x$. Certainly $x_1 < z < x_2$ because both endpoints are points of continuity; also for every

$$\delta_1 < \delta_2 < \min\{z - x_1, x_2 - z\}$$

at this point $g(z - h) < 0$ for all but denumerably many $\delta_2 > h > \delta_1$ while there are uncountably many $h < \delta_2$ with $0 < h < \delta_2$ and $g(z+h) > 0$. Translating this in terms of the function f this means that there are arbitrarily small positive numbers h with

$$\frac{f(z + h) - f(z - h)}{2h} > M'$$

which contradicts the main assumption of the theorem. This contradiction completes the proof.

As a corollary we obtain Wolibner's result by setting $M = 0$ in the theorem.

Corollary 2.20 (Wolibner) *Let the function f satisfy at every point x the condition $SD\, f(x) = 0$. Then f is constant on its set of points of continuity.*

Of course we know from Theorem 2.9 that this set of points of continuity necessarily includes all points excepting only a scattered set. If we combine what we now know with the Theorem 1.16 of Auerbach then we can obtain the following. Note that here we no longer need to assume in advance that the function is measurable.

Corollary 2.21 *Let f be a real valued function on the interval $[a, b]$ that satisfies the condition*

$$\limsup_{t \to 0} \left| \frac{f(x + t) - f(x - t)}{2t} \right| < +\infty$$

at every point of (a, b) and suppose that $SD\, f(x) = 0$ at almost every point of (a, b). Then f is constant in this interval off a scattered set.

A final remark of Szpilrajn [42, p. 225] is needed in order to close this discussion. Theorem 2.19 asserts that a function f that satisfies the condition

$$\limsup_{h \to 0} \left| \frac{f(x + h) - f(x - h)}{2h} \right| \le M$$

at every point agrees with some continuous Lipschitz function at the points of continuity of f. If we weaken this to the condition

$$\limsup_{h\to 0} \left| \frac{f(x+h)-f(x-h)}{2h} \right| < \infty \qquad (2.31)$$

then f need not agree closely with any continuous function. The simple example $f(x) = \cos x^{-1}$ satisfies property (2.31) at every point but allows agreement with no everywhere continuous function on any set containing points close to 0.

2.5.1 A Monotonicity Theorem

The methods we have just studied can be used to obtain a monotonicity theorem, one that was overlooked for some time. Freiling [111] has pointed this out and explored the consequences in great detail. We shall return to his ideas in Chapter 5 but for now we prove this result using the classical ideas we have already seen. This result combines the monotonicity and continuity ideas in a fruitful way.

Theorem 2.22 *Let the function f satisfy at every point x the condition*

$$0 \le \underline{SD} f(x) \le \overline{SD} f(x) < +\infty.$$

Then f is nondecreasing off a scattered set.

Proof. The proof is again an elementary compactness argument nearly identical to that used in Theorem 2.19. By Charzyński's theorem the function f is continuous off a scattered set S; we show that it is nondecreasing off S. If this is false then there are points x_1 and x_2 with $x_1 < x_2$ at which the function is continuous with nevertheless $f(x_1) > f(x_2)$. Define the function

$$g(x) = f(x) + \eta \left(x - \frac{x_1 + x_2}{2} \right) - \frac{f(x_1) + f(x_2)}{2}$$

where $\eta > 0$ is chosen sufficiently small that $g(x_1) > 0 > g(x_2)$.

Note that g is just f plus a linear function so that they have the same points of continuity. Let z be the supremum of those numbers $x_1 < x < x_2$ with $g(t) > 0$ for all but denumerably many $x_1 \le t < x$. Certainly $x_1 < z < x_2$ because both endpoints are points of continuity; also for every

$$\delta_1 < \delta_2 < \min\{z - x_1, x_2 - z\}$$

at this point $g(z-h) > 0$ for all but denumerably many $\delta_2 > h > \delta_1$ while there are uncountably many $h < \delta_2$ with $0 < h < \delta_2$ and $g(z+h) < 0$. Translating this in terms of the function f this means that there are arbitrarily small positive numbers h with

$$\frac{f(z+h) - f(z-h)}{2h} < -\eta$$

so $\underline{SD} f(z) \le -\eta$. This contradicts the main assumption of the theorem and this contradiction completes the proof.

2.6 Jurek–Szpilrajn Example

We have seen, in Corollary 2.21, that for the condition SD $f(x) = 0$ to hold everywhere the function f must be constant off some scattered set. This is the best possible statement that can be made. Szpilrajn [279] published in 1933 the following construction showing that any scattered set may appear as the exceptional set outside of which f is constant. He acknowledged in a note in the next volume of Fundamenta Mathematicae that the first proof must be credited to Jurek.

Theorem 2.23 (Jurek–Szpilrajn) *Let S be scattered. Then there is a function f that is positive on S and vanishes otherwise but has a zero symmetric derivative everywhere.*

Proof. Let S be scattered. Thus S is both left scattered and right scattered and we may use Lemma A.7 (in the Appendix) to produce functions g_1 and g_2 both vanishing off S, both positive on S and such that $g_1(x - h) \leq h^2$ and $g_2(x + h) \leq h^2$ for all x and all sufficiently small $h > 0$. Take $f = g_1 + g_2$. Then, at every point x and for all sufficiently small $h > 0$,

$$-h^2 \leq -g_1(x - h) \leq f(x + h) - f(x - h) \leq g_2(x + h) \leq h^2.$$

Evidently SD $f(x) = 0$ everywhere. The theorem is then proved.

The arguments of Szpilrajn [279] were used by Evans and Larson [96] to show that Theorem 2.17 cannot be further improved. Since their construction is so intimately related to that for Theorem 2.23 and is the even version of that theorem we present it here as a corollary.

Corollary 2.24 (Evans–Larson) *Let S be a scattered subset of the interval $(0, 1)$. Then there is a measurable function g defined on that interval such that*

$$\limsup_{h \to 0} \left| \frac{g(x + h) + g(x - h) - 2g(x)}{h^2} \right| < +\infty \qquad (2.32)$$

at every point x and g is discontinuous precisely at the points in the set S.

Proof. Let $s_1, s_2, s_3 \ldots$ be an enumeration of the set S and let f be the function defined in the theorem. We define a function H on the interval $(0, 1)$ by writing

$$H(x) = \sum_{s_i < x} 2^{-i-1} f(s_i).$$

This function is a saltus function that has jump discontinuities exactly at the points of S. We can adjust this function so as to satisfy (2.32) by defining $g(x) = H(x)$ if $x \notin S$ and at any point x in S we write

$$g(x) = \frac{H(x+0) + H(x-0)}{2}.$$

We claim that g satisfies (2.32) and that its set of discontinuity points is precisely S. The latter fact is clear. To see the former take any point $x \in (0,1)$. If $x \notin S$ then for sufficiently small h

$$|g(x+h) + g(x-h) - 2g(x)| \le |g(x+h) - g(x)| + |g(x) - g(x-h)|$$
$$\le \sup\{f(s_i) : s_i \in (x-h, x)\} + \sup\{f(s_i) : s_i \in (x, x+h)\} \le 2h^2.$$

If $x \in S$ then

$$|g(x+h) + g(x-h) - 2g(x)| \le |g(x+h) - g(x+0)| + |g(x-0) - g(x-h)|$$

and again $|g(x+h) + g(x-h) - 2g(x)| \le 2h^2$.

In each case $|g(x+h) + g(x-h) - 2g(x)| = O(h^2)$ as $h \to 0$ as required for (2.32). Since g is monotone it is measurable and the proof is complete.

Freiling [111] produces a more refined version of the Jurek–Szpilrajn example using semi-scattered sets; this is useful in setting a limit as to how far a monotonicity theorem can go for the lower symmetric derivative.

Corollary 2.25 (Freiling) *Given any disjoint sets R and L such that R is right scattered and L is left scattered there is a function f positive on R, negative on L and elsewhere zero such that $\underline{SD} f(x) = 0$ everywhere.*

Proof. Suppose that L is left scattered and R right scattered with $R \cap L = \emptyset$; again we use Lemma A.7 to produce functions g_1 and g_2 vanishing off of R and L respectively, both positive on these sets and such that $g_1(x-h) \le h^2$ and $g_2(x+h) \le h^2$ for all x and all sufficiently small $h > 0$. Take $f = g_1 - g_2$. Then, at every point x and for all sufficiently small $h > 0$,

$$-2h^2 \le -g_1(x-h) - g_2(x+h) \le f(x+h) - f(x-h).$$

Evidently $\underline{SD} f(x) \ge 0$ everywhere; but f vanishes off a countable set and so $\underline{SD} f(x) \le 0$ too.

2.7 Pesin–Preiss Theorem

As a special case of the Stein and Zygmund Theorem (Theorem 2.1) we know that for a measurable function f that is everywhere symmetrically continuous the set of points $D(f)$ where f is discontinuous in the ordinary sense is of measure zero. Since this set is necessarily a Borel set of type F_σ (this is true for any function) this means that a measurable, symmetrically continuous function

f must have its set of discontinuity points $D(f)$ both of measure zero and of the first category.

The arguments in Charzyński's original paper show that the set $D(f)$ must be first category without assuming measurability for the function f. (This was reproduced by Fried [119] with a direct acknowledgment that the ideas were due to Charzyński; unfortunately the literature more often cites Fried than the original paper of Charzyński to whom the ideas are properly attributed.) It remained unknown for some time whether a symmetrically continuous function would have to be measurable. This was proved to be the case by Pesin [220] and, independently, by Preiss [230]; Preiss shows that Charzyński's (or Fried's) methods together with some further computations will prove measurability. The Stein–Zygmund Theorem can be used to avoid even these computations.

Theorem 2.26 (Pesin–Preiss) *Let f be a function that is everywhere symmetrically continuous. Then f is measurable and its set of points of discontinuity is of measure zero and of the first category.*

Proof. Because of the Stein–Zygmund Theorem it would be enough to show that such a function is measurable; it follows then that the set of points of discontinuity is a measure zero F_σ-set and so also first category. However to achieve this we shall need to show that f is a.e. continuous in any case.

We shall use the methods of Charzyński to show that a function that is everywhere symmetrically continuous must be continuous on a dense (in fact residual) set. It is enough to show that, for every interval (a, b) and every positive number ϵ, there is a subinterval $(c, d) \subset (a, b)$ so that $\omega_f(x) \leq \epsilon$ at every point x of (c, d). Here ω_f denotes the oscillation of the function f defined as usual:

$$\omega_f(x) = \limsup_{h \to 0+} \{ |f(x_1) - f(x_2)| \; : \; x_1, x_2 \in (x - h, x + h) \}.$$

For each $x \in (a, b)$ there is a $\delta(x)$ so that

$$0 < t < \delta(x) \implies |f(x + t) - f(x - t)| < \epsilon/8.$$

By our standard methods there is a sequence of sets $\{E_n\}$ covering (a, b) so that

$$x \in E_n, 0 < t < 1/n \implies |f(x + t) - f(x - t)| < \epsilon/8.$$

By the Baire category Theorem there is a set E_n and a subinterval (c, d) of (a, b) so that E_n is dense in (c, d) and $d - c < 1/2n$. We merely show that for any two points $x_1, x_2 \in (c, d)$ the inequality $|f(x_1) - f(x_2)| < \epsilon$ must hold and the theorem is proved.

Let x_0 be a point of $E_n \cap (c, d)$ and let x_1 be any point of the interval (c, d); set $2t = x_1 - x_0$ and select t' so that $x_0 + t' \in E_n$ is sufficiently close to $x_0 + t/2$ so that

$$|f(x_0 + 2t') - f(x_0 + 2t - 2t')| < \epsilon/8.$$

Here we have used the symmetric continuity of f at the point $x_0 + t$. Now by reflections about the points $x_0 + t'$, $x_0 + t$, x_0 and $x_0 + t'$ we have

$$
\begin{aligned}
|f(x_0) - f(x_0 + 2t')| &< \epsilon/8 \\
|f(x_0 + 2t') - f(x_0 + 2t - 2t')| &< \epsilon/8 \\
|f(x_0 + 2t - 2t') - f(x_0 + 2t' - 2t)| &< \epsilon/8 \\
|f(x_0 + 2t' - 2t) - f(x_0 + 2t)| &< \epsilon/8
\end{aligned}
$$

which together give

$$|f(x_0) - f(x_1)| = |f(x_0) - f(x_0 + 2t)| < \epsilon/2.$$

Consequently, for any two points x_1 and x_2 of this interval,

$$|f(x_2) - f(x_1)| < \epsilon$$

as required. It follows that f is continuous at the points of a dense set. By Theorem 2.5 there is a measurable function that is everywhere symmetrically continuous and has the same set of points of discontinuity. We know that such a function is a.e. continuous and so f must be a.e. continuous too.

2.8 Local Symmetry

A set E is exactly symmetric at a point x if $x + h \in E$ whenever $x - h \in E$. A function f is exactly symmetric at a point x if $f(x + h) = f(x - h)$ for all h.

In this section we study local versions of these. We can say that a function f is *exactly locally symmetric* if at each point x there is a $\delta > 0$ so that $f(x + h) = f(x - h)$ holds for all $0 < h < \delta$. A set is *exactly locally symmetric* if its characteristic function is exactly locally symmetric. If f is the characteristic function of a set then the notions of exact local symmetry and symmetric continuity evidently coincide. It is for this reason that we place these considerations in this chapter.

It was posed as a query in an early issue of the Real Analysis Exchange (see [78]) whether such a set (or function) need be measurable. Of course the theorem of Pesin–Preiss, just established, shows that the answer is yes. It was also indicated by Davies [60], in a subsequent issue of the Exchange, that the arguments of Charzyński's Theorem allowed a sharper statement. In the same issue Ruzsa [250] obtained the identical result by similar means. Since this result responds to the classical methods already studied we shall include it here to help further illustrate the methods.

Theorem 2.27 (Davies–Ruzsa) *Let f be a function such that at each point x there is a positive number $\delta(x)$ so that*

$$0 < t < \delta(x) \implies f(x+t) - f(x-t) = 0.$$

Then f is constant off a closed countable set.

Proof. It is enough to show that f is constant on at least one open interval (c, d); for if so and f assumes the value k there then consider the set

$$X = \mathbb{R} \setminus \text{int } f^{-1}(k).$$

If Y is the set of condensation points of X then either Y is empty (which establishes the theorem) or else Y is perfect. If the latter is true then, since we know Y cannot contain all real numbers there is an interval (a, b) complementary to Y with one endpoint finite. But for this endpoint the symmetry condition would require it to be isolated in Y which is impossible.

Thus the proof reduces to obtaining f constant on some open interval. The condition of the theorem translates by our usual methods (cf. the proof of Theorem 2.26) to the existence of a sequence of sets $\{E_n\}$ covering the line so that

$$x \in E_n, \ 0 < t < 1/n \implies f(x+t) - f(x-t) = 0.$$

A Baire category argument provides an interval (c, d) and a set E_n dense in that interval and the series of reflections from the proof of Theorem 2.26 show that f is constant there.

If f is the characteristic function of a set E then the theorem reduces to the following.

Corollary 2.28 *Let E be locally symmetric at every point of the real line. Then either E or its complement has countable closure.*

The even analogue of Theorem 2.27 is obtained by Kostyrko [154] and [157] (who calls such functions *locally Jensen*).

Theorem 2.29 (Kostyrko) *Let f be measurable and suppose that at each point x there is a positive number $\delta(x)$ so that*

$$0 < t < \delta(x) \implies f(x+t) + f(x-t) - 2f(x) = 0.$$

Then there exists a closed locally symmetric set S and a constant m such that in each interval complementary to S the graph of f is a straight line with slope m.

Proof. The reader may consult [154, Theorem 2, p. 66] for details. Very much the the same proof as that given for Theorem 2.27 may be given. One obtains by a Baire category argument and a series of reflections as in the proof of Theorem 2.26 that there is an interval in which the condition

$$f(x+t) + f(x-t) - 2f(x) = 0$$

holds uniformly. For measurable functions this means that f is of the form $f(x) = mx + b$ in that interval. (See Section 4.2.3 below for the history of this last assertion and a proof.)

We conclude this section with a remark on a related problem. One might ask, as did Marcus [192], whether a set S can exist with the property that at every point x there is a $\delta > 0$ so that $x + h \in S$ if and only if $x - h \notin S$ ($0 < |h| < \delta$). Such a set would be the opposite of locally symmetric, say locally *antisymmetric*. It is easy to suspect that no such set exists; if a measurable set S had this property it would clearly have density $1/2$ at every point, which is impossible. Kostyrko [157] has shown that no such set, measurable or not, can exist. The methods are exactly the methods of Charzyński that we have used already in this section. See Section 6.4.4 for further developments that continue this theme.

2.8.1 Points of Local Symmetry

The set of points of local symmetry of a set will play a role in many investigations. We shall write $E^\star$ for the set of all points at which a set is locally symmetric; thus $x \in E^\star$ if and only if there is a $\delta > 0$ so that $x + h \in E$ implies $x - h \in E$ for all $0 < |h| < \delta$. Note that, in this language, a set E is exactly locally symmetric if $E^\star = \mathbb{R}$.

Having just discussed the case of sets everywhere locally symmetric it is natural to turn to an investigation of the structure of the sets $E^\star$. We will return to this topic in Chapter 6. Here are two elementary results on points of symmetry; the second is from Hammill [128]. See also Corollary 2.45 for a further property of such sets.

Theorem 2.30 *Let C denote the Cantor ternary set. Then $C \cap C^\star = \emptyset$.*

Proof. Suppose that C is symmetric about a point $z \in C$ inside some interval $(z - \delta, z + \delta)$. Let us call any interval complementary to C in $[0, 1]$ an nth stage interval if it has length 3^{-n}. For large enough N we may choose the right most Nth stage interval (say I) in $(z - \delta, z)$. If I' is the reflection of I about z then I' is also an Nth stage interval and there can be no other Nth stage intervals between them. By the way in which the Cantor set is constructed this places z at the center of an interval that should have been removed at an earlier stage in the construction and this contradicts the fact that z must belong to C.

Theorem 2.31 (Hammill) *If S is open (or closed) then $S^\star$ is the union of an open set and a countable, nowhere dense set.*

Proof. Let $S \neq \mathbb{R}$ be open. Clearly $S^* \supset S \cup (\mathrm{int}(\mathbb{R} \setminus S))$. Write

$$F = S^* \setminus (S \cup \mathrm{int}(\mathbb{R} \setminus S))$$

and let $x \in F$. Since $x \in S^*$ there is an open interval $I(x)$ of length ≤ 1 centered at x such that $y \in S$ if and only if $2x - y \in S$ for all $y \in I(x)$. Since $x \notin S \cup (\mathrm{int}(\mathbb{R} \setminus S))$ there are points of S in $I(x)$. Then, S being open, there are component intervals, J_x^+ and J_x^-, of S such that $J_x^+ \cap I(x)$ and $J_x^- \cap I(x)$ are nonempty and symmetric about x. Associate with x the largest such pair. In this way we can map F one to one into the set of pairs of component intervals of S. Since this set is countable F is also countable.

Now we need show only that F is nowhere dense. But this is clear since S is disjoint from F and has component intervals arbitrarily close to any point of F.

As a corollary, also from [128], we have the following which enlarges on Theorem 2.30. We remark too that [128] supplies an example of a perfect, nowhere dense set S in which $S^* \cap S$ is dense.

Corollary 2.32 (Hammill) *If S is a perfect, nowhere dense set then $S^* \cap S$ is countable.*

2.9 Points of Discontinuity

A symmetrically continuous function f can evidently have points of discontinuity. We have seen that the set $D(f)$ of points of discontinuity is both of measure zero and first category. On the other hand, as indicated previously, any countable set can appear as the set of discontinuities of a symmetrically continuous function. Hausdorff posed in *Fundamenta Mathematicae* the problem of whether the set of points of discontinuity can be uncountable. Preiss [230] has answered this by showing how to construct a function f, everywhere symmetrically continuous, such that the set $D(f)$ is uncountable. A complete characterization of the set $D(f)$ for symmetrically continuous functions f is not currently known.

We shall express this example in the language of N–sets. A set E is said to be an N–set if there is a trigonometric series

$$a_0/2 + \sum_{n=1}^{\infty} (a_n \cos nx + b_n \sin nx) \qquad (2.33)$$

with

$$\sum_{n=1}^{\infty} (|a_n| + |b_n|) = +\infty$$

and yet the series (2.33) converges absolutely at every point $x \in E$. From the theorem of Lusin and Denjoy (Lemma 2.34 proved below) we know that an

N-set must have measure zero and be of the first category. That there are uncountable N-sets is well known: for example Zygmund [313, vol. I, p. 250] gives the series

$$\sum_{n=1}^{\infty} n^{-1} \sin n! x$$

as an example of a trigonometric series which converges absolutely on an uncountable set without being absolutely convergent everywhere. (His suggestion is to consider the graphs of the curves $y = \sin n! x$.) Thus an N-set may be uncountable. No characterization of such sets is known.

The theorem shows that given any N-set E there is a symmetrically continuous function f that is discontinuous at every point in E. (It may be discontinuous at other points too.)

Theorem 2.33 (Preiss) *Let E be an N-set. Then there is a bounded, 2π-periodic function f that is everywhere symmetrically continuous and discontinuous at each point in E.*

For the proof we require a number of elementary lemmas. The first of these gives necessary conditions for an N-set. This is due, independently, to Lusin [181] and Denjoy [65]. We reproduce the standard proof (cf. [313, vol. I, pp. 232–233] or [11, vol. I, pp. 173–174]).

Lemma 2.34 (Lusin–Denjoy) *Let $a_0/2 + \sum_{n=1}^{\infty} (a_n \cos nx + b_n \sin nx)$ be a trigonometric series with*

$$\sum_{n=1}^{\infty} |a_n| + |b_n| = +\infty.$$

Then the set

$$E = \{x : \sum_{n=1}^{\infty} |a_n \cos nx + b_n \sin nx| < +\infty\}$$

is of type F_σ and of measure zero.

Proof. Let us simplify the writing with a standard manipulation: let

$$\rho_n = \sqrt{(a_n)^2 + (b_n)^2}, \quad a_n = \rho_n \cos \alpha_n \text{ and } b_n = \rho_n \sin \alpha_n.$$

Then

$$a_n \cos nx + b_n \sin nx = \rho_n \cos(nx - \alpha_n)$$

and E is exactly the set of points x at which

$$g(x) = \sum_{n=1}^{\infty} \rho_n |\cos(nx - \alpha_n)| < +\infty.$$

Suppose, in order to obtain a contradiction, that $|E| > 0$. Then there is a measurable set $P \subset E$ of positive on which g is bounded and so

$$\sum_{n=1}^{\infty} \rho_n \int_P |\cos(nx - \alpha_n)| \, dx = \int_P g(x) \, dx < \infty.$$

From this we conclude that $\sum_{n=1}^{\infty} \rho_n < +\infty$ which contradicts the assumption in the theorem that

$$\sum_{n=1}^{\infty} (|a_n| + |b_n|) = +\infty.$$

This is because the limit

$$\int_P |\cos(nx - \alpha_n)| \, dx \geq \int_P \cos^2(nx - \alpha_n) \, dx \to |P|/2$$

can be established from the Riemann-Lebesgue Lemma (see [313, vol. I, pp. 45–46]).

We have established thus far that E has measure zero. To see its structure write

$$E_n = \left\{ x : \sum_{n=1}^{\infty} |a_n \cos nx + b_n \sin nx| \leq n \right\}.$$

Each E_n is evidently closed and $E = \bigcup_{n=1}^{\infty} E_n$. This expresses E as a Borel set of type F_σ and completes the proof.

The set of points of absolute convergence of a trigonometric series has curious symmetry properties in general. We need one simple property (others are given in [313, vol. I, Theorem 1.10, p. 233]. This is attributed to Fatou in [11, vol. I, p. 308]).

Lemma 2.35 (Fatou)   *The set of points*

$$A = \left\{ x : \sum_{n=1}^{\infty} |a_n \cos nx + b_n \sin nx| < +\infty \right\}$$

is symmetric about each of its points.

Proof. To simplify the writing let

$$A_n(x) = a_n \cos nx + b_n \sin nx.$$

An elementary computation with trigonometric identities shows that

$$A_n(x + h) + A_n(x - h) = 2A_n(x) \cos nh.$$

Consequently if $x \in A$ then $x + h \in A$ if and only if $x - h \in A$.

Now we show how to construct a symmetrically continuous function that is discontinuous at a given N–set.

Lemma 2.36 *Let*

$$a_0/2 + \sum_{n=1}^{\infty} (a_n \cos nx + b_n \sin nx)$$

be a trigonometric series with $\sum_{n=1}^{\infty} (|a_n| + |b_n|) = +\infty$. Then the function

$$f(x) = \lim_{N \to \infty} \left(1 + \sum_{n=1}^{N} |a_n \cos nx + b_n \sin nx| \right)^{-1} \qquad (2.34)$$

is everywhere symmetrically continuous and upper semicontinuous.

Proof. The limit is seen to be the limit of a decreasing sequence of continuous functions and hence is upper semicontinuous (eg. see Natanson [210, vol. II, p. 153]). Also from the semicontinuity it is evident that f is continuous at every point x at which $f(x) = 0$.

From the preceding lemma we know that the set

$$A = \{x : \sum_{n=1}^{\infty} |a_n \cos nx + b_n \sin nx| < +\infty\}$$

is symmetric about each of its points. Thus if $x_0 \in A$ then $f(x_0 + h) = 0$ if and only if $f(x_0 - h) = 0$.

Suppose, on the other hand, that $x_0 \in A$ and $f(x_0 + h) > 0$. An elementary computation with trigonometric identities shows that

$$\Big| |A_n(x + h)| - |A_n(x - h)| \Big| \leq 2|A_n(x)|$$

and

$$\Big| |A_n(x + h)| - |A_n(x - h)| \Big| \leq 2(|a_n| + |b_n|)|\sin nh|.$$

For $\alpha_i \geq 0$ and $\beta_i \geq 0$ it is easy to see that

$$\left| \left(1 + \sum_{i}^{\infty} \alpha_i \right)^{-1} - \left(1 + \sum_{i}^{\infty} \beta_i \right)^{-1} \right| \leq \sum_{i}^{\infty} |\alpha_i - \beta_i|.$$

Using these inequalities we see that

$$|f(x_0 + h) - f(x_0 - h)| \leq \sum_{n=1}^{\infty} \Big| |A_n(x_0 + h)| - |A_n(x_0 - h)| \Big|$$

$$\leq \sum_{n=1}^{N} 2(|a_n| + |b_n|)|\sin nh| + \sum_{n=N+1}^{\infty} |A_n(x_0)|.$$

From this it is easy to establish the symmetric continuity of f at every point in A. At the remaining points $x \notin A$ we know that $f(x) = 0$ and can see that f is continuous there. Consequently f is symmetrically continuous everywhere.

Finally let us return to the proof of Theorem 2.33. Let E be an N–set. If

$$a_0/2 + \sum_{n=1}^{\infty} (a_n \cos nx + b_n \sin nx)$$

is a trigonometric series with $\sum_{n=1}^{\infty} |a_n| + |b_n| = +\infty$ then the set

$$P = \left\{ x : \sum_{n=1}^{\infty} |a_n \cos nx + b_n \sin nx| < +\infty \right\}$$

is a set of type F_σ and of measure zero. Since E is an N–set there is some suitable choice of the coefficients so that $E \subset P$. Then, as proved in Lemma 2.36, the function f defined by (2.34) is everywhere symmetrically continuous and upper semicontinuous. Since P is precisely the set of points where f is positive and since f vanishes on a dense set it must be discontinuous at every point of the set P. This completes the proof of Theorem 2.33.

It follows too, from Theorem 2.33, that if a set E is a countable union of N–sets then there is a 2π–periodic function f that is everywhere symmetrically continuous and discontinuous at each point in E. For if

$$E = \bigcup_{k=1}^{\infty} E_k$$

where each E_k is an N–set then take f_k as in the proof of the theorem for the set E_k. We have each f_k symmetrically continuous, $0 \le f_k \le 1$, $f_k(x) > 0$ for $x \in E_k$ and $f_k(x) = 0$ outside of a set of measure zero. If we take

$$f(x) = \sum_{k=1}^{\infty} 2^{-k} f_k(x)$$

then f is the uniform limit of symmetrically continuous functions, f is positive on E and zero outside of a set of measure zero. Evidently f is symmetrically continuous and discontinuous at every point in E.

Countable unions of N–sets probably do not provide a characterization of the discontinuity points of symmetrically continuous 2π–periodic functions. They do provide at least an indication of sufficient conditions and they can dispense with some naive conjectures. For example it is known that an N–set can have Hausdorff dimension arbitrarily close to unity; accordingly there must exist a function f that is everywhere symmetrically continuous and discontinuous at the points of a set of Hausdorff dimension one.

Also one might, perhaps, have expected a symmetrically continuous function to be discontinuous only at a σ-porous set. Again this is not so because of the following theorem of Konjagin (cited in Zajíček [310]).

Theorem 2.37 (Konjagin) *The set*

$$E = \left\{ x \; : \; \sum_{n=1}^{\infty} \frac{|\sin n!\pi x|}{n} \leq 1 \right\}$$

is a closed and non–σ–porous N–set.

2.9.1 Theorem of Tran

Tran [292] proves the following assertion. His construction is complicated and makes no appeal to ideas of the preceding section. It is reasonable to conjecture that there is an even analogue of Theorem 2.33 which would provide the same conclusion but this is not known.

Theorem 2.38 (Tran) *There is a bounded, measurable function f that is everywhere symmetric and is discontinuous at each point of an uncountable set.*

2.9.2 Theorems of Ponomarev and Chlebík

We conclude our discussion of discontinuity points of symmetrically continuous functions with an elementary theorem of Marcus [188], and extensions of Ponomarev [227] and Chlebík [43]. Marcus' theorem asserts that the classical Cantor set cannot be precisely the set of points of discontinuity of such functions. This already answers a problem of Hausdorff [129] as to whether, given an arbitrary measure zero set of type F_σ, there can exist a symmetrically continuous function that is discontinuous exactly at the points of that set.

Theorem 2.39 (Marcus) *Let $C \subset [0,1]$ be the usual Cantor ternary set. Then there cannot exist a symmetrically continuous function f so that*

$$C = \{x \; : \; f \text{ is discontinuous at } x\}.$$

Proof. Suppose that such a function f did exist. By Theorem 2.6 we can suppose that f is upper semicontinuous, that $0 \leq f \leq 1$ and that it is continuous precisely when it vanishes. As such an f is Baire 1 there must exist a point $z \in C \cap (0,1)$ so that f is continuous at z relative to the set C.

We claim first that such a point must be a point of local symmetry for C. If z is not a point of local symmetry of C then there is a sequence $x_n \in C$, $x_n \to z$ and $2z - x_n \notin C$. As f is supposed continuous outside of C it must vanish outside of C and so $f(2z - x_n) = 0$ for all n. On the other hand $f(z) > 0$ since $z \in C$ is a point of discontinuity and hence $f(x_n) \to f(z) > 0$. These two assertions are incompatible with the symmetric continuity of f at z since that must require $f(2z - x_n) - f(x_n) \to 0$.

But, by Theorem 2.30, C has no points of local symmetry; this contradiction completes the proof.

This theorem immediately raises more questions than it answers however. Is it just the structure of the Cantor set that makes this so or would other perfect nowhere dense sets have the same property? Is there a symmetrically continuous function that is discontinuous at the points of the Cantor set (and some other points too)? The theorems of Ponomarev and Chlebík answer this. While a complete characterization for the problem of discontinuity points for symmetrically continuous functions still eludes us these theorems point a possible direction.

Theorem 2.40 (Ponomarev) *Let $P \subset [0,1]$ be a perfect set. Then there cannot exist a symmetrically continuous function f so that*

$$P = \{x : \ f \text{ is discontinuous at } x\}.$$

Proof. Suppose that such a function f did exist. Certainly P must be nowhere dense. Exactly as in the proof of Theorem 2.39 we can suppose that f is upper semicontinuous, that $0 \leq f \leq 1$ and that it is continuous precisely when it vanishes. As such an f is Baire 1 there must exist a set Q residual in P so that f is continuous at each $z \in Q$ relative to the set P.

Again, as in the proof of Theorem 2.39, each point $z \in Q$ must belong to $P^\star$. But, by Corollary 2.32, $P^\star \cap P$ is countable; this contradiction completes the proof.

Chlebík [43] has given a broader attack on the same problem; we state and prove his theorem here. He shows that a certain type of perfect set is too large to allow the existence of a symmetrically continuous function that is discontinuous on the set. For any set E let the sets $S_k(E)$ be defined inductively by writing $S_0(E) = E$ and

$$S_k(E) = \{2x - y : \ x \in E, \ y \in S_{k-1}(E)\}.$$

A perfect set P will be called a *Chlebík set* if there is an integer k so that for every portion P' of P the set $S_k(P')$ contains an interval. The original proof appears not to have been published ([43] contains only the announcement); the one we present here is almost surely less elegant than what Chlebík had in mind.

Theorem 2.41 (Chlebík) *Let P be a Chlebík set and f a symmetrically continuous function. Then f is continuous at the points of a set residual in P.*

Proof. Let g denote the oscillation of f as used in the proof of Theorem 2.5. Recall the special features of g that we have already used; in particular g itself is everywhere symmetrically continuous and upper semicontinuous and g vanishes on a dense set.

For any $\epsilon > 0$ write

$$P(\epsilon) = \{x \in P : g(x) \geq \epsilon\}.$$

Each such set $P(\epsilon)$ is closed and

$$\{x \in P : f \text{ is discontinuous at } x\} = \bigcup_{n=1}^{\infty} P(1/n). \tag{2.35}$$

To prove the theorem it is enough, because of (2.35), if we prove that each set $P(\epsilon)$ is nowhere dense in P. To obtain a contradiction let us suppose that $\epsilon > 0$ and that there is an interval I_1 with $P(\epsilon) \cap I_1 = P \cap I_1 \neq \emptyset$.

Since P is a Chlebík set there is an integer k so that every portion of P has the property stated after the k reflections. Choose numbers

$$\epsilon/2 = r_k < r_{k-1} < \ldots < r_2 < r_1 = \epsilon.$$

For each natural number m let A_m denote the set of points $x \in \mathbb{R}$ such that

$$i = 1, 2, \ldots, k, \ 0 \leq |t| \leq 1/m, \ g(x+t) \geq r_i \implies g(x-t) > r_{i+1}. \tag{2.36}$$

Since g is symmetrically continuous at each point and the sequence of numbers $i = 1, 2, \ldots, k$ in the definition of the sets is finite it can be seen that $\bigcup_{m=1}^{\infty} A_m = \mathbb{R}$.

By the Baire category theorem there is a subinterval I_2 of I_1 and an integer m so that A_m is dense in $P \cap I_2 \neq \emptyset$. We may evidently suppose that $|I_2| < 1/(km)$.

Write $Q_0 = P \cap I_2$ and, for $i = 1, 2, \ldots$, let $Q_i = 2Q - Q_{i-1}$. Since P is a Chlebík set there is an interval J so that $Q_k \supset J$ where the integer k is as above. Every point z in the interval J has the following property: there are points x, x_1, x_2, $\ldots x_k$ in $P \cap I_2$ so that x reflected in turn about the points x_1, x_2, $\ldots$, x_k goes to z.

Since $x \in P \cap I_2 \subset P(\epsilon)$ we know that $g(x) \geq r_1$. As $x_1 \in P \cap I_2$ and A_m is dense there we can choose a sequence of points $x_{1p} \to x_1$ so that each $x_{1p} \in A_m$ and $|x_{1p} - x| < 1/m$; it follows that $g(2x_{1p} - x) > r_2$ for each p. By semicontinuity properties of g we then have at the limit point that $g(2x_1 - x) \geq r_2$. Repeating the same argument at the point x_2 would then give that $g(2x_2 - (2x_1 - x)) \geq r_3$. Continuing the series of reflections and limits through points in A_m we obtain finally that $g(z) \geq r_k = \epsilon/2$.

But this means that g is positive at every point of the interval J which contradicts the fact that g should vanish on a dense set. The statement of the theorem follows.

By a familiar argument (cf. [313, Vol. I, p. 235]) one can easily show that for the Cantor set C the set $\{2x - y : x, y \in C\}$ contains an interval; by the self-similarity of C this argument applies to every portion and it is then easy to see that the Cantor set is a Chlebík set. Consequently, as a corollary, we obtain an improved version of the Marcus theorem just proved.

Corollary 2.42 *Let $C \subset [0,1]$ be the Cantor ternary set. Then there cannot exist a symmetrically continuous function f that is discontinuous at every point in a subset of C that is second category in C.*

As a further corollary we obtain, because of Theorem 2.5, an identical version for symmetric functions.

Corollary 2.43 *Let $C \subset [0,1]$ be the Cantor ternary set. Then there cannot exist a measurable, symmetric function f that is discontinuous at every point in a subset of C that is second category in C.*

There is nothing special about the Cantor set here in these last two corollaries. Let $C(\xi)$ denote *any* symmetric perfect set with constant ratio ξ. Choose an integer p so that $1/p < \xi$. Then every $x \in [0,1]$ can be written as $x = \sum_{i=1}^{2p} x_i$ for some choice of $x_i \in C(\xi)$ (see [11, Theorem 4, p. 300]). Choose any $t \in [0,1]$, $z \in C(\xi)$ with $z \leq t$ and write

$$\tfrac{1}{2}(t - z) = \sum_{i=1}^{2p} x_i$$

in exactly this way. This gives

$$t = \sum_{i=1}^{4p} 2(-1)^{i+1} y_i + z$$

where $y_{2i+1} = x_i$ and $y_{2i} = 0$. This shows that $[0,1] \subset S_{4p+1}(C(\xi))$. As any portion of $C(\xi)$ behaves similarly, we have proved that any symmetric perfect set with constant ratio is a Chlebík set. Thus in the statements of Corollaries 2.42 and 2.43 any symmetric perfect set with constant ratio can be used in place of the Cantor set.

2.9.3 Points of Even/Odd Continuity

Is it possible to characterize the set of points at which a function can be symmetrically continuous? Is it possible to characterize the set of points at which a measurable function can be symmetric?

These are natural enough problems given that it is entirely elementary to characterize the set of points at which an arbitrary function may be continuous in the ordinary sense. This was achieved long ago by W. H. Young who showed that a set is of this type if and only it is a G_δ (or, in the language that he invented, "an inner limiting set"). We will return to problems of this kind in Chapter 6. In this section we present a few interesting facts.

Ponomarev [225] noted that the set O_f of points at which a function f can be symmetrically continuous need not be measurable if f is not measurable and need not be Borel if f is merely measurable. Jaskula and Szkopińska [140, p. 4] point this out too, as do Kostyrko, Neubrunn, Šalát and Smítal [158]. These observations are easy to make from the perspective of subgroups of the reals; if G is an additive subgroup of $\mathbb{R}$ then χ_G is symmetrically continuous precisely at the points in $\frac{1}{2}G$. See Section 6.2.1 for a discussion of additive subgroups.

It is fairly easy to see directly from the definition that the set of points at which a Borel measurable function may be symmetrically continuous [symmetric] is, however, at least co-analytic. We reproduce a proof from Darji [57].

Theorem 2.44 *For an arbitrary function f let O_f denote the set of points at which f is symmetrically continuous and let E_f denote the set of points at which f is symmetric. If f is Borel measurable then O_f and E_f are co-analytic. If f is Baire class 1 then O_f and E_f are $G_{\delta\sigma\delta}$.*

Proof. We consider first the case for the set O_f. For each integer m let $\mathcal{B}_m$ denote the class of all intervals with rational endpoints and length less than $1/m$. Write for $n \in \mathbb{N}$, $r \in \mathbb{Q}$, $I \in \mathcal{B}_m$

$$A(n,r,I) = \frac{1}{2}\left[f^{-1}\left((-\infty, r) \cap I \right) + f^{-1}\left((r+1/n, +\infty) \cap I \right) \right].$$

(Here $\frac{1}{2}[S_1 + S_2]$ is used to denote the set of points $\frac{1}{2}[s_1 + s_2]$ for $s_1 \in S_1$, $s_2 \in S_2$.)

We claim that

$$\mathbb{R} \setminus O_f = \bigcup_{n=1}^{\infty} \bigcap_{m=1}^{\infty} \bigcup_{I \in \mathcal{B}_m} \bigcup_{r \in \mathbb{Q}} A(n,r,I). \tag{2.37}$$

Suppose that $x \notin O_f$. There is a natural number k so that for every $\delta > 0$ there is a $0 < h < \delta$ such that $|f(x+h) - f(x-h)| > 2/k$. Fix any $m \in \mathbb{N}$. There is an $I \in \mathcal{B}_m$ with $x \in I$. Choose $h > 0$ with $x+h$, $x-h \in I$ and $|f(x+h) - f(x-h)| > 2/k$. Choose $r \in \mathbb{Q}$ so that r and $r+1/k$ lie between $f(x+h)$ and $f(x-h)$. Then $x \in A(k,r,I)$. Thus, since it is true for each m that such an I and r can be found, x is in the set

$$= \bigcap_{m=1}^{\infty} \bigcup_{I \in \mathcal{B}_m} \bigcup_{r \in \mathbb{Q}} A(k,r,I)$$

and it follows that x is in the set on the right hand side of (2.37). This proves set inclusion in one direction; the other direction is immediate.

It remains only to analyze the set in (2.37). If f is Borel measurable then each $A(n,r,I)$ is analytic (it is the sum of two Borel sets) and so $\mathbb{R} \setminus O_f$ is analytic and hence O_f is co-analytic. If f is Baire 1 then each $A(n, r, I)$ is an F_σ (it is the sum of two F_σ sets) and so $\mathbb{R} \setminus O_f$ is $F_{\sigma\delta\sigma}$ and hence O_f is $G_{\delta\sigma\delta}$.

The proof for the set E_f is similar but requires quite different arithmetic. Write this time

$$A(n, (r,s), (t,u), I) = \frac{1}{2} \left[f^{-1} ((r,s) \cap I) + f^{-1} ((t,u) \cap I) \right]$$

$$\cap f^{-1} \left(\mathbb{R} \setminus [-n^{-1} + (r+t)/2, n^{-1} + (s+u)/2] \right)$$

and show that

$$\mathbb{R} \setminus E_f = \bigcup_{n=1}^{\infty} \bigcap_{m=1}^{\infty} \bigcup_{J \in \mathcal{B}_m} \bigcup_{K \in \mathcal{B}_m} \bigcup_{I \in \mathcal{B}_m} A(n, J, K, I). \tag{2.38}$$

The details are left to the reader (who can cheat by consulting [57]). Once again it is clear that if f is Borel measurable then E_f is co-analytic and if f is Baire 1 then E_f is $G_{\delta\sigma\delta}$.

Corollary 2.45 *If a set E is a G_δ or an F_σ then E^* is a $G_{\delta\sigma\delta}$. If E is Borel then E^* is co-analytic.*

Proof. Apply the theorem to the function $\chi_E = f$. Then E^* is O_f.

One might hope for more than this last theorem asserts. Darji [59] has shown that there is no better topological statement available. The methods utilize the fact that there is a Hamel basis containing a perfect set (see Section A.6 in the Appendix) and use some techniques developed by Laczkovich [170] that we will return to in Chapter 6. We state this theorem, without proof, just for the two notions studied here but the theorem in [59] applies equally to the set of points of just about any specified symmetric behavior (points of smoothness, points of symmetric differentiability, etc.).

Theorem 2.46 (Darji) *Let M be any zero-dimensional co-analytic subset of $\mathbb{R}$. Then there is a Baire 2 function f such that the sets O_f and E_f of points where f is symmetrically continuous/symmetric are homeomorphic to M.*

In the other direction we ask what examples *can* be constructed. Marcus [187] has a preliminary result in this direction; for any G_δ set E there is a function that is symmetrically continuous at exactly the points of E. Jaskula and Szkopińska [140] also discuss this problem and give a necessary condition and a sufficient condition for a set E to be exactly the set of points of symmetric continuity of some function; as their ideas are technical and do not completely solve the problem we shall not reproduce them here. We do present a more elementary construction from Darji [59] who shows the following as a first step in his proof of Theorem 2.46.

Theorem 2.47 (Darji) *Let P be any perfect, nowhere dense subset of $\mathbb{R}$. Then there is a Baire 2 function f with $0 \leq f \leq 1$, with $\{x : f(x) > 0\}$ countable and such that $O_f = E_f = P$.*

Proof. We can assume that P is compact. Let H be a Hamel basis for the real numbers over the rationals (see Section 4.2.2). Choose a decreasing sequence of open sets $\{G_n\}$ and a disjoint sequence $\{A_n\}$ of countable subsets of H such that

(*i*) $\overline{G}_{n+1} \subset G_n$.

(*ii*) $\mathbb{R} \setminus G_n$ is perfect.

(*iii*) $P = \bigcap_{n=1}^{\infty} G_n$.

(*iv*) A_n is a dense subset of $\mathbb{R} \setminus G_n$.

Define ϵ_n as the distance from P to $\mathbb{R} \setminus G_n$. Since these sets are disjoint each ϵ_n is positive and $\epsilon_n \to 0$ by property (iii).

Define $f(x)$ to be $\min\{1, \epsilon_n{}^2\}$ for each $x \in A_n$ and as 0 if x belongs to no A_n. Clearly $0 \leq f \leq 1$, the set $\{x : f(x) > 0\}$ is countable and, hence, f is a Baire 2 function. At each point $x \in P$ a direct computation shows that $f'(x) = 0$. In particular, then, both sets O_f and E_f include P.

Now suppose that $x \notin P$. Then x does not belong to G_n and so is a two-sided limit point of A_n. We claim x cannot be in O_f or E_f. Suppose that x belongs to some A_m. There are $t_k \to 0$ with $x + t_k \in A_n$ and each $x - t_k$ cannot belong to any A_m because of the linear independence of H. But then $f(x - t_k) = 0$ and so f cannot be symmetric or symmetrically continuous at x. On the other hand if x belongs to no A_m then $f(x) = 0$. Again there are $t_k \to 0$ with $x + t_k \in A_n$ and so

$$|f(x + t_k) + f(x - t_k) - 2f(x)| \geq f(x + t_k)$$

which is fixed and so does not tend to zero. Hence $x \notin E_f$. Also $|f(x + t_k) - f(x - t_k)|$ cannot tend to zero unless $x - t_k$ belongs to A_n for large k. But, again using the linear independence, we see that $x - t_k$ cannot belong to A_n for two distinct k. Hence $x \notin O_f$. Finally then $O_f = E_f = P$ as required.

To conclude let us report one further observation of Darji [57]. One might think that if a set E is too complicated topologically then it could immediately be ruled out as a possible candidate for an O_f or an E_f. Certainly, as we have seen, this is so if we insist that f be a Baire function. A Baire 1 function places strong requirements; even just a Baire function means the sets must be at least co-analytic. Darji shows, in a result related to Theorem 2.46, that given any zero dimensional subset M of $\mathbb{R}$ there is a function f so that O_f (or E_f) is homeomorphic to M. This applies as well to the sets of points of even/odd differentiability.

3

Covering Theorems

3.1 Introduction

The notion of a covering theorem and its role in the study of the differentiation properties of functions are now widely known. In many parts of modern analysis these concepts are now central. The same program can be applied equally well to the study of functions of a single variable with a resulting clarification of techniques. The goal is to focus on what might be called the geometric properties of a differentiation basis.

Let us begin with some remarks on coverings in the simplest setting, in the study of the ordinary derivative.

We define a *full cover* of a set E as a collection $\mathcal{C}$ of closed intervals such that for every $x \in E$ there is a $\delta(x) > 0$ so that $[y, z] \in \mathcal{C}$ if $x \in [y, z]$ and $z - y < \delta(x)$. We define a *fine cover* of a set E as a collection $\mathcal{C}$ of closed intervals such that for every $x \in E$ and every $\epsilon > 0$ there is an interval $[y, z] \in \mathcal{C}$ with $x \in [y, z]$ and $z - y < \epsilon$.

To see how these arise naturally in the study of the ordinary derivative notice that the collection

$$\mathcal{C} = \left\{ [y, z] : \frac{f(z) - f(y)}{z - y} > r \right\}$$

is full cover of the set

$$\{ x : \underline{D}f(x) > r \}$$

and a fine cover of the (larger) set

$$\{ x : \overline{D}f(x) > r \}.$$

In fact the notions of full and fine covers characterize completely the ordinary derivates: the upper derivate of a function can be defined in terms of the full covers as

$$\overline{D}f(x) = \inf_{C \in \text{full}} \sup \left\{ \frac{f(z) - f(y)}{z - y} : x \in [y, z], \ [y, z] \in C \right\}.$$

or equally in terms of the fine covers as

$$\overline{D}f(x) = \sup_{C \in \text{fine}} \inf \left\{ \frac{f(z) - f(y)}{z - y} : x \in [y, z], \ [y, z] \in C \right\}.$$

Thus in fact all of the information about the nature of the process of ordinary differentiation is contained in these dual concepts of a full cover and a fine cover. The program we wish to follow is to determine geometric properties of the covers that express differentiation properties of functions.

By a *covering theorem* we shall mean any theorem asserting that a subcover (in some sense) with certain properties may be extracted from a cover (in some sense). This necessarily vague notion is best realized with two famous examples which are our models in the theory. We express the Vitali Theorem in an imprecise form in order to illustrate it as belonging to the same family of ideas as the more elementary theorem which follows.

Theorem 3.1 (Vitali) *Suppose that C is a fine cover of a bounded set E and $\epsilon > 0$. Then one can extract from C a collection*

$$\{I_1, I_2, \ldots, I_n\}$$

of nonoverlapping intervals so that

$$\left| \sum_{i=1}^{n} |I_i| - |E| \right| < \epsilon.$$

Theorem 3.2 (Cousin) *Suppose that C is a full cover of the interval $[a, b]$. Then one can extract from C a collection*

$$\{I_1, I_2, \ldots, I_n\}$$

of nonoverlapping intervals that forms a partition of $[a, b]$.

The second theorem is now commonly called Cousin's Lemma after Pierre Cousin, a student of Henri Poincaré, who essentially formulated it in his thesis [46] in 1895. Over the years it has been subjected to almost constant rediscovery: Goursat [126] in 1900, Lusin [182] in 1912, G. C. Young and

W. H. Young [308] in 1915, Tonelli [291] in 1936 and many later authors including an almost periodic recurrence in the American Mathematical Monthly
(for example [17], [73], [109], [146], [203] and [254]).

Applications of the Vitali Theorem are well known. For an elementary application of the Cousin covering Theorem consider how simple it is to prove the
following monotonicity theorem: *If f has a positive lower derivate $\underline{D}f(x) > 0$
everywhere then f is increasing.* For a proof we construct the collection

$$\mathcal{C} = \left\{ [y, z] : \frac{f(z) - f(y)}{z - y} > 0 \right\}$$

which is evidently a full cover of any interval $[a, b]$. By Cousin's covering Theorem one can extract from $\mathcal{C}$ a collection

$$\{[a_0, a_1], [a_1, a_2], \ldots, [a_{n-1}, a_n]\}$$

of nonoverlapping intervals that forms a partition of $[a, b]$ giving

$$f(b) - f(a) = \sum_{i=1}^{n} f(a_i) - f(a_{i-1}) > 0.$$

Since a and b are arbitrary points with $a < b$ and we have shown that $f(a) < f(b)$
we have established that f is increasing.

Of course a simple compactness proof can be constructed for this monotonicity theorem. However this proof, based directly on a covering argument,
better reveals the true nature of the theorem: a local property of a function can
be used to provide a global property. In fact Cousin's Theorem is equivalent to
the Bolzano-Weierstrass Theorem on the real line and can be used in a variety
of elementary situations where a global statement must flow from a local one.

There is a direct analogue of Cousin's Theorem that is equally useful in
establishing elementary properties of the symmetric derivative. If we wish to
apply this same approach to the study of the symmetric derivative we are led
immediately to this definition: by a *symmetric full cover* on the real line is meant
a collection $\mathcal{S}$ of closed intervals with the property that for every real x there
is a $\delta(x) > 0$ so that

$$[x - h, x + h] \in \mathcal{S}$$

for every $0 < h < \delta(x)$. It should be clear that symmetric covers will play the
same role in the study of symmetric continuity and symmetric derivatives that
ordinary covers play in the study of ordinary continuity and ordinary derivatives.

We shall prove the following theorem in Section 3.3.3.

Theorem 3.3 *Let $\mathcal{S}$ be a symmetric full cover on the real line. Then there is
a denumerable set N so that $\mathcal{S}$ contains a partition of every interval neither of
whose endpoints belongs to N.*

As an immediate application we have, just by using the same methods as above, an elementary monotonicity theorem that includes, for example, Theorem 1.4 on page 6 but whose proof is now simpler, more transparent and more general: *if f has a positive lower symmetric derivate $\underline{SD}f(x) > 0$ everywhere then f is increasing off some countable set.*

Our goal in this chapter is to develop the basic covering theorems that should be associated with any study of the various symmetry properties of real functions. As we shall see there is a considerable economy afforded by this approach and many proofs are clarified. The covering theorems discussed in this chapter are reproduced mostly from [111], [231], [232], [284], [285] and [116].

3.2 Symmetric Covering Relations

We have defined a *symmetric full cover* on a set E to be a collection $\mathcal{S}$ of closed intervals with the property that for every $x \in E$ there is a $\delta(x) > 0$ so that

$$[x - h, x + h] \in \mathcal{S}$$

for every $0 < h < \delta(x)$. Such a notion evidently arises naturally in a study of symmetric limits. For example, if a function f is symmetrically continuous at each point of E then the collection of intervals

$$\{[a, b] : |f(b) - f(a)| < \epsilon\}$$

is, for every $\epsilon > 0$, a symmetric full cover of E.

There are some technical problems in adopting this definition that are resolved by a slight shift of viewpoint. It is common in derivation theory to study interval functions by considering an interval function $[a, b] \rightarrow h([a, b])$ as instead a function of two variables $h(a, b)$ defined in the plane $\mathbb{R}^2$. Then the interval $[a, b]$ corresponds to the planar point (a, b) and limit processes on the line correspond to limit process in the plane. Since we focus only on symmetric type limit processes it is more natural for our purposes to associate the interval $[a, b]$ more closely with its midpoint by selecting the point $((a + b)/2, (b - a)/2)$ to represent this interval. Thus we think of the scheme

$$(x, h) \longleftrightarrow [x - h, x + h]$$

or

$$[a, b] \longleftrightarrow \left(\frac{a + b}{2}, \frac{b - a}{2} \right).$$

The point (x, h) is the vertex of an isoceles right-angled triangle whose hypotenuse is the interval $[x - h, x + h]$ on the x–axis. The figure illustrates the picture that should be kept in mind.

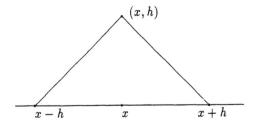

Fig. 3.1. The vertex for the interval $[x - h, x + h]$.

We say that (x, h) is the *vertex* for the interval $[x - h, x + h]$ and that the interval $[x - h, x + h]$ and the vertex (x, h) are *associated*.

Thus we convert from collections of intervals to subsets of $\mathbb{R}^2$ but retain the notion of coverings. We refer to subsets of $\mathbb{R}^2$ as *symmetric covering relations* in keeping with language which is standard in some parts of derivation theory. We shall allow covering relations to contain points below the x–axis; the point (x, h) with $h < 0$ can still play a role in some of our definitions and will usually indicate that the interval $[x + h, x - h]$ is traversed in the reverse direction in some sense. The definitions explain the language to be used.

Definition 3.4 Let $E \subset \mathbb{R}$. A set $V \subset E \times \mathbb{R}$ is said to be a *symmetric covering relation* on E. A point (x, h) is said to belong to V if $(x, h) \in V$; an interval $[c - h, c + h]$ is said to "belong to V" if $(c, h) \in V$.

Of course it is an abuse of language to say that an interval belongs to V when we mean really that the associated vertex of the interval belongs to V. But this looseness is convenient. We also require a language in which to express Theorem 3.3 in terms of the covering relation.

Definition 3.5 We say that a covering relation V *allows a partition of the interval* $[a, b]$ if there is a finite sequence of numbers

$$a = x_0 < x_1 < \ldots < x_{n-1} < x_n = b$$

so that each pair

$$\left(\frac{x_{i+1} + x_i}{2}, \frac{x_{i+1} - x_i}{2} \right) \in V$$

(i.e. each interval $[x_i, x_{i+1}]$ belongs to V).

For a number of our other covering theorems in this chapter we need a more refined language. In order to piece together two or more adjacent intervals from covering relations we adopt the following notation.

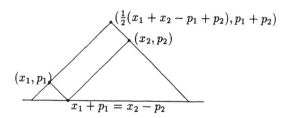

Fig. 3.2. Illustration of $V_1 \times V_2$.

Definition 3.6 Let V_1 and V_2 be symmetric covering relations. Then by $V_1 \times V_2$ we denote the symmetric covering relation consisting of all pairs

$$\left(\frac{x_1 + x_2 - p_1 + p_2}{2}, p_1 + p_2 \right)$$

where $(x_1, p_1) \in V_1$, $(x_2, p_2) \in V_2$ and $x_1 + p_1 = x_2 - p_2$.

The figure illustrates how the two intervals

$$[x_1 - p_1, x_1 + p_1] \quad \text{and} \quad [x_2 - p_2, x_2 + p_2]$$

in the definition are joined. As well, of course, other possible configurations are possible if the points in V_1 or V_2 lie below the axis.

It is convenient to have some language for a covering relation that is closed under iteration of this operation.

Definition 3.7 A symmetric covering relation V is said to be *transitive* if $V \supset V \times V$.

Note that if V is transitive and allows a partition of an interval $[a, b]$ then the interval $[a, b]$ belongs to V.

3.3 Basic Covering Theorems

All of our covering theorems concern covering relations $V \subset \mathbb{R}^2$ usually with some local property such as

$$x \in E, \;\; 0 < t < \delta(x) \Longrightarrow (x, t) \in V.$$

Recall that the pairs (x, h) in V correspond to the intervals $[x - h, x + h]$ and while we can, in most cases, rephrase all statements in the language of interval coverings, we prefer to maintain throughout the two dimensional picture.

The first few of our covering lemmas are just extractions of the underlying geometry of the arguments used in Chapter 1.

3.3.1 Uniformization

Our first elementary covering theorem allows us to codify a well known and much used device in analysis. The idea is to convert a local property to one that holds uniformly on certain sets.

Theorem 3.8 *Suppose that the covering relation $V \subset I\!\!R^2$ has the property that for every $x \in E$ there is a positive number $\delta(x)$ so that*

$$x \in E, \ 0 < t < \delta(x) \Longrightarrow (x, t) \in V.$$

Then there is a denumerable partition $\{E_n\}$ of E so that

$$x \in E_n, \ 0 < t < 1/n \Longrightarrow (x, t) \in V.$$

Proof. To prove this it is enough to define the sequence of sets

$$A_n = \{x \in E : \delta(x) < 1/n\}$$

and then, inductively, $E_1 = A_1$ and, for $n \geq 2$,

$$E_n = A_n \setminus \bigcup_{k < n} A_k.$$

In general, even if E is measurable, there is no guarantee that the sets E_n of the partition will be measurable. In some cases, though, if these sets arise from a consideration of some measurable function then it is not difficult to arrange for the partition of E to be measurable.

3.3.2 Charzyński's Covering Theorems

Here we present a covering version of the argument used in Lemma 2.10, due originally to Charzyński.

Theorem 3.9 *Let x_1, $x_2 \in I\!\!R$ with $h = x_2 - x_1 > 0$. Suppose that the covering relations V_1 and V_2 have the property that there are positive numbers $\delta_1 > h$ and $\delta_2 < h$ so that*

$$0 < t < \delta_1 \Longrightarrow (x_1, t) \in V_1$$

and

$$0 < t < \delta_2 \Longrightarrow (x_2, t) \in V_2.$$

If $x_1 < x < x_1 + \delta_2$ then $(x, h) \in V_1 \times V_2$.

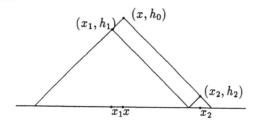

Fig. 3.3. Two rotations in Theorem 3.9.

Proof. The proof requires only the geometric argument that the translation $x - h \to x + h$ can be obtained by two rotations, one about the point x_1 and one about the point x_2 using points in V. In order to use the second rotation the first rotation must carry $x - h$ sufficiently close to x_2.

Here are the algebraic computations necessary to justify the notions; the figure illustrates the two rotations in an obvious manner. For any $x_1 < x < x_1 + \delta_2$ we may choose h_1 and h_2 so that $x - h = x_1 - h_1$ and $x_2 + h_2 = x + h$. Evidently then $0 < h_1 < \delta_1$ since $h_1 = x_1 - x + h$, $x_1 < x < x_1 + \delta_2$ and $\delta_2 < h < \delta_1$. Similarly $0 < h_2 < \delta_2$. Thus $(x_1, h_1) \in V_1$ and $(x_2, h_2) \in V_2$. Then

$$x - h = x_1 - h_1 < x_1 + h_1 = x_2 - h_2 < x_2 + h_2 = x + h$$

exhibits the two rotations about x_1 and x_2 and the proof is complete.

A simple variant on this theorem is useful. Here we need to pick points on the other side and so one of the rotations goes backwards. The second figure illustrates this situation and the proof is similarly obtained.

Theorem 3.10 *Let x_1, $x_2 \in \mathbb{R}$ with $h = x_2 - x_1 > 0$. Suppose that the covering relations V_1 and V_2 have the property that there are positive numbers $\delta_1 > h$ and $\delta_2 < h$ so that*

$$0 < t < \delta_1 \Longrightarrow (x_1, t) \in V_1$$

and

$$-\delta_2 < t < 0 \Longrightarrow (x_2, t) \in V_2.$$

If $x_1 - \delta_2 < x < x_1$ then $(x, h) \in V_1 \times V_2$.

In the same spirit as the preceding theorems we can prove also a covering version of the argument of Charzyński in Lemma 2.12 (cf. also the proof of Theorem 2.26).

Theorem 3.11 *Suppose that the covering relations V and W have the following property: there is a set E, a positive number δ and an interval (c, d) so that*

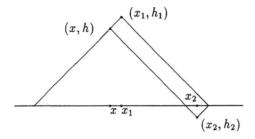

Fig. 3.4. Two rotations in Theorem 3.10.

$d - c < \delta$, E *is dense in* (c, d),

$$x \in E, \, 0 < |t| < \delta \Longrightarrow (x, t) \in V$$

and for each $c < x < d$ *there is a* $\delta(x) > 0$ *so that*

$$0 < |t| < \delta(x) \Longrightarrow (x, t) \in W.$$

Then, for any two points $c < x_1 < x_2 < d$, *the interval* $[x_1, x_2]$ *belongs to*

$$V \times V \times W \times V \times V \times W \times V \times V.$$

Proof. Let x_0 be a point of $E \cap (c, d)$ and let x_1 be any point of the interval (c, d); set $2t = x_1 - x_0$ and select t' so that $x_0 + t' \in E$ is sufficiently close to $x_0 + t/2$ so that the vertex corresponding to the interval $[x_0 + 2t', x_0 + 2t - 2t']$ belongs to W, i.e. so that $(x_0 + t, t - 2t') \in W$. This is just the local symmetric condition at the point $x_0 + t$.

Now by reflections about the points $x_0 + t'$, $x_0 + t$, x_0 and $x_0 + t'$ we have the reflection sequence

$$x_0, \, x_0 + 2t', \, x_0 + 2t - 2t', \, x_0 + 2t' - 2t, \, x_0 + 2t = x_1$$

showing that V leads from x_0 to x_1 by a sequence of four reflections, first in V, then W then twice again in V. Evidently then for any pair of points x_1 and x_2 in this interval V and W carry x_1 to x_2 by a sequence of eight reflections in the order stated in the theorem. The figure below illustrates the proof; here we have chosen $x_0 = 0$ and $x_1 = 2t$.

3.3.3 Fundamental Covering Theorem for Full Symmetric Covers

The covering relation which is associated with ordinary symmetric notions (symmetric continuity, symmetric differentiability, etc.) can be defined as follows.

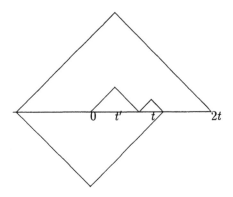

Fig. 3.5. Four rotations in Theorem 3.11.

Definition 3.12 A covering relation V is a *full symmetric covering relation* on a set $E \subset \mathbb{R}$ provided that for every $x \in E$ there is a positive number $\delta(x)$ so that

$$0 < t < \delta(x) \Longrightarrow (x, t) \in V.$$

Our fundamental elementary result can be considered the symmetric analogue of Cousin's Theorem (Lemma 3.2). It asserts that a full symmetric cover of the real line permits partitions

$$a = x_0 < x_1 < \ldots < x_{n-1} < x_n = b$$

for every interval $[a, b]$ with only a countable exceptional set; this has obvious applications to the study of symmetric properties that are assumed to hold everywhere and shall be used in Chapter 9 as a basis for defining a Riemann type integral relative to full symmetric covers. This covering theorem was observed independently in Preiss and Thomson [231] and in Freiling [111]. The latter article goes on to characterize the countable set as splattered.

Theorem 3.13 *Suppose that V is a full symmetric covering relation on an interval (a, b). Then there is a right scattered set R and a left scattered set L such that V allows a partition of every interval $[c, d] \subset (a, b)$ with $c \notin R$, $d \notin L$.*

The "countable" part of the proof follows in three simple steps the first of which is useful on its own; the "splattered" part of the proof is given in the next subsection. Fragments of this proof can be found in a number of situations (eg. McGrotty [199], Kurzweil and Jarník [168], and Davies [60]).

Lemma 3.14 *Let c denote any point of (a, b). Then there is a closed denumerable set $C_c \subset (c, b)$ so that V allows a partition of $[2c - x, x]$ for every $x \in (c, b) \setminus C_c$ with $2c - x > a$.*

Proof. The proof is a simple compactness argument. Let Z be the set of all points $z \in (c, b)$ with the property that V allows a partition of $[2c - x, x]$ for all $x \in [c, z]$ but a closed denumerable subset. Let $z_0 = \sup Z$. The proof is complete if $z_0 = b$.

Evidently $z_0 > c$ simply because for all sufficiently small t we have $(c, t) \in V$ so that V includes $[2c - x, x]$ for $x = c + t$. Also $z_0 < b$ yields a contradiction: for then there is a $\delta > 0$ so that $(z_0, t), (2c - z_0, t) \in V$ for $0 < t < \delta$ and this requires $z_0 + \delta$ to belong to Z. Thus $z_0 = b$ and we are done.

Lemma 3.15 *For every $0 < h < b - a$ there is a denumerable set T_h so that V allows a partition of $[x, x + h]$ if $x \in (a, b) \setminus T_h$ and $x + h < b$ and for every $0 > h > -(b - a)$ there is a denumerable set T_h so that V allows a partition of $[x + h, x]$.*

Proof. For $h > 0$ write

$$T_h = \bigcup_{r \in Q \cap (a, b)} (2r - C_r) \cup (C_{r + h/2} - h) \tag{3.1}$$

where the union is taken over all rationals Q and C_r has been defined in the statement of Lemma 3.14. If x does not belong to T_h then choose a rational number $r \in (x, x + h/2)$. Using the centers r and $r + h/2$ we see that V allows a partition of $[x, 2r - x]$ because

$$x \notin (2r - C_r)$$

and V allows a partition of $[2r - x, x + h]$ because

$$x \notin (C_{r + h/2} - h).$$

This gives a partition of $[x, x + h]$ as required. A similar argument works for $h < 0$.

Lemma 3.16 *There is a denumerable set $N \subset (a, b)$ so that V allows a partition of $[x, y]$ if $[x, y] \subset (a, b)$ and $x, y \notin N$.*

Proof. Let $N = \bigcup_{r \in Q} T_r$ where T_r is defined in (3.1) and where the union is taken over all nonzero rationals. If neither x nor y belongs to N then choose a rational s so that $x + s \in (m - \delta(m), m)$ where $m = (x + y)/2$. Evidently V allows a partition of $[x, x + s]$ and of $[y - s, y]$; since it also contains the interval $[x + s, y - s]$ it allows a partition of $[x, y]$ as required. This completes the proof of Lemma 3.16 and hence of the main theorem for just the case of a countable exceptional set.

We have still to prove that in the assertion of the covering Theorem 3.13 the exceptional sets R and L can be taken as semi-scattered. We give here the proof from Freiling and Thomson [117] showing that once the countable exceptional set is granted an application of the "scattered Baire" Theorem completes the argument. (The proof uses the material in Section A.2 in the Appendix.) The first proof of this fact appears in [111] and is expressed in the language of games. Later Humke and Laczkovich [135] provided a category proof of this same result; note that the proof here could be considered as just an interpretation of their proof within the setting of the scattered Baire Theorem. It has the advantage that it is constructive rather than providing only an indirect proof of the scattered nature of the exceptional sets. We express it here in the language of intervals rather than covering relations.

Lemma 3.17 *Let C be a collection of closed intervals with the following properties:*

(1) if $[a, b]$, $[b, c] \in C$ then $[a, c] \in C$.
(2) for each x there is a $\delta(x) > 0$ so that $[x - t, x + t] \in C$ for all $0 < t < \delta(x)$.
(3) there is a countable set N so that if a, $b \notin N$ then $[a, b] \in C$.

Then there is a right scattered set S_1 and a left scattered set S_2 so that if $x < y$, $x \notin S_1$ and $y \notin S_2$ then $[x, y] \in C$.

Proof. Let Z denote the set of points of the form $(x + y)/2$ for x, $y \in N$ and write $Z \cup N = \{z_1, z_2, z_3, \ldots\}$. To start our Baire category argument write

$$E_n = \{z_1, z_2, z_3, \ldots z_n\} \cup \{x \notin Z \cup N : [x - t, x + t] \in C \text{ if } 0 < t < n^{-1}\}.$$

These sets cover the real line and are expanding so the scattered Baire Theorem may be applied. Let G_η denote the open sets and let R and L denote the set of right and left endpoints from that application. We claim that with $S_1 = L$ and $S_2 = R$ if $x < y$, $x \notin S_1$ and $y \notin S_2$ then $[x, y] \in C$.

To show this it is enough to show that for any $x_1 < x < x_2$ with x_1, $x_2 \notin N$ and $x \notin L$ (resp. $x \notin R$) $[x, x_2] \in C$ (resp. $[x_1, x] \in C$). This is already true if $x \notin N$ so we may suppose that $x \in N$. There is a first ordinal so that $x \in G_{\eta+1}$ but $x \notin G_\eta$. Let E_i be dense in $(a_\eta, b_\eta) \setminus G_\eta$. Then for any $\delta > 0$ the set $(x, x + \delta) \setminus G_\eta$ must be infinite and so also E_i is infinite in $(x, x + \delta)$.

Choose a point z from $E_i \cap (x, x + \frac{1}{2}(x_2 - x))$ so that $z \neq z_j$ for any $j \leq i$. As $x \in N$ and $z \neq z_j$ it follows that $2z - x$ cannot be in N so $[2z - x, x_2] \in C$ and, as $z \in E_i$, that $[x, 2z - x] \in C$. By (1) then $[x, x_2] \in C$ as required. (The similar statement for $[x_1, x] \in C$ admits the same proof.)

These four steps complete the proof of the main theorem. The corollaries are immediate and just express the theorem in a form that is easier to use for many applications.

Corollary 3.18 *Let V be a transitive, full symmetric cover of the real line. Then there is a right scattered set R and a left scattered set L so that V contains every interval $[a, b]$ with $a < b$ and $a \notin R$, $b \notin L$.*

Corollary 3.19 *Let V be a transitive, full symmetric cover of the real line. Then there is a splattered set S so that V contains every interval $[a, b]$ with $a < b$ and $a, b \notin S$.*

3.3.4 Other Variants

A variant on Theorem 3.13 allows a countable exceptional set from the beginning. We suppose that there is a countable set C so that except for intervals with an endpoint in C we are given a symmetric cover of the line. Then by exactly the same arguments as in Section 3.3.3 we can establish the following theorem.

Theorem 3.20 *Let C be a countable set. Suppose that the covering relation $V \subset \mathbb{R}^2$ has the property that for every x in an interval (a, b) there is a positive number $\delta(x)$ so that $x \in (a, b)$, $0 < t < \delta(x)$, $x + t \notin C$, $x - t \notin C$ implies that $(x, t) \in V$. Then there is a countable set $N \subset (a, b)$ so that V allows a partition of every subinterval $[c, d]$ of (a, b) whose endpoints do not belong to N.*

One might have hoped that in the statement of the theorem the countable exceptional set C could be replaced by a countable set C_x that varies from point to point; in Section 3.3.5 and Section 6.2.2 we shall see why this would be impossible.

Symmetric partitions can often be conveniently adjusted by employing a simple device that in [107] is called "jiggling" and in [169] is called an "h–modification" of a partition. Suppose that points $\{x_i\}$,

$$a < x_1 < x_2 < x_3 < \ldots < x_n < b,$$

have been chosen so as to be δ–fine at the centers, i. e. so that each

$$|x_{i+1} - x_i| < \delta\left(\tfrac{1}{2}(x_{i+1} + x_i)\right) \qquad (i = 1, 2, \ldots n - 1). \tag{3.2}$$

Just write $z_i = x_i + (-1)^i h$ for $i = 1, 2, \ldots n$ and $h > 0$. For all sufficiently small h, we still have $a < z_1 < z_2 < \ldots < z_m < b$, the center of each interval $[z_i, z_{i+1}]$ remains as the center of $[x_i, x_{i+1}]$ and the new intervals are also δ–fine at the centers.

This device along with the basic covering theorem allows us to prove a variant that will be useful in Chapter 9.

Theorem 3.21 *Let G be an open dense set and let δ be a gauge on $[a, b] \setminus G$ where $[a, b]$ is an interval with endpoints in G. There exists a partition $\mathcal{P}$ of $[a, b]$ into subintervals with endpoints in G so that each interval $I \in \mathcal{P}$ is one of the following types:*

(i) $I \subset G$, or

(ii) I is centered at a point $x \in [a, b] \setminus G$ and $|I| < \delta(x)$.

Proof. Extend the gauge δ to all of $[a, b]$ by writing

$$\delta(x) = \text{dist}(x, [a, b] \setminus G) \quad (x \in G).$$

By applying our basic covering theorem we may evidently determine points $\{x_i\}$ so that

$$a < x_1 < x_2 < x_3 < \ldots < x_n < b,$$

so that $[a, x_1], [x_n, b] \subset G$ and so that (3.2) holds. By the construction each interval of the partition is of type (i) or (ii) but the endpoints may not be in G. Since G is both open and dense a finite number of small jiggles on the points $\{x_i\}$ suffices to place the endpoints in G without altering (3.2).

3.3.5 Freiling's Negligent Version

We can continue the theme of the preceding section by investigating in more detail what happens if we neglect sets from some σ–ideal. Theorem 3.20 merely neglected a single countable set. As already mentioned we cannot hope for a partitioning theorem if we neglect arbitrary countable sets. Freiling [110] has provided an analysis of this situation by exploring its set-theoretic foundations.

Let $\mathcal{N}$ be a σ–ideal of sets which satisfies the following properties for some cardinal κ:

(i) $\mathcal{N}$ is translation invariant.

(ii) $\mathcal{N}$ is reflection invariant.

(iii) $\mathcal{N}$ contains all countable sets.

(iv) $\mathbb{R}$ is not the union of fewer than κ many sets in $\mathcal{N}$.

(v) every set not in $\mathcal{N}$ contains a subset of cardinality $< \kappa$ which is also not in $\mathcal{N}$.

It is by no means clear that such a σ-ideal of sets can exist that fulfills the conditions (i)—(v). It is known that it is *consistent* in ZFC that the σ-ideal of measure zero sets has these properties. Also if there exists an uncountable cardinal $\alpha < 2^{\aleph_0}$ then the σ-ideal of sets possessing cardinality $< \alpha$ will have these five properties (with $\kappa = \alpha^+$). Thus if one assumes the negation of the continuum hypothesis then the σ-ideal of countable sets satisfies (i)—(v).

Theorem 3.22 (Freiling) *Let $\mathcal{N}$ be a σ-ideal of sets that fulfills the conditions (i)—(v). Suppose that the covering relation $V \subset I\!\!R^2$ has the property that for every x in an interval (a,b) there is a positive number $\delta(x)$ and a set $N_x \in \mathcal{N}$ so that $x \in (a,b)$, $0 < t < \delta(x)$, $x + t \notin N_x$, $x - t \notin N_x$ implies that $(x,t) \in V$. Then there is a set $N \in \mathcal{N}$ so that V allows a partition of every subinterval $[c,d]$ of (a,b) whose endpoints do not belong to N.*

Proof. The proof requires a repetition of the same arguments as in Section 3.3.3. See [110] for details.

From Theorem 3.22 and the Sierpiński theorems of Section 6.2.2 Freiling [110] deduces the following. This gives an observation of Sierpiński: Theorem 3.20 cannot be improved without invoking additional set-theoretic assumptions.

Theorem 3.23 (Freiling) *The following assertion is equivalent to the negation of the continuum hypothesis:*

> *Suppose that the covering relation $V \subset I\!\!R^2$ has the property that for every x in an interval (a,b) there is a positive number $\delta(x)$ and a countable set C_x so that $x \in (a,b)$, $0 < t < \delta(x)$, $x + t \notin C_x$, $x - t \notin C_x$ implies that $(x,t) \in V$. Then there is a countable set $N \subset (a,b)$ so that V allows a partition of every subinterval $[c,d]$ of (a,b) whose endpoints do not belong to N.*

3.4 Khintchine Covering Theorem

The theorem of Khintchine (Theorem 1.23), which forms the centerpiece of Chapter 1, was the first in a series of studies devoted to studying the fundamental properties of measurable functions on which a symmetry condition is imposed. The geometrical arguments involved in the proof of that theorem may be (oversimply) described as noting that a symmetric covering relation that holds uniformly on some set E imposes strong conditions at the density points of E. We can present the essence of these arguments as a single covering theorem; because the idea is implicit in the article of Khintchine [151], we shall attribute the theorem to him. This material is reproduced from [285].

As in Sections A.2 and A.4 of the Appendix, for any set E let $b(E)$ denote either $d(E)$ the set of density points of E, or alternatively, $c(E)$ the set of second category points of E. With either interpretation we have the following theorem.

Theorem 3.24 *Let $E \subset \mathbb{R}$, $V \subset \mathbb{R}^2$ and $\eta > 0$. We suppose that*

$$x \in E, \ 0 < t < \eta \Longrightarrow (x, t) \in V$$

and that $z \in b(E)$. Then there is a positive number δ so that for every point $0 < |z - x| < \delta$ there is a set $A = A(x)$ with $z \in b(A)$ such that, for all $a \in A$,

$$z < a < x \Longrightarrow [a, x] \ \text{belongs to } V \times V$$

and

$$z > a > x \Longrightarrow [x, a] \ \text{belongs to } V \times V.$$

Proof. Let us take $b(E) = d(E)$ and give the language of the density proof. (For convenience the single point at which a translation for the $b(E) = c(E)$ proof must be made is indicated.) Also we will shift everything to the origin and assume that the point z is 0. We argue just on the right at 0 as the situation on the left is similarly handled.

We use Lemma A.11(9) to choose $\delta < \eta$ so that if $0 < x < \delta$ then the set

$$2b(E) \cap (2E - x) \cap (0, x)$$

is nonempty. This just uses the sets $A_1 = 2b(E)$ and $A_2 = 2E$ and the numbers $\alpha_1 = 0$ and $\alpha_2 = -x$; since $0 \in 2b(E)$ and $0 \in b(2E)$ the property of Lemma A.11(9) even shows that the set $2b(E) \cap (2E - x) \cap (0, x)$ can be arranged to have positive measure [to be of the second category]. If $0 < x < \delta$ then we choose

$$u \in b(E) \cap (E - \tfrac{1}{2}x) \cap (0, \tfrac{1}{2}x).$$

Note that $u \in b(E)$ so that, by Lemma A.11(6), we know that $0 \in b(2E - 2u)$. Thus if we define

$$A = (2E - 2u) \cap (-2u, 2u)$$

then we have $0 \in b(A)$ as we require.

For any point $a \in A$, $0 < a < x$ then we can exhibit that $[a, x]$ is in $V \times V$ by a scheme of two reflections

$$a < 2u < x$$

(see the Figure) and we have only to check that the vertex points are in V. But the midpoint of the first reflection is $(a + 2u)/2 = u + \tfrac{1}{2}a$ and $a \in A \subset 2E - 2u$ so that $u + \tfrac{1}{2}a \in E$. Certainly then $(u + \tfrac{1}{2}a, (2u - a)/2) \in V$ since $(2u - a)/2 < \eta$.

The midpoint of the second reflection is $(x + 2u)/2 = u + \tfrac{1}{2}x$ but by choice $u \in E - \tfrac{1}{2}x$ so this too is in E and again $(u + \tfrac{1}{2}x, (x - 2u)/2) \in V$ since $(x - 2u)/2 < \eta$.

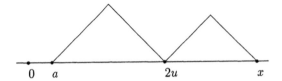

Fig. 3.6. Two reflections in the proof of Theorem 3.24.

3.5 Uher Covering Lemma

The covering argument of Theorem 3.24 has been used, in some form, frequently in the literature. For the most part it is limited in usefulness to obtaining the properties of measurable functions. Few genuinely new techniques have arisen in this subject since the original article of Khintchine. In order to remove the measurability assumption far more subtle geometrical arguments are needed. The most penetrating analysis has been provided by Uher in his articles [293] and [294]. In this section we express the main argument of Uher [294, Lemma 1, pp. 423–426] in the form of covering theorems. This was first done in Thomson [285]. Subsequently Humke and Laczkovich [136] simplified the proofs; it is their presentation that we give here.

Uher's original application of his ideas expressed them in a form using parallel notions of density and category. It is a bit simpler to split the two ideas into separate theorems. The category version, as it is the easier of the two, is given first. Recall that the covering relation is expressed in terms of membership as either $(x, t) \in V$ or the phrase "the interval $[x - t, x + t]$ belongs to V" and it is convenient to use both.

Theorem 3.25 *Let $E \subset \mathbb{R}$ have the Baire property and suppose that $V \subset \mathbb{R}^2$ is a covering relation on E. We suppose that for every point $x \in E$ there is a positive number $\delta(x)$ so that either*

$$0 < t < \delta(x) \Longrightarrow (x, t) \in V$$

or alternatively

$$0 > t > -\delta(x) \Longrightarrow (x, t) \in V.$$

Then there is an open set G such that $E \setminus G$ is of the first category and, for every interval $[x - t, x + t]$ that is contained entirely in G, either (x, t) or $(x, -t)$ belongs to V^5. Where $V^5 = V \times V \times V \times V \times V$.

Proof. Let $\mathcal{I}$ denote the family of all open intervals I with the property that for any $x, y \in I$ there is a chain $x = x_0, x_1, x_2, x_3, x_4, x_5 = y$ or $y = x_0, x_1, x_2, x_3, x_4, x_5 = x$ with each pair satisfying

$$\left(\frac{x_{i+1} + x_i}{2}, \frac{x_{i+1} - x_i}{2}\right) \in V \qquad (i = 0, 1, 2, 3, 4). \tag{3.3}$$

We show that if E is residual in (a, b) then there is a subinterval of (a, b) belonging to $\mathcal{I}$. The statement of the theorem then follows by taking the set G as a union of a maximal disjoint subfamily of $\mathcal{I}$.

Let E be residual in (a, b). Write E^+ for the set of points in E at which there is a positive number $\delta(x)$ so that $0 < t < \delta(x) \implies (x, t) \in V$ and write E^- for the set of points in E at which there is a positive number $\delta(x)$ so that $0 > t > -\delta(x) \implies (x, t) \in V$. Then $E = E^+ \cup E^-$ and so one of these sets is second category in (a, b). For the remainder of the proof we assume it is E^+ that has this property. The same proof works for the other case by making the appropriate directional changes.

For each natural number n let

$$E_n^+ = \{x \in E^+ : 0 < t < 1/n \implies (x, t) \in V\}$$

and

$$E_n^- = \{x \in E^- : 0 > t > -1/n \implies (x, t) \in V\}.$$

Then $E^+ = \bigcup_{n=1}^{\infty} E_n^+$ and hence, since E^+ is second category in (a, b), there is an open interval $I_1 \subset (a, b)$ and an n such that $|I_1| < 1/n$ and E_n^+ is second category in every subinterval of I_1. Let I_2 be that open subinterval of I_1 concentric with I_1 and of $1/10$ its length. It is this interval I_2 that we shall show belongs to $\mathcal{I}$.

Let $x, y \in I_2$, $x < y$. To simplify the notation suppose that $x = 0$. We shall obtain a subinterval $J \subset (0.9y, y)$ such that for every $z \in J$ there is a four element chain $0 = x_0, x_1, x_2, x_3, x_4 = z$ of the type (3.3) that we require. If this interval J is found then choose $w \in (0.9y, y) \cap E_n^+$ such that $2w - y \in J$; we would then obtain first our promised four element chain $x = 0 = x_1, x_2, x_3, x_4 = 2w - y$ and then note that this expands to five elements $x = 0 = x_1, x_2, x_3, x_4, x_5 = y$ satisfying (3.3) since $\left(w, \frac{1}{2}(y - w)\right) \in V$.

In the interval $(0.9y, y)$ the set E is residual and the set $E_n^+ + \frac{3}{4}y$ is second category; hence $E \cap (E_n^+ + \frac{3}{4}y)$ is also second category in $(0.9y, y)$. It follows, from the fact that $E = \bigcup_{m=1}^{\infty} E_m^+ \cup E_m^-$, that there is an integer m and an open interval $J \subset (0.9y, y)$ so that one of the sets

$$E_m^+ \cap (E_n^+ + \tfrac{3}{4}y) \quad \text{or} \quad E_m^- \cap (E_n^+ + \tfrac{3}{4}y) \tag{3.4}$$

is dense in J.

Let $z \in J$. We shall construct four points p, q, r, s so that a sequence of reflections about these points in turn takes 0 to z.

Since E is residual and E_n^+ is second category in every subinterval of I_1 the set

$$A = (\tfrac{1}{2}E + \tfrac{3}{4}y) \cap (\tfrac{1}{2}E_n^+ + \tfrac{1}{4}z + \tfrac{3}{8}y)$$

is dense in J. Therefore we can choose a point $t \in A \cap J$ so that either $z - 1/m < t < z$ or alternatively $z + 1/m > t > z$ depending on which of the two sets in (3.4) is dense in J. We then put $q = 2t - \frac{3}{2}y$, $r = 2t - \frac{1}{2}z - \frac{3}{4}y$. Note that $q \in E$, $r \in E_n^+$ and $0 < q < r < t$.

Since one of the two sets in (3.4) is dense in J we can choose a point $s \in J$ in that set and arbitrarily close to t. Let $p = s - \frac{3}{4}y$, then $p \in E_n^+$.

If s is chosen close enough to t then each of the following inequalities can be satisfied: $0 < p < q$, $2q - 2p < r$, $|2p - q| < \delta(q)$, $2p < q$ if $q \in E^+$ and $2p > q$ if $q \in E^-$, $|s - z| < 1/m$ and $s < z$ or $s > z$ depending respectively on which of the two sets in (3.4) is dense in J. Note also the identities $2q - 2r = z - \frac{3}{2}y$ and $2s - 2p = \frac{3}{2}y$.

This gives then our final chain

$$
\begin{aligned}
x_0 &= 0 \\
x_1 &= 2p \\
x_2 &= 2q - 2p \\
x_3 &= 2s - z = 2r - 2q + 2p \\
x_4 &= z = 2s - 2r + 2q - 2p = 2w - y \\
x_5 &= y
\end{aligned}
$$

Since the centers of these are (in order) $p \in E_n^+$, $q \in E$, $r \in E_n^+$, $s \in E_m^+$ or $s \in E_m^-$, and $w \in E_n^+$ each element of the chain can be seen to be in V. This completes the proof.

We measure theoretic version of Theorem 3.25 is similar. Note that the statement that "$(x + t, t)$ is in V^5" for $t > 0$ means, in our other language that the interval $[x, x + 2t]$ belongs to V^5.

Theorem 3.26 *Let $E \subset \mathbb{R}$ be measurable and suppose that $V \subset \mathbb{R}^2$ is a covering relation on E. We suppose that for every point $x \in E$ there is a positive number $\delta(x)$ so that either*

$$0 < t < \delta(x) \Longrightarrow (x, t) \in V$$

or alternatively

$$0 > t > -\delta(x) \Longrightarrow (x, t) \in V.$$

Then, for almost every point $x \in E$, there is a neigbourhood U_x of x such that, for each $x + 2t \in U_x$, either $(x + t, t)$ or $(x + t, -t)$ is in V^5.

Proof. The proof repeats much of the proof of Theorem 3.25. We may assume, by removing a set of measure zero if necessary, that every point of E is a density point. Let E^+, E^-, E_n^+, E_n^- be as before. We show that every point x that belongs to the set

$$\bigcup_{n=1}^{\infty} (d(E_n^+) \cup d(E_n^-)),$$

where $d(S)$ denotes the set of density points of a set S, has a neighbourhood satisfying the assertion of the theorem.

Let x be a point of density of E_n^+. (The arguments for x a point of density of E_n^- are similar.) We may assume for notational convenience that $x = 0$. There is a $0 < \eta < 1/n$ so that for every $|y| < \eta$ each of the sets $2E_n^+ - y$, $\frac{1}{2}E_n^+ + \frac{3}{4}y$, $\frac{2}{3}E_n^+ + \frac{1}{2}y$ and $E \cap (E_n^+ + \frac{3}{4}y)$ intersects the interval $(0.9y, y)$ in a set of outer measure greater than $0.09|y|$.

Fix any point $0 < y < \eta$. We prove that there is a chain

$$0 = x_0, x_1, x_2, x_3, x_4, x_5 = y$$

with each pair satisfying (3.3). (A similar argument would show that for every $0 > y > -\eta$ there is a chain $y = x_0, x_1, x_2, x_3, x_4, x_5 = 0$ with each pair satisfying (3.3).)

Let $B = \{z \in (0.9y, y) : [z, y] \text{ belongs to } V\}$. If $C = E_n^+ \cap (0.95y, y)$ then $B \supset 2C - y$ and so $|B| \geq |2C - y| > 0.09y$.

It will be enough to prove the theorem if we show that there is a measurable set $F \subset (0.9y, y)$ such that $|F| > 0.01y$ and so that for every $z \in F$ there is a four element chain $0 = x_0, x_1, x_2, x_3, x_4 = z$ satisfying (3.3). Indeed if this is true then $F \cap B \neq \emptyset$. Choose any $z \in F \cap B$; we would then obtain first our promised four element chain $x = 0 = x_1, x_2, x_3, x_4 = z$ and then note that this expands to five elements $x = 0 = x_1, x_2, x_3, x_4, x_5 = y$ satisfying (3.3) since $[z, y]$ belongs to V.

Observe that

$$\left| E \cap (E_n^+ + \tfrac{3}{4}y) \cap (0.9y, y) \right| > 0.09y$$

It follows then that there is an integer m so that one of the sets

$$E_m^+ \cap (E_n^+ + \tfrac{3}{4}y) \cap (0.9y, y) \quad \text{or} \quad E_m^- \cap (E_n^+ + \tfrac{3}{4}y) \cap (0.9y, y) \qquad (3.5)$$

has measure exceeding $0.04y$. Take A_m as the set in (3.5) that has this property. Write

$$F = (\tfrac{1}{2}E + \tfrac{3}{4}y) \cap d(\tfrac{2}{3}E_n^+ + \tfrac{1}{2}y) \cap d(A_m).$$

We show that the set F has the property claimed for it. By the choice of η, $|F| > 0.02y$. Let $z \in F$. We shall, as before, construct four points p, q, r, s so that a sequence of reflections about these points in turn takes 0 to z.

Since $z \in d(\tfrac{2}{3}E_n^+ + \tfrac{1}{2}y)$ it follows that $z \in d(\tfrac{1}{2}E_n^+ + \tfrac{1}{4}z + \tfrac{3}{8}y)$. Let

$$A = (\tfrac{1}{2}E + \tfrac{3}{4}y) \cap (\tfrac{1}{2}E_n^+ + \tfrac{1}{4}z + \tfrac{3}{8}y) \cap d(A_m).$$

Then $z \in d(A)$. Therefore we can choose a point $t \in A$ so that either $z - 1/m < t < z$ or alternatively $z + 1/m > t > z$ depending on which of the two sets in (3.5) has has measure exceeding $0.04y$. As before we then put $q = 2t - \tfrac{3}{2}y$, $r = 2t - \tfrac{1}{2}z - \tfrac{3}{4}y$; note that $q \in E$, $r \in E_n^+$ and $0 < q < r < t$.

Since t is a density point of A_m it is also a bilateral limit point of that set; consequently we can choose a point s in that set and arbitrarily close to t. Let $p = s - \frac{3}{4}y$, then $p \in E_n^+$. If s is chosen close enough to t then each of the following inequalities can be satisfied: $0 < p < q$, $2q - 2p < r$, $|2p - q| < \delta(q)$, $2p < q$ if $q \in E^+$ and $2p > q$ if $q \in E^-$, $|s - z| < 1/m$ and $s < z$ or $s > z$ depending respectively on which of the two sets in (3.5) has the required property. Note also the identities $2q - 2r = z - \frac{3}{2}y$ and $2s - 2p = \frac{3}{2}y$.

This gives then, much as before, our final chain

$$
\begin{aligned}
x_0 &= 0 \\
x_1 &= 2p \\
x_2 &= 2q - 2p \\
x_3 &= 2s - z = 2r - 2q + 2p \\
x_4 &= z = 2s - 2r + 2q - 2p \\
x_5 &= y
\end{aligned}
$$

Since the centers of these are (in order) $p \in E_n^+$, $q \in E$, $r \in E_n^+$, $s \in E_m^+$ or $s \in E_m^-$, and $\frac{1}{2}(z+y) \in E_n^+$ each element of the chain can be seen to be in V. This completes the proof.

3.6 Approximate Symmetric Covering Relations

For the study of the approximate symmetric derivative one naturally considers covering relations $V \subset \mathbb{R}^2$ so that V contains pairs (x,t) for $t > 0$ in a set having density 1 at 0. The quantity that measures this density at a point x is given by the formula

$$\varrho(V,x) = \limsup_{h \searrow 0} |\{t \in (0,h) : (x,t) \notin V\}|/h \tag{3.6}$$

where as usual $|E|$ denotes the Lebesgue outer measure of the set E.

There are a number of definitions and notions which we summarize in the following assertions.

Definition 3.27 A symmetric covering relation V is said to be a *approximate symmetric covering relation in the large sense* on a set E if for every $x \in E$

$$|\{t \in (0,h) : (x,t) \notin V\}| = o(h)$$

as $h \searrow 0$ (i.e. using *outer* measure).

Definition 3.28 A symmetric covering relation V is said to be a *approximate symmetric covering relation* on a set E if for every $x \in E$

$$|\{t \in (0,h) : (x,t) \notin V\}|_i = o(h)$$

as $h \searrow 0$ (i.e. using *inner* measure).

Definition 3.29 A symmetric covering relation V is said to be an *essential symmetric covering relation* on a set E if for every $x \in E$ there is a $t_0 > 0$ so that the set

$$\{t \in (0,t_0) : (x,t) \notin V\}$$

has zero measure.

Definition 3.30 A symmetric covering relation V is said to be *measurable* if V is measurable (as a subset of $\mathbb{R}^2$).

3.6.1 A Lemma for Measurable Covering Relations

The measurability assumption (Definition 3.30) on a covering relation will play an important role in our deliberations. We need this to obtain the following lemma which is used in the next section. This is reproduced from Preiss and Thomson [232].

Lemma 3.31 *If V is a measurable, approximate symmetric cover then for almost every point x*

$$\limsup_{h \searrow 0} |\{t \in (0,h) : (x + t/2, t/2) \in V\}| / h = 1 \qquad (3.7)$$

and

$$\limsup_{h \searrow 0} |\{t \in (0,h) : (x - t/2, t/2) \in V\}| / h = 1. \qquad (3.8)$$

Proof. Let χ be the characteristic function of $\mathbb{R} \times (0,\infty) \setminus V$, and define the sequences of functions

$$f_n(x) = n\,|\{t \in (0,1/n) : (x,t) \notin V\}|$$

and

$$g_n(x) = n\,|\{t \in (0,1/n) : (x+t,t) \notin V\}|.$$

We compute on any interval $[a,b]$, using the Fubini Theorem and a change of variables,

$$\int_a^b g_n(x)\,dx = n \int_a^b \int_0^{1/n} \chi(x+t,t)\,dt\,dx$$
$$\leq n \int_a^{b+1} \int_0^{1/n} \chi(u,v)\,dv\,du = \int_a^{b+1} f_n(u)\,du.$$

As $0 \leq f_n \leq 1$ and $f_n \to 0$ (because of the density requirements on V) we conclude that $\int_a^b g_n(x)\,dx \to 0$ so that g_n converges to 0 in measure. Consequently we may pass to a subsequence converging almost everywhere on $[a, b]$. This gives equation (3.7) at almost every point of $[a, b]$ and hence almost everywhere on $\mathbb{R}$; assertion (3.8) is similarly proved.

3.6.2 Analysis of the Covering Properties

The general study of the approximate symmetric derivative was impeded for some time by the lack of a monotonicity theorem. Although the literature contained some accounts all of the proofs were flawed. It was not until the work of Freiling and Rinne [112] that the techniques were developed. They showed in a later paper [113] that the monotonicity result was equivalent to a formulation in terms of partitioning properties of approximate symmetric covering relations. It is this equivalence to which we now turn. The proof of the covering theorem can then be obtained by a number of strategies.

Theorem 3.32 (Freiling–Rinne) *The following assertions are equivalent (can be deduced from each other):*

(i) *if V is an approximate symmetric covering relation on an interval (a, b) and $c = (a + b)/2$ then there is a measure zero set so that V allows a partition of each interval $[2c - x, x]$ for $x \in (c, b) \setminus N$.*

(ii) *if V is a measurable, approximate symmetric covering relation on an interval (a, b) then there is a measure zero set so that V allows a partition of each interval $[c, d]$ with $c, d \in (a, b) \setminus N$.*

(iii) *if f is a measurable function with $\underline{ASD}\,f(x) \geq 0$ at every point of an interval (a, b) then f is equivalent to a nondecreasing function.*

(iv) *if W and B are measurable subsets of an interval (a, b) whose union has full measure in (a, b) and such that for each $a < x < b$ the set*

$$\{h > 0 : x - h \in W,\ x + h \in B\} \tag{3.9}$$

has density zero on the right at 0 then each such set in (3.9) in fact has measure zero.

Proof. We show that each may be deduced from the other by proving that $(i) \Longrightarrow (ii)$, $(ii) \Longrightarrow (iii)$, $(iii) \Longrightarrow (iv)$, and $(iv) \Longrightarrow (i)$ in that order.

$(i) \Longrightarrow (ii)$. Because we assume that V is measurable we may apply Lemma 3.31 to obtain a set of full measure $B \subset (a, b)$ so that for all points $x \in B$ the sets

$$R_x = \{y > x : [x, y] \text{ belongs to } V\}$$

and

$$L_x = \{y < x : [y, x] \text{ belongs to } V\}$$

are measurable and have upper density 1 at x on the right and left respectively. In particular they have positive measure. So given any pair of points $c < d$ from B there must be points $c < c' < d' < d$ so that c' is a density point of R_c and d' is a density point of L_d. Let z be the midpoint of (c', d') and apply the hypothesis (i) to obtain a set of measure zero N so that V allows a partition of each interval $[2z - x, x]$ for $x \in (z, d') \setminus N$.

Consequently, because of the density assumptions at c' and d', there are points $c'' \in R_c$ and $d'' \in L_d$ with V allowing a partition of $[c'', d'']$. But $[c, c'']$ and $[d'', d]$ are in V and so V allows a partition of $[c, d]$. As this holds for any such pair in B we have proved (ii) as required.

$(ii) \implies (iii)$. Under the assumptions on f in (iii) with any integer n the set

$$V_n = \{(x, h) : f(x + h) - f(x - h) > -h/n\}$$

is a measurable, approximate symmetric covering relation on (a, b). Applying (ii) we have a set E_n of measure zero and a partition from V for any subinterval $[c, d]$ of (a, b) whose endpoints do not belong to E_n. For such intervals evidently

$$f(d) - f(c) > -2(d - c)/n.$$

Let $E = \bigcup_{n=1}^{\infty} E_n$ and then f is nondecreasing off of E.

$(iii) \implies (iv)$. Apply (iii) to the function $f = \chi_W$.

$(iv) \implies (i)$. Assume that V is an approximate symmetric covering relation on interval $(-b, b)$ and with center $c = 0$. then we shall show that there is a measure zero set so that V allows a partition of each interval $[-x, x]$ for $x \in (0, b) \setminus N$. For sets C and D we use the notation

$$CD_x = \{h > 0 : x - h \in W, x + h \in B\} \tag{3.10}$$

and we write

$$V_x = \{h > 0 : (x, h) \in V\}.$$

Let W' denote the set of points $x \in (0, b)$ with the property that V allows a partition of the interval $[-x, x]$. Write $B' = (0, b) \setminus W'$. Let B be a measurable cover of B' and write $W = (0, b) \setminus B$. We know that $W' \supset V_0$ and V_0 has inner density 1 at 0 so W' has positive inner measure and so $|W| > 0$. Our goal is to show that $|B| = 0$ and hence that $|B'| = 0$.

Consider the sets WB_x. Each such set has density zero at 0. If not then for some $x \in (0, b)$, some $\delta > 0$ and some sequence $t_n \searrow 0$

$$|WB_x \cap V_x \cap V_{-x} \cap (0, t_n)| \geq \delta t_n.$$

Since $W \subset W'$ and $|B| = |B'|$ we have too that

$$|W'B'_x \cap V_x \cap V_{-x} \cap (0, t_n)| \geq \delta t_n.$$

Interpreting this we find that $h \in W'B'_x$ means that V partitions $[-(x-h), x-h]$ but fails to partition $[-x-h, x+h]$. But $h \in V_x \cap V_{-x}$ means that both intervals $[x-h, x+h]$ and $[-x-h, -x+h]$ are in V. Thus there can be no points h in the intersection $W'B'_x \cap V_x \cap V_{-x} \cap (0, t_n)$ and this is a contradiction.

From the contradiction we now know that each set WB_x has density zero at 0. If we apply the hypothesis (iv) we obtain that $|WB_x| = 0$ for all x. It follows that $|B| = 0$. For if not then choose points $x_0 > y_0 > 0$ so that x_0 is a point of density of B and y_0 is a point of density of W (recall that W has positive measure in each interval to the right of 0). Set

$$T = (x_0 - B) \cap (W - y_0).$$

With $x = (x_0 + y_0)/2$, $h_0 = (x_0 - y_0)/2$ we see that $WB_x \supset h_0 - T$ since if $t \in T$ then $x_0 - t \in B$, $y_0 + t \in W$, $x + (h_0 - t) \in B$ and $x - (h_0 - t) \in W$. But T has positive measure and this contradicts the fact that $|WB_x| = 0$.

Now B, and hence also B', has measure zero and so V allows a partition of the interval $[-x, x]$ for each $x \in (0, b) \setminus B'$ as required to complete the proof.

Some final remarks are in order. In Theorem 3.32 the following assertion is also equivalent to the others (i.e. they can be deduced from each other). This is proved in Freiling–Rinne [113]:

(v) if W and B are open subsets of an interval (a, b) whose union has full measure in (a, b) and such that for each $a < x < b$ the set

$$\{h > 0 : x - h \in W, x + h \in B\} \tag{3.11}$$

has density zero on the right at 0 then each such set in (3.11) is empty.

From (v) it is possible to deduce directly (see [112]) that if f is a *continuous* function on a interval $[a, b]$ with $\underline{ASD} f(x) \geq 0$ at every point of (a, b) then f is nondecreasing. Freiling and Rinne [112] give a complete proof that statement (v) holds and so, consequently, all of the assertions of Theorem 3.32 follow. In the next section, in Theorem 3.33, we shall give a direct proof of (ii).

Note that in (ii) it is assumed that the covering relation is *measurable*. Can this be dropped and our theorem proved without a measurability assumption? After all notice that statement (i) does not require measurability.

We shall see in Section 6.2.2 that in ZFC + CH the statement is false for nonmeasurable relations. However it does follow from Theorem 3.22 and Theorem 3.32(i) by using the σ–ideal of measure zero sets that the partitioning property for nonmeasurable approximate symmetric covering relations is *consistent* with ZFC. For more details see Freiling [110].

Thus, while such a partitioning theorem cannot be established in ZFC and is even false in ZFC + CH no counterexample can be produced and proven in ZFC alone. In this sense an analyst perhaps need never worry about the measurability assumption.

3.6.3 Partitioning Theorem

This section is dedicated to the proof of the main covering theorem available for approximate symmetric covers. As we have seen in the preceding section this may be obtained directly from a monotonicity theorem or from a variety of equivalent assertions. The first proof is then the monotonicity theorem of Freiling and Rinne [112] as we mentioned at the close of the preceding section. The proof we give here is obtained directly as a partitioning lemma and is reproduced from Preiss and Thomson [232]. This was obtained by utilizing ideas from [112].

Theorem 3.33 *For every measurable, approximate symmetric covering relation V there is a set N of measure zero such that V allows a partition of every interval with endpoints in* $\mathbb{R} \setminus N$.

The proof is obtained by a series of lemmas containing the main computations. Theorem 3.39 gives a covering property for approximate symmetric covers that are not necessarily measurable and then the proof of Theorem 3.33 for the measurable case follows almost immediately by using Lemma 3.31.

We begin with some notations needed just for the proofs in this section. The interval concentric with a given bounded interval I and having length $\kappa|I|$ is denoted $\kappa * I$.

Definition 3.34 Suppose that E is a measurable subset of the real line, $x \in \mathbf{R}$ and $\kappa \in (0, 1)$.

(a) We denote by $\partial_+(E)$ the set of all $x \in \mathbf{R}$ such that every neighbourhood of x contains points $u < v < w$ with $|(u, v) \cap E| > 0$ and $|(v, w) \setminus E| > 0$.

(b) We denote by $\partial_-(E)$ the set of all $x \in \mathbf{R}$ such that every neighbourhood of x contains points $u < v < w$ with $|(u, v) \setminus E| > 0$ and $|E \cap (v, w)| > 0$.

(c) We denote by $\delta(E, x, \kappa)$ the supremum of the lengths of all those open intervals J containing x for which $|J \setminus E| < \kappa|J|$. If there is no such interval, we let $\delta(E, x, \kappa) = 0$.

(d) We denote by $\Delta_+(E, x, \kappa)$ the supremum of all $t > 0$ such that

$$|[(x, x + h) \cap (2x - E)] \setminus E| < \kappa h$$

for every $h \in (0, t)$. If there is no such t, we let $\Delta_+(E, x, \kappa) = 0$.

(e) We denote by $\Delta_-(E, x, \kappa)$ the supremum of all $t > 0$ such that

$$|[(x, x - h) \cap (2x - E)] \setminus E| < \kappa h$$

for every $h \in (0, t)$. If there is no such t, we let $\Delta_-(E, x, \kappa) = 0$.

Lemma 3.35 *Suppose that E is a measurable subset of the real line. Then*

(1) $\partial_-(E) = -\partial_+(-E)$,

(2) *the sets $\partial_+(E)$ and $\partial_-(E)$ are closed subsets of $\mathbf{R}$,*

(3) *for every open interval I on the real line $I \cap \partial_+(E) = I \cap \partial_+(I \cap E)$ and $I \cap \partial_-(E) = I \cap \partial_-(I \cap E)$,*

(4) *for every interval $I = (a, b)$ on the real line $I \cap \partial_+(E) = \emptyset$ if and only if there are $c \in [a, b]$ and a set $N \subset I$ of Lebesgue measure zero such that*

$$(c, b) \setminus N \subset E \subset (c, b) \cup N,$$

(5) *for every interval $I = (a, b)$ on the real line $I \cap \partial_-(E) = \emptyset$ if and only if there are $c \in [a, b]$ and a set $N \subset I$ of Lebesgue measure zero such that*

$$(a, c) \setminus N \subset E \subset (a, c) \cup N,$$

(6) *for every $\kappa \in (0, 1)$ and every $\delta \in \mathbf{R}$ the set $\{x \in \mathbf{R}; \delta(E, x, \kappa) > \delta\}$ is open,*

(7) $\Delta_-(E, x, \kappa) = \Delta_+(-E, -x, \kappa)$ *for every $\kappa \in (0, 1)$ and every $x \in \mathbf{R}$. and*

(8) $\Delta_+(\mathbf{R} \setminus E, x, \kappa) = \Delta_-(E, x, \kappa)$ *and $\Delta_-(\mathbf{R} \setminus E, x, \kappa) = \Delta_+(E, x, \kappa)$ for every $\kappa \in (0, 1)$ and every $x \in \mathbf{R}$.*

Proof. All statements of the lemma can be verified directly.

Lemma 3.36 *Suppose that A and B are measurable subsets of bounded open intervals I and J, respectively, and that $\kappa \in (0, 1)$ and $\delta \in (0, |I|)$. Then there is $t \in J - I$ such that*

$$|[A + s] \cap B| \geq \left[\frac{|B|}{|I| + |J|} - \frac{|s - t|}{\delta} - 2\kappa \right] |\{x \in I; \delta(A, x, \kappa) > \delta\}|$$

for every $s \in \mathbf{R}$.

Proof. Since the statement is clear if $|A| = 0$ or if $|B| = 0$ we shall assume that the sets A and B have positive measure. Let $\mathcal{G}$ be the family of all open subintervals K of I such that $|K| > \delta$ and $|K \setminus A| \le \kappa |K|$, and let G be the union of this family. Then G is an open subset of I each of whose components has length at least δ. Moreover, there is a (necessarily finite) subfamily of $\mathcal{G}$ covering G such that every point of I belongs to at most two of its members. Hence $|G \setminus A| \le 2\kappa |G|$. Since $[G + t] \cap B = \emptyset$ if $t \notin J - I$, and since

$$\int_{-\infty}^{\infty} |[G + t] \cap B| \, dt = |G||B|,$$

there is $t \in J - I$ such that

$$|[G + t] \cap B| \ge |G||B|/(|I| + |J|).$$

Since each component of G has length at least δ,

$$|[G + t] \setminus [G + s]| \le |s - t||G|/\delta$$

for every $s \in \mathbf{R}$. We also observe that

$$G \supset \{x \in A;\ \delta(A, x, \kappa) > \delta\},$$

since, if K is an interval, $K \setminus I \ne \emptyset$, and if $\check{K}$ is a subinterval of I such that $K \cap I \subset \check{K}$ and $|\check{K}| = \min[|K|, |I|]$, then $|K \setminus A|/|K| \le |\check{K} \setminus A|/|\check{K}|$.

Combining the previous estimates, we see that

$$
\begin{aligned}
|[A + s] \cap B| &\ge |[G + t] \cap B| - |[G + t] \setminus [G + s]| - |G \setminus A| \\
&\ge \left[\frac{|B|}{|I| + |J|} - \frac{|s - t|}{\delta} - 2\kappa \right] |G| \\
&\ge \left[\frac{|B|}{|I| + |J|} - \frac{|s - t|}{\delta} - 2\kappa \right] |\{x \in A;\ \delta(A, x, \kappa) > \delta\}|
\end{aligned}
$$

for every $s \in \mathbf{R}$.

Lemma 3.37 *Suppose that I is a bounded open interval, E is a measurable subset of $\mathbf{R}$ and $\varepsilon \in (0, 1/2)$. Suppose further that the center w of I belongs to the closure of the set*

$$Q = \{x \in I; \Delta_+(E, x, \varepsilon) > 3|I|\},$$

that $|J \cap E| \le |J|/2$ whenever J is a subinterval of $5 * I$ with left end point w, and that $|J \cap E| \ge |J|/2$ whenever J is a subinterval of $5 * I$ with right end point w.

Then the following statements hold.

(1) *Whenever w belongs to the closure of a subinterval J of $5 * I$ then*

$$|J|/2 - \varepsilon |J| \le |J \cap E| \le |J|/2 + \varepsilon |J|.$$

(2) *Whenever J is a subinterval of* $5 * I$ *then*

$$|J|/2 - 2\varepsilon \operatorname{diam}(\{w\} \cup J) \le |J \cap E| \le |J|/2 + 2\varepsilon \operatorname{diam}(\{w\} \cup J).$$

(3) *Each component of* $I \setminus \partial_+(E)$ *has length at most* $8\varepsilon|I|$.

(4) *Whenever* $x \in Q$ *then*

$$|(x - 2|I|, x + 2|I|) \cap [(2x - E) \setminus E]| \le 16\varepsilon|I|.$$

(5) *Whenever* $x, y \in Q$ *then*

$$|[2(y - x) + (I \cap E)] \setminus E| \le 28\varepsilon|I|.$$

Proof. (1) Assume first that $J = (a, w)$. For every $\tau \in (0, \min(|I|, |J|)/2)$ we find a point $x \in (w - \tau, w + \tau) \cap Q$, we let $\tilde{J} = (a, x)$, and we use the inequality

$$|[(2x - \tilde{J}) \cap (2x - E)] \setminus E| < \varepsilon|\tilde{J}|$$

to conclude that

$$\begin{aligned}
|J|/2 \ &\le\ |J \cap E| \le |\tilde{J} \cap E| + \tau = |(2x - \tilde{J}) \cap (2x - E)| + \tau \\
&\le\ |[(2x - \tilde{J}) \cap (2x - E)] \setminus E| + |(2x - \tilde{J}) \cap E| + \tau \\
&\le\ |(2x - J) \cap E| + \varepsilon|\tilde{J}| + 2\tau \le |J|/2 + \varepsilon|J| + 3\tau.
\end{aligned}$$

Hence

$$|J|/2 \le |J \cap E| \le |J|/2 + \varepsilon|J| + 3\tau,$$

and

$$|J|/2 \ge |(2x - J) \cap E| \ge |J|/2 - \varepsilon|J| - 3\tau,$$

which shows that the inequality required in (1) holds provided that w is an endpoint of J. The general case follows easily by a decomposition.

(2) If w belongs to the closure of J, this follows from (1). Otherwise we write $J = J_1 \setminus J_2$, where J_1 and J_2 are intervals with endpoint w, and we estimate

$$|J \cap E| = |J_1 \cap E| - |J_2 \cap E|$$

$$\le |J|/2 + \varepsilon(|J_1| + |J_2|) \le |J|/2 + 2\varepsilon \operatorname{diam}(\{w\} \cup J)$$

and

$$|J \cap E| = |J_1 \cap E| - |J_2 \cap E|$$

$$\ge |J|/2 - \varepsilon(|J_1| + |J_2|) \ge |J|/2 - 2\varepsilon \operatorname{diam}(\{w\} \cup J).$$

(3) According to Lemma 3.35 (4) every component L of $I \setminus \partial_+(E)$ contains an interval J such that $|J| \ge |L|/2$ and $|J \cap E| = 0$ or $|J \setminus E| = 0$. Hence (2) implies that $|L| \le 2|J| \le 8\varepsilon|I|$.

(4) Since the interval $(x - |I|, x + |I|)$ contains w, (2) implies that

$$|(x, x + 2|I|) \cap E| \leq (1 + 6\varepsilon)|I|$$

and

$$|(x, x + 2|I|) \cap (2x - E)| \geq (1 - 6\varepsilon)|I|.$$

Recalling that $\Delta_+(E, x, \varepsilon) > 3|I|$ implies

$$|[(x, x + 2|I|) \cap (2x - E)] \setminus E| \leq 2\varepsilon|I|,$$

we combine all these inequalities to get

$$|(x, x + 2|I|) \cap E \cap (2x - E)| = |(x, x + 2|I|) \cap (2x - E)|$$
$$- |[(x, x + 2|I|) \cap (2x - E)] \setminus E| \geq (1 - 6\varepsilon)|I| - 2\varepsilon|I| = (1 - 8\varepsilon)|I|.$$

Hence

$$|(x - 2|I|, x) \cap [(2x - E) \setminus E]| = |[(x, x + 2|I|) \cap E] \setminus (2x - E)|$$
$$= |(x, x + 2|I|) \cap E| - |(x, x + 2|I|) \cap E \cap (2x - E)|$$
$$\leq (1 + 6\varepsilon)|I| - (1 - 8\varepsilon)|I| = 14\varepsilon|I|,$$

which immediately implies (4).

(5) From (4) we see that there is a set N_1 with measure at most $14\varepsilon|I|$ such that

$$2x - (I \cap E) \subset E \cup N_1.$$

Using (4) once more, we find a set N_2 with measure at most $14\varepsilon|I|$ such that

$$(y - 2|I|, y + 2|I|) \cap (2y - E) \subset E \cup N_2.$$

Finally, we use that $2x - (I \cap E) \subset (y - 2|I|, y + 2|I|)$ to infer that

$$2(y - x) + E = 2y - (2x - E)$$
$$\subset [(y - 2|I|, y + 2|I|) \cap (2y - E)] \cup (2y - N_1)$$
$$\subset E \cup N_2 \cup (2y - N_1).$$

Lemma 3.38 *Suppose that E is a measurable subset of the real line, I is a bounded open interval with center w belonging to $\partial_+(E)$, $\varepsilon \in (0, 1/2)$, and that the set*

$$Q = \{x \in I \cap \partial_+(E); \; \Delta_+(E, x, \varepsilon) > 3|I|\}$$

is dense in $I \cap \partial_+(E)$. Let γ be defined as the smallest nonnegative number such that

$$|\{z \in I; \; \mathrm{dist}(z, I \cap \partial_+(E)) \leq \gamma\}| \geq 3|I|/4.$$

Then the following statements hold.

(i) *For every s in* $(-|I|/2, |I|/2)$ *and every* $\tau > 0$ *there are* $x, y \in I \cap Q$ *such that* $|(y - x) - s| < 2\gamma + \tau.$

(ii) *For every* $\kappa \in (0, 1/4)$ *either*

$$|\{z \in I; \delta(I \cap E, z, \kappa) > \gamma/2\}| \geq |I|/8$$

or

$$|\{z \in I; \delta(I \setminus E, z, \kappa) > \gamma/2\}| \geq |I|/8.$$

(iii) *Whenever* $\delta \in (0, |I|/8)$, $\kappa \in (0, 1/2)$, *and* $x \in (1 - 8\delta/|I|) * I$, *then*

$$\delta(I \cap E, x, \kappa + \varepsilon/2) \geq 2(1 - \kappa)\delta$$

provided that $\delta(I \cap E, x, \kappa) > \delta$, *and*

$$\delta(I \setminus E, x, \kappa + \varepsilon/2) \geq 2(1 - \kappa)\delta$$

provided that $\delta(I \setminus E, x, \kappa) > \delta$.

(iv) *If* $j = 1, 2, \ldots$ *is such that* $\varepsilon \leq j^{-2}$, *if* $\delta \in (0, 2^{-j-3}|I|)$, *and if* $x \in (1 - 2^{j+3}\delta/|I|) * I$, *then*

$$\delta(I \cap E, x, (j + 1)\varepsilon/2) > 2^{j-1}\delta$$

provided that $\delta(I \cap E, x, \varepsilon/2) > \delta$, *and*

$$\delta(I \setminus E, x, (j + 1)\varepsilon/2) > 2^{j-1}\delta$$

provided that $\delta(I \setminus E, x, \varepsilon/2) > \delta$.

(v) *If* $j = 1, 2, \ldots$ *is such that* $\varepsilon \leq j^{-2}$ *and such that* $\gamma \leq 2^{-j-7}|I|)$ *then*

$$|\{z \in I; \delta(I \cap E, z, (j + 1)\varepsilon/2) > 2^{j-2}\gamma\}| \geq |I|/16$$

or

$$|\{z \in I; \delta(I \setminus E, z, (j + 1)\varepsilon/2) > 2^{j-2}\gamma\}| \geq |I|/16.$$

Proof. (i) Since the set $M = \{z \in I; \text{dist}(z, I \cap Q) \leq \gamma\}$ has measure at least $3|I|/4$, the assumption that $M \cap (s + M) = \emptyset$ would imply

$$3|I|/2 = |M \cup (s + M)| \leq |I \cup (s + I)| < 3|I|/2.$$

Hence there is $u \in M \cap (s + M)$ and it suffices to consider $x, y \in I \cap Q$ such that $|u - y| < \gamma + \tau/2$ and $|(u - s) - x| < \gamma + \tau/2$.

(ii) Since this statement is clear if $\gamma = 0$ (one just notes, that the sets whose measure we are estimating contain all density points of $I \cap E$ and $I \setminus E$, respectively), we shall assume that $\gamma > 0$. Let G be the union of all bounded components J of $I \setminus \partial_+(E)$ having length at least γ and satisfying $|J \cap E| \geq |J|/2$ and let H be the union of all bounded components J of $I \setminus \partial_+(E)$ having length at least γ and satisfying $|J \setminus E| \geq |J|/2$. Noting that γ is the smallest nonnegative number such that

$$|\{z \in I; \; \mathrm{dist}(z, (I \cap Q)) \leq \gamma\}| \geq 3|I|/4$$

and that $\gamma > 0$, we see that $|G \cup H| \geq |I|/4$. Moreover, for every component J of G we may use that $J \cap \partial_+(E) = \emptyset$ and Lemma 3.35 (4) to conclude that

$$\delta(I \cap E, x, \kappa) > |J \cap E| \geq |J|/2$$

for every $x \in J \cap E$ and every $\kappa > 0$. Hence the first inequality from (ii) holds in case $|G| \geq |G \cup H|/2$. A similar argument shows that the second inequality from (ii) holds in case $|H| \geq |G \cup H|/2$.

(iii) We prove the first inequality only. The second inequality may be proved in a similar way, or one can note that Lemma 3.35 (7) and (8) imply that each of the inequalities in question follows from the other one.

Since the statement is clear in case $\delta((I \cap E), x, \kappa) > 2(1 - \kappa)\delta$ we may assume that $\delta((I \cap E), x, \kappa) \in (\delta, 2(1 - \kappa)\delta]$.

Let $\tau \in (0, \kappa)$ be such that $\delta((I \cap E), x, \kappa - \tau) > \delta$. We let

$$\sigma = \min[\tau \delta/4, \delta((I \cap E), x, \kappa - \tau) - \delta],$$

and we find an open interval $J = (a, b)$ containing x such that $|J \setminus E| < (\kappa - \tau)|J|$ and $|J| > \delta((I \cap E), x, \kappa - \tau) - \sigma$. Noting that

$$|J| \leq \delta(I \cap E, x, \kappa) \leq 2(1 - \kappa)\delta$$

implies $2(b - a + \sigma) \leq 4(1 - \kappa)\delta + 2\kappa\delta < 4\delta$, and that $x \in [(1 - 8\delta/|I|) * I] \cap J$, we infer that $(a, 2b - a + \sigma) \subset I$.

Let c denote the smallest number from $[a, b]$ such that

$$|\{y \in E; \; c < y < b\}| = 0.$$

Since $|J \setminus E| < (\kappa - \tau)|J|$, it follows that $c > a + (1 - (\kappa - \tau))|J|$. Hence $c - \sigma > a$ and

$$(c - \sigma, c + \sigma) \subset (a, 2b - a + \sigma) \subset I.$$

We claim that $(c - \sigma, c + \sigma) \cap \partial_+(E) \neq \emptyset$. Indeed, otherwise we could use Lemma 3.35 (4) and the fact that $|(c - \sigma, c) \cap E| > 0$ to infer that E contains almost all of $(c, c + \sigma)$. But this would imply that $c = b$ and

$$|(a, b + \sigma) \setminus E| \leq |J \setminus E| < (\kappa - \tau)|J| < (\kappa - \tau)((b + \sigma) - a).$$

Consequently

$$\delta((I \cap E), x, \kappa - \tau) \geq (b + \sigma) - a = |J| + \sigma > \delta((I \cap E), x, \kappa - \tau),$$

which would be a contradiction.

Thus our claim is proved and we may use it together with the above observation that $(c - \sigma, c + \sigma) \subset I$ to find a point $z \in (c - \sigma, c + \sigma) \cap Q$. Then

$$|(a, 2z - a)| = 2(z - a) > 2(|J| - (\kappa - \tau)|J| - \sigma) > 2(1 - \kappa)\delta,$$

$$\begin{aligned}
|(a, 2z - a) \setminus E| &= |(a, z) \setminus E| + |(z, 2z - a) \setminus E| \leq |(a, z) \setminus E| \\
&+ |(z, 2z - a) \setminus (2z - E)| + |[(z, 2z - a) \cap (2z - E)] \setminus E| \\
&= 2|(a, z) \setminus E| + |[(z, 2z - a) \cap (2z - E)] \setminus E| \\
&< 2((\kappa - \tau)|J| - (b - c) + \sigma) + \varepsilon(|J| - (b - c) + \sigma) \\
&\leq (2\kappa + \varepsilon)(z - a) + 2((1 + \kappa + \varepsilon)\sigma - \tau|J|) \\
&\leq (2\kappa + \varepsilon)(z - a),
\end{aligned}$$

and

$$x \in (a, b) \subset (a, 2z - a) \subset I,$$

which implies that

$$\delta(I \cap E, x, \kappa + \varepsilon/2) \geq |(a, 2z - a)| > 2(1 - \kappa)\delta.$$

(iv) Using (iii) inductively, we see that

$$\delta(I \cap E, x, (i + 1)\varepsilon/2) > 2^i(1 - i^2\varepsilon/2)\delta$$

for each $i = 1, 2, \ldots, j$, provided that $\delta(I \cap E, x, \varepsilon/2) > \delta$. The second inequality is handled similarly.

(v) This statement follows immediately from (iv) with $\kappa = \varepsilon/2$ and from (ii) with $\delta = \gamma/2$.

Theorem 3.39 *There is a constant $\varepsilon > 0$ with the following property:*
Whenever V is a covering relation on an interval $(a, b) \subset \mathbf{R}$ such that $\varrho(V, x) < \varepsilon$ for every $x \in (a, b)$, and whenever E is a measurable subset of (a, b), then for almost every $v \in (a, b)$ for which $|(a, v) \cap E| > 0$ there is $u \in (a, v) \cap E$ such that the interval $[u, v]$ admits a partition from V.

Proof. We prove that the statement holds with $\varepsilon = 2^{-18}$. We may clearly assume that the set E is open in the density topology. (A set is density open if it is measurable and has density 1 at each of its points.) Also, we may find sets $T_x \subset (0, \infty)$ open in the density topology such that, for each $x \in \mathbb{R}$ the right lower density of T_x at 0 is greater than $1 - \varepsilon$ and $(x, t) \in V$ for every $t \in T_x$.

Instead of V it will be more convenient to consider the covering relation $\tilde{V}$ defined by the requirement $(x, t) \in \tilde{V}$ if and only if $t \in T_x$. Then we easily see that the set $\tilde{E}$ of all $v \in (a, b)$ for which there is $u \in E \cap (a, v)$ such that the interval $[u, v]$ admits a partition from $\tilde{V}$ is density open. Consequently, $\tilde{E}$ is measurable. We also observe that $x + t \in \tilde{E}$ whenever $x \in (a, b)$, $t \in T_x \cap (0, \min[b - x, x - a])$, and $x - t \in \tilde{E}$. Thus $\Delta_+(\tilde{E}, x, \varepsilon) > 0$ for every $x \in (a, b)$.

From Lemma 3.35 (4) we see that the statement of the proposition is equivalent to $\partial_+(\tilde{E}) = \emptyset$. Thus, in order to find a contradiction, we shall assume that $\partial_+(\tilde{E}) \neq \emptyset$. Then we may use Lemma 3.35 (2) and the Baire Category Theorem to find $\Delta > 0$ and an open interval $I_0 \subset (a, b)$ such that $I_0 \cap \partial_+(\tilde{E}) \neq \emptyset$ and the set

$$\{x \in I_0 \cap \partial_+(\tilde{E}); \Delta_+(\tilde{E}, x, \varepsilon) > \Delta\}$$

is dense in $I_0 \cap \partial_+(\tilde{E})$.

Since $I_0 \cap \partial_+(\tilde{E}) \neq \emptyset$, there are points $u_0 \in I_0 \cap \tilde{E}$ and $v_0 \in I_0 \setminus \tilde{E}$ such that v_0 is a density point of $I_0 \setminus \tilde{E}$ and $u_0 < v_0 < u_0 + \Delta$. Let $w \in [u_0, v_0]$ be a point at which the function $x \mapsto |(u_0, x) \cap \tilde{E}| - x/2$ attains its maximum on $[u_0, v_0]$. Noting that

$$|(w, x) \cap \tilde{E}| \leq (x - w)/2 \text{ whenever } w \leq x \leq v_0$$

and

$$|(x, w) \cap \tilde{E}| \geq (w - x)/2 \text{ whenever } u_0 \leq x \leq w,$$

we use the facts that u_0 is a density point of $\tilde{E}$ and v_0 is a density point of $I_0 \setminus \tilde{E}$ to deduce that $w \in (u_0, v_0) \cap \partial_+(\tilde{E})$.

Let I be an open interval with center w such that $|I| < \Delta/3$ and $5 * I \subset (u, v)$. The above inequalities imply that the assumptions of Lemma 3.37 hold. As in Lemma 3.38 and Lemma 3.37 we denote

$$Q = \{x \in I; \Delta_+(\tilde{E}, x, \varepsilon) > 3|I|\},$$

and we let γ be the smallest nonnegative number such that

$$|\{z \in I; \mathrm{dist}(z, I \cap \partial_+(\tilde{E})) \leq \gamma\}| \geq 3|I|/4.$$

From Lemma 3.37(3) we immediately see that $\gamma \leq 8\varepsilon|I|$. Letting $j = 8$ and noting that $2^{j+7}\gamma \leq |I|$, we deduce from Lemma 3.38 (v) that either

$$|\{z \in I; \delta(I \cap \tilde{E}, z, (j + 1)\varepsilon/2) > 2^{j-2}\gamma\}| \geq |I|/16$$

or

$$|\{z \in I; \delta(I \setminus \tilde{E}, z, (j + 1)\varepsilon/2) > 2^{j-2}\gamma\}| \geq |I|/16.$$

Assume, for example, that the first of these alternatives holds. (To handle the second case, one just exchanges the notation A and B in the following argument.) Then we let $A = I \cap \tilde{E}$, $B = I \setminus \tilde{E}$, and we observe that

$$|\{z \in I; \delta(I \cap A, z, 1/64) > 64\gamma\}| \geq |I|/16.$$

Thus we may use Lemma 3.36 to infer that there is $t \in (-|I|, |I|)$ such that

$$|[A+s] \cap B| \geq \left[\frac{|B|}{2|I|} - \frac{|s-t|}{64\gamma} - 1/32\right] |I|/16$$

for every $s \in \mathbb{R}$. Consequently, if $|s-t| < 5\gamma$ then

$$|[A+s] \cap B| \geq [1/4 - \varepsilon/2 - 5/64 - 1/32] |I|/16 > 2^{-7}|I|.$$

On the other hand, Lemma 3.38(i) implies that there are $x, y \in Q$ such that $|(y-x) - t/2| < 5\gamma/2$. Hence, letting $s = 2(y-x)$, we use Lemma 3.37(5) to conclude that

$$|[A+s] \cap B| \leq 28\varepsilon|I| < 2^{-7}|I|.$$

Since $|s-t| < 5\gamma$, this gives the required contradiction.

We are now ready to complete the proof of Theorem 3.33. Let V be a measurable approximate symmetric cover. We note first that for every $r > 0$ and for almost every x the sets

$$A(x) = \{y \in (x-r, x); (y, (x-y)/2) \in V\}$$

and

$$B(x) = \{y \in (x, x+r); ([x, (y-x)/2) \in V\}$$

are measurable and have positive measure. Their measurability follows immediately from the measurability of V and the fact that they have, for almost every point x, positive measure follows from Lemma 3.31.

Then, to finish the proof, we just observe that, whenever $x < y$, $B(x)$ is measurable and has positive measure, and $A(y)$ is measurable and has positive measure, we may use Theorem 3.39 with $E = (x, (x+y)/2) \cap B(x)$ to find $v \in ((x+y)/2, y) \cap A(x)$ and $u \in (x, (x+y)/2) \cap B(x)$ such that the interval $[u,v]$ admits a partition from V. This partition can be, in an obvious way, extended to a partition of the whole interval $[x,y]$.

3.7 2–Interval Partitions

In this section we shall develop a covering theorem that expresses the geometry of the Schwarz theorem (Theorem 1.12) of Chapter 1.

One might protest that the original Schwarz Theorem has a sufficiently elementary proof and that our covering arguments are really very subtle and involved. The goal however is not to simplify the proof but to extract some geometrical information from that result in a form that can be applied to a variety of concerns. The origins of the covering theorem here are in a paper of Ash, Freiling and Rinne [8] where, with the help of such arguments, they are able to prove a new uniqueness theorem for multiple trigonometric series. Explicitly the theorem is first stated and proved in Freiling, Rinne and Thomson [116] where it is used to develop an integral (that we shall see in Chapter 9).

We wish a covering argument that would allow us to conclude that if F is a continuous function then, from the condition

$$F(x + h) + F(x - h) - 2F(x) = 0$$

holding locally, one can conclude that this condition holds globally (so that F is linear). This is exactly analogous to the situation that, for a continuous function f, the condition $f(x + h) - f(x - h) = 0$ locally implies the same globally (so that f is constant). In the latter we can collect intervals $[x - h, x + h]$ for all x and small h and piece them together to form partitions of larger intervals. The identity $f(x + h) - f(x - h) = 0$ on each of the smaller intervals carries over by additivity and with the covering theorems in Section 3.3.3 to the same identity on larger intervals.

Here the configuration must be more complicated than simply partitions of intervals with subintervals. We can associate with the identity

$$F(x + h) + F(x - h) - 2F(x) = 0$$

the ordered pair $([x - h, x + h], [x, x])$, called a 2–interval, and in turn associate the 2–interval with a square in the plane. The partitions required then would be something of the nature of partitions of squares into smaller subsquares. But even this does not quite work because squares are too special. Instead we consider the closely related identity

$$F(x + h) + F(x - h) - F(x + k) - F(x - k) = 0.$$

We associate with this identity the ordered pair $([x - h, x + h], [x + k, x - k])$, again called a 2–interval, and associate the 2–interval with a rectangle in the plane. We seek then a partition of rectangles into subrectangles satisfying some local condition.

With this sketchy introduction the reader should be encouraged to accept the study of 2–intervals and rectangles which follows. This material is all reproduced from [116].

3.7.1 Rectangles and 2–Intervals

3.7.1.1 Rectangles By a *rectangle* R we mean a product of two closed intervals

$$R = [a, b] \times [c, d]. \tag{3.12}$$

The *eccentricity* of a rectangle R is the maximum ratio of its two sides

$$e(R) = \max\left\{\frac{d-c}{b-a}, \frac{b-a}{d-c}\right\}. \tag{3.13}$$

If $R = \bigcup_{i=1}^{n} R_i$ where $R, R_1, R_2, \ldots, R_n$ are rectangles and the $\{R_i\}$ do not overlap then we call the finite sequence $(R_1, R_2, \ldots, R_n)$ a *partition* of R. Any subsequence of $(R_1, R_2, \ldots, R_n)$ shall be called a *subpartition* of R.

3.7.1.2 2–Intervals A *2–interval* K is an ordered pair of intervals of the form

$$K = ([a, b], [a + p, b - p]) \quad (a < b, \ 0 < p \le \frac{a+b}{2}). \tag{3.14}$$

Note that this includes the possibility that the second interval of the pair is degenerate. The *centers* of the two intervals are the same and we write as $c(K)$. For any 2–interval K with center c we have then

$$K = ([c - h, c + h], [c - k, c + k])$$

for some $0 \le k < h$. We write

$$|K| = h \quad \text{and} \quad \|K\| = h^2 - k^2 \tag{3.15}$$

and refer to $|K|$ as the *length* of K and to $\|K\|$ as the *area* of K. We refer to $c \pm h$ as the *outer endpoints* of K and $c \pm k$ as the *inner endpoints* of K. The 2–interval K is said to be *r–regular* if $0 \le k \le rh$. For all of our applications $r = \frac{1}{3}$ and so the condition will read $0 \le 3k \le h$.

Every 2–interval K may be regarded as the "projection" of a rectangle R by the following scheme. Write $\pi : (x, y) \to (x + y)$ for the projection from $\mathbb{R}^2$ to $\mathbb{R}$. Then if $R = [a, b] \times [c, d]$ is a rectangle the four corners project to the four points $\pi(a, c), \pi(b, c), \pi(b, d)$ and $\pi(a, d)$, which can be viewed as the four vertices of a 2–interval. The points $\pi(a, c)$ and $\pi(b, d)$ form its outer endpoints and the points $\pi(b, c)$ and $\pi(a, d)$ its inner endpoints. We write $\pi([a, b] \times [c, d])$ for the 2–interval. For example if $b + c < a + d$ then

$$\pi([a, b] \times [c, d]) = ([a + c, b + d], [b + c, a + d]). \tag{3.16}$$

See the figure.

Naturally there is a close connection between properties of the rectangle R and its associated 2–interval $\pi(R)$. Notice that the center of the 2–interval is given by

$$c(\pi([a, b] \times [c, d])) = \frac{a + b + c + d}{2} \tag{3.17}$$

and the area by

$$\|\pi([a, b] \times [c, d])\| = (b + d - (a + b + c + d)/2)^2 - (b + c - (a + b + c + d)/2)^2$$

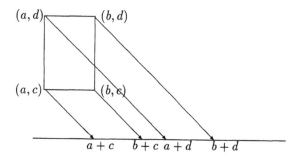

Fig. 3.7. A rectangle projecting to a 2–interval.

$$= bd - bc - ad + ac = (b - a)(d - c)$$

which latter is exactly the area of the rectangle. Similarly the length of a 2–interval is

$$|\pi\left([a, b] \times [c, d]\right)| = \frac{1}{2}(b + d - a - c) \qquad (3.18)$$

which is one quarter of the perimeter of the rectangle. The regularity of a 2–interval is related to the eccentricity of a rectangle that projects to it; if R has eccentricity no more than e then $\pi\left(R\right)$ is $(e - 1)/(e + 1)$–regular. We mainly require $\frac{1}{3}$–regularity and this holds for $e \le 2$.

3.7.1.3 2–Partitions Given any 2–interval K we define a partition into "smaller" 2–intervals by means of the projections. If R is any rectangle with $\pi\left(R\right) = K$ and $(R_1, R_2, \ldots, R_n)$ is a partition of R then we shall call

$$\left(\pi\left(R_1\right), \pi\left(R_2\right), \ldots, \pi\left(R_n\right)\right) \qquad (3.19)$$

a *2–partition* of the 2–interval K. If $(R_1, R_2, \ldots, R_n)$ is a subpartition of R then we shall call (3.19) a *2–subpartition* of K. Thus a finite sequence $(K_1, K_2, \ldots, K_n)$ of 2–intervals is a 2–partition or 2–subpartition of K if it can be so expressed. While the members of the sequence $(R_1, R_2, \ldots, R_n)$ must be distinct in order for them to form a partition of R there can easily be duplications in the sequence (3.19). We also need to be able to refer to the position of a member in the sequence; for these reasons we must use the language of sequences here rather than sets.

Let δ be a gauge (i.e. a positive function) defined on (a, b), let

$$K = ([a, b], [a + p, b - p])$$

be a 2–interval and suppose that

$$(K_1, K_2, \ldots, K_n)$$

is a 2–partition or a 2–subpartition of K. We say that such a partition or subpartition is δ–*fine at the centers* if

$$|K_i| < \delta(c(K_i)) \quad (i = 1, 2, \ldots, n).$$

We shall also measure δ–fineness at the outer endpoints in some situations. The partition is r–*regular* if each member K_i is r–regular.

3.7.2 Geometrical Arguments

The geometrical arguments collected here are needed in an induction step in the final covering theorem.

Lemma 3.40 *Let δ be a gauge in the plane which is constant on lines with slope -1. Let L be the union of a closed set of lines with slope -1 and let D denote the complement of L. Suppose that L' is a union of some subcollection of the lines in L such that L' is dense in L. Then, for any $a < b$ and any y there is a c such that $\frac{1}{2}(b - a) \leq c - y \leq 2(b - a)$ and the rectangle $R = [a, b] \times [y, c]$ can be partitioned into subrectangles such that each rectangle in the partition is of one of the following types:*

(α) *its center is in L' and its eccentricity is less than or equal to 2,*

(β) *its interior is in the open set D,*

(γ) *it is a δ–fine square with center in L.*

Proof. We may assume without loss of generality that $a = 0$, $b = 1$ and that $y = 0$.

Case 1. Some line ℓ in L (and hence also then some line ℓ' in L) intersects the y–axis in the interval $(\frac{3}{4}, \frac{3}{2})$ (see Figure 2.).

Then ℓ' also intersects the vertical line $x = \frac{1}{2}$ at some point $(\frac{1}{2}, p')$ where $\frac{1}{4} < p' < 1$. Then if we let $c = 2p'$ the rectangle $R = [0, 1] \times [0, c]$ has center $(\frac{1}{2}, p')$ and is of type α so we are done.

For the remaining cases we shall take $c = \frac{1}{2}$ so that $R = [0, 1] \times [0, \frac{1}{2}]$.

Case 2. Some line ℓ in L intersects the y–axis in the interval $(\frac{1}{2}, \frac{3}{4}]$ but not in $(\frac{3}{4}, \frac{3}{2})$ as for Case 1. Let ℓ be that line in L with the greatest y–intercept in $(\frac{1}{2}, \frac{3}{4}]$. Since ℓ has a y–intercept exceeding $\frac{1}{2}$ it must intersect the horizontal line $y = \frac{1}{4}$ at some point $(p, \frac{1}{4})$ where $\frac{1}{4} < p \leq \frac{1}{2}$ (see Figure 3.). Then, since L' is dense in L, there is a line ℓ' in L' that is sufficiently close to ℓ that it intersects

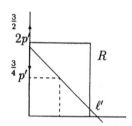

Fig. 3.8. Case 1 in Lemma 3.40

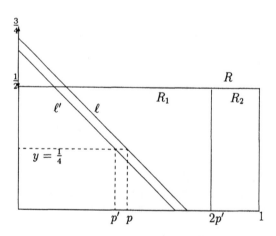

Fig. 3.9. Case 2 in Lemma 3.40.

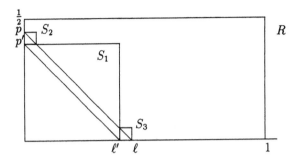

Fig. 3.10. Case 3 in Lemma 3.40.

the horizontal line $y = \frac{1}{4}$ at some point $(p', \frac{1}{4})$ where $\frac{1}{2}p + \frac{1}{8} < p' \le p$. Note that $2p' > p + \frac{1}{4}$ which is the x–intercept of ℓ. Thus R can be partitioned into the two rectangles $R_1 = [0, 2p'] \times [0, \frac{1}{2}]$ and $R_2 = [2p', 1] \times [0, \frac{1}{2}]$ with R_1 of type (α) since its center is on the line ℓ' and R_2 of type (β) by the assumptions on L and the maximality of ℓ.

Case 3. Some line ℓ in L intersects the y–axis in the interval $(0, \frac{1}{2}]$ but not in $(\frac{1}{2}, \frac{3}{2})$ as for Cases 1 and 2. Let ℓ be that line in L with the greatest y–intercept $(0, p)$ in $(0, \frac{1}{2}]$. We may choose ℓ' in L' so that ℓ' has y–intercept $(0, p')$ with $p - p'$ as small as we please (see Figure 4.).

Note that for any choice of ℓ' the square $S_1 = [0, p'] \times [0, p']$ is of type (α), while the squares $S_2 = [0, p - p'] \times [p', p]$ and $S_3 = [p', p] \times [0, p - p']$ have centers on the line ℓ. (If $p = p'$ then the squares are degenerate and can be left out.) The remainder of the rectangle, i.e. $R \setminus (S_1 \cup S_2 \cup S_3)$, has its interior in D and so can be partitioned into subintervals of type (β). We have only then to choose p' sufficiently close to p so that squares S_2 and S_3 are δ–fine at their centers (recall that δ is constant on ℓ). Accordingly we have partitioned R as required.

Case 4. In the final case no line ℓ in L intersects the y–axis in the interval $(0, \frac{3}{2})$. Then L does not meet the interior of the rectangle R and so R itself if of type (β) and we are done.

Lemma 3.41 *Let δ, L, L' and D be as in the preceding lemma and let S denote the square $[a, b] \times [a, b]$. Then there is a subrectangle $R = [a, d] \times [a, b]$ of S where $d - a \ge \frac{1}{12}(b - a)$ and R can be partitioned into rectangles each of type (α), (β) or (γ).*

Proof. As before we may assume that $a = 0$ and $b = 1$ so that S is the unit square $[0, 1] \times [0, 1]$.

Case 1. There is a line ℓ in L (and hence also a line ℓ' in L') such that its y–intercept is in $(\frac{3}{4}, 1)$ (see Figure 5.). Such an ℓ' intersects the horizontal line $y = \frac{1}{2}$

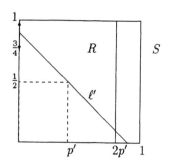

Fig. 3.11. Case 1 in Lemma 3.41.

at some point $(p', \frac{1}{2})$ where $\frac{1}{4} < p' < \frac{1}{2}$. Then the rectangle $R = [0, 2p'] \times [0, 1]$ is of type (α) with center $(p', \frac{1}{2})$ in L' and $2p' \geq \frac{1}{12}$ so that R satisfies the required conditions.

Case 2. Suppose that L does not intersect the y–axis between $\frac{3}{4}$ and 1 (see Figure 6.). Let $d = \frac{1}{12}$ so that R is the rectangle $[0, \frac{1}{12}] \times [0, 1]$. By the preceding lemma there is a partition of the desired type for the rectangle $R_1 = [0, \frac{1}{12}] \times [0, c_1]$ for some choice of $\frac{1}{24} \leq c_1 \leq \frac{1}{6}$. Applying the lemma again we partition $R_2 = [0, \frac{1}{12}] \times [c_1, c_2]$ for some choice of $\frac{1}{24} \leq c_2 - c_1 \leq \frac{1}{6}$. Continue inductively partitioning the interval $R_i = [0, \frac{1}{12}] \times [c_{i-1}, c_i]$ for some choice of $\frac{1}{24} \leq c_i - c_{i-1} \leq \frac{1}{6}$ until some $c_n \in [\frac{3}{4}, \frac{11}{12}]$. This must occur at some stage since $\frac{11}{12} - \frac{3}{4} = \frac{1}{6}$. In this way we have obtained a partition of the required form for the rectangle

$$[0, \tfrac{1}{12}] \times [0, c_n] = \bigcup_{i=1}^{n} R_i.$$

We will be done if we can also partition the rectangle $R' = [0, \frac{1}{12}] \times [c_n, 1]$

If R' has its interior entirely in D then immediately R' itself is of type (β) and we are done. If not then there is at least one line ℓ in L which intersects the y–axis at a point $(0, p)$ where $1 \leq p < \frac{13}{12}$.

Let ℓ be that line in L with the least y–intercept $(0, p)$ in $[1, \frac{13}{12})$. We may choose ℓ' in L' so that ℓ' has y–intercept $(0, p')$ with $p' - p$ as small as we please.

Note that for any choice of ℓ' the square $S_1 = [p' - 1, \frac{1}{12}] \times [p' - \frac{1}{12}, 1]$ is of type (α), while the squares $S_2 = [p - 1, p' - 1] \times [1 - p' + p, 1]$ and $S_3 = [\frac{1}{12} - p' + p, \frac{1}{12}] \times [p - \frac{1}{12}, p' - \frac{1}{12}]$ have centers on the line ℓ. (Again see Figure 6; this argument is a repetition of the ideas found in Case 3 of the preceding lemma. Note that if $p = p'$ then the squares S_2 and S_3 are degenerate and can be left out.))

The remainder of the rectangle, i.e. $R' \setminus (S_1 \cup S_2 \cup S_3)$, has its interior in D and so can be partitioned into subintervals of type (β). We have only then to choose p' sufficiently close to p so that squares S_2 and S_3 are δ–fine at their centers (recall that δ is constant on ℓ). Accordingly we have partitioned R' and hence also R as required.

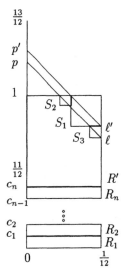

Fig. 3.12. Case 2 in Lemma 3.41.

Lemma 3.42 *Let δ, L, L' and D be as in the preceding lemmas. Let R be any rectangle, x any point inside $\pi(R)$ and N any neighbourhood of x. Then R can be partitioned into rectangles each of type (α), (β) or (γ) together with up to four rectangles with eccentricity not exceeding 2 that project into N and have an outer corner that projects to x.*

Proof. Quadrasect R by a horizontal line and a vertical line so that $P = \pi(x)$ is the common corner of the four resulting rectangles. We show that each of these four can be partitioned into rectangles each of type (α), (β) or (γ) together with one further rectangle with eccentricity not exceeding 2, that projects into N and has an outer corner that projects to x. The lemma then follows.

Without loss of generality we may assume that $|N| < \delta(x)$ if $x \in \pi(L)$ and $N \subset \pi(D)$ if $x \in \pi(D)$.

We illustrate the arguments with the T chosen as that subrectangle with lower right corner P (i.e. T is the northwest rectangle). The southeast rectangle admits an identical treatment and the remaining two rectangles (northeast and southwest) are slightly easier.

Let S_1 denote the largest square contained in T containing the corner of T that is opposite to P. If S_1 contains the left edge of T we can, by applying Lemma 3.41, obtain a partition of a rectangle $T_1 \subset S_1$ where T_1 contains at least the left $\frac{1}{12}$-th of S_1. If S_1 contains the top edge of T we can similarly obtain a partition of a rectangle $T_1 \subset S_1$ where T_1 contains at least the top $\frac{1}{12}$-th of S_1. After choosing T_1 we let S_2 be the largest square contained in the rectangle

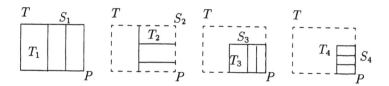

Fig. 3.13. Partition of T in Lemma 3.42.

$\overline{T} \setminus T_1$ again containing the opposite corner to P. By the same procedure there is subrectangle $T_2 \subset S_2$ where T_2 contains at least the left or the top $\frac{1}{12}$-th of S_2 (see Figure 7.).

Continuing in this manner we get a sequence of rectangles $T_1, T_2, T_3, \dots$ each of which may be partitioned as required. At some stage we will arrive at a decomposition so that

$$T' = \overline{T} \setminus (T_1 \cup T_2 \cup \dots T_n)$$

is a rectangle with eccentricity not more than 2, with a lower right corner P, and that projects entirely into the neighbourhood N. Note that in this case P is an *inner* corner of T' and so we are not done. However had the original rectangle T been chosen as the northeast or southwest rectangle then this same argument can terminate since P would, in those cases, be an outer corner of T' and so T' is of the type we require and thus a partition of T itself of the desired type is obtained.

We continue then. Let S' be the largest square contained in T' and with corner P. Note that the center of the square also projects to x. If $x \in \pi(L)$ then, since $|N| < \delta(x)$, S' is of type (γ) while if $x \in \pi(D)$, since $N \subset \pi(D)$, S' is of type (β). The remainder of the rectangle T', namely $T' \setminus S'$, now has an outer corner projecting to x and can be treated as either a northeast or a southwest rectangle. But we already know how to partition such a rectangle in the manner required.

Corollary 3.43 *Let δ, L, L' and D be as in the preceding lemmas. Let R be any rectangle which intersects the open set D. Then R can be partitioned into rectangles each of type (α), (β) or (γ).*

Proof. Apply the lemma with P any point in the rectangle that lies in D. Then the four subrectangles of the lemma that have corners at P can be chosen inside D and so are of type (β).

3.7.3 The Covering Theorem

We now state and prove our covering theorem.

Theorem 3.44 *Let δ be a gauge on $\mathbb{R}$. Then there is a countable set E with the following property: for any 2–interval*

$$K = ([a, b], [a + p, b - p]) \quad (0 < p \leq \tfrac{1}{2}(a + b))$$

and for any gauge γ on $E \times \mathbb{N}$ there is a sequence

$$(I_1, I_2, \ldots, I_n)$$

forming a 2–partition of the 2–interval K that is $\frac{1}{3}$–regular, such that each I_i is either δ–fine at the center or else I_i has an outer endpoint x in E with $|I_i| < \gamma(x, i)$.

Proof. For each natural number i let $A_i = \{x : \delta(x) > i^{-1}\}$ and apply the splattered Baire category Theorem, Lemma A.8 to obtain a transfinite sequence of open sets $\{G_\eta\}$ with the properties stated. Let E_0 be the splattered set of the endpoints of the $\{G_\eta\}$ and let E be E_0 *together with* the addition of any countable dense set. Since E is countable we may write $E = \{x_1, x_2, x_3, \ldots\}$.

It is more convenient to translate the theorem to a statement about rectangles. Let $\delta'(x, y) = \delta(x + y)$; then δ' is a gauge on $\mathbb{R}^2$ that is constant on lines with slope -1 (so that the lemmas of the preceding sections may be applied). We shall show that any rectangle R that projects to a subset of· a set G_η has the following property:

($*$) for any gauge γ on $\mathbb{N} \times \mathbb{N}$ there is a sequence

$$(R_1, R_2, \ldots, R_n)$$

of subrectangles forming a partition of R such that, for each $i = 1, 2 \ldots n$, the rectangle R_i has eccentricity no more than 2 and either

(*i*) R_i is δ'–fine at its center, or

(*ii*) R_i has an outer corner that projects to a point x_j in E with $|R_i| < \gamma(i, j)$.

Since $G_\eta = \mathbb{R}$ at some stage this proves the theorem.

The property ($*$) can be more conveniently stated in an equivalent form:

($**$) for any gauge γ on $\mathbb{N} \times \mathbb{N}$ there is a sequence

$$(R_1, R_2, \ldots, R_n)$$

of subrectangles forming a partition of R such that, for each $i = 1, 2 \ldots n$, the rectangle R_i has eccentricity no more than 2 and either (*i*) R_i is δ'–fine at its center, or

(*ii*) R_i has an outer corner that projects to a point x_j in E with $|R_i| < \gamma(i,j)$, or

(*iii*) R_i itself satisfies (*).

We prove that (*) and (**) are equivalent. It is clear that (*) $\Longrightarrow$ (**). Conversely suppose R satisfies (**) and γ is a gauge on $\mathbb{N} \times \mathbb{N}$. Then there is a sequence $(R_1, R_2, \ldots, R_n)$ of subrectangles forming a partition of R such that each R_i has eccentricity no more than 2 and each is of type (i), (ii) or (iii). We show how to construct a new sequence $(R'_1, R'_2, \ldots, R'_m)$ forming a partition of R with each R'_i having eccentricity no more than 2 and each of type (i) or (ii). It will follow that R satisfies (*) and we are done. Each R_i in the original sequence is treated, in turn, in the following manner. If R_i is of type (i) or (ii) then set $R'_i = R_i$. If R_i is of type (iii) then choose an appropriate gauge γ_i and apply (*) with that gauge to obtain a partition $(R_{i1}, R_{i2}, \ldots, R_{in_i})$ of R_i. Replace $R'_i = R_{i1}$ and place $(R_{i2}, R_{i3}, \ldots, R_{in_i})$ at the end of the sequence so far obtained. It is easy to select each γ_i so that the resulting sequence satisfies (ii) with respect to the original gauge γ.

Now we observe some facts about (*) (or equivalently (**)). The property (*) is additive. If $R = \bigcup_{i=1}^{n} R_i$ is a partition of R and each R_i has property (*) then so too does R. This is clear since if this is so then R has property (**) trivially and this has been seen to be equivalent to (*).

Note too that (*) has a compactness property. If for every $\alpha < \beta$ the property (*) holds for each rectangle that projects into the open set G_α then (*) holds for each rectangle R that projects into the open set $\bigcup_{\alpha < \beta} G_\alpha$. By compactness R projects into some finite union $G_{\alpha_1} \cup G_{\alpha_2} \ldots \cup G_{\alpha_n}$. R may be then partitioned into subrectangles each of which projects into a single G_{α_i}. Since each of these has property (*) it follows, from the additivity property, that R does too.

And finally note that (*) has a closure property. Let (a, b) be an open interval with endpoints $a, b \in E$. If every rectangle that projects into (a, b) has (*) then so too does every rectangle R that projects into $[a, b]$. Let such a rectangle R and a gauge γ on $\mathbb{N} \times \mathbb{N}$ be given. If an outer corner of R projects to a or b it can be handled as follows. If $R = [c_1, d_1] \times [c_2, d_2]$ then it may be partitioned into $(S_1, S_2, T_1, T_2, T_3)$ where $S_1 = [c_1, c_1 + \epsilon] \times [c_2, c_2 + \epsilon]$ and $S_2 = [d_1 - \epsilon, d_1] \times [d_2 - \epsilon, d_2]$, for arbitrarily small $\epsilon > 0$, are two squares and T_1, T_2, T_3 are three rectangles which project entirely into (a, b). The subrectangles T_i satisfy (**)(*iii*) (since they project to subsets of (a, b)) and S_1, S_2 will satisfy (**)(*ii*) if ϵ is made sufficiently small since the outer corners project to points in E. By definition R must have property (**) (and hence (*) as well) and we are done.

We show now that for each ordinal η and any rectangle R that projects to a subset of G_η the assertion (*) holds. For the initial member of the sequence, $G_0 =$

$\emptyset$ there is nothing to prove. For λ, a limit ordinal, the compactness property supplies the statement.

Now suppose that we have verified $(*)$ for all rectangles that project into G_n and we have a rectangle R that projects into G_{n+1}. Recall that for some i the set A_i is dense in $G_{n+1} \setminus G_n$. By the additivity property we may assume that $|R| < i^{-1}$.

Since E is dense, choose a point P in the rectangle R such that $\pi(P) = x_j \in E$. Let N be a neighbourhood of x_j smaller in length than $\gamma(i,j)$ for $i = 1,2,3,4$. Use Lemma 3.42 with the gauge δ', $D = \pi^{-1}(G_n)$ and $L' = L \cap \pi^{-1}(A_i)$ where π is, as before, the projection map $\pi(x,y) = x + y$. In this way we can partition R into rectangles of type (α), (β) or (γ) together with up to four rectangles of eccentricity no more than 2 which project into N and have an outer corner projecting to x_j. These four rectangles are satisfactory for the first four elements of a partition of R by the definition of N. The remaining rectangles, since they are of type (α), (β) or (γ), satisfy $(*)$. Accordingly R satisfies $(**)$ and hence $(*)$ as required. The theorem now follows by induction.

We should mention that the original article [116] also proves a version under the assumption that the gauge δ has the Baire property. In that case the exceptional countable set of the theorem may be chosen to be splattered.

3.7.4 Partitions into Squares

Let us make a further remark on this partitioning theorem. The second symmetric derivative is more closely associated with squares than with rectangles having some regularity condition. That is, in the language of 2–intervals, we would prefer a partitioning into 2–subintervals of the form $([x - h, x + h], [x, x])$ rather than the, more general, regular ones. Accordingly one might have hoped for a partitioning of rectangles into squares rather than subrectangles. This would make our subsequent work rather easier since we would not then have to introduce the modification of the second symmetric derivative that we shall see in Chapter 8 and 9.

It has been long known, however, that not all rectangles can be partitioned by squares. Here is a simple proof (from Pokrovskii [223]) of a theorem of Dehn [63] that any rectangle whose sides have irrational ratio cannot be so partitioned. For any rectangle $R = [a,b] \times [c,d]$ write

$$F(R) = f(a + c) + f(b + d) - f(b + c) - f(a + d)$$

where f is an arbitrary real function. It is easy to check that F is an additive rectangle function. Suppose that α/β is irrational and consider the rectangle R_0 centered at the origin and with northeast corner at the point (α, β). Choose a Hamel basis containing both $\alpha + \beta$ and $\alpha - \beta$. Let $f(x) = r^2$ where r is the coefficient corresponding to $\alpha - \beta$ in the Hamel expansion of x. Then, with F as above, note that

$$F(R_0) = f(\alpha + \beta) + f(-\alpha - \beta) - f(\alpha - \beta) - f(-\alpha + \beta) = -2.$$

Observe for a square S that $F(S)$ is of the form $f(x+h)+f(x-h)-2f(x)$. Write $x+h = r(\alpha-\beta)+\ldots$ and $x-h = s(\alpha-\beta)+\ldots$ so that $x = \frac{1}{2}(r+s)(\alpha-\beta)+\ldots$ all in the Hamel expansions. Then $f(x+h) = r^2$, $f(x-h) = s^2$ and $f(x) = \frac{1}{4}(r+s)^2$. Hence we observe that $F(S) \geq 0$ for every square S. By the additivity of F it follows that R_0 cannot be partitioned into squares.

4

Even Properties

4.1 Introduction

In the first two chapters we have studied the symmetry properties of real functions following mostly the themes of continuity and differentiability as they arise in the properties of the expressions

$$\Delta^1{}_s f(x,t) = f(x+t) - f(x-t)$$

and

$$\Delta^2{}_s f(x,t) = f(x+t) + f(x-t) - 2f(x).$$

As we have now most of the technical tools needed, this chapter will attempt a rather more systematic study of just the even properties. This will repeat to a small degree material we have seen but will also delve much more deeply into these matters.

Recall that the even properties of a function reflect in some way the odd symmetric structure of the function. At the simplest level a function is odd at a point if its even part there vanishes. If the even part of a function f vanishes everywhere then one expects f to be odd everywhere and that seems to suggest that f must be linear. (Things are not, as we shall see, quite this simple.) In general the conditions that we place on the even part can be viewed as imposing greater or lesser odd symmetric structure on the function.

Historically the most important studies of the expression $\Delta^2{}_s f(x,t)$, arising from the even part of f, are those of Riemann developed more than a century ago, Jensen in 1905 and 1906, and Zygmund in 1945. The notion of smoothness that we discuss in Section 4.8 is from Riemann's study. The fundamental paper that initiated a serious investigation of smoothness as a study in its own right, rather than for specific applications in trigonometric series, is that of Zygmund [312]. The Jensen study is connected with convexity, especially with the condition that $\Delta^2_s f(x,t) \geq 0$. This is the notion of midpoint-convexity that is studied in Section 4.3.

The rest of the chapter is devoted to the study of local growth conditions of the form

$$\Delta^2_s f(x,h) = O(\phi(h)) \quad \text{or} \quad \Delta^2_s f(x,h) = o(\phi(h)$$

as $h \to 0$ where ϕ is a monotonic function (most frequently $\phi(h) = h^\alpha$). Loosely speaking we shall call all such conditions "smoothness" conditions. In the literature a function f is said to be *symmetric* if $\Delta^2_s f(x,h) = o(1)$, *smooth* if $\Delta^2_s f(x,h) = o(h)$, and *quasi-smooth* if $\Delta^2_s f(x,h) = O(1)$. At the risk of abusing this language we shall call functions satisfying a condition $\Delta^2_s f(x,h) = O(\phi(h))$ for functions ϕ tending faster to zero, *super-smooth*. We shall investigate the structure of functions satisfying these and other related properties.

4.2 Midpoint-Linear Functions

We begin our study with the most severe of the even symmetry conditions that can be imposed. Let us introduce the following classes of functions. A function f is *additive* if

$$f(x+y) = f(x) + f(y)$$

for all real x and y. A function f is *linear* if

$$f(\lambda x + (1-\lambda)y) = \lambda f(x) + (1-\lambda)f(y)$$

for all real x and y and all $0 \leq \lambda \leq 1$. Finally a function f is *midpoint-linear* if

$$f\left(\frac{x+y}{2}\right) = \frac{f(x) + f(y)}{2}$$

for all real x and y.

It is only the last of these classes that relates directly to our concerns since only this last is a property expressed by the even part of f. In particular note that a midpoint-linear function satisfies the relation

$$f(z+h) + f(z-h) = 2f(z)$$

for all z and h which is a more familiar expression from our viewpoint. (Simply replace $z + h = x$ and $z - h = y$.)

The class of linear functions is transparent: the definition just asserts that the graph is a straight line between any two points $(x, f(x))$ and $(y, f(y))$ and it is easy to check that this must require $f(x) = mx + c$ for some numbers m and c. One might be led to expect that the other classes can be handled in as trivial a fashion. Are not all additive functions simply of the form $f(x) = mx$? Are not all midpoint-linear functions merely linear? An elementary result of Sierpiński [256] shows that the classes of additive functions and midpoint-linear functions are closely related. The existence then of nonmeasurable additive functions shows that the answer is not so clear.

Lemma 4.1 (Sierpińksi) *The function f is midpoint-linear if and only if the function $g(x) = f(x) - f(0)$ is additive.*

Proof. If $g(x) = f(x) - f(0)$ is additive then certainly

$$
\begin{aligned}
f(z + h) + f(z - h) &= g(z + h) + f(0) + g(z - h) + f(0) \\
&= 2g(z) + 2f(0) = 2f(z)
\end{aligned}
$$

and hence f is midpoint-linear. Conversely, if f is midpoint-linear, then

$$2f(x) = f(2x) + f(0) \text{ and } 2f(y) = f(2y) + f(0)$$

for any x and y. Using the midpoint-linearity again this gives

$$2f(x + y) = f(2x) + f(2y) = 2f(x) + 2f(y) - 2f(0)$$

and hence that $g(x + y) = g(x) + g(y)$ so that g is additive.

4.2.1 Additive Functions

Because of Lemma 4.1 we may reduce the study of midpoint-linear functions to the study of additive functions. The additive functions, i.e. those functions f that satisfy the Cauchy functional equation

$$f(x + y) = f(x) + f(y),$$

have now a long history with contributions tracing back to Legendre, Gauss and Cauchy. The existence of nonmeasurable solutions was established by Hamel [127] in 1905. Since then it has been shown that a variety of weak regularity conditions imposed on the solution will force the solution to assume the form $f(x) = mx$ for some real number m. We present just the simplest of these observations.

Lemma 4.2 *Let f be additive. Then for any real numbers $x_1, x_2, \ldots x_n$ and rational numbers $r_1, r_2, \ldots r_n$,*

$$f\left(\sum_{i=1}^{n} r_i x_i\right) = \sum_{i=1}^{n} r_i f(x_i). \tag{4.1}$$

Proof. If f is additive then we can establish, by induction, that

$$f\left(\sum_{i=1}^{n} x_i\right) = \sum_{i=1}^{n} f(x_i) \tag{4.2}$$

and so also that $f(mx) = mf(x)$ for real x and natural numbers m. This then extends to $f(mx/n) = mf(x)/n$ for natural numbers m and n by writing $mf(x) = f(mx) = f(n(mx/n)) = nf(mx/n)$. Thus $f(px) = pf(x)$ for positive rational p. For $p = 0$ the same identity holds because, by additivity $f(0 + 0) = f(0) + f(0)$ so $f(0) = 0$. Also for p negative we have again the same identity as the computation

$$0 = f(px - px) = f(px) + f(-px) = f(px) + (-p)f(x)$$

shows. Finally this fact and the relation (4.2) combine to give relation (4.1) completing the proof.

Theorem 4.3 *Let f be additive. Then either f is of the form $f(x) = mx$ for some real number m or else the graph of f is everywhere dense in $\mathbb{R}^2$.*

Proof. If f is not of the form $f(x) = mx$ for some real number m then we may select nonzero numbers x_1 and x_2 so that $f(x_1)/x_1 \neq f(x_2)/x_2$. This can be interpreted as asserting that the vectors $(x_1, f(x_1))$ and $(x_2, f(x_2))$ are linearly independent in $\mathbb{R}^2$. Consequently the set of vectors

$$\{r_1(x_1, f(x_1)) + r_2(x_2, f(x_2)) \,:\, r_1, r_2 \in \mathbb{Q}\}$$

is dense in $\mathbb{R}^2$. But

$$r_1(x_1, f(x_1)) + r_2(x_2, f(x_2)) = (r_1 x_1 + r_2 x_2, f(r_1 x_1 + r_2 x_2))$$

is a point on the graph of f and so that graph must also be dense in $\mathbb{R}^2$.

Corollary 4.4 *Let f be a real-valued, additive function. Then f must be of the form $f(x) = mx$ for some real number m should any one of the following conditions hold:*

1. *f has a point of continuity.*

2. *f is monotonic on some interval.*

3. *f is bounded above or below on some interval.*

4. *f is bounded above or below on some measurable set of positive measure.*

5. *f is measurable with respect to some set of positive measure.*

Proof. For the first three of these items it is clear that such a function cannot have a dense graph; accordingly Theorem 4.3 asserts that the other alternative, that $f(x) = mx$, must be realized.

For the fourth case we need a measure-theoretic fact: if E is a measurable set of positive measure then the set

$$E + E = \{x + y : x, y \in E\}$$

contains an interval. Consequently if f is bounded above (say) on E and $E + E$ contains an interval I then for every $z \in I$, $z = x + y$ for some $x, y \in E$ and

$$f(z) = f(x + y) = f(x) + f(y) \leq 2 \sup_{e \in E} f(e).$$

Once again Theorem 4.3 shows that f must assume the form $f(x) = mx$. The final case follows too since, if f is measurable relative to a measurable set E of positive measure, some set $\{x \in E : f(x) < n\}$ or $\{x \in E : f(x) > n\}$ must have positive measure.

Corollary 4.5 *Let f be real-valued and midpoint-linear. Then f must be of the form $f(x) = mx + c$ for some real numbers m and c should any one of the following conditions hold:*

1. *f has a point of continuity.*

2. *f is monotonic on some interval.*

3. *f is bounded above or below on some interval.*

4. *f is bounded above or below on some measurable set of positive measure.*

5. *f is measurable with respect to some set of positive measure.*

4.2.2 Hamel Theorem

Having seen that additive and midpoint-linear functions are generally very simple provided some regularity condition is met, we now indicate the worst situation. The proof follows directly from the existence of a Hamel basis for the real numbers. (This material is presented in Section A.6 in the Appendix where additional information about such bases appears.) In particular we can show that the nonmeasurable, additive function of the theorem can be chosen so as to vanish on a dense set or to have the Darboux property. In this latter case the graph of the function is not merely dense (as Theorem 4.3 would require) but the function must assume every value on every interval.

Theorem 4.6 (Hamel) *There exist nonmeasurable functions that are additive and midpoint-linear.*

Proof. If B is a Hamel basis then choose an arbitrary function f on B and extend to the rest of $\mathbb{R}$ by writing $f(x) = \sum_{i=1}^{n} r_i f(h_i)$ where $x = \sum_{i=1}^{n} r_i h_i$ is the Hamel expansion of x relative to the basis B. Clearly f is additive. The nonmeasurability can follow from properties of Hamel bases but is evident directly from Theorem 4.3 and Corollary 4.4: since f may be chosen freely on B it is easy to avoid the form $f(x) = mx$.

4.2.3 Close to Midpoint-Linear Functions

Suppose that a function f is very close to being midpoint-linear in the sense that for some $\epsilon > 0$ the inequality

$$|f(x + h) + f(x - h) - 2f(x)| \leq \epsilon$$

holds for all x, $h \in \mathbb{R}$. Then it might be expected that f is not too distant from some midpoint-linear function. This is the content of the next theorem. This is closely related to material in [161, Ch. 17] on "approximately additive functions".

Theorem 4.7 *Let $\epsilon > 0$ and suppose that the inequality*

$$|f(x + h) + f(x - h) - 2f(x)| \leq \epsilon \tag{4.3}$$

holds for all x, $h \in \mathbb{R}$. Then there is a midpoint-linear function F for which

$$|F(x) - f(x)| \leq \epsilon \tag{4.4}$$

holds for all $x \in \mathbb{R}$. The function F is unique up to an additive constant.

Proof. We write
$$F(x) = \lim_{n \to \infty} 2^{-n} f(2^n x) + f(0) \tag{4.5}$$

and show that this limit exists, that F is midpoint-linear and that (4.4) holds. Write

$$G_n(x) = 2^{-n} (f(2^n x) - f(0)) + f(0).$$

We prove, inductively, for all integers n that

$$|G_n(x) - f(x)| \leq (1 - 2^{-n})\epsilon. \tag{4.6}$$

The case $n = 0$ is trivial. Suppose that (4.6) holds for some integer n. By (4.3) we know that

$$|f(2 \cdot 2^n x) + f(0) - 2f(2^n x)| \leq \epsilon. \tag{4.7}$$

Now, using (4.7) and the induction hypothesis, we compute

$$
\begin{aligned}
|G_{n+1}(x) - f(x)| &= \left|2^{-(n+1)}\left(f(2^{n+1}x) - f(0)\right) + f(0) - f(x)\right| \\
&\leq \left|2^{-n}\left(f(2^n x) - f(0)\right) + f(0) - f(x)\right| + 2^{-n-1}\left|f(2 \cdot 2^n x) + f(0) - 2f(2^n x)\right| \\
&\leq 2^{-(n+1)}\epsilon + (1 - 2^{-n})\epsilon = (1 - 2^{-(n+1)})\epsilon.
\end{aligned}
$$

Hence (4.6) follows.

An elementary computation shows that

$$
\begin{aligned}
G_{n+m}&(x) - G_m(x) \\
&= \left\{2^{-(n+m)}\left(f(2^m \cdot 2^n x) - f(0)\right) + f(0)\right\} - \left\{2^{-n}\left(f(2^n x) - f(0)\right) + f(0)\right\} \\
&= 2^{-n}\left\{2^{-m}\left(f(2^m \cdot 2^n x) - f(0)\right) + f(0) - f(2^n x)\right\} \\
&= 2^{-n}\left(G_m(2^n x) - f(2^n x)\right).
\end{aligned}
$$

Now, using (4.6), we find that

$$
|G_{n+m}(x) - G_m(x)| \leq 2^{-n}[(1 - 2^{-m})\epsilon] < 2^{-n}\epsilon
$$

and so the sequence $\{G_n(x)\}$ is Cauchy.

It follows that the limit defining F in (4.5) exists and that $F(x) = \lim_{n\to\infty} G_n(x)$. The inequality (4.4) follows immediately now from (4.6). Using (4.3) we obtain that

$$
\begin{aligned}
|G_n(x + h) &+ G_n(x - h) - 2G_n(x)| \\
&= \left|2^{-n}(f(2^n(x + h)) + f(2^n(x - h)) - 2f(2^n x)\right| \leq 2^{-n}\epsilon.
\end{aligned}
$$

Finally, letting $n \to \infty$ in this inequality, we obtain that F is midpoint-linear.

It remains only to show that the function F is unique. Let F be any midpoint-linear function satisfying (4.4). For any $x \in \mathbb{R}$ and any integer n, the midpoint-linearity property provides $F(nx) - F(0) = n(F(x) - F(0))$. Hence, from (4.4), we can obtain that

$$
F(x) - F(0) = \lim_{n\to\infty} \frac{1}{n}(f(nx) - f(0))
$$

which shows that F is uniquely determined by f up to an additive constant.

In Theorem 4.7 it can be shown that most regularity assumptions on f carry over to F; for example if f is measurable then so too is F so that F is in this case a linear function.

4.2.4 Locally Midpoint-Linear Functions

Recall that a function f is midpoint-linear if

$$f\left(\frac{x+y}{2}\right) = \frac{f(x) + f(y)}{2}$$

for all real x and y. Let us say that f is *locally midpoint-linear* if this holds merely locally, i.e. if for each x there is a $\delta(x) > 0$ so that

$$f(x) = \frac{f(x+h) + f(x-h)}{2}$$

for all $0 < h < \delta(x)$. Because of what we have seen in the preceding sections we shall study only measurable functions that satisfy such a condition. Note that this is the even analogue of the locally symmetric functions of Section 2.8 and, not surprisingly, much the same techniques will provide the structure of such functions.

A simple example illustrates that when this condition is imposed locally, rather than globally, the function need not be linear. Write $g(x) = x - [x]$ for each x not an integer, where $[x]$ denotes the integer part of x (the greatest integer less than x) and set $g(x) = \frac{1}{2}$ at the integers. It is easy to see that this function is locally midpoint-linear. While it is certainly not linear it is in a sense just "fragments" of the linear function $f(x) = x$. This example suggests what all such functions will look like.

Our study of these functions is based on the work of Kostyrko [154] and [157] (who calls such functions *locally Jensen*). This reproduces Theorem 2.29 from Chapter 2 and we refer to that material for an indication of the method of proof.

Theorem 4.8 (Kostyrko) *Suppose that f is measurable and locally midpoint-linear. Then there exists a closed, countable, locally symmetric set S and a constant m such that in each interval complementary to S the graph of f is a straight line with slope m.*

In the original [154, Theorem 2, p. 66] this theorem is stated in the more general form that, without assuming measurability, there is a single, globally midpoint-linear function F and f is a translate of F in each of the intervals complementary to S. Of course if f is measurable then F must be a linear function and this is the version we have preferred to state. The corollary follows from the theorem since, if f has no jump discontinuities, then the set S of the theorem must be empty.

Corollary 4.9 *Suppose that f is a measurable, locally midpoint-linear function that has the Darboux property. Then f is linear.*

Kostyrko [154, p. 68] goes on to show that the class of locally midpoint-linear functions is not closed under uniform limits. We state this without proof.

Theorem 4.10 (Kostyrko) *There is a uniformly convergent sequence of functions $\{f_n\}$, each locally midpoint-linear, whose limit f is not locally midpoint-linear.*

4.3 Midpoint-Convex Functions

A function f is said to be *midpoint-convex* if

$$f\left(\frac{x+y}{2}\right) \leq \frac{f(x)+f(y)}{2}$$

for all real x and y. We can read this as asserting that

$$f(x+t) + f(x-t) \geq 2f(x)$$

which states that the point $(x, f(x))$ lies below the chord joining the point $(x+t, f(x+t))$ to the point $(x-t, f(x-t))$. This should be contrasted with the definition of a convex function: a function f is said to be *convex* if

$$f\left(\lambda x + (1-\lambda)y\right) \leq \lambda f(x) + (1-\lambda)f(y)$$

for all real x and y and all $0 \leq \lambda \leq 1$.

There are some facts that we should recall about convex (rather than midpoint-convex) functions. If f is convex on a interval $[a, b]$ it must be continuous on (a, b) and differentiable at every point there with the possible exception of a denumerable set. It has right and left hand derivatives everywhere there and so the points where the derivative fails to exist are just those (countably many) points where the right and left hand derivative fail to agree. The derivative f' is continuous and nondecreasing on the set where it exists. These facts can be found in Roberts and Varberg [245, Chap. I] for example. (Their Chapter VII also contains an excellent account of midpoint-convex functions.)

In the presence of a regularity condition (for example measurability) the two concepts, convexity and midpoint convexity, are equivalent and so in the literature either definition may be found labeled simply as "convex". Here we shall be more careful with the terminology. The definition can be restricted to an interval with no changes in the arguments throughout.

We notice that a midpoint-linear function is necessarily midpoint-convex too so that, because of Theorem 4.6, nonmeasurable midpoint-convex functions do exist. Here, as in the preceding sections, the thrust of our work is to show that under mild assumptions on the function the pathological case cannot occur. There is a considerable literature devoted to this problem.

A brief survey might be interesting. The first study of such functions is that of Jensen [143] in 1905. He showed that these notions coincide if f is bounded above. Bernstein and Doetsch [15] in 1915 refined this by showing that the same is true if f is merely bounded above in a neighbourhood of a single point. Blumberg [16] and Sierpiński [256] independently in 1919 and 1920 showed that a measurability assumption is enough for this. Other refinements on this same theme can be found in Ostrowski [218], Kurepa [167], Kemperman [150], and Mehdi [200] among others. Of course the literature of this subject is wider still.

There does not appear to be too much to be said about general nonmeasurable, midpoint convex functions apart from the fact that every midpoint linear function is also midpoint convex. Bruckner, Ceder and Weiss [19] have proved the following theorem. Recall that a midpoint convex function itself need not be Darboux.

Theorem 4.11 (Bruckner–Ceder–Weiss) *Any midpoint convex function is the uniform limit of a sequence of Darboux functions.*

4.3.1 Theorem of Blumberg–Sierpiński

The main result that we wish to establish is that, under a measurability assumption, midpoint-convex functions are convex. This theorem is due, independently, to Blumberg [16] and Sierpiński [256].

Theorem 4.12 (Blumberg–Sierpiński) *Suppose that f is a measurable, midpoint-convex function. Then f is convex.*

The proof of the theorem follows from a series of lemmas each of interest on its own. The first lemma is due to Jensen; it asserts that for a midpoint-convex function there is a natural inequality for all *rational* convex combinations of points.

Lemma 4.13 (Jensen's inequality) *Let f be midpoint-convex, let x_1, x_2, $\ldots x_n$ be real numbers and suppose that r_1, r_2, $\ldots r_n$ are positive rational numbers with $\sum_{i=1}^{n} r_i = 1$. Then*

$$f\left(\sum_{i=1}^{n} r_i x_i\right) \le \sum_{i=1}^{n} r_i f(x_i). \tag{4.8}$$

Proof. This clever proof is mostly due to Cauchy. For the first step we shall show that

$$f\left(\frac{x_1 + x_2 + \ldots x_n}{n}\right) \le \frac{f(x_1) + f(x_2) + \ldots f(x_n)}{n} \tag{4.9}$$

which is evidently a special case of (4.8). For $n = 2$ this is immediate. For $n = 4$ this follows from two applications of the $n = 2$ case. Thus we have from one application of this inequality

$$f\left(\frac{1}{2}\left[\frac{x_1 + x_2}{2}\right] + \frac{1}{2}\left[\frac{x_3 + x_4}{2}\right]\right) \leq \frac{1}{2}\left[f\left(\frac{x_1 + x_2}{2}\right) + f\left(\frac{x_3 + x_4}{2}\right)\right]. \quad (4.10)$$

The inequality (4.10) together with the $n = 2$ case again supplies the $n = 4$ case. In this manner the situation for $n = 2^k$ is handled.

For other values of n we need only some arithmetic: if $2^{k-1} < n < 2^k$ then an application of the above inequality for the 2^k case gives

$$f\left(\frac{x_1 + x_2 + \ldots x_n}{n}\right)$$

$$= f\left(\frac{1}{2^k}\left(x_1 + x_2 + \ldots x_n + (2^k - n)\frac{x_1 + x_2 + \ldots x_n}{n}\right)\right)$$

$$\leq \frac{1}{2^k}\left(f(x_1) + f(x_2) + \ldots f(x_n) + (2^k - n)f\left(\frac{x_1 + x_2 + \ldots x_n}{n}\right)\right).$$

The inequality (4.9) now follows after some obvious algebraic manipulation.

Finally to establish (4.8) in general let d be the least common denominator of the r_i. Set $s_i = r_i d$ so that the s_i are integers. Since $\sum_{i=1}^{n} r_i = 1$ we have $\sum_{i=1}^{n} s_i = d$. Thus the expression $\sum_{i=1}^{n} r_i x_i$ reduces to

$$\frac{1}{d}\left(\left(\sum_{i=1}^{s_1} x_1\right) + \left(\sum_{i=1}^{s_2} x_2\right) + \ldots \left(\sum_{i=1}^{s_n} x_n\right)\right).$$

This is now in a form that allows us to apply the inequality (4.9) and the inequality (4.8) must follow.

Lemma 4.14 *If f is continuous and midpoint-convex then f is convex.*

Proof. This follows directly from Lemma 4.13 since, if f is continuous, the inequality (4.8) must hold for all *real* convex combinations and this is the definition of a convex function.

Lemma 4.15 (Sierpiński) *If f is midpoint-convex and has a point of discontinuity x_0 then*

$$\limsup_{x \to x_0} f(x) = +\infty.$$

Proof. This elegant proof is due to Sierpiński. Let $\epsilon = \omega_f(x_0)/2$. In any neighbourhood of the point x_0 there must be a point x with

$$|f(x) - f(x_0)| > \epsilon. \quad (4.11)$$

We reflect x about x_0 to obtain $x' = 2x_0 - x$ and the condition for midpoint convexity gives

$$[f(x) - f(x_0)] + [f(x') - f(x_0)] \geq 0. \tag{4.12}$$

From (4.11) and (4.12) we can see that either $f(x) - f(x_0) > \epsilon$ or else $f(x') - f(x_0) > \epsilon$. Thus every neighbourhood of x_0 contains at least one point x_1 with $f(x_1) - f(x_0) > \epsilon$.

Now reflect x_0 about x_1 to obtain $x_2 = 2x_1 - x_0$ and hence, from the midpoint-convexity

$$f(x_2) - f(x_0) \geq 2\,(f(x_1) - f(x_0)) > 2\epsilon.$$

Continuing this process inductively we can evidently obtain in any neighbourhood of x_0 a sequence of points $x_1, x_2, \ldots, x_n$ with

$$f(x_n) - f(x_0) > 2^{n-1}\epsilon$$

and the proof is complete.

Lemma 4.16 (Sierpiński) *A function that is measurable and midpoint-convex is continuous.*

Proof. If, contrary to the lemma, there is a point of discontinuity then by the preceding lemma there must exist an interval (a, b) containing a sequence of points $\{\xi_k\}$ with $f(\xi_k) \geq k$. Let

$$A_k = \{x \in (a, b) \,:\, f(x) \geq k\}$$

and let $B_k = 2\xi_k - A_k$ which is just the set A_k reflected about the point ξ_k. We claim that $A_k \cup B_k \supset (a, b)$. Indeed if $x \in (a, b)$ then either $x \in A_k$ or else $f(x) < k$. If the latter is true then $x' = 2\xi_k - x \in B_k$ and by the midpoint-convexity $f(x') \geq 2f(\xi_k) - f(x) \geq k$. This means that $x \in B_k$.

Now the situation is evidently impossible. The sets A_k and B_k are measurable and have the same measure, $A_k \cup B_k \supset (a, b)$ and yet $|A_k| \to 0$. From this contradiction the lemma follows.

4.3.2 Convex Sets, Anticonvex Sets

These notions can be applied to certain classes of sets of real numbers. Following Steinhaus [276], Ruziewicz [249], Luxemburg [183] and Marcus [189] [193], [190], [191] and [192] we make these definitions: a set S is *midpoint convex* (or *Jensen convex*) if

$$x, y \in S \Longrightarrow \tfrac{1}{2}(x + y) \in S$$

S is *anticonvex* if

$$x, y \in S, \ x \neq y \implies \tfrac{1}{2}(x+y) \notin S.$$

This latter notion can be expressed in a number of forms. A set is evidently anticonvex provided it contains no three points x, $x + h$, $x + 2h$ in arithmetic progression, or if it contains no point equidistant from any pair in the set.

About midpoint convex sets Marcus [193], [190] proves the following theorem. Such sets only become interesting if they contain no interval or, equivalently, they have a dense complement. For examples of midpoint convex sets that are of this type one need only take a nontrivial additive subgroup of the reals. We discuss these in some detail in Section 6.2.

Theorem 4.17 (Marcus) *Let S be a midpoint convex set such that $\mathbb{R} \setminus S$ is dense. Then every measurable subset of S has measure zero and every subset of S with the Baire property is first category.*

We turn now to the anticonvex sets. It is immediately clear that any Hamel basis is anticonvex, so that there exist anticonvex sets that are not measurable and do not have the Baire property. We recall that a measurable Hamel basis must have measure zero. Luxemburg [183] asks whether every measurable anticonvex set must have measure zero; the category analogue would be that every anticonvex set with the Baire property is first category. These questions are easily dispensed with by a simple density argument (cf. [190]); Ruziewicz [249] also gives an argument based on distance sets.

Theorem 4.18 *Let S be an anticonvex set. Then every measurable subset of S has measure zero and every subset of S with the Baire property is first category.*

4.4 Uniform Smoothness Conditions

We turn now to the study of local growth conditions. A condition of the form

$$|f(x+h) + f(x-h) - 2f(x)| = O(\phi(h))$$

for some function ϕ decreasing to zero will be said (loosely) to be a "smoothness" condition. Such a condition holds uniformly if there are positive numbers C and η so that

$$|f(x+h) + f(x-h) - 2f(x)| \leq C\phi(h)$$

for all x and all $0 < h < \eta$.

More specifically the condition

$$|f(x+h) + f(x-h) - 2f(x)| = o(h)$$

is known in the literature as *smoothness* while the weaker condition

$$|f(x + h) + f(x - h) - 2f(x)| = O(h)$$

has been called *quasi-smoothness*. The next few sections examine some consequences if these conditions hold uniformly.

Note that there is no parallel story for uniform *odd* conditions. For example if the condition $f(x+h)-f(x-h) = O(h)$ holds uniformly in x then immediately, with $c = (x + y)/2$ and $h = (y - x)/2$, we have

$$|f(y) - f(x)| = |f(c + h) - f(c - h)| < Kh$$

for small h and so f is Lipschitz in any case.

4.4.1 Modulus of Continuity

Let f be defined in an interval $[a, b]$. The function

$$\omega(\delta) = \sup \{|f(x_1) - f(x_2)| : x_1, x_2 \in (a, b), |x_1 - x_2| < \delta\}$$

is called the *modulus of continuity* of the function f. The modulus of continuity measures the degree of continuity of f; f is clearly continuous if and only if $\omega(\delta) \to 0$ as $\delta \to 0$. If $\omega(\delta) \leq C\delta^\alpha$ for some $0 < \alpha \leq 1$ then f is Lipschitz of order α.

We say f is *uniformly quasi–smooth* if the condition

$$|f(x + h) + f(x - h) - 2f(x)| = O(h) \tag{4.13}$$

holds uniformly. For our first result we show that uniformly quasi–smooth functions provide an estimate on the modulus of continuity.

The first observations of this type can be found in Zygmund [312, Theorem 10, p. 52]. It arises from some trigonometric estimates; simpler real-variable proofs appear in [313, vol. I, p. 44]. Note that a function f that has a modulus of continuity satisfying $\omega(\delta) = O(\delta \log \delta)$ must also satisfy a Lipschitz condition

$$|f(y) - f(x)| \leq C|x - y|^{1-\epsilon}$$

for sufficiently small $y - x$ and any choice of $0 < \epsilon < 1$. This is because the inequality $|\log h| \leq h^{-\epsilon}$ holds for small enough h. In general such functions need not satisfy a stronger Lipschitz condition

$$|f(y) - f(x)| \leq C|x - y|$$

because, as we shall see in Section 4.7.2, there are continuous, uniformly quasi-smooth functions that are nowhere differentiable. This is in marked contrast to the odd version: if a function satisfies $|f(x + h) - f(x - h)| \leq Ch$ even locally then it differs from a Lipschitz function only on a small set (see Section 2.5).

Theorem 4.19 *Let f be a bounded function defined on an interval $[a, b]$ and let ω denotes the modulus of continuity of f on that interval. If f satisfies the condition*

$$|f(x + h) + f(x - h) - 2f(x)| = O(h) \tag{4.14}$$

uniformly then $\omega(\delta) = O(\delta \log \delta)$.

Proof. (The proof is from [313, vol. I, p. 44].) We suppose that

$$|f(x + t) + f(x - t) - 2f(x)| \leq Ct$$

for $0 < t < \gamma$ and any x in the interval. We argue at a fixed x and use the notation $g(t) = f(x + t) - f(x)$. We have then the inequality

$$\left| g(t) - 2g\left(\frac{t}{2}\right) \right| \leq Ct$$

for $0 < t < \gamma$. Consequently for $i = 1, 2, \ldots n$, by replacing t successively by τ, $\tau/2$, $\tau/2^2$, ... we obtain

$$\left| 2^{i-1} g\left(\frac{\tau}{2^{i-1}}\right) - 2^i g\left(\frac{\tau}{2^i}\right) \right| \leq C\tau.$$

Summing these inequalities for $i = 1, 2, \ldots n$ we obtain

$$\left| g(\tau) - 2^n g\left(\frac{\tau}{2^n}\right) \right| \leq Cn\tau. \tag{4.15}$$

For any $0 < h < \gamma/2$ there is an integer n so that $\tau = 2^n h \in (\frac{1}{2}\gamma, \gamma)$. Let M be a bound for the values of $|f(x)|$; then $2M$ is a bound for the values of $|g(t)|$. Since $2^n h < \gamma$ there is a number K with $n \leq K \log |h|$. Now from (4.15) we see that

$$|g(h)| \leq \frac{2M}{2^n} + \frac{Cn\tau}{2^n} \leq \frac{4Mh}{\gamma} + CKh \log |h|.$$

Since $g(h) = f(x + h) - f(x)$ this gives $\omega(\delta) = O(\delta \log \delta)$ as required.

The same proof applies to functions satisfying stronger conditions. For example if f satisfies the condition

$$|f(x + h) + f(x - h) - 2f(x)| = o(h)$$

uniformly then $\omega(\delta) = o(\delta \log \delta)$. Again, by the same methods, if f satisfies the condition

$$|f(x + h) + f(x - h) - 2f(x)| = O(h^\alpha)$$

uniformly for some $0 < \alpha < 1$ then $\omega(\delta) = O(\delta^\alpha)$. This cannot be extended to $\alpha = 1$ (i.e. a Lipschitz condition). Instead we have the following which, in the extreme case $\beta = 1$, gives a "smoothness" condition that implies a Lipschitz condition.

Theorem 4.20 *Let f be a bounded function defined on an interval $[a, b]$, let $0 < \beta \leq 1$ and let ω denotes the modulus of continuity of f on that interval. If f satisfies the condition*

$$|f(x + h) + f(x - h) - 2f(x)| = O\left(\frac{h}{|\log h|^\beta}\right) \qquad (4.16)$$

uniformly as $h \to 0+$ then

$$\omega(\delta) = O\left(\delta(\log \delta)^{1-\beta}\right).$$

Proof. (The proof again follows [313, vol. I, p. 44].) We suppose that

$$|f(x + h) + f(x - h) - 2f(x)| \leq Ch(\log |h|)^{-\beta}$$

for $0 < h < \gamma$ and any x in the interval. We argue at a fixed x and use the notation $g(t) = f(x + t) - f(x)$. We then have the inequality

$$\left|g(\tau) - 2g\left(\frac{\tau}{2}\right)\right| \leq \tfrac{1}{2}C\tau(\log|\tfrac{1}{2}\tau|)^{-\beta}$$

for $0 < \tau < \gamma$. Consequently for $i = 1, 2, \ldots n$ we obtain

$$\left|2^{i-1}g\left(\frac{\tau}{2^{i-1}}\right) - 2^i g\left(\frac{\tau}{2^i}\right)\right| \leq C\tau(\log|\tau|)^{-\beta}.$$

Summing these inequalities for $i = 1, 2, \ldots n$ we obtain

$$\left|g(\tau) - 2^n g\left(\frac{\tau}{2^n}\right)\right| \leq Cn\tau\left(\log\left|\frac{\tau}{2^n}\right|\right)^{-\beta}. \qquad (4.17)$$

For any $0 < h < \gamma/2$ there is an integer n so that $\tau = 2^n h \in (\tfrac{1}{2}\gamma, \gamma)$. Let M be a bound for the values of $|f(x)|$; then $2M$ is a bound for the values of $|g(t)|$. Since $2^n h < \gamma$ there is a number K with $n \leq K \log |h|$. Now we obtain from (4.17) that

$$|g(h)| \leq \frac{2M}{2^n} + \frac{Cn\tau}{2^n}\left(\log\left|\frac{\tau}{2^n}\right|\right)^{-\beta} \leq \frac{4Mh}{\gamma} + CKh(\log |h|)^{1-\beta}.$$

Since $g(h) = f(x + h) - f(x)$ this gives

$$\omega(\delta) = O\left(\delta(\log \delta)^{1-\beta}\right)$$

as required.

4.4.2 Theorem of Anderson and Pitt

The estimate on the growth of uniformly quasi-smooth functions expressed in Theorem 4.19 can be improved if we ask for an estimate only almost everywhere.

Theorem 4.21 (Anderson–Pitt) *Let f be a continuous function defined on an interval $[a, b]$ such that*

$$|f(x + h) + f(x - h) - 2f(x)| = O(h) \qquad (4.18)$$

uniformly as $h \to 0$. Then for almost every point x

$$\limsup_{h \to 0} \frac{|f(x + h) - f(x)|}{|h|\sqrt{\log \frac{1}{|h|} \log \log \log \frac{1}{|h|}}} < +\infty.$$

The paper of Anderson and Pitt [1] should be consulted for a proof; they employ martingale methods that are beyond the scope of this text.

There are other studies of uniform quasi-smoothness that the reader might wish to see. Timan [289] and [290] obtains a refinement of Theorem 4.19: *If a continuous function f satisfies*

$$|f(x + h) + f(x - h) - 2f(x)| \leq 2M|h|$$

uniformly on an interval $[a, b]$ with $f(a) = f(b) = 0$ then

$$\omega(h) \leq \frac{M}{\log 2} h \log \frac{1}{h} + Ah$$

where A depends on f and $[a, b]$ and $\frac{M}{\log 2}$ is best possible. Also Brudnyĭ [20] continues problems of this type by finding $\sup f$ for functions satisfying the conditions above.

Morozov [202] gives some results on approximation to quasi-smooth functions by trigonometric polynomials. Timan and Dzyadyk [77] do some approximation studies under assumptions that a function has a r-th order derivative that is uniformly quasi-smooth. Kotljar [159] shows that for functions of two variables in very simple domains a quasi-smoothness condition in each variable separately implies a quasi-smoothness condition in the two variables together. Kārkliņa [148] studies higher order uniform quasi-smoothness conditions (in one and several variables) using the higher order symmetric differences.

4.4.3 Continuously Differentiable Functions

We pass now to stronger uniform smoothness conditions. By raising the strength of the conditions we can ensure that the function is continuously differentiable.

Theorem 4.22 *Let f be a continuous function and suppose, for some $\beta > 1$, that*

$$f(x+t) + f(x-t) - 2f(x) = O\left(\frac{t}{|\log t|^\beta}\right) \tag{4.19}$$

uniformly in x as $t \to 0+$. Then f is continuously differentiable.

Here is another variant that we can prove by the same elementary methods.

Theorem 4.23 *Let f be a continuous function and suppose, for some $1 < \alpha \le 2$, that*

$$f(x+t) + f(x-t) - 2f(x) = O\left(t^\alpha\right) \tag{4.20}$$

uniformly in x as $t \to 0+$. Then f is continuously differentiable and the derivative f' has a modulus of continuity $\omega(\delta) = O(\delta^{\alpha-1})$.

Proof. Under either of the assumptions in Theorems 4.22 or 4.23 we know, from Theorem 4.20, that f is Lipschitz. Hence f is the indefinite integral of a bounded function g. We need to show that g is equivalent to a continuous function.

We assume f is a continuous function for which an inequality

$$|f(x+t) + f(x-t) - 2f(x)| \le t\phi(t) \tag{4.21}$$

holds for every x and all $0 < t < \eta$ and we assume at first only that $\phi(t)$ decreases monotonically to 0 with t. (Later on in the proof we shall take $\phi(t) = Ct^{\alpha-1}$ or $\phi(t) = C|\log t|^{-\beta}$.)

From (4.21) we have

$$\left|\int_x^{x+h} g(t)\,dt - \int_{x-h}^x g(t)\,dt\right| \le h\phi(h) \tag{4.22}$$

if $0 < h < \eta$. Take any points z_1, z_2 that are Lebesgue points for g and for which $0 < h = z_2 - z_1 < \eta$. We obtain an estimate for $|g(z_2) - g(z_1)|$ and hence show that g is uniformly continuous restricted to its set of Lebesgue points.

Construct the sequence $x_n = z_1 + 2^{-n}(z_2 - z_1)$ so that $x_0 = z_2$, $x_n \searrow z_1$ and $x_{n-1} - x_n = 2^{-n}h$. Apply (4.22) at each point x_n for $n = 1, 2, 3, \ldots$ to obtain

$$\left|\int_{z_1}^{x_{n-1}} g(t)\,dt - 2\int_{z_1}^{x_n} g(t)\,dt\right| = \left|\int_{x_n}^{x_{n-1}} g(t)\,dt - \int_{z_1}^{x_n} g(t)\,dt\right|$$

$$\le \frac{h}{2^n}\phi\left(\frac{h}{2^n}\right).$$

Multiply each of these inequalities by 2^n and add for $n = 1, 2, \ldots, N$ to obtain

$$\left|\int_{z_1}^{z_2} g(t)\,dt - 2^N \int_{z_1}^{x_N} g(t)\,dt\right| \le \frac{h}{2}\sum_{n=1}^{N} \phi\left(\frac{h}{2^n}\right).$$

Now divide by h and use the fact that $x_N - z_1 = 2^{-N} h$ to obtain

$$\left| \frac{1}{z_2 - z_1} \int_{z_1}^{z_2} g(t)\, dt - \frac{1}{x_N - z_1} \int_{z_1}^{x_N} g(t)\, dt \right| \leq \psi(h)$$

where we write

$$\psi(h) = \tfrac{1}{2} \sum_{n=1}^{\infty} \phi\left(\frac{h}{2^n} \right).$$

Since z_1 is a Lebesgue point of g

$$\frac{1}{x_N - z_1} \int_{z_1}^{x_N} g(t)\, dt \to g(z_1)$$

as $N \to \infty$ and so we have obtained

$$\left| \frac{1}{z_2 - z_1} \int_{z_1}^{z_2} g(t)\, dt - g(z_1) \right| \leq \psi(h).$$

In exactly the same manner one proves that

$$\left| \frac{1}{z_2 - z_1} \int_{z_1}^{z_2} g(t)\, dt - g(z_2) \right| \leq \psi(h)$$

and hence $|g(z_2) - g(z_1)| \leq 2\psi(h)$.

Now by choosing $\phi(t) = C t^{\alpha - 1}$ or $\phi(t) = C|\log t|^{-\beta}$ we evidently obtain uniform continuity of g on the set of Lebesgue points with the desired modulus of continuity. So g may be replaced by an everywhere continuous function and $f'(x) = g(x)$ everywhere. Thus both theorems have been proved.

Before we leave this discussion a simple example suffices to show that the conditions of these theorems may not be much relaxed. The function

$$f(x) = \sum_{n=1}^{\infty} \frac{\cos 2^n x}{n 2^n}$$

is an example of a function that satisfies the hypotheses of Theorem 4.23 with $\alpha = 1$ and yet is not continuously differentiable; it fails to have a derivative

$$f'(x) = \sum_{n=1}^{\infty} \frac{\sin 2^n x}{n}$$

at a dense set of points as an inspection of this latter series will show.

4.4.4 Theorem of M. Weiss and Zygmund

In the Section 4.4.3 we have shown that a strong type of uniform smoothness condition implies the existence of a continuous derivative. In order to provide a little more insight into these matters let us cite, without proof, a result from 1959 due to Weiss and Zygmund [307] showing that a weakening of this condition still allows a statement about the derivative. Once again the condition is assumed to hold uniformly but note that it is weaker than the assertion in Theorem 4.22. Later on (in Section 4.9) we shall report on the situation where a local, rather than uniform, condition of this type is employed.

Theorem 4.24 (Weiss–Zygmund) *Let f be 2π-periodic and continuous. Suppose, for some $\beta > \frac{1}{2}$, that*

$$f(x+t) + f(x-t) - 2f(x) = O\left(\frac{t}{|\log t|^\beta}\right) \tag{4.23}$$

uniformly in x as $t \to 0+$. Then f is absolutely continuous and its derivative f' is in $L_p[0, 2\pi]$ for every $p > 1$.

The theorem is false for $\beta = \frac{1}{2}$. Indeed the function

$$f(x) = \sum_{n=1}^{\infty} \frac{\cos 2^n x}{2^n \sqrt{n}}$$

satisfies the condition (4.23) uniformly and is almost nowhere differentiable. (See Section 4.4.7.)

The original proof of Theorem 4.24 in [307] is fairly short but utilizes a number of deep results including a theorem of Littlewood and Paley. A proof in John and Nirenberg [144] is more direct, more general and as well shows that the derivative f' is not merely in each L_p space but also of bounded mean oscillation. A related result is sketched in [272, p. 164].

In fact the special form of the growth condition in assertion (4.23) is unnecessary. It can be replaced by the following version. Let $\alpha(t)$ be a monotonic nondecreasing function on an interval $(0, \eta)$ such that

$$\int_0^\eta \frac{\alpha^2(t)}{t}\, dt < +\infty. \tag{4.24}$$

If the continuous function f satisfies an inequality

$$|f(x+t) + f(x-t) - 2f(x)| \le t\alpha(t)$$

for all x and all $0 < t < \eta$ then f is absolutely continuous and its derivative f' is in $L_p[0, 2\pi]$ for every $p > 1$.

We shall not prove these facts in full generality but we will report on a special case that is remarkably easy to prove. Here we establish merely that the derivative is in L_2. The elementary proof is from [255] where it is attributed to Carleson. It uses nothing more than familiar methods of Fourier series.

Theorem 4.25 *Suppose that f is continuous and 2π–periodic and that*

$$|f(x+t) + f(x-t) - 2f(x)| \le w(t)$$

where

$$\int_0^1 t^{-3}[w(t)]^2\, dt < +\infty. \tag{4.25}$$

Then f is absolutely continuous and f' is in L_2.

Proof. Let f have the Fourier expansion $\sum_{-\infty}^{\infty} c_n e^{inx}$. In order to see that f is absolutely continuous and that f' is in L_2 it is enough, by familiar arguments in Fourier series, to establish that

$$\sum_{-\infty}^{\infty} n^2|c_n|^2 < +\infty. \tag{4.26}$$

But this follows from (4.25) by some direct computations. We see first that

$$\int_0^1 \int_0^{2\pi} \frac{|f(x+t) + f(x-t) - 2f(x)|^2}{t^3}\, dx\, dt < +\infty \tag{4.27}$$

and (4.27) in turn gives

$$16 \sum_{-\infty}^{\infty} |c_n|^2 \int_0^1 t^{-3} \sin^4(nt/2)\, dt < +\infty. \tag{4.28}$$

Since

$$\int_0^1 t^{-3} \sin^4(nt/2)\, dt = \frac{n^2}{4} \int_0^{n/2} u^{-3} \sin^4 u\, du \ge Bn^2$$

for some positive constant B and all n we see that the convergence of the series in (4.28) provides the convergence of the series in (4.26) as required.

4.4.5 Almost Nowhere Differentiable Functions

The uniform smoothness condition (4.24) is just sharp enough to prove almost everywhere differentiability. Let us reproduce the following theorem from Stein and Zygmund [274, Theorem D, p. 250] showing that any weaker growth condition on a function f allows f to be almost nowhere differentiable. The theorem was first stated, without proof, in Stein and Zygmund [275, p. 297] and then a proof sketched in [274, pp. 250–251]. The condition $\alpha(2t)/\alpha(t) \to 1$ in the statement of the theorem is not essential; evidently some kind of regularity is needed though.

Theorem 4.26 (Stein–Zygmund) *Let $\alpha(t)$ be a monotonic, nondecreasing function on an interval $(0,\eta)$ such that $\alpha(t) \to 0$ and $\alpha(2t)/\alpha(t) \to 1$ as $t \to 0$ with*

$$\int_0^\eta \frac{\alpha^2(t)}{t}\, dt = +\infty.$$

Then there is a continuous function f such that

$$|f(x+t) + f(x-t) - 2f(x)| \le t\alpha(t) \tag{4.29}$$

for all x and all $0 < t < \eta$ and f' exists only in a set of measure zero.

Proof. For f we take $f(x) = Cg(x)$ for a suitable constant C where

$$g(x) = \sum_{n=N}^{\infty} \alpha\left(\frac{1}{2^n}\right) \frac{\sin 2^n x}{2^n}$$

with N chosen sufficiently large so that $2^{-N} < \eta$. That the function g defined by this series is differentiable only in a set of measure zero follows from familiar arguments in the study of Fourier series. The hypotheses on α require that the series $\sum \alpha^2(2^{-n})$ diverge. This means that the differentiated series

$$\sum_{n=N}^{\infty} \alpha\left(\frac{1}{2^n}\right) \cos 2^n x$$

is not summable by any linear method of summation on a set of positive measure (see [313, Vol. I, p. 203]). In particular g' cannot exist on a set of positive measure.

It remains only to show that g satisfies

$$g(x+t) + g(x-t) - 2g(x) = O\left(t\alpha(t)\right) \tag{4.30}$$

uniformly so that a choice of C may be made in such a way that the inequality (4.29) holds.

The following arguments are also familiar in the study of Fourier series. We compute

$$g(x+t) + g(x-t) - 2g(x) = -4 \sum_{n=N}^{\infty} \alpha\left(\frac{1}{2^n}\right) \frac{\sin 2^n x \sin^2 \frac{1}{2} 2^n t}{2^n}.$$

For $0 < t < 2^{-N}$ there is an $M \ge N$ so that $2^{-M-1} \le t < 2^{-M}$. Then the sum above may be split into two parts, say $-4P$ and $-4Q$ where P is the sum $\sum_{n=N}^{M}$ and Q is the sum $\sum_{n=M+1}^{\infty}$ of the same expression. Now

$$|Q| \le \sum_{n=M+1}^{\infty} \alpha\left(\frac{1}{2^n}\right) \frac{1}{2^n} \le \alpha\left(\frac{1}{2^{M+1}}\right) \frac{1}{2^M} \le 2t\alpha(t) \tag{4.31}$$

and

$$|P| \le t^2 \sum_{n=N}^{M} \alpha\left(\frac{1}{2^n}\right) 2^n \le t^2 \sum_{n=N}^{M} \alpha\left(\frac{1}{2^n}\right) \left(\frac{3}{2}\right)^n \left(\frac{4}{3}\right)^n.$$

Since

$$\left(\frac{3}{2}\right)^n \alpha\left(\frac{1}{2^n}\right) \leq \left(\frac{3}{2}\right)^{n+1} \alpha\left(\frac{1}{2^{n+1}}\right)$$

for sufficiently large n we may choose a constant A that does not depend on M so that

$$\left(\frac{3}{2}\right)^n \alpha\left(\frac{1}{2^n}\right) \leq A\left(\frac{3}{2}\right)^M \alpha\left(\frac{1}{2^M}\right)$$

for each $n = N, N+1, \ldots M$. Thus

$$|P| \leq K t^2 \alpha\left(\frac{1}{2^M}\right) 2^M$$

for some constant K and so finally $|P| = O\left(t\alpha(t)\right)$. Together with (4.31) this shows that (4.29) must hold for $f(x) = Cg(x)$ a suitable choice of constant C. This completes the proof.

4.4.6 Monotonic, Uniformly Smooth Functions

It is perhaps remarkable that there exist monotonic singular functions that are uniformly smooth. Such functions form the basis for a counterexample in the theory of conformal mappings, playing a role in approximation theory and polynomial expansions. Duren, Shapiro and Shields [72] show that a certain construction for rectifiably bounded Jordan domains is in fact equivalent to the problem of constructing a monotonic, singular function that is uniformly quasi-smooth. Thus an example of the former (by Keldyš and Lavrentiev in 1937) provides the existence of the latter.

Because of that a number of authors have given direct constructions of such singular functions by real variable methods. The first is due to Piranian [221]. We cite his theorem here and refer the reader to the original paper for details of the construction.

Theorem 4.27 (Piranian) *There exists a continuous, strictly increasing, singular function that is uniformly smooth.*

Kahane [147] and Shapiro [255] have also given constructions that allow a more detailed statement. This next theorem shows that in Theorem 4.26 the function constructed can be taken as monotonic and singular. Again, too, Theorem 4.24 of Weiss and Zygmund is seen to be sharp even in the case of monotonic functions.

Theorem 4.28 *Let $\alpha(t)$ be a positive, monotonic nondecreasing function on an interval $(0, \eta)$ such that $\alpha(4t) \leq 2\alpha(t)$ and with*

$$\int_0^\eta \frac{\alpha^2(t)}{t}\, dt = +\infty.$$

Then there is a nondecreasing, singular and continuous function f whose support is a compact set of measure zero such that

$$f(x+t) + f(x-t) - 2f(x) = O(t\alpha(t))$$

uniformly as $t \to 0$.

4.4.7 Some Examples

Examples of almost nowhere differentiable functions satisfying some uniform smoothness condition can be readily obtained from lacunary Fourier series. Consider the functions of the form

$$f(t) = \sum_{k=1}^{\infty} \frac{a_k}{n_k} \cos(n_k t + \theta_k) \tag{4.32}$$

where $n_k \in \mathbb{N}$ and $n_{k+1}/n_k \geq q > 1$.

The conditions on the $\{a_k\}$ that are of interest are:

(α) $\sup |a_k| < +\infty$.

(β) $\lim_{k\to\infty} a_k = 0$.

(γ) $\sum_{k=1}^{\infty} |a_k|^2 = +\infty$.

If (γ) holds then f in (4.32) is almost nowhere differentiable. (If (γ) fails, that is if $\sum_{k=1}^{\infty} |a_k|^2$ converges, then it is easy to check that f is absolutely continuous with a derivative in L_2.) It is known (see [313, Vol. I, p. 206, p. 321]) that f is uniformly quasi-smooth if and only if (α) holds; f is uniformly smooth if and only if (β) holds.

A further family of examples from [313, Vol. I, p. 73] can be used to illustrate the scope of a number of the theorems of the preceding sections.

Theorem 4.29 *Let* $0 < \alpha < 1$ *and* $-\infty < \beta < +\infty$. *The function*

$$f_{\alpha,\beta}(x) = \sum_{n=1}^{\infty} 2^{-n\alpha} n^{-\beta} \cos 2^n x$$

has modulus of continuity

$$\omega(\delta) = O\left(\delta^\alpha \left(\log \frac{1}{\delta}\right)^{-\beta}\right)$$

and the function

$$f_\beta(x) = \sum_{n=1}^{\infty} n^{-\beta} \cos 2^n x$$

has modulus of continuity

$$\omega(\delta) = O\left(\left(\log \frac{1}{\delta}\right)^{-(\beta-1)}\right).$$

This theorem provides a family of examples that will satisfy various smoothness conditions; the differentiability properties are investigated by looking at the differentiated series. For example we have already mentioned that the function

$$f_{1,1}(x) = \sum_{n=1}^{\infty} \frac{\cos 2^n x}{n2^n}$$

is an example of a function that satisfies the hypotheses of Theorem 4.23 with $\alpha = 1$ and yet is not continuously differentiable; it fails to have a derivative at certain points because of the divergence of the differentiated series

$$f'_{1,1}(x) = \sum_{n=1}^{\infty} \frac{\sin 2^n x}{n}$$

at a dense set of points. Similarly the example

$$f_{1,1/2}(x) = \sum_{n=1}^{\infty} \frac{\cos 2^n x}{\sqrt{n}2^n}$$

has been mentioned to illustrate that the Weiss–Zygmund Theorem (Theorem 4.24) is sharp. That $f'_{1,1/2}(x)$ exists almost nowhere follows from the arguments in Theorem 4.26.

4.5 Boundedness

We pass now from a study of uniform conditions to more local conditions. The question we first address is to what degree an even boundedness condition imposes some degree of ordinary boundedness on a function. Of course a function can be unbounded and yet have an even part that is locally bounded at every point; even for measurable functions this can easily occur. Take $f(x) = x^{-1}$ for $x \neq 0$ and $f(0) = 0$.

The first elementary result was observed (in greater generality) in Marcinkiewicz and Zygmund [186, p. 13]. The method of proof is taken from Zygmund [313, Vol. II, Lemma 4.34, p. 79] and repeats ideas we have already seen and will see again many times.

Lemma 4.30 *Let f be measurable and suppose that*

$$f(x+t) + f(x-t) - 2f(x) = O(1)$$

as $t \to 0$ *at every point* x *of a measurable set* E. *Then* f *is bounded in a neighbourhood of almost every point in* E.

Proof. Since f is measurable we may assume that f is approximately continuous at each point of E. For any pair of integers m and n define the set

$$A(m, n) = \{x \in E : |f(x + t) + f(x - t) - 2f(x)| < m, \ 0 < t < 1/n\}.$$

Since $E = \bigcup_{m,n} A(m, n)$ it is enough if we show that, whatever be m and n, f is bounded in a neighbourhood of almost every point of the set $A(m, n)$.

Let x_0 be a point of density of $A(m, n)$, which we take for simplicity as 0. Since f is approximately continuous at 0 there is a measurable set B with 0 as a point of density and a number $C > 0$ so that $|f(x)| \leq C$ for all $x \in B$.

There is a positive number η so that for each $0 < u < \eta$ there is a point v in the intersection

$$B \cap (2B - u) \cap (2A(m, n) - u) \cap (u, 2u).$$

This just uses the density computations in Lemma A.10. (Here the set B is measurable but we do not need to check that $A(m, n)$ is measurable.) Since $(u + v)/2 \in A(m, n)$ this gives that

$$|f(u) + f(v) - 2f((u + v)/2)| < m.$$

But also since $v, (u + v)/2 \in B$ we have $|f(v)| \leq C$ and $|f((u + v)/2)| \leq C$. Consequently $|f(u)| < m + 3C$ for all u sufficiently close to x_0. We see that f is bounded in a neighbourhood of every density point of $A(m, n)$ and hence at almost every point of the given set E.

Important Remark. In the argument here we have avoided assuming the measurability of the set $A(m, n)$. The proofs in the literature (eg. [313]) that employ this technique generally take measurability for granted. That some caution is warranted is provided by the following example due to K. Ciesielski. Let C_0 denote the usual Cantor ternary set and write $C = C_0 \cap [0, \frac{1}{2}]$. Since $\frac{1}{2}(C + C) = [0, \frac{1}{3}]$ there must exist a set $S \subset C$ so that $\frac{1}{2}(S + S)$ is not measurable. Write f for the characteristic function of S; as S has measure zero f is a measurable function. But the set

$$E = \{x : f(x + h) + f(x - h) - 2f(x) < 2 \text{ for all } 0 < h < \frac{1}{2}\}$$

is *not* measurable. Indeed

$$\left((\mathbb{R} \setminus E) \cap [0, \tfrac{1}{2}]\right) \setminus C$$
$$= \{x \in [0, \tfrac{1}{2}] \setminus C : x + h \in S \text{ and } x - h \in S \text{ for some } h\}$$

and this latter set is not measurable.

It is also useful to know the nature of the set of points of unboundedness for a function with an everywhere bounded even part.

Lemma 4.31 *Let f be measurable and suppose that*

$$f(x+t) + f(x-t) - 2f(x) = O(1)$$

as $t \to 0$ at every point. Then the set

$$Z = \{x : \omega_f(x) = +\infty\}$$

is a closed, countable and locally symmetric set.

Proof. We show firstly that Z is locally symmetric. At any point x choose C and $\delta > 0$ so that

$$|f(x+t) + f(x-t) - 2f(x)| < C \qquad (4.33)$$

if $0 < t < \delta$. Then if $x + h \in Z$ and $0 < |h| < \delta$ there must exist a sequence $h_n \to h$ so that $|f(x+h_n)| \to +\infty$. In that case, in view of the inequality (4.33), $|f(x-h_n)| \to +\infty$ too. This shows that $x - h \in Z$. Consequently Z is locally symmetric about every point x.

The set Z is closed since

$$Z = \bigcap_{n=1}^{\infty} \{x : \omega_f(x) \geq n\} \, .$$

Theorem 2.27 shows that either Z or its complement has countable closure; by Lemma 4.30 it must be Z that is countable.

4.6 Symmetric Functions

Let us turn now to one of the most interesting and simplest of the even symmetry growth conditions but one that has acquired by now a fairly large literature. A function is said to be *symmetric* at a point x if

$$f(x+t) + f(x-t) - 2f(x) = o(1)$$

as $t \to 0$. Thus a function f is symmetric at a point x if the even part of f at that point is continuous.

We have already remarked in Section 2.2 on the analogies between even continuity (symmetry) and odd continuity (symmetric continuity). Some of the basic continuity properties have already been established and will be reviewed here. In particular recall that an everywhere symmetric function may have any countable set as its set of discontinuity points and not much more about characterizing this is known. We shall see necessary conditions but no more sufficient conditions.

As usual throughout our investigations we require some kind of hypothesis such as measurability or boundedness; this cannot be removed in this setting since there are nonmeasurable midpoint-linear functions f and, of course, any such function is also symmetric at every point. In the next few sections we shall see that symmetric functions which are measurable (or bounded or have the Baire property) have strong continuity properties.

4.6.1 Continuity Properties of Symmetric Functions

4.6.1.1 Stein–Zygmund Theorem. The most important continuity property of symmetric functions was first established in Mazurkiewicz[198]; he showed that a bounded symmetric function is almost everywhere continuous. Much later Stein and Zygmund [274, Lemma 9, p. 266] and, independently, Neugebauer [213] extended this to measurable symmetric functions. We repeat for reference the Theorem of Stein and Zygmund that was proved in Section 2.2.1. Recall that the only technique needed to effect this proof is a density argument similar to that originated by Khinchine that we have seen in Chapter 1.

Theorem 4.32 (Stein–Zygmund) *Let f be measurable and symmetric at each point x of a measurable set E. Then f is continuous at almost all points of E.*

Note in particular that a measurable function that is everywhere symmetric is continuous at every point excepting only a set that is of measure zero; since the set of discontinuities of any function is a set of type F_σ this set must be also first category. We can express this as a corollary.

Corollary 4.33 *Let f be a measurable function that is symmetric at every point x. Then f is continuous at all points excepting a set of measure zero and first category.*

The exact characterization of the exceptional set of the corollary is unknown. Tran [292] constructs a measurable, symmetric function that is discontinuous at the points of an uncountable set.

By the same methods used to establish the Stein–Zygmund Theorem one can prove also the following variants, due to Darji [58].

Theorem 4.34 (Darji) *Let f be a measurable function, let S_f denote the set of points at which f is symmetric and let C_f denote the set of points at which f is continuous. Then S_f is measurable and $S_f \setminus C_f$ has measure zero.*

The category analogue may be similarly proved (again see [58]).

Theorem 4.35 (Darji) *Let f have the Baire property, let S_f denote the set of points at which f is symmetric and let C_f denote the set of points at which f is continuous. Then S_f has the Baire property and $S_f \setminus C_f$ is first category.*

To show the limitations Darji [58] adapts some existing constructions to show that, under the continuum hypothesis, *there is a function f with the Baire property so that $S_f \setminus C_f$ has positive measure and there is a measurable function f so that $S_f \setminus C_f$ has second category.* (His statements are somewhat stronger.)

4.6.1.2 Bounded, Symmetric Functions. A measurable symmetric function is continuous almost everywhere. In an early paper Mazurkiewicz [198] showed that this is true for an arbitrary bounded function without an *a priori* measurability assumption. This was improved by Auerbach [9] who showed that it was not necessary to assume the symmetry condition at every point if the boundedness assumption is made.

Theorem 4.36 *Let f be a bounded function that is symmetric at almost every point x in an interval (a, b). Then f is measurable and continuous at almost all points of (a, b).*

Proof. (There is an oversight in the original proof which is amended here.) Since the function f is bounded we may write

$$G(x) = \overline{\int_a^x} f(t)\, dt \quad \text{and} \quad H(x) = \underline{\int_a^x} f(t)\, dt$$

where these are interpreted as Darboux upper and lower integrals. We expect that for an integrable f an even symmetry condition would translate into some kind of odd symmetry condition for G and H; here the fact that we have only upper and lower integrals just requires a rather more careful attention to detail.

Let $\epsilon > 0$. At any point x at which f is symmetric we may choose a $\delta > 0$ so that if $0 < t < \delta$ then

$$2f(x) - f(x - t) - \epsilon \le f(x + t) \le 2f(x) - f(x - t) + \epsilon. \tag{4.34}$$

If we integrate the inequalities (4.34) (as a function of t) we obtain

$$[2f(x) - \epsilon]h - \underline{\int_0^h} f(x - t)\, dt \le \overline{\int_0^h} f(x + t)\, dt \le [2f(x) + \epsilon]h - \underline{\int_0^h} f(x - t)\, dt.$$

We can write this in the form

$$|G(x + h) - G(x) + H(x) - H(x - h) - 2f(x)h| \le \epsilon h. \tag{4.35}$$

Similarly, rewriting (4.34) as

$$2f(x) - f(x+t) - \epsilon \leq f(x-t) \leq 2f(x) - f(x+t) + \epsilon \qquad (4.36)$$

and integrating the inequalities (4.36), we can obtain

$$|G(x) - G(x-h) + H(x+h) - H(x) - 2f(x)h| \leq \epsilon h. \qquad (4.37)$$

Now set $K = \frac{1}{2}(G + H)$. Then K is a continuous function and we have, on adding (4.35) and (4.37),

$$\left| \frac{K(x+h) - K(x-h)}{2h} - f(x) \right| \leq \epsilon/2$$

for $0 < h < \delta$. Thus SD $K(x) = f(x)$ at every point x at which f is symmetric. As this holds almost everywhere we see that f is measurable and, by Theorem 2.3, it is continuous almost everywhere as required.

(Incidentally, instead of appealing to Theorem 2.3, one could continue the proof to show that G and H are identical. Since G and H are Lipschitz functions it would be enough to show that SD $G(x) =$ SD $H(x) = f(x)$ almost everywhere and apply Theorem 1.16. This means in particular that f is Riemann integrable on $[a, b]$ because its upper and lower Darboux integrals agree, and so f is almost everywhere continuous.)

Before we leave the proof note that if f is symmetric almost everywhere then the proof expresses f as almost everywhere the symmetric derivative of a continuous function and so, in particular,

$$f(x) = \lim_{n \to \infty} n(K(x + 1/n) - K(x - 1/n))/2.$$

If f is symmetric *everywhere* then we have just shown that f is in the first Baire class; this will be carried further in Section 4.6.2.

One last comment is in order too. A bounded, symmetric function is measurable. But one can employ weaker boundedness assumptions. If we remember that any nonmeasurable midpoint-linear function f has the property that $\{x : f(x) \geq \alpha\}$ has full exterior measure for every α then we might anticipate that we can make do with some weak boundedness assumption that does not allow such nonmeasurable functions to be considered. This was given by Ponomarev [224].

Theorem 4.37 (Ponomarev) *Let f be a function that is symmetric at every point of an interval (a, b) and for which*

$$\lim_{n \to +\infty} |\{x \in (a, b) : f(x) \geq n\}| = 0.$$

Then f is measurable.

4.6.1.3 Evans–Larson Theorem. Either of the assumptions of boundedness or measurability suffices to establish strong regularity properties for symmetric functions. This situation was clarified further in a work of Evans and Larson [96] who show that functions with the Baire property also have this property.

Theorem 4.38 (Evans–Larson) *Let f be a function that possesses the Baire property and is symmetric at every point. Then f is measurable and almost everywhere continuous.*

Rather than reproduce the arguments in [96] we argue from some of our earlier results. We need only a simple lemma that is a category analogue of Lemma 4.30 and can be proved in exactly the same way by using the category density notions (see Section A.5).

Lemma 4.39 *Let f have the Baire property and suppose that*

$$f(x+t) + f(x-t) - 2f(x) = O(1)$$

as $t \to 0$ at every point x of a set E that has the Baire property. Then f is bounded in a neighbourhood of every point in E excepting only a first category set.

Now we are in a position to prove the theorem. By Lemma 4.39 f is bounded on a dense set of intervals. Each such interval must, by Theorem 4.36, contain a point of continuity of f. We know then from Theorem 2.5 that there is a measurable, symmetrically continuous function g with the same set of points of continuity as f. Hence g is almost everywhere continuous by Theorem 2.3. This shows that f too is continuous almost everywhere as required.

4.6.2 Baire Class

The arguments of Section 4.6.1.2 already show that a bounded function that is everywhere symmetric must be in the first Baire class. This is first proved in Mazurkiewicz [198]. Auerbach [9] extended this to integrable functions and Neugebauer [213] to arbitrary measurable symmetric functions. Auerbach also shows that under a slightly stronger condition more can be said. The condition that a function f is Baire 1 is that it can be expressed as a pointwise limit of continuous functions; the condition[1] that it be Baire*1 is that there is a sequence of closed sets $\{E_n\}$ covering the real line and f is continuous relative to each set E_n. Such a function must be continuous at the points of an open dense set, as an application of the Baire category Theorem easily shows.

---
[1] The notation Baire*1 is due to Richard O'Malley; H. W. Ellis earlier used the notation [CG] for the same concept. An equivalent concept has been termed Baire $\frac{1}{2}$.

This fact allows us to show that the cases $\alpha = 0$ and $\alpha > 0$ in Lemma 4.40 are distinct. It is easy to construct a measurable, symmetric function with a specified countable set as its set of discontinuity points. Since that set may be dense such functions, while Baire 1, need not be Baire∗1. Even under an extra hypothesis the discontinuities may be dense: Larson [176] gives an example of a measurable, symmetric function f that is also everywhere approximately continuous and such that the set of points where f is discontinuous is precisely the set of rationals.

Lemma 4.40 (Auerbach) *Let f be integrable on an interval (a, b). Suppose, at every point $x \in (a, b)$, that*

$$f(x + t) + f(x - t) - 2f(x) = o(1)$$

as $t \to 0$. Then f is Baire 1. Suppose $\alpha > 0$ and, at every point $x \in (a, b)$, that

$$f(x + t) + f(x - t) - 2f(x) = O(t^\alpha)$$

as $t \to 0$. Then f is Baire∗1.

Proof. If the inequality $|f(x + t) + f(x - t) - 2f(x)| < \epsilon$ holds for $0 < t < \delta$ and F is an indefinite integral of f then, for $0 < h < \delta$, we have

$$\left| \frac{F(x + h) - F(x - h)}{2h} - f(x) \right|$$

$$= \left| \frac{1}{2h} \int_0^h |f(x + t) + f(x - t) - 2f(x)| \, dt \right| < \frac{1}{2h} \int_0^h \epsilon \, dt < \epsilon.$$

Thus f is everywhere the symmetric derivative of a continuous function F and so f is Baire 1.

If the inequality $|f(x + t) + f(x - t) - 2f(x)| < Ct^\alpha$ holds instead then the same reasoning gives

$$\left| \frac{F(x + h) - F(x - h)}{2h} - f(x) \right| \leq \frac{1}{2h} \int_0^h Ct^\alpha \, dt < Ch^\alpha.$$

Thus we get that for every x there is an $N(x)$ so that

$$|f_n(x) - f(x)| < C/n^\alpha$$

if $n \geq N(x)$ by setting $f_n(x) = n(F(x + n^{-1}) - F(x - n^{-1})/2$. This expresses f as a pointwise limit of a sequence of continuous functions with enough extra structure that we can show f is Baire∗1.

Now let

$$E_m = \{x : N(x) \leq m\} \cap [-m, m].$$

The sets $\{E_m\}$ cover the real line and $f_n \to f$ uniformly on each set E_m so that f is continuous on each such set. We need continuity on the closures too and an elementary argument will give this. Fix m, fix a point x in the closure of E_m, let $\epsilon > 0$ and choose $n > m$ so that $C/n < \epsilon/3$ and $|f_n(x) - f(x)| < \epsilon/3$. Then by the continuity of f_n choose η so that $y \in E_m$ and $|y - x| < \eta$ requires $|f_n(x) - f_n(y)| < \epsilon/3$. Now if $y \in E_m$ and $|y - x| < \eta$ then

$$|f(x) - f(y)| \leq |f(x) - f_n(x)| + |f_n(x) - f_n(y)| + |f_n(y) - f(y)| < \epsilon.$$

It is now easy to argue that the continuity of f over each set E_m extends as well to the closures and so f is Baire∗1 as required.

We are now ready to give the extension of Neugebauer.

Theorem 4.41 (Auerbach-Neugebauer) *Let the function f be measurable and everywhere symmetric. Then f is in the first Baire class.*

Proof. Neugebauer's proof consists just of pushing the Auerbach result. We know from Lemma 4.31 that the set

$$Z = \{x : \omega_f(x) = +\infty\}$$

is closed and countable. Let G be the (open) complement of this set and (a, b) any component interval. Since $\omega_f(x) < +\infty$ at every point of (a, b) the function f is bounded on any compact subinterval. By Lemma 4.40 f is Baire 1 on any such interval. This extends in a simple manner to (a, b) itself and hence also to the whole set G. Finally since Z is countable this extends further to $\mathbb{R} = G \cup Z$ and so f is of the first Baire class on the line as required. (We leave the details of each of these extensions to the reader.)

4.6.3 Approximately Symmetric Functions

Theorem 4.41 can be generalized further still by generalizing the concept of a symmetric function. Larson [175] has shown that this same theorem is true for measurable functions that are "approximately" symmetric. We say that f is *approximately symmetric* at a point x if

$$\operatorname{app} \lim_{t \to 0} f(x + t) + f(x - t) - 2f(x) = 0.$$

That is to say the set

$$\{t : |f(x + t) + f(x - t) - 2f(x)| \geq \epsilon\}$$

must have zero density at 0 for every $\epsilon > 0$.

If f is measurable, bounded and approximately symmetric then very nearly the same proof as given in Lemma 4.40 shows that f is Baire 1. The boundedness assumption can be removed but requires different techniques; apparently no elementary proof is available. Our proof is from [175, pp. 89–91].

We remark that while a measurable, approximately symmetric function must be Baire one it may be almost nowhere continuous. Zahorski [309] provides an example of a bounded, measurable, approximately continuous function f such that the set $\{x \; : \; f(x) = 0\}$ is dense and has measure zero. Such a function f is measurable, approximately symmetric and can only be continuous at points in the measure zero set where f vanishes.

Theorem 4.42 (Larson) *Let f be a measurable function that is approximately symmetric at every point. Then f is in the first Baire class.*

The proof is given in a series of lemmas that are of independent interest. We write

$$Q(x,t) = \frac{f(x+t) + f(x-t)}{2} \tag{4.38}$$

and note that at every point x the approximate limit of $Q(x,t)$ as $t \to 0$ is $f(x)$.

Lemma 4.43 *Let M be a bounded measurable set and let $\epsilon > 0$ and $\delta > 0$. Then there is a number $r > 0$ so that*

$$|\{t \in M \; : \; |Q(x,t) - Q(y,t)| < \delta\}| > |M| - \epsilon$$

if $|x - y| < r$.

Proof. Since f is measurable and M is bounded and measurable there is a compact set $P \subset M$ so that f is continuous relative to P and $|P| > |M| - \epsilon/2$. Choose $r_1 > 0$ so that $|f(x') - f(x'')| < \delta$ if $x', x'' \in P$ and $|x' - x''| < r_1$.

Choose $r_2 > 0$ so that

$$|(P - x) \cap (P - y)| > |M| - \epsilon/2 \tag{4.39}$$

if $|x - y| < r_2$. For example to justify this observe that the measure of $(P - x) \cap (P - y)$ can be estimated from

$$\left| |(P - x) \cap (P - y)| - |P| \right|$$
$$= \left| \int \chi_P(t + h)\chi_P(t)\,dt - |P| \right| \leq \int_P |\chi_P(t + h) - \chi_P(t)|\,dt$$

where $h = x - y$ and this expression tends to 0 with h.

Now with $r = \min\{r_1, r_2\}$ we easily verify that

$$|\{t \in M \; : \; |f(x + t) - f(y + t)| < \delta\}| > |M| - \epsilon/2.$$

for this set includes the set $(P - x) \cap (P - y)$. A similar argument provides

$$|\{t \in M \ : \ |f(x - t) - f(y - t)| < \delta\}| > |M| - \epsilon/2$$

and together these assertions prove the lemma since

$$2(Q(x, t) - Q(y, t)) = f(x + t) - f(y + t) + f(x - t) - f(y - t).$$

Lemma 4.44 *Let* $t > 0$, $0 < \rho < 1$ *and* c *a real number. Then the set*

$$G(t, \rho, c) = \{x \ : \ |\{\tau \in (0, t) \ : \ Q(x, \tau) > c\}| > \rho t\}$$

is open.

Proof. Let $x \in G(t, \rho, c)$. Then by definition the set

$$\{\tau \in (0, t) \ : \ Q(x, \tau) > c\}$$

must have measure exceeding ρt. We can choose $\delta > 0$ sufficiently small so that the set

$$M = \{\tau \in (0, t) \ : \ Q(x, \tau) > c + \delta\}$$

also has measure exceeding ρt. Let $\epsilon < |M| - \rho t$. We apply Lemma 4.43 with M, δ, and ϵ as here to obtain a number $r > 0$ so that

$$|\{t \in M \ : \ |Q(x, t) - Q(y, t)| < \delta\}| > |M| - \epsilon$$

if $|x - y| < r$.

We show that $y \in G(t, \rho, c)$ for every $y \in (x - r, x + r)$ and thus we have proved that $G(t, \rho, c)$ is open. Now

$$\{\tau \in (0, t) \ : \ Q(y, \tau) > c\}$$
$$\supset \ \{\tau \in (0, t) \ : \ Q(x, \tau) > c + \delta\} \cap \{\tau \in (0, t) \ : \ |Q(x, \tau) - Q(y, \tau)| < \delta\}$$
$$= \ \{\tau \in M \ : \ |Q(x, \tau) - Q(y, \tau)| < \delta\}.$$

But this last set has measure exceeding $|M| - \epsilon > \rho t$. It follows that $y \in G(t, \rho, c)$ and the proof is complete.

Lemma 4.45 *For any* $\frac{1}{2} < \rho < 1$,

$$\{x \ : \ f(x) \geq c\} = \bigcap_{m=1}^{\infty} \bigcap_{k=1}^{\infty} \bigcup_{n=k}^{\infty} G\left(\frac{1}{n}, \rho, c - \frac{1}{m}\right).$$

Proof. Let us first show that any point x that belongs to the set on the right of the identity of the lemma must have $f(x) \geq c$. Any such point belongs to

$$\bigcap_{k=1}^{\infty} \bigcup_{n=k}^{\infty} G\left(\frac{1}{n}, \rho, c - \frac{1}{m}\right).$$

for all m. Accordingly there must be an increasing sequence of integers $\{n_k\}$ so that

$$x \in G\left(\frac{1}{n_i}, \rho, c - \frac{1}{m}\right).$$

for every i and m. By definition this means that the set

$$\{\tau \in (0, 1/n_i) : Q(x, \tau) > c - 1/m\}$$

has measure exceeding ρ/n_i. But $Q(x, t) \to f(x)$ at every point x as $t \to 0$ through a set of density 1 at 0. Consequently because of the above measure estimate we must have $f(x) \geq c - 1/m$ for all m. So $f(x) \geq c$ as required.

In the other direction suppose $f(x) \geq c$; then for each m and sufficiently large n the set

$$\{\tau \in (0, 1/n) : Q(x, \tau) > c - 1/m\}$$

must have measure exceeding ρ/n. By definition then

$$x \in \bigcup_{n=k}^{\infty} G\left(\frac{1}{n}, \rho, c - \frac{1}{m}\right).$$

for all k and all m. But this is exactly the condition that x belongs to the set on the right of the identity of the lemma and so the proof is complete.

Now the proof of the theorem is clear. From Lemmas 4.44 and 4.45 we see that the set $\{x : f(x) \geq c\}$ is a Borel set of type G_δ. The same arguments (obtained by replacing f by $-f$) would show that $\{x : f(x) \leq c\}$ is also of type G_δ. Thus f is in the first Baire class and the theorem is proved.

4.6.4 Points of Symmetry

Let f be an arbitrary function and let E_f denote the set of points at which f is symmetric. It is natural to ask for the character of such sets. The following results are pointed out in Darji [57] and [58]. Note that the first two assertions follow from Theorem 4.34 and Theorem 4.35; the third and fourth are proved in Theorem 2.44.

Theorem 4.46 *Let f be arbitrary function and let E_f denote the set of points at which f is symmetric.*

1. *if f is measurable then E_f is a measurable set.*

2. *if f has the Baire property then the set E_f has the Baire property.*

3. if f is Borel measurable then E_f is a co-analytic set.

4. if f is Baire 1 then E_f is a Borel set of type $G_{\delta\sigma\delta}$.

The exact character of the set E_f in general is unknown; any characterization must be of a geometric or algebraic nature. In order to show that the set of points E_f cannot be given a topological characterization [58] establishes that, for any zero dimensional subset M of $\mathbb{R}$, there is a function f such that E_f is topologically equivalent to M.

4.6.5 Points of Approximate Symmetry

Evans and Humke [91] and Evans [87] and [86] study the relations among the sets of points where a function is continuous, is approximately continuous or is approximately smooth. We report on this work here.

Let f be a measurable function defined on an interval (a,b). We write

(i) C_f for the set of points at which f is continuous,

(ii) AC_f for the set of points at which f is approximately continuous,

(iii) AS_f for the set of points at which f is approximately symmetric, and

(iv) L_f for the set of points at which f is the derivative of its integral.

If f is everywhere approximately symmetric then, from Section 4.6.2, f is Baire 1 and so C_f is residual. For a function f that is approximately symmetric at most points but not everywhere there may be no points of continuity. (Let f be the characteristic function of the rationals; then f is approximately symmetric at every irrational point and, of course, nowhere continuous.) The natural question seems to be to ask for a relation between AS_f and AC_f, i.e. between the sets of points of approximate continuity and approximate *even* continuity.

We cite these results without proof.

Theorem 4.47 (Evans) *For a measurable function f the set $AS_f \setminus AC_f$ is measure zero and first category.*

Theorem 4.48 (Evans–Humke) *For a measurable, locally integrable function f the set $AS_f \setminus L_f$ is measure zero and first category. If f is bounded then this set is σ-porous.*

4.6.6 L_p-Symmetric Functions

As a corollary to Theorem 4.42 we have a similar assertion for functions that are L_p-symmetric. This is from Neugebauer [212, Theorem 8, p. 90].

Theorem 4.49 (Neugebauer) *Let $p \geq 1$ and suppose that f is a measurable function such that, at every point x,*

$$\lim_{h \to 0} \frac{1}{h} \int_0^h |f(x+t) + f(x-t) - 2f(x)|^p \, dt = 0.$$

Then f is in the first Baire class.

Proof. To obtain this as a corollary to Theorem 4.42 we shall show that a function f that satisfies the hypotheses here must be approximately symmetric. Fix x, let $\epsilon > 0$ and let

$$E_h = \{0 < t < h \, : \, |f(x+t) + f(x-t) - 2f(x)| \geq \epsilon\}.$$

We compute

$$\frac{1}{h} \int_0^h |f(x+t) + f(x-t) - 2f(x)|^p \, dt \geq \frac{1}{h} \int_{E_h} \epsilon^p \, dt \geq \epsilon^p \frac{|E_h|}{h}.$$

Thus $|E_h|/h \to 0$ as $h \to 0+$ and this is exactly the requirement that f is approximately symmetric.

One might have suspected that a function satisfying the hypotheses of this theorem would also be continuous almost everywhere; a measurable, symmetric function has this property. Neugebauer [212, p. 90] points out that an example of Zahorski [309] (already cited in Section 4.6.3) provides an example of a measurable, L_p-symmetric function that is almost nowhere continuous.

4.6.7 Determining Sets of Symmetric Functions

A determining set (see [18, p. 199]) for a class $\mathcal{F}$ of real functions is a set E such that if two members f_1 and f_2 of the class $\mathcal{F}$ agree on E then $f_1 = f_2$. As part of his study of symmetric functions Neugebauer [213, p. 24] characterized the determining sets for the symmetric, measurable functions.

We begin with an observation for general (not necessarily measurable) symmetric functions.

Lemma 4.50 (Neugebauer) *If f and g are symmetric and the set*

$$\{x \, : \, f(x) = g(x)\}$$

is residual in an interval (a, b) then $f = g$ on that interval.

Proof. It is enough to show that, for any symmetric function f for which $E = \{t : f(t) = 0\}$ is residual, f must vanish everywhere. Fix any point $x \in (a, b)$; we show that $f(x) = 0$. The set $2x - E$, which is the reflection of E about x, is also residual in a neighbourhood of x and, accordingly, so too is the intersection $E \cap (2x - E)$. Take any $x_n \to x$ from this intersection and then $f(x_n) = f(2x - x_n) = 0$. But, by the fact that f is symmetric, we know that

$$f(x_n) + f(2x - x_n) - 2f(x) \to 0.$$

Hence $f(x) = 0$ as we require.

We remark that it is not enough in the proof of Lemma 4.50 that the set

$$\{x : f(x) = g(x)\}$$

be dense. There exist discontinuous additive function (i.e. solutions of the Cauchy equation $f(x + y) = f(x) + f(y)$) that are zero on a dense set (see Section 4.2.2).

Our main theorem shows that in order for a set to be determining for the class of measurable, symmetric functions it is necessary and sufficient that it be dense. The necessity is, of course, obvious.

Theorem 4.51 (Neugebauer) *If f and g are measurable and symmetric and the set*

$$\{x : f(x) = g(x)\}$$

is dense in an interval (a, b) then $f = g$ on that interval.

Proof. As before it is enough to establish that a measurable, symmetric function f with $E = \{x : f(x) = 0\}$ dense in an interval vanishes everywhere there. Note first that such a function must vanish at any point of continuity. By Theorem 2.3 the set C_f of continuity points of f is dense and, at each point x in C_f, $f(x) = 0$. For any function f the set C_f is always a G_δ and so $E \supset C_f$ is residual. The proof is completed by appealing now to Lemma 4.50.

There are some immediate corollaries that follow from this theorem.

Corollary 4.52 *Suppose that f is measurable and symmetric. If f is bounded on a set dense in an interval then f is bounded on that interval.*

Corollary 4.53 *Suppose that f is measurable and symmetric. If f is nondecreasing on a set dense in an interval then f is nondecreasing on that interval.*

4.6.8 Derivates of Symmetric Functions

There are by now a host of relations that are known to hold among the Dini derivatives. The most famous collection of such relations is that known as the Denjoy–Young–Saks Theorem. A simpler set of relations for continuous functions was given in Neugebauer [211], rediscovering an early result of W. H. Young: for a continuous function f the set of points

$$\left\{x \ : \ \overline{D}^+ f(x) \neq \overline{D}^- f(x) \ \text{or} \ \underline{D}^+ f(x) \neq \underline{D}^- f(x)\right\}$$

is of the first category. Neugebauer [213, p. 30] noted that it is enough for the even part of f to be continuous in this statement. In this context we should mention too Pu and Pu [238] and [233] who extend some of these ideas involving relations for symmetric functions, employing approximate Dini derivatives and extreme symmetric derivates.

Theorem 4.54 (Neugebauer) *Let the function f be measurable and symmetric. Then the set*

$$\left\{x \ : \ \overline{D}^+ f(x) \neq \overline{D}^- f(x) \ \text{or} \ \underline{D}^+ f(x) \neq \underline{D}^- f(x)\right\}$$

is of the first category.

Proof. It is enough to show that the set

$$\left\{x \ : \ \overline{D}^+ f(x) > \overline{D}^- f(x)\right\}$$

is first category. As usual in such problems this reduces in turn to showing that each set of the form

$$A_{rs} = \left\{x \ : \ \overline{D}^+ f(x) > s > r > \overline{D}^- f(x)\right\}$$

for $s > r$ rational is first category. Once again this reduces to showing that each set

$$A_{rsk} = \left\{x \ : \ \overline{D}^+ f(x) > s\right\}$$
$$\cap \ \{x \ : \ f(x) - f(y) < r(x - y) \ \text{for} \ x - 1/k < y < x\}$$

for $k = 1, 2, \ldots$ is first category. By Theorem 4.41 the function f is Borel measurable and so standard arguments (cf. [18, Chap. 4]) show that each set in this intersection is Borel. Consequently to show that each such set is first category it is enough to show that no set of that form may be residual in an interval.

In order to obtain a contradiction let us suppose that a set A_{rsk} is residual in an interval (c, d). Choose any point $x \in A_{rsk} \cap (c, d)$ and let y be an arbitrary point with $x < y < x + 1/k$. Since A_{rsk} is residual in a neighbourhood of y the reflected set $2y - A_{rsk}$ is too. Choose a sequence of points $\{y_n\}$ from A_{rsk}, $x < y_n < x + 1/k$ so that $y_n \to y$ and $2y - y_n \in A_{rsk}$.

We have then
$$f(y_n) - f(x) < r(y_n - x)$$
and
$$f(2y - y_n) - f(x) < r(2y - y_n - x).$$

Because f is symmetric at y we know that $f(2y - y_n) + f(y_n) \to 2f(y)$. We conclude then that $f(y) - f(x) \le r(y - x)$. As this holds for all $y > x$ sufficiently close to x we must have $\overline{D}^+ f(x) \le r$; but, since $x \in A_{rsk}$ we have too that $\overline{D}^+ f(x) > s$ and this is a contradiction. This completes the proof.

4.6.9 Typical Symmetric Functions

There is a large literature devoted to the study of what has been called "typical" (or sometimes "generic") properties of functions. One introduces a Banach space of functions and, relative to that space, a property of functions is said to be *typical* if the set of all functions possessing that property is residual in the space. This allows the construction of examples of functions possessing perhaps strange properties along with an observation that, moreover, this strange property is shared by almost all functions in the space. The best known example is nondifferentiability: the set of functions in C[a, b], the Banach space of continuous functions on [a, b] with the supremum norm, that do not possess a derivative at any point is residual. Hence nowhere differentiability is typical (at least in that space and with that metric).

The terminology is somewhat vague and could be subject to abuse; accordingly some authors have not approved of it. Even so it has been widely adopted nowadays and has proved itself invaluable in its ability to communicate rapidly the essence of this idea. Let us then seek for "typical" properties of symmetric functions. Neugebauer [213, p. 26] first addressed this problem with respect to how discontinuous a symmetric function might be. We know that we can construct a measurable, symmetric function with a dense set of discontinuities. The theorem shows that this is typical. This is in contrast to functions that satisfy a slightly sharper symmetry condition.

Theorem 4.55 (Neugebauer) *Let* BS[a, b] *be the set of all bounded, measurable, symmetric functions equipped with the supremum metric. Then the typical function f $\in$BS[a, b] has a dense set of discontinuities.*

Proof. Let A_n denote the set of functions in BS[a, b] for which there is an interval $(c, d) \subset [a, b]$ with $d - c \ge 1/n$ and f continuous on (c, d). It is easy to see that each set A_n is closed in BS[a, b]. If $\{f_k\}$ is a sequence of functions from

A_n converging uniformly to a function f then there are intervals $(c_k, d_k) \subset [a, b]$ with $d_k - c_k \geq 1/n$ and f_k continuous on (c_k, d_k). We may suppose that (by passing to a subsequence if necessary) $c_n \to c$ and $d_n \to d$. An elementary compactness argument shows this. Evidently $(c, d) \subset [a, b]$ with $d - c \geq 1/n$. Every point of (c, d) lies inside an interval (c_k, d_k) for large enough k and f_k is continuous there. Since f is the uniform limit of the f_k we see that f too is continuous on (c, d). Thus by definition $f \in A_n$.

We now show that each A_n is nowhere dense. Let $f \in A_n$ and let $\epsilon > 0$. Suppose that $I_1, I_2, \ldots I_k$ are the components of the set of continuity points of f that are at least $1/n$ in length. For each interval $I_i = (c, d)$ we can construct a function h_i that is measurable, symmetric and for which $h_i(x) = 0$ if $x \leq c$ or $x \geq d$, and such that $0 \leq h_i(x) \leq \epsilon$ everywhere and with a dense set of discontinuity points in (c, d). The function $g = f + \sum_{i=1}^{k} h_i$ cannot belong to A_n and is within ϵ of f everywhere. This proves that each A_n is nowhere dense.

Finally the set of functions $f \in BS[a, b]$ that have a dense set of discontinuities can be expressed as the complement of the first category set $\bigcup_{n=1}^{\infty} A_n$ and this is what we were required to prove.

4.7 Quasi-Smooth Functions

A continuous function f is said to be *smooth* at a point x if

$$f(x + h) + f(x - h) - 2f(x) = o(h) \quad (\text{as } h \to 0). \tag{4.40}$$

A continuous function f is said to be *quasi-smooth* at a point x if

$$f(x + h) + f(x - h) - 2f(x) = O(h) \quad (\text{as } h \to 0). \tag{4.41}$$

The graph of a smooth function can evidently have no corners and this is the source of the terminology. The graph of a quasi-smooth function may have corners but can have no cusps. The terminology here has evolved by a familiar second order process. The same phenomenon occurs in other languages: for example some French authors refer to smooth and quasi-smooth functions as *lisse* and *presque lisse*.

4.7.1 Continuity Properties

We have already mentioned the continuity properties of quasi-smooth functions in Section 2.3.5 of Chapter 2. We derived Theorem 2.17 from the Charzyński theorem. For reference we include the statements here. See also Section 5.3.4 in Chapter 5 for a different proof, based on monotonicity arguments, and a slightly more general statement.

Theorem 4.56 *Let f be a measurable function that satisfies the condition*

$$\limsup_{h \to 0} \left| \frac{f(x+h) + f(x-h) - 2f(x)}{h} \right| < +\infty \qquad (4.42)$$

at every point x. Then f is continuous at every point with the exception only of a scattered set.

The fact that this theorem actually characterizes the set of discontinuity points of quasi-smooth functions (and smooth functions) is due to Evans and Larson [96]. A proof has been given in Section 2.6.

Theorem 4.57 (Evans–Larson) *Let S be a scattered subset of the interval (0,1). Then there is a smooth, measurable function f defined on that interval that is discontinuous precisely at the points in the set S.*

4.7.2 A Nowhere Differentiable Quasi-Smooth Function

We shall see that a continuous, smooth function must have many points of differentiability. Here we show that a continuous, quasi-smooth function may have none.

Theorem 4.58 *There is a continuous, quasi-smooth function f that is nowhere differentiable.*

Proof. The classical Weierstrass function

$$f(x) = \sum_{k=1}^{\infty} \frac{\cos b^k x}{b^k}$$

for $b > 1$ provides a well-known example. This function is uniformly quasi-smooth. See the discussion in Section 4.4.1 and in [313, p. 48 and p. 206].

4.7.3 Derivates of Quasi–Smooth Functions

We have just seen that a continuous, quasi–smooth function may be nowhere differentiable. It may seem that, accordingly, nothing much positively may be said about the differentiability properties of such functions.

But smoothness conditions obviously place some local restrictions on the disposition of the Dini derivatives, and these restrictions have interesting global implications. If f is smooth at a point x_0 then

$$\frac{f(x_0+h) - f(x_0)}{h} - \frac{f(x_0-h) - f(x_0)}{-h} \to 0 \qquad (4.43)$$

as $h \to 0$. This means that the sets of right hand derived numbers and left hand derived numbers are identical. This identity forces the existence of the derivative on a c–dense set (as we see in Section 4.8.2 below). If f is quasi–smooth at a point x_0 then for some $C > 0$, $\delta > 0$

$$\left| \frac{f(x_0 + h) - f(x_0)}{h} - \frac{f(x_0 - h) - f(x_0)}{-h} \right| \leq C \qquad (4.44)$$

for all $0 < h < \delta$. This means that for every right (left) hand derived number α there is a corresponding left (right) hand derived number β with $\beta \in [\alpha - C, \alpha + C]$. This too has differentiability implications.

Garg [121] has given an analysis of the level structure of continuous functions that yields an interesting and curious observation for quasi-smooth functions. He calls a number $\alpha \in \mathbb{R} \cup \{\pm\infty\}$ a *knotted lower derivative* of f at x_0 if

$$\overline{D}^+ f(x_0) = +\infty, \ \underline{D}^- f(x_0) = -\infty, \ \overline{D}^- f(x_0) = \underline{D}^+ f(x_0) = \alpha. \qquad (4.45)$$

For example the function $f(x) = \sqrt{|x|}\sin^2 1/x$ has a knotted lower derivative 0 at 0. Evidently a smooth or quasi–smooth function cannot have at any point a knotted lower derivative. Garg [121] then shows that if f is a continuous function on an interval I and E is the set of points at which f has a derivative, finite or infinite, or has a knotted lower derivative then E is c–dense in I and dense in $f^{-1}(y)$ for almost every $y \in \mathbb{R}$.

The following theorem is now immediate. Our proof does not, however, invoke this analysis and just applies standard material on the level structure of continuous functions.

Theorem 4.59 (Garg) *Let f be a continuous, quasi–smooth function on an interval I, let E be the set of points at which f has a derivative, finite or infinite and for each $\lambda \in \mathbb{R}$ let $E_\lambda = \{x \in E : f(x) = \lambda\}$. Then there is a set N of measure zero so that for every $\lambda \in \mathbb{R} \setminus N$ each point of E_λ is isolated and E_λ is dense in $f^{-1}(\lambda)$. (Indeed every point of $f^{-1}(\lambda) \setminus E_\lambda$ is a bilateral limit of points in E_λ.)*

Proof. Let $N_1 = \{x \in I : f'(x) = 0\}$. By [251, Thm 6.5, p. 227] we know that $|f(N_1)| = 0$. Let

$$N_2 = \{x \in I \setminus E : \overline{D} f(x) \text{ or } \underline{D} f(x) \text{ is finite}\}.$$

By [251, Thm 4.2, p. 270] we know that $|f(N_2)| = 0$.

Suppose that λ does not belong to $f(N_1) \cup f(N_2)$. Then each point x_0 in E_λ is isolated in $f^{-1}(\lambda)$; if not then $f'(x_0) = 0$ and this contradicts the fact that λ does not belong to $f(N_1)$.

Now consider a point x_0 that is in the level $f^{-1}(\lambda)$ and which is isolated on the right or the left. Suppose that x_0 is isolated on the right. If the graph of f lies above the line $y = \lambda$ in a neighbourhood $(x_0, x_0 + \delta)$ then

$$0 \leq \underline{D}^+ f(x_0) \leq \overline{D}^+ f(x_0) \leq +\infty.$$

But at x_0 the function f is quasi-smooth so there is a $C_x > 0$ with

$$|f(x_0 + h) + f(x_0 - h) - 2f(x_0)| \leq C_x|h|$$

for all small h. It follows that

$$\underline{D}^+ f(x_0) - C_x \leq \underline{D}^- f(x_0) \leq \overline{D}^- f(x_0) \leq \overline{D}^+ f(x_0) + C_x.$$

In particular $-\infty < -C_x \leq \underline{D} f(x_0) \leq +\infty$. But as λ does not belong to $f(N_2)$ it follows that the point x_0 must be in E_λ. Similar arguments come to the same conclusion if x_0 is isolated on the other side or if the graph of f is on the other side of $y = \lambda$.

The level $f^{-1}(\lambda)$ is closed and must be nowhere dense (otherwise it contains a point with a zero derivative). Consequently the points in $f^{-1}(\lambda)$ that are isolated on one side at least are dense in $f^{-1}(\lambda)$ and we know that these points are all in E_λ. This completes the proof.

Corollary 4.60 *Let f be a continuous, quasi–smooth function on an interval I and let E be the set of points at which f has a derivative, finite or infinite. Then E is c–dense in I and dense in $f^{-1}(\lambda)$ for almost every $\lambda \in \mathbb{R}$.*

Corollary 4.61 *Let f be a continuous, quasi–smooth function on an interval I and suppose that f maps the set of points at which f has an infinite derivative into a set of measure zero. Then f has a finite derivative on a dense set in I.*

Corollary 4.62 *Let f be a continuous, quasi–smooth function on an interval $[a, b]$. Let*

$$P = \{x \in [a, b] : 0 < f'(x) \leq +\infty\} \quad and \quad N = \{x \in [a, b] : 0 > f'(x) \geq -\infty\}.$$

Then

$$-|f(N)| \leq f(b) - f(a) \leq |f(P)|.$$

Proof. This follows easily from the theorem using the methods of [251, pp. 280–281].

Corollary 4.63 *Let f be a continuous, quasi–smooth function on an interval $[a, b]$. Let E be a measurable set with the property that every point $x \in E$ is isolated on one side at least in the level $f^{-1}[f(x)]$. Then f is almost everywhere differentiable in E.*

Proof. One argues, precisely as in the theorem, that $\underline{D} f(x) > -\infty$ or $\overline{D} f(x) < +\infty$ at each point $x \in E$. Then the a.e. differentiability of f on E follows from [251, Theorem 10.1, p. 234].

Recall the example in Section 4.7.2 of a continuous, quasi–smooth function f that is nowhere differentiable. We see now that f possesses an infinite derivative on a dense set and f must map that set into a set of positive measure.

These results seem to suggest that a continuous, quasi–smooth function should satisfy Banach's condition T_2. Indeed Anderson and Pitt [1] have shown that if the quasi-smoothness holds *uniformly* then this is the case. It is apparently unknown whether the pointwise version has this property. We shall record this as an open problem.

Another related result of Garg [121, p. 71] says more about the Dini derivatives of a semi-continuous, quasi–smooth function. Let us say that a function f is *nonangular* if

$$\max\{\underline{D}^+ f(x), \underline{D}^- f(x)\} \leq \min\{\overline{D}^+ f(x), \overline{D}^- f(x)\} \qquad (4.46)$$

at each point x. We say that a number $\alpha \in \mathbb{R} \cup \{\pm\infty\}$ is a *lower gradient* of f at x_0 if

$$\overline{D}^- f(x_0) \leq \alpha \leq \underline{D}^+ f(x_0). \qquad (4.47)$$

Garg shows that every lower semicontinuous function on an interval has a finite lower gradient at a dense set of points; if the function is moreover nonangular then it has a finite lower gradient at a c–dense set of points. For a quasi-smooth function the existence of a finite lower gradient at a point forces all four Dini derivatives at that point to be finite. Hence we obtain the following (first observed in Zygmund [312, p. 55]).

Theorem 4.64 *If f is lower semi-continuous and quasi-smooth in an interval (a, b) then then the set of points at which all four Dini derivatives of f are finite is dense in (a, b).*

For smooth functions more can be said. First of all it is easy to check that a smooth function is nonangular; accordingly the set of points where it has a finite lower gradient is c–dense. But a smooth function has a finite lower gradient at a point if and only if it has a finite derivative there. Thus the smooth version of Theorem 4.64 is that a lower semi-continuous, smooth function has a finite derivative on a c–dense set. This is given in Section 4.8.2.

4.7.4 A Tauberian Theorem

In this section we establish a relationship that must hold between the ordinary derivative and the approximate derivative for quasi-smooth functions. This theorem is due to Marcinkiewicz and Zygmund [186]; a proof also appears in Marcinkiewicz [185, Lemma 2, p. 67]. Stein [272, p. 261] proves a similar theorem using a related derivative; he suggests that this result may be viewed as a kind of "Tauberian" theorem in that it allows a weaker form of derivative (the approximate derivative) to be refined in the presence of quasi-smoothness to a stronger form of derivative (the ordinary derivative). For a well known example recall that for monotonic functions approximate differentiability at any point implies ordinary differentiability. Here we have the same implication but only almost everywhere. We shall discuss too a variant on this due to Neugebauer [214]; Starcev [271] has a closely related result.

Theorem 4.65 *Let f be a measurable function and E a measurable set. Suppose that*

$$|f(x+h) + f(x-h) - 2f(x)| = O(h) \qquad (4.48)$$

as $h \to 0$ at every point $x \in E$. Then f is differentiable for almost every point $x \in E$ at which the approximate derivative of f exists.

Proof. For any pair of integers m and n define the set

$$E_{mn} = \{x \in E : |f(x+t) + f(x-t) - 2f(x)| < mt, \ 0 < t < 1/n\}.$$

These sets cover E. We show, at any density point z of a set E_{mn}, that should the approximate derivative exist then so too does $f'(z)$. Evidently the theorem follows since almost every point of E can be realized as a density point of one of these sets.

Let z be a density point of E_{mn}, choose a measurable set G having density 1 at z and so that

$$\lim_{y \to z, \, y \in G} \frac{f(y) - f(z)}{y - z} = \mathrm{AD}\, f(z).$$

We may suppose that $z = f(z) = \mathrm{AD}\, f(z) = 0$ to simplify the computations. Thus we have

$$\lim_{y \to 0, \, y \in G} \frac{f(y)}{y} = 0.$$

We will argue just on the right.

Let $\epsilon > 0$. For any $x \in G$ there is a $\eta = \eta(\epsilon)$ with $0 < \eta < 1/n$ and so that $0 < |x| < \eta$, $x \in G$ requires $|f(x)| \leq \epsilon|x|$. Because $z = 0$ is a point of density of G and of E_{mn} and because G is measurable, for any $p > 0$, there is a $0 < \delta < \eta$ so that the set

$$G \cap (2G - x) \cap (2E_{mn} - x) \cap \left(\frac{p-1}{p}x, x\right)$$

is nonempty for every $0 < x < \delta$. (The density computations used here are set out in Section A.4.)

For any $0 < x < \delta$ choose y in the set above. Then $(x+y)/2 \in G$, $y \in G$, and $|y - x| < |x|/p$. Because of this we have $|f(y)| \leq \epsilon|x|$ and $|f((x+y)/2)| \leq \epsilon|x|$. Because $(x+y)/2 \in E_{mn}$ we have

$$\left| f(y) + f(x) - 2f\left(\frac{x+y}{2}\right) \right| \leq \frac{m}{p}|x|.$$

Putting these together we obtain

$$|f(x)| \leq |f(y)| + 2\left|f\left(\frac{x+y}{2}\right)\right| + \frac{m}{p}|x|$$

$$\leq |x|\left(3\epsilon + \frac{m}{p}\right).$$

Consequently

$$\limsup_{x \to 0+}\left|\frac{f(x)}{x}\right| = 0.$$

since ϵ and p are arbitrary. This translates into the existence and vanishing of the right hand derivative of f at $z = 0$; similarly the left hand derivative also vanishes. The proof is now complete.

A number of applications of this theorem come to mind. One need only combine a condition that implies the existence of an approximate derivative with quasi-smoothness to obtain a differentiation theorem. For example the following easy application was noted in [1]. It asserts that an occupation measure $\mu(A) = |f^{-1}(A)|$ is singular for certain functions f.

Theorem 4.66 *Let f be a continuous, almost nowhere differentiable, quasi-smooth function. Then every level $f^{-1}(\lambda)$ has measure zero.*

Proof. If any level $f^{-1}(\lambda)$ has positive measure then AD $f(x) = 0$ for each density point x of $f^{-1}(\lambda)$. By Theorem 4.65 then $f'(x) = 0$ a.e. in $f^{-1}(\lambda)$ which violates the fact that f is almost nowhere differentiable.

There is a variant on Theorem 4.65 that may be useful. We can compute derivatives and assign local growth conditions in a way that ignores sets of measure zero. We first show, following Neugebauer [214, p. 1206], that a function that is quasi-smooth ignoring sets of measure zero locally is in fact quasi-smooth ignoring a single set of measure zero.

Lemma 4.67 *Let f be a measurable function and E a measurable set. If for each $x \in E$ there is a set N_x of measure zero so that*

$$\limsup_{t \to 0+,\ x+t \notin N_x}\left|\frac{f(x+t) + f(x-t) - 2f(x)}{t}\right| < +\infty$$

then there is a set N of measure zero so that

$$\limsup_{t \to 0+,\ x+t \notin N}\left|\frac{f(x+t) + f(x-t) - 2f(x)}{t}\right| < +\infty$$

Proof. It is enough to take for N the set of points at which f is not approximately continuous.

We now can prove the variant on Theorem 4.65 that allows a set of measure zero to be ignored at each point.

Theorem 4.68 *Let f be a measurable function and E a measurable set. Assume that for each $x \in E$ there is a set N_x of measure zero so that*

$$\limsup_{t \to 0+, \ x+t \notin N_x} \left| \frac{f(x+t) + f(x-t) - 2f(x)}{t} \right| < +\infty$$

and that f has an approximate derivative at every point of E. Then f is equivalent to a function g that has a derivative at almost every point of E.

Proof. By Lemma 4.67 there is a single set of measure zero N so that f is quasi-smooth at each point of E provided N is ignored. Essentially the same arguments as used in the proof of Theorem 4.65 now will establish that f is differentiable in the ordinary sense at a.e. point of E provided that points in N are ignored. Finally a simple argument (cf. Neugebauer [214, Lemma 1, p. 1206]) shows that f must then be equivalent to a function g with the required property.

4.8 Smooth Functions

A function f is *smooth* at a point x provided

$$f(x+t) + f(x-t) - 2f(x) = o(t)$$

as $t \to 0$. We might have called this "even differentiability" since this condition is equivalent to the even part of f at x having a derivative. The choice of terminology is due to Zygmund and is explained below. We will discuss some variants on this notion as well. Higher order smoothness properties have also been investigated by a number of authors (eg. [30], [85], [74], and [76]); we shall not pursue these ideas ourselves.

We have already seen in Section 4.7.1 the complete situation for the continuity properties of smooth functions. Our interest accordingly will shift towards the rather curious differentiability properties of smooth functions.

4.8.1 Elementary Properties

If the derivative $f'(x_0)$ exists and is finite then certainly f is smooth at x_0, but smoothness cannot imply differentiability. If, however, the right and left hand derivatives of f exist at that point and f is smooth there then it is not difficult to see that f must be differentiable since the derivates on the two sides agree. This excludes a smooth function from having any "angular" points and explains partly the choice of terminology.

We should recall that the smoothness of a function f at a point x_0 really means the differentiability of the even part of f at that point; thus if a function is smooth and also symmetrically differentiable at a point then both even and odd parts are differentiable so that, of course, the function is differentiable too. In Theorem 4.69 below a general relation between the ordinary and symmetric derivates of smooth functions is given. Further differentiability properties of smooth functions are developed in the next few sections.

The notion of smoothness was introduced by Riemann in his famous memoir on trigonometric series; we shall see in later sections some of his applications. The first modern study of smoothness to go substantially beyond the original work of Riemann is that of Zygmund [312] from 1944 entitled simply "Smooth functions". We shall reproduce some of the more immediate classical results both from Zygmund and earlier. As Zygmund [312, p. 48] explains it, the notion of a smooth function can be expected to arise in problems "where by rough extrapolation of known results we should have (wrongly) anticipated functions with continuous derivatives".

There is a close connection with the first order symmetric continuity in that if a function f satisfies

$$f(x_0 + t) - f(x_0 - t) \to 0$$

as $t \to 0$ and F is a primitive of f then, by integrating

$$\int_0^h \left(f(x_0 + t) - f(x_0 - t) \right) \, dt = F(x_0 + h) + F(x_0 - h) - 2F(x_0),$$

we see that F must be smooth at x_0. Thus, again as pointed out by Zygmund, "smooth functions may be regarded as a generalization of the integral of continuous functions".

In Section 4.8.2 we present some simple results from Zygmund [312, pp. 54–55] illustrating the remarkable derivation properties of smooth functions. These theorems show just how close a continuous smooth function is to having the familiar properties of a continuously differentiable function; in particular the mean-value theorem (Theorem 4.72) and the Darboux property (Theorem 4.75) should be somewhat surprising especially in view of the fact that the derivative of a continuous smooth function may exist only on a set of measure zero.

The arguments are mostly elementary and flow just from the observation that smooth functions do not allow angular points. More precisely we note that the condition of smoothness at a point x_0 may be rewritten as

$$\lim_{t \to 0} \left(\frac{f(x_0 + t) - f(x_0)}{t} - \frac{f(x_0) - f(x_0 - t)}{t} \right) = 0. \qquad (4.49)$$

This makes some observations really quite transparent. Firstly if $f'(x_0)$ exists then evidently f is smooth at x_0. If a one-sided derivative of f exists at x_0 and f is smooth there then the derivative on the other side also exists and with the same value. If f has a local extremum at a point x_0 and is smooth there then necessarily $f'(x_0) = 0$. This same nonangularity forces symmetric derivatives and ordinary derivatives of smooth functions to coincide. For reference let us state this as a theorem.

Theorem 4.69 *Let f be smooth at a point x_0. Then*

$$\overline{SD}\, f(x_0) = \overline{D}\, f(x_0) = \overline{D}^+ f(x_0) = \overline{D}^- f(x_0)$$

and

$$\underline{SD}\, f(x_0) = \underline{D}\, f(x_0) = \underline{D}^+ f(x_0) = \underline{D}^- f(x_0).$$

Proof. This follows almost immediately from the expression in (4.49).

Corollary 4.70 *Let f be smooth. Then the ordinary derivative of f exists at any point at which the symmetric derivative exists.*

This is a convenient spot to mention one other elementary fact about smooth functions. The product of two smooth functions need not be smooth: take $F(x) = \sqrt{|x|}\sin x^{-1}$ which is smooth at $x = 0$ and compare with $F^2(x) = |x|\sin^2 x^{-1}$ which is not. However a product FG is smooth if F is continuous and smooth and if G is differentiable.

4.8.2 Continuous, Smooth Functions

The differentiability properties of continuous, smooth functions were first explored in Rajchman [242]; he showed that the derivative must exist on a dense set. This was refined by Zalcwasser to show that this dense set has continuum many points in any interval (cf. [312, p. 48]). Some of this material was also anticipated by G. C. Young and W. H. Young who studied the differential properties of continuous functions whose right and left derivates are identical (as is the case for smooth functions in view of Theorem 4.69).

The following theorem may also be viewed as a special case of a more general result of Garg [121, p. 71] and as the specification of Theorem 4.64 to smooth functions. See the discussion in Section 4.7.3.

Theorem 4.71 (Rajchman-Zalcwasser) *Let g be a continuous function that is smooth in (a,b). Then the derivative $g'(x)$ exists and is finite at a c–dense set of points of that interval.*

Proof. (Our proof is from [312, p. 54] and was repeated in [313, Vol I, p. 43]). In any subinterval (c, d) of (a, b) let $L(x)$ be the linear function which coincides with $g(x)$ at the points c and d. Then $f(x) = g(x) - L(x)$ is continuous, smooth and vanishes at the two endpoints c and d. If x_0 is a point in (c, d) where f attains a maximum then all the derived numbers on the right are nonpositive and all the derived numbers on the left are nonnegative. The expression for the smoothness of f at x_0 in the form (4.49) shows then that the derivative $f'(x_0)$ must exist and vanish. The same argument applies at a minimum. Thus in any case there is a point x_0 with $g'(x_0) = (g(d) - g(c))/(d - c)$. Thus we have shown that the derivative $g'(x)$ exists and is finite at an everywhere dense set of points of (a, b).

This also establishes a mean-value theorem for continuous, smooth functions and the rest of the proof exploits this. Inside every interval $(c, z) \subset (c, d)$ there is a point x_z where the derivative $g'(x_z)$ is equal to the slope of the line joining $(c, g(c))$ and $(z, g(z))$. If the slopes for two different values of z are different then the corresponding points x_z must also differ. Either g is linear in (c, d) or else there are continuum many points z here with differing slopes. In either case then there are continuum many points in (c, d) at which the derivative of g exists. This completes the proof.

We have just seen that continuous, smooth functions have a mean-value property. We express this as a theorem.

Theorem 4.72 *If f is continuous and smooth in $[a, b]$ then*

$$\frac{f(b) - f(a)}{b - a} = f'(\xi)$$

for some point $\xi \in (a, b)$.

This mean value theorem gives an obvious monotonicity result.

Theorem 4.73 *Let the function f be continuous and smooth in (a, b). If the derivative $f'(x)$ is nonnegative at every point at which it exists then f in non-decreasing in that interval.*

As a corollary we obtain a result clarifying the nature of smooth functions that are differentiable only on sets having small measure. We shall need this later in Section 4.8.6.

Corollary 4.74 *Let f be continuous and smooth on an interval (a, b) and suppose that D_f, the set where $f'(x)$ exists, has full measure in no subinterval of (a, b). Then every set*

$$\left\{ x \in (a, b) : \overline{D}^+ f(x) \le c \right\}$$

and

$$\left\{x \in (a,b) \ : \ \underline{D}^+ f(x) \geq c\right\}$$

is dense in (a, b).

Proof. If not then the set $\left\{x \in (a,b) \ : \ \overline{D}^+ f(x) \leq c\right\}$ misses some subinterval $(c, d) \subset (a, b)$. This means that $f'(x) > c$ at every point in (c, d) at which the derivative exists. By Theorem 4.73 the function $f(x) - cx$ must be monotone in that interval and hence almost everywhere differentiable there. But this contradicts the assumption that D_f has full measure in no subinterval of (a, b). This contradiction completes the proof.

This mean value property also extends to the full Darboux property. Recall that a function g defined on a set E has the *Darboux property* if given any two points c, d in E the function g takes on the set $(c, d) \cap E$ every value between $g(c)$ and $g(d)$.

Theorem 4.75 (Zygmund) *If f is continuous and smooth in (a, b) and D_f denotes the set of points at which f is differentiable (finitely) then f' has the Darboux property on D_f.*

Proof. In any subinterval (c, d) of (a, b) with f differentiable at c and d let us suppose that $f'(c) < 0 < f'(d)$. We shall find a point $z \in (c, d)$ with $f'(z) = 0$. The general case would then follow from the same arguments. The strategy is simply to find two points on the same level of f: between these two points there is a point where the derivative vanishes. (This is the same argument as in the proof of Theorem 4.71).)

Consider the function $g(x) = [f(x + h) - f(x)]/h$ where $0 < h < d - c$ is sufficiently small that $g(c) < 0$ and $g(d - h) > 0$. As g is continuous there must be a point x_0 inside the interval $(c, d - h)$ at which $g(x_0) = 0$. Evidently $f(x_0 + h) = f(x_0)$. Now there must be a point z inside the interval $(x_0, x_0 + h)$ at which f attains a maximum or minimum; at this point the derivative $f'(z)$ vanishes and we are done.

As Zygmund remarks ([313, Vol I, p. 44]) this argument shows somewhat more. For convenient reference this can be expressed as a theorem too.

Theorem 4.76 (Zygmund) *Let f be continuous and smooth. If $A < C < B$,*

$$\liminf_{h \to 0} (f(\alpha + h) - f(\alpha))/h \leq A$$

and

$$\limsup_{h \to 0} (f(\beta + h) - f(\beta))/h \geq B$$

then there is a point γ between α and β such that $f'(\gamma) = C$.

The following related observation is from Zygmund [312, p. 55].

Theorem 4.77 *If f is continuous and smooth in an interval (a, b) then the set of points at which all four Dini derivatives of f are finite is dense in (a, b) and each of the four Dini derivatives of f has the Darboux property (a, b) .*

The results above, showing that continuous smooth functions have interesting differentiation properties, must be balanced by the fact that the set of points of differentiability may be small. The function

$$f(x) = \sum_{n=1}^{\infty} \frac{1}{\sqrt{n}2^n} \cos 2^n x$$

(see Section 4.4.7) is continuous, smooth (even uniformly smooth) and yet differentiable only in a set of measure zero. In fact the function f here satisfies the rather stronger condition

$$|f(x+t) + f(x-t) - 2f(x)| = O\left(\frac{t}{\sqrt{|\log t|}}\right).$$

We state these observations as a theorem.

Theorem 4.78 *There is a continuous, smooth, almost nowhere differentiable function.*

4.8.3 Makarov's Theorems

The differentiablity properties of continuous, smooth functions have been subjected to further study by Makarov [184]. It was not known until very recently whether a *complex-valued* smooth function had to possess any points of differentiability. Of course if $F = f + ig$ is continuous and smooth then both functions f and g are themselves continuous and smooth. Since the sets of points where f' and g' exist can be small (measure zero) it seems immediately likely that f and g can be chosen in such a way that the two derivatives do not exist simultaneously. But this was not known. The problem is mentioned in a 1989 review of Zygmund's works [314, p. xxix] prepared by Fefferman, Kahane and Stein. Makarov solves the problem. A more elementary attack is due to Ullrich [296] and we outline here his ideas.

Lemma 4.79 (Ullrich) *Let E be a measure zero set of type F_σ. Then there is a 2π-periodic, continuous, uniformly smooth function f so that $f'(x) = +\infty$ for every $x \in E$.*

This lemma is proved in [296] using complex methods. A real variable proof would be welcome. The next lemma, from the same source, follows using familiar techniques for lacunary series.

Lemma 4.80 (Ullrich) *Let $a_j \to 0$ with $a_j \geq 0$ and $\sum_{j=1}^{\infty} a_j^2 = +\infty$. Then*

$$f(x) = \sum_{j=1}^{\infty} 2^{-j} a_j \cos 2^j x \qquad (4.50)$$

is continuous and uniformly smooth and the set

$$E = \left\{ x \ : \ \sup_{h>0} |f(x+h) - f(x)|/h < +\infty \right\} \qquad (4.51)$$

is a measure zero set of type F_σ.

Now if we take f and E as in (4.50) and (4.51) there is, by Lemma 4.79, a continuous, uniformly smooth function g so that $g'(x) = +\infty$ for every $x \in E$. The complex function $F = f + ig$ is uniformly smooth and nowhere differentiable. In fact $|F(x+h) - F(x)|/h \to +\infty$ as $h \to 0+$ at every point x. Thus *complex* smooth functions may not have a derivative at any point in contrast to the situation for real functions. This was first proved by Makarov [184] using martingale methods. Ullrich [296] gives the proof we have sketched here.

Theorem 4.81 (Makarov) *There is a continuous, complex-valued, uniformly smooth function on $\mathbb{R}$ that is nowhere differentiable.*

The second result of Makarov that we wish to cite here concerns the size of the set of points of differentiability. In [184] he proves the following theorem using martingale methods. A real variable proof might be instructive.

Theorem 4.82 (Makarov) *A continuous real function on $\mathbb{R}$ that is uniformly smooth is differentiable on a set of Hausdorff dimension one.*

4.8.4 Measurable, Smooth Functions

Let us now drop the assumption that our smooth functions are also continuous. This then would disallow most of the arguments used so far. Fortunately some of these ideas are salvaged by the fact that a measurable smooth function must be continuous on a dense open set, on each component of which the earlier properties would then hold.

The following theorem summarizes some of the continuity and differentiability properties of measurable, smooth functions.

Theorem 4.83 *Let f be measurable and smooth in an interval (a, b), let C_f denote the set of points at which f is continuous and let D_f denote the set of points at which f is differentiable (finitely). Then*

1. $(a, b) \setminus C_f$ is scattered (and so, in particular, nowhere dense and countable).

2. C_f contains an open set dense in (a, b).

3. f is in the first Baire class on (a, b).

4. D_f is of the power of the continuum in every subinterval of (a, b).

Proof. We already know assertions (1), (2) and (3). Assertion (4) follows from Theorem 4.71: since such a function is continuous on an open dense set G this theorem can be applied to show that D_f is of the power of the continuum in every component of G.

4.8.5 Neugebauer's Example

Let us reproduce an example of Neugebauer [213, p. 27] illustrating a measurable, smooth function f such that, even though f' exists almost everywhere, f' assumes only two distinct values. This then shows that an arbitrary measurable, smooth function does not necessarily have the property of Theorem 4.75 that continuous, smooth functions possess. A smooth continuous function can have no "corners" so that its graph cannot be made up of line segments unless it is itself linear. If we drop the assumption of continuity this can occur as long as we require at each of the jumps that

$$f(x) = \frac{f(x + 0) + f(x - 0)}{2} \tag{4.52}$$

and maintain the same slopes for each piece.

Define the sequence of points $x_1 = 1$, $x_{n+1} = x_n(1 - x_n/2)$. Define the function f so that (i) $f(x) = 0$ if $x \leq 0$, (ii) $f(x) = 2(x - 1)$ if $x > 1$, (iii) $f(x) = x_n^2/2$ if $x = x_n$, and (iv) $f(x) = 2(x - x_{n+1})$ if $x_{n+1} < x < x_n$. Note that the function has one horizontal piece on $(-\infty, 0)$ and line segments with slope 2 in each interval $(1, +\infty)$, (x_2, x_1), (x_3, x_2), (x_4, x_3) .... and satisfies (4.52) at the jumps. Thus f is smooth at every point $x \neq 0$. The segments all lie between the x-axis and the parabola $y = x^2$ and so f must be smooth too at $x = 0$. Finally only the values 0 and 2 occur for the derivative of f even though the derivative exists everywhere except at the points in the sequence $\{x_n\}$.

The figure illustrates the initial steps of the construction.

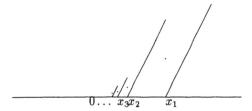

Fig. 4.1. Neugebauer's example.

4.8.6 Neugebauer Alternative

We recall that a continuous, smooth function f must have a derivative on a set D_f that is c–dense and that f' has the Darboux property on this set. In view of the fact that D_f may well have measure zero this fact may seem rather surprising. Neugebauer [213, p. 29] in his 1964 study of smooth functions has clarified this considerably. He shows that if $|D_f| = 0$ then the set

$$\left\{ x \ : \ \overline{D}^+ f(x) = \overline{D}^- f(x) = +\infty \ \text{and} \ \underline{D}^+ f(x) = \underline{D}^- f(x) = -\infty \right\}$$

must be residual and accordingly, by Theorem 4.76, the derivative f' assumes every value on every interval; in this case the function is wildly oscillatory. Thus if the set D_f appears (naively) too small to carry the Darboux property for f' it is only because the function f is so highly oscillatory that the Darboux property holds, but in this sense: f' takes on every value in every interval.

The theorem that we now prove shows that for smooth functions one of these two situations must occur: either the derivative exists almost everywhere in some interval or else the oscillatory behavior is evident. Note that this salvages the Darboux property for measurable (not continuous) smooth functions; while the Darboux property can fail (as in Section 4.8.5) it must hold in the oscillatory case.

Theorem 4.84 (Neugebauer) *Let f be measurable and smooth, let D_f denote the set of points at which f is differentiable and let (a, b) be an arbitrary interval. Then one or other of the following alternatives must hold: either*

(1) *the set D_f has full measure in some subinterval of (a, b), or*

(2) *the set*

$$\left\{ x \ : \ \overline{D}^+ f(x) = \overline{D}^- f(x) = +\infty \ \text{and} \ \underline{D}^+ f(x) = \underline{D}^- f(x) = -\infty \right\}$$

is residual in (a, b) and f' assumes every real value in every subinterval of (a, b).

For the proof of this we require an elementary lemma on the behavior of the Dini derivatives ([213, p. 28]).

Lemma 4.85 *Let f be continuous on an interval (a, b) and suppose that the set*

$$\left\{ x \in (a, b) : \overline{D}^+ f(x) \leq r \right\}$$

is dense in that interval. Then the set

$$\left\{ x \in (a, b) : \underline{D}^+ f(x) > r \right\}$$

is first category.

Proof. To show that the set $\left\{ x \in (a, b) : \underline{D}^+ f(x) > r \right\}$ is first category it is enough to show that for all rationals $s > r$ and all integers i each set

$$E_{si} = \left\{ x \in (a, b) : \frac{f(y) - f(x)}{y - x} > s \text{ if } 0 < y - x < 1/i \right\}$$

is nowhere dense.

To obtain a contradiction suppose that a set E_{si} is dense in a subinterval $(c, d) \subset (a, b)$. Since the set $\left\{ x : \overline{D}^+ f(x) \leq r \right\}$ is dense in that interval we may choose a point x there for which $\overline{D}^+ f(x) \leq r$. If $x < y < x + 1/i$ choose a sequence of points $x_n \searrow x$ with each $x_n \in E_{si} \cap (x, y)$. We have then that $f(y) - f(x_n) > s(y - x_n)$. Since f is continuous this gives $f(y) - f(x) \geq s(y - x)$. As this holds for all $x < y < x + 1/i$ we have proved that $\underline{D}^+ f(x) \geq s > r$ which is the desired contradiction.

Now we return to the proof of Theorem 4.84. Lemma 4.85 and Corollary 4.74 together show that any smooth, measurable function f for which the set D_f has full measure in no subinterval of (a, b) must have each set

$$A_n = \left\{ x \in (a, b) : \underline{D}^+ f(x) > n \right\}$$

and

$$B_n = \left\{ x \in (a, b) : \overline{D}^+ f(x) < n \right\}$$

first category for $n = 1, 2, 3, \ldots$. But then the set

$$\left\{ x \subset (a, b) : \overline{D}^+ f(x) = +\infty \text{ and } \underline{D}^+ f(x) = -\infty \right\}$$

$$= (a, b) \setminus \bigcup_{n=1}^{\infty} (A_n \cup B_n)$$

is residual in (a, b). A similar argument works for the left hand Dini derivatives. Thus if assertion (1) fails then the first part of assertion (2) must be true. Finally should the first part of assertion (2) hold then Theorem 4.76 shows that on each subinterval of (a, b) on which f is continuous (remember that these intervals are dense) the derivative f' assumes every real value and so the second part of assertion (2) holds. This completes the proof of the theorem.

Corollary 4.86 (Neugebauer) *Let f be measurable and smooth. If either of the sets*

$$\left\{ x \ : \ \overline{D}^+ f(x) = +\infty \ \text{and} \ \underline{D}^+ f(x) = -\infty \right\}$$

or

$$\left\{ x \ : \ \overline{D}^- f(x) = +\infty \ \text{and} \ \underline{D}^- f(x) = -\infty \right\}$$

is of the first category in an interval (a, b) then f is differentiable almost everywhere on a dense open subset of (a, b).

Proof. This follows easily from the theorem. If these sets are first category then every subinterval of (a, b) contains a further subinterval in which f is a.e. differentiable.

4.8.7 Approximately Smooth Functions

We say that a measurable function f is *approximately smooth* at a point x if

$$\operatorname{app} \lim_{h \to 0} \frac{f(x+t) + f(x-t) - 2f(x)}{t} = 0.$$

That is to say the set

$$\{t \ : \ |f(x+t) + f(x-t) - 2f(x)| \geq \epsilon |t|\}$$

must have zero density at 0 for every $\epsilon > 0$. We might anticipate that approximately smooth functions would share some of the properties of smooth functions. The first study of such functions is that of Neugebauer [212]. The analogue of Theorem 4.71 is proved in much the same way (see [212, Lemma 4, p. 82]) and the analogue of Theorem 4.75 as well (see [212, Remark, p. 82 and Theorem 3, p. 85] and O'Malley [215, p. 80]).

The approximate continuity hypothesis that we impose in the statement of the next theorem allows us to use the fact that such functions have approximate maxima and minima ([215, Theorem 1, p. 76]) and then the methods we have already employed in Theorem 4.75 can be used with little modification. It is not known whether this assumption might be relaxed in the first part of this theorem. If it is true that a measurable, approximately smooth function must be approximately continuous on a dense open set then we would know that the approximate derivative must exist on a c–dense set of points. In regards to this, note that Evans and Humke [94] give an example of a measurable, approximately smooth function that is approximately discontinuous at an uncountable set of points.

Theorem 4.87 *Let f be a function that is approximately continuous and approximately smooth at each point in (a, b). Then the approximate derivative $AD f(x)$ exists and is finite at a c-dense set of points of that interval. If AD_f denotes the set of points at which f is approximately differentiable (finitely) then $AD f$ has the Darboux property on AD_f.*

As before the set of points AD_f must be c-dense but may well have measure zero. Indeed take a continuous, smooth function f that is almost nowhere differentiable. Then, by Theorem 4.65, the function f is also almost nowhere approximately differentiable.

One additional remark might be in order. Since continuous, approximately smooth functions have some degree of regularity as regards approximate differentiability properties one would expect that the typical continuous function is nowhere approximately smooth (typical in the sense of category as in Section 4.6.9). Evans [86] shows much more. (The odd analogue is given in the same paper and reproduced in Section 7.4.6 in Chapter 7.)

Theorem 4.88 (Evans) *The typical continuous function f in $C[0, 1]$ must satisfy*

$$\text{ap} \lim_{h \to 0} \left| \frac{f(x + t) + f(x - t) - 2f(x)}{t} \right| = +\infty$$

for each $0 < x < 1$.

4.8.8 L_p-Smooth Functions

We say that a measurable function f is L_p-smooth $(p \geq 1)$ at a point x if

$$\left\{ \frac{1}{h} \int_0^h |f(x + t) + f(x - t) - 2f(x)|^p \, dt \right\}^{1/p} = o(h)$$

as $h \to 0$. Again we might anticipate that L_p-smooth functions would share some of the properties of approximately smooth and smooth functions. The first suggestion that this should be the case is apparently due to Zygmund and is reported in Neugebauer [212]. The appropriate sense in which to interpret the derivative is in the L_p-sense. We say that f has an L_p-derivative $(p \geq 1)$ at a point x_0 if

$$\left\{ \frac{1}{2h} \int_{-h}^h |f(x_0 + t) - a_0 - a_1 t|^p \, dt \right\}^{1/p} = o(h)$$

as $h \to 0$. The numbers a_0 and a_1 in this case would be unique and we will write L_p-D $f(x_0) = a_1$.

The connection between this derivative and the approximate derivative is given in [212, Lemma 7, p. 85]).

Theorem 4.89 *Let f be measurable. Then at almost all points x at which the L_p-derivative of f exists, L_p-$Df(x) = ADf(x)$. If f is moreover continuous then at every x at which the L_p-derivative of f exists, L_p-$Df(x) = ADf(x)$.*

Neugebauer obtains the following analogue of Theorem 4.71 ([212, Theorem 2, p. 84]).

Theorem 4.90 *Let f be a measurable function that is L_p-smooth on an interval (a, b). Then f is continuous on a dense open subset of (a, b) and the L_p-derivative of f exists and is finite at a c–dense set of points of that interval.*

Similarly the analogue of Theorem 4.75 is obtained ([212, Theorem 3, p. 85]).

Theorem 4.91 *Suppose that f is continuous and L_p-smooth in (a, b) and let S denotes the set of points at which f has an L_p-derivative. Then L_p-$Df(x)$ has the Darboux property on S.*

More of the results of Section 4.8.4 may be extended to this setting; see [212, Theorem 4, p. 86]).

We conclude with two further extensions of ideas we have seen to the study of L_p-smooth functions from O'Malley [216, p. 191]. The first is the analogue of the Auerbach Theorem (Theorem 4.41) and can be proved in much the same manner. The second is a version of Theorem 4.75 and, by familiar methods (along with a property of Darboux Baire∗1 functions [216, p. 189]), extends Theorem 4.91 just mentioned.

Theorem 4.92 (O'Malley) *Suppose that f is a measurable function that is L_p-smooth in (a, b). Then f is Baire∗1.*

Theorem 4.93 (O'Malley) *Suppose that f is a measurable function, Darboux continuous and L_p-smooth in (a, b). Let S denote the set of points at which f has an L_p-derivative. Then L_p-$Df(x)$ has the Darboux property on S.*

4.8.9 Discontinuities of L_p-Smooth Functions

We recall that measurable, smooth functions are mostly continuous; the set of discontinuities of such a function must be scattered. We know from Theorem 4.90 that the set of discontinuities of an L_p-smooth function must be nowhere dense. Neugebauer [212, Theorem 5, p. 87]) shows that this is the best that can be said.

Theorem 4.94 (Neugebauer) *Let C be a closed, nowhere dense subset of $(0,1)$ and let $p \geq 1$. Then there exists a bounded, measurable function that is L_p-smooth in $(0,1)$ and f is discontinuous precisely at the points in C.*

Perhaps a more natural question to ask in this context is not how small the set of discontinuities of a measurable function that is L_p-smooth must be but how small is the set of L_p-discontinuities. We say that f is L_p-continuous $(p \geq 1)$ at a point x_0 if

$$\left\{ \frac{1}{2h} \int_{-h}^{h} |f(x_0 + t) - f(x_0)|^p \, dt \right\}^{1/p} = o(1)$$

as $h \to 0$. One might expect that the set L_p-discontinuities of an L_p-smooth function must be countable but this is not so. Evans and Humke [91] have exhibited an L_p-smooth function that is approximately discontinuous at an uncountable set. See also the discussion in Neugebauer [212, pp. 88–89] for some related ideas.

4.8.10 Discontinuities of approximately smooth functions

We know that a measurable, approximately smooth function is Baire 1. It follows that the set of discontinuities is small in the sense of category. Can more be said?

Because of Theorem 4.94 we see that no measure theoretic assertion is possible. Given any nowhere dense, closed set C (even of positive measure) there is a bounded, measurable, approximately smooth function discontinuous precisely at the points in C. But, along with Evans [84], we might ask whether a measurable, approximately smooth function f is necessarily Baire∗1. This is strongly suggested by the fact that every smooth function, every symmetrically differentiable function and every approximately differentiable function has this property. While the answer is not yet known Larson [176] has shown that this is so under the extra hypothesis that f is approximately continuous.

We state and prove the theorem for approximate smoothness but it can easily be amended to cover the case that $f(x+h) + f(x-h) - 2f(x)$ is approximately $O(h^\alpha)$ for any $\alpha > 0$ as in the Auerbach Lemma (Lemma 4.40). The remarks following the proof show that the case $\alpha = 0$ cannot be included.

Theorem 4.95 (Larson) *Let f be approximately continuous and approximately smooth at each point. Then f is Baire∗1.*

Proof. To prove this we shall exhibit a sequence of closed sets $\{C_n\}$ so that $\mathbb{R} = \bigcup_{n=1}^{\infty} C_n$ and f is continuous relative to each C_n. Write Q_n for the set of all $x \in \mathbb{R}$ so that $0 < \eta < 1/n$ implies that

$$\left| \{ h \in (0, \eta) \; : \; |f(x+h) + f(x-h) - 2f(x)| \geq 2h \} \right| < \eta/2. \tag{4.53}$$

By the smoothness assumption on f it is clear that $\mathbb{R} = \bigcup_{n=1}^{\infty} Q_n$. Note that, as a particular consequence of (4.53), for any $x \in Q_n$ and all sufficiently small $\eta > 0$,

$$\left| \{ h \in (0, \eta) \; : \; f(x+h) - f(x) \leq -\eta \text{ and } f(x-h) - f(x) \leq -\eta \} \right| < \eta/2 \tag{4.54}$$

since for any h in the set in (4.54) one has

$$f(x+h) - f(x) + f(x-h) - f(x) \leq -2\eta < -2h$$

and such an h belongs to the set in (4.53).

Let $C_n = \overline{Q}_n$. Then $\mathbb{R} = \bigcup_{n=1}^{\infty} C_n$ and each C_n is closed. We show that f is continuous relative to each set C_n. For this it is sufficient that we establish

$$\lim_{x \to x_0, \; x \in Q_n} f(x) = f(x_0) \tag{4.55}$$

for each $x_0 \in C_n$.

If (4.55) fails then there exists a positive number δ and a sequence $x_m \to x_0$, $x_m \in Q_n$, so that

$$f(x_m) - f(x_0) > 2\delta \qquad (\text{or} < -2\delta) \tag{4.56}$$

for all m. Of the two possibilities for the inequality let us assume that it is the former and derive a contradiction. In the case of the opposite inequality a similar proof is available (but using a different version of (4.54) of course).

Write $\eta_m = |x_m - x_0|$ so that $\eta_m \to 0$. For any $\eta > 0$ write

$$A(\eta) = \{ y \; : \; |y - x_0| < \eta, \; |f(y) - f(x_0)| \geq \delta \}. \tag{4.57}$$

Since f is approximately continuous at each point we know that $|A(\eta)| = o(\eta)$ as $\eta \to 0$.

Suppose that $0 < h < \eta_m < \delta$. Note that if

$$f(x_m + h) > f(x_m) - \eta_m$$

then, by (4.56),

$$f(x_m + h) > f(x_0) + 2\delta - \eta_m > f(x_0) + \delta$$

so that $x_m + h \in A(\eta_m)$. Similarly if

$$f(x_m - h) > f(x_m) - \eta_m$$

then $x_m - h \in A(\eta_m)$. For large enough m, $|A(\eta_m)| < \frac{1}{4}\eta_m$ and consequently, for such m,

$$\left|\{h \in (0, \eta_m) : f(x_m + h) - f(x_m) > -\eta_m \text{ or } f(x_m - h) - f(x_m) > -\eta_m\}\right|$$

$$< \eta_m/4. \tag{4.58}$$

But each $x_m \in Q_n$ and (4.58) directly contradicts (4.54) since together the two sets make up all of $(0, \eta_m)$. From this contradiction the proof follows.

Larson [176] gives an example to show that the theorem is false for functions that are approximately continuous and symmetric; he produces a simple construction of a measurable, symmetric function f that is everywhere approximately continuous and such that the set of points where f is discontinuous is precisely the set of rationals. Thus, while f is Baire 1, it cannot be Baire∗1 as it has a dense set of points of discontinuity.

4.9 Super–Smooth Functions

The condition that $f(x + t) + f(x - t) - 2f(x) = o(t)$ was termed smoothness by Zygmund. In order to have some suitable language for the weaker condition

$$f(x + t) + f(x - t) - 2f(x) = O(t) \quad (h \to 0)$$

we have followed a number of authors in calling this quasi-smoothness. We now proceed to investigate stronger conditions of the form

$$f(x + t) + f(x - t) - 2f(x) = O(t\delta(t))$$

where $\delta(t)$ tends monotonically to 0. Again in order to have a suitable language to express our problems let us say that f is *super-smooth* at x if such a condition holds for a function δ tending to zero sufficiently rapidly that

$$\int_0^1 \frac{\delta^2(t)}{t}\, dt < +\infty.$$

Note that $\delta(t) = t^\epsilon$ for $\epsilon > 0$ or $\delta(t) = (\log|t|)^{-\beta}$ for $\beta > \frac{1}{2}$ are functions with this property. We have already seen a condition of this kind in Section 4.4.4 although there it was required to hold uniformly. The terminology is introduced solely for the purpose of collecting together some related results and the reader is not encouraged to use the terminology elsewhere.

Conditions of this type have been much studied in the literature. While we know that smooth functions may be nowhere differentiable the present condition is sufficient to guarantee the differentiability of f almost everywhere at which the condition holds.

4.9.1 Differentiability a.e.

As we have seen in Section 4.8 a function may be measurable and smooth and yet have a derivative only on a set of measure zero. This contrasts sharply with the situation for odd differentiability; by Khintchine's Theorem a function that is differentiable in the odd sense (i.e. symmetrically differentiable) is almost everywhere differentiable. For the same conclusion among the even growth conditions we need something a bit stronger. We shall prove by elementary means the following theorem in this section. Sharper results will be cited later.

Theorem 4.96 *Let f be a measurable function and suppose, at every point x in a measurable set E, that*

$$\lim_{h \to 0} \left| \frac{f(x+t) + f(x-t) - 2f(x)}{t^\alpha} \right| < +\infty. \tag{4.59}$$

where $\alpha > 1$. Then f is differentiable at almost every point of E.

The proof of Theorem 4.96 is given in a series of lemmas.

Lemma 4.97 *Let f be a measurable function such that the condition (4.59) holds at every point x in a measurable set E. Then f satisfies the condition*

$$\limsup_{t \to 0} \left| \frac{f(x+2t) - 2f(x+t) + f(x)}{t^\alpha} \right| < +\infty \tag{4.60}$$

at almost every point of E.

Proof. The proof follows techniques in [313, Vol. II, pp. 78–79]. There, however, the measurability questions have not been addressed (cf. the remark on page 136). We handle this with a device from Fejzić and Weil [101] who have clarified these issues.

The sets

$$E_{mn} = \{x \in E : |f(x+t) + f(x-t) - 2f(x)| < mt^\alpha,\ 0 < t < 1/n\}$$

cover E. Thus it is enough to show that (4.60) holds at every (outer) density point of a set E_{mn}. Let us simplify by taking 0 as that density point and arguing just on the right.

Define the set

$$S_1 = \{u \in [t, 2t] : |f(t+u) - 2f(\tfrac{1}{2}u + \tfrac{1}{2}t) + f(0)| \le m|t|^\alpha\} \tag{4.61}$$

Since f is meaurable and t fixed the set S_1 is measurable. Moreover, as E_{mn} has zero as a point of density the set $(2E_{mn} - t) \cap (t, 2t)$ has measure approaching t as $t \searrow 0$. But notice that

$$S_1 \supset (2E_{mn} - t) \cap (t, 2t)$$

for small enough t since any point $u \in (2E_{mn} - t) \cap (t, 2t)$ has $\frac{1}{2}(u + t) \in E_{mn}$ and hence the inequality in (4.61) must hold. Thus S_1 is a measurable subset of $[t, 2t]$ and has measure approaching t for small values of t.

Similarly write

$$S_2 = \{u \in [t, 2t] : |f(2u) - 2f(u + t) + f(2t)| \le m|t|^\alpha\} \qquad (4.62)$$

and

$$S_3 = \{u \in [t, 2t] : |f(u) - 2f(\tfrac{1}{2}u + \tfrac{1}{2}t) + f(t)| \le m|t|^\alpha\}. \qquad (4.63)$$

Again, much as before, for small enough t we have the inclusions

$$S_2 \supset (E_{mn} - t) \cap [t, 2t]$$

and

$$S_3 \supset (2E_{mn} - t) \cap [t, 2t].$$

Each of these sets is a measurable subset of $[t, 2t]$ and has measure approaching t for small values of t. Consequently for small t the set $S_1 \cap S_2 \cap S_3$ is nonempty.

Let u be a point in this intersection. We now express

$$
\begin{aligned}
f(2t) - 2f(t) + f(0) = {} & 2\left[f(u + t) - 2f\left(\frac{u + t}{2}\right) + f(0)\right] \\
& + [f(2u) - 2f(u + t) + f(2t)] \\
& - 2\left[f(t) - 2f\left(\frac{u + t}{2}\right) + f(u)\right].
\end{aligned}
$$

Each of the three terms on the right of this identity is bounded in absolute value by $m|t|^\alpha$; thus

$$|f(2t) - 2f(t) + f(0)|$$

has a bound of the form $C|t|^\alpha$ for small enough t and this is exactly the condition (4.60) that we wished to establish. At least this is the right hand version; as the left hand version is similar the lemma is proved.

Lemma 4.98 *Let f be a function that is bounded in a neighbourhood of a point x and satisfies the condition (4.60) at the point x for some $\alpha > 1$. Then*

$$\limsup_{h \to 0} \left|\frac{f(x + h) - f(x)}{h}\right| < +\infty. \qquad (4.64)$$

Proof. Suppose that at a point x_0 we have

$$|f(x_0 + 2t) - 2f(x_0 + t) + f(x_0)| \le C|t|^\alpha \qquad (4.65)$$

for all $0 < |t| < 2\delta$ and f is bounded by M in $(x_0 - \delta, x_0 + \delta)$. To simplify the computations let us suppose that $x_0 = 0 = f(x_0)$ and argue just on the right. Then

$$\left| f(t) - 2f\left(\frac{t}{2}\right) \right| \leq C' |t|^\alpha \tag{4.66}$$

for a constant C' and all $0 < t < \delta$. Thus for any $i = 0, 1, 2, \ldots n - 1$ we have

$$\left| f\left(\frac{h}{2^i}\right) - 2f\left(\frac{h}{2^{i+1}}\right) \right| \leq C' |h|^\alpha 2^{-i\alpha}. \tag{4.67}$$

Consequently

$$\left| f(h) - 2^n f\left(\frac{h}{2^n}\right) \right| \leq \sum_{i=0}^{n-1} 2^i \left| f\left(\frac{h}{2^i}\right) - 2f\left(\frac{h}{2^{i+1}}\right) \right|$$

$$\leq C' |h|^\alpha \frac{1}{1 - 2^{-\alpha}}.$$

Let t be any number in $(0, \delta/2)$. Then $t = h/2^n$ for some $h \in (\delta/2, \delta)$ and some n and this gives, from the above computations,

$$\left| \frac{f(t)}{t} \right| \leq \frac{2M}{\delta} + C' \delta^{\alpha-1} \frac{1}{1 - 2^{-\alpha}}.$$

which is the boundedness conclusion that we require.

Now the proof of Theorem 4.96 is easily completed. If the condition (4.59) holds everywhere in E then f is bounded in a neighbourhood of almost every point x (Lemma 4.30) and hence satisfies (4.64) at almost every point in E. Now by classical material (Saks [251, p. 230 and p. 234]) the function f must be a.e. differentiable in E as required.

4.9.2 Integral of Marcinkiewicz

The condition of super-smoothness is very close to being a characterization of a.e. differentiability. The true picture is due to Zygmund and his collaborators starting with an idea of Marcinkiewicz. Following the work that Zygmund did with Mary Weiss (sketched in Section 4.4.4) he and Stein, in a now celebrated paper [274], gave a complete solution to this problem; the earlier related paper [275] should be consulted too. This has now generated a large literature and given insight into many areas. The ideas can be traced also in Calderón and Zygmund [39]. The treatise of Stein [272] contains a full account; the historical survey Stein [273] explains well the source of these ideas and places them in a larger context.

We content ourselves with citing just one version of these theorems from [272, p. 262]; the version there is stated in higher dimensions and the version in [274] is stated for higher order symmetric differences as well.

Theorem 4.99 *Let f be a measurable function defined in an open neighbourhood of every point of a set E. Then f has a derivative in the ordinary sense at almost every point of E if and only if f is quasi-smooth at almost every point of E and for almost every $x \in E$*

$$\int_0^\eta \frac{|f(x+t) + f(x-t) - 2f(x)|^2}{t^3} \, dt < +\infty$$

where η is a sufficiently small positive number (depending on x).

4.9.3 Theorem of Denjoy

We have just seen in Section 4.9.1 that, $\alpha > 1$, the condition

$$f(x+t) + f(x-t) - 2f(x) = O(t^\alpha) \tag{4.68}$$

as $t \to 0$ is enough for the a.e. differentiability of f. This condition with $\alpha = 2$ was first studied by Denjoy. He announced [67, p. 1220] in 1921 that a measurable function f possessing (4.68) at every point of a measurable set E must have, at almost every point x in E, a second order Peano derivative $f_{(2)}(x)$ that is the approximate derivative of f'. A complete proof was published in his memoir [68]. Before the appearance of that proof a more general theorem for all higher order symmetric derivates was proved by Marcinkiewicz and Zygmund [186]. In some ways this may be thought of as a second order analogue of the Khintchine Theorem.

The clearest exposition of the proof appears in Zygmund [313, Vol II, pp. 78–80] and it is this that we mainly follow. The proof in [186] handles the full general case and so requires a deeper argument.

We say that a measurable function f has a *second order Peano derivative* at a point x_0 if there are numbers $f_{(0)}(x_0)$, $f_{(1)}(x_0)$ and $f_{(2)}(x_0)$ so that

$$f(x_0 + t) = f_{(0)}(x_0) + f_{(1)}(x_0)t + \left(f_{(2)}(x_0) + \epsilon_{x_0}(t)\right)\frac{t^2}{2!}$$

where $\epsilon_{x_0}(t)$ tends to 0 as $t \to 0$. If f is continuous at x_0 then certainly the number $f_{(0)}(x_0)$ is just $f(x_0)$. If in addition f is differentiable at x_0 then the number $f_{(1)}(x_0)$ is $f'(x_0)$. Finally if f is twice differentiable at this point then $f_{(2)}(x_0)$ is $f''(x_0)$. Since f is assumed measurable all the functions which appear may be shown to be measurable too.

Theorem 4.100 (Denjoy) *Suppose that f is measurable and that*

$$\limsup_{h \to 0} \left| \frac{f(x+h) + f(x-h) - 2f(x)}{h^2} \right| < +\infty$$

at every point x of a measurable set E of positive measure. Then at almost every point x of E the function f is differentiable, the second order Peano derivative $f_{(2)}(x)$ exists and

$$AD f'(x) = f_{(2)}(x).$$

We break the proof into a series of lemmas that reveal the structure of the argument. The first lemma we do not prove; it is a special case of a result of Marcinkiewicz and Zygmund. A proof can be found in [185] and in [313, Vol. II, pp. 73–80]. With this lemma it is clearly just a matter of translating the symmetric condition to this unsymmetrical one.

Lemma 4.101 *Let f be measurable and suppose that at each point x in a measurable set E*

$$f(x+t) = f(x) + f'(x)t + O(t^2)$$

as $t \to 0$. Then at almost every point x of E the second order Peano derivative $f_{(2)}(x)$ exists.

Lemma 4.102 *Let f be measurable and suppose that, at each point x in a measurable set E, the derivative $f'(x)$ and the second order Peano derivative $f_{(2)}(x)$ exist. Then at almost every point $x \in E$*

$$AD f'(x) = f_{(2)}(x).$$

Proof. The argument is taken mostly from [186, p. 19]. (A simpler argument is given in [313, vol. II, p. 77] but it appeals to a "splitting" theorem of Marcin-kiewicz.) We may suppose that at every point $x \in E$

$$f(x+t) = f(x) + f'(x)t + \left(f_{(2)}(x) + \epsilon_x(t) \right) \frac{t^2}{2!}$$

where $\epsilon_x(t)$ tends to 0 as $t \to 0$. Let $P \subset E$ be a subset chosen so that $f_{(2)}(x)$ is continuous relative to P and so that $\epsilon_x(t) \to 0$ uniformly for $x \in P$. It is enough to show that, at any density point x of P, $AD f'(x) = f_{(2)}(x)$. Almost every point of E is a density point of such a set P.

Suppose x is a density point of P and that $x + t \in P$. Then

$$f(x+t) = f(x) + f'(x)t + f_{(2)}(x)t^2/2 + \epsilon_x(t)t^2/2, \qquad (4.69)$$

$$f(x+2t) = f(x+t) + f'(x+t)t + f_{(2)}(x+t)t^2/2 + \epsilon_{x+t}(t)t^2/2, \qquad (4.70)$$

and
$$f(x + 2t) = f(x) + 2f'(x)t + 2f_{(2)}(x)t^2 + 2\epsilon_x(t)t^2. \tag{4.71}$$

Equations (4.69) and (4.71) together imply that
$$\frac{f(x + 2t) - 2f(x + t) + f(x)}{t^2} \to f_{(2)}(x).$$

Equations (4.69) and (4.70) imply that
$$\frac{f'(x + t) - f'(x)}{t} = \frac{f(x + 2t) - 2f(x + t) + f(x)}{t^2}$$
$$+ \left(f_{(2)}(x + t) - f_{(2)}(x) \right) + (\epsilon_{x+t}(t) - \epsilon_x(t)).$$

Thus the limit of the quotient $(f'(x + t) - f'(x))/t$ as $t \to 0$ with $x + t \in P$ must be $f_{(2)}(x)$ by the way in which P has been chosen. This completes the proof of the lemma.

The third lemma is just a repeat of Lemma 4.97.

Lemma 4.103 *Let f be a measurable function and E a measurable set such that*
$$f(x + t) + f(x - t) - 2f(x) = O(t^2)$$
holds for each $x \in E$ as $t \to 0$. Then
$$f(x + 2t) - 2f(x + t) + f(x) = O(t^2)$$
at almost every point of E.

The final lemma now allows us to convert the condition of the theorem to that of Lemma 4.101.

Lemma 4.104 *Let f be a measurable function and E a measurable set such that*
$$f(x + 2t) - 2f(x + t) + f(x) = O(t^2)$$
holds for each $x \in E$ as $t \to 0$. Then at almost every point x in the set E
$$f(x + t) = f(x) + f'(x)t + O(t^2)$$
as $t \to 0$.

Proof. We can argue very much as in the proof of Lemma 4.98. As there suppose that at a point x_0 we have
$$|f(x_0 + 2t) - 2f(x_0 + t) + f(x_0)| \le C|t|^2$$

for all $0 < |t| < 2\delta$. We know that f is almost everywhere differentiable so let us suppose that $f'(x_0)$ exists. To simplify the computations let us suppose that $x_0 = 0 = f(x_0)$ and argue just on the right.

Then

$$\left| f(t) - 2f\left(\frac{t}{2}\right) \right| \le C'|t|^2$$

for a constant C' and all $0 < t < \delta$. Thus, for any $i = 0, 1, 2, \ldots n-1$, we have

$$\left| f\left(\frac{h}{2^i}\right) - 2f\left(\frac{h}{2^{i+1}}\right) \right| \le C'|h|^2 2^{-2i}.$$

Thus we have again, exactly as in the proof of Lemma 4.98, an inequality

$$\left| f(h) - 2^n f\left(\frac{h}{2^n}\right) \right| \le Kh^2$$

for some constant K, all sufficiently small h and all n. Since f is differentiable at $x_0 = 0$ and $f(x_0) = 0$ we see that

$$\lim_{n\to\infty} 2^n f\left(\frac{h}{2^n}\right) = hf'(x_0).$$

Letting $n \to \infty$ in the above inequality and interpreting this, we have shown that

$$|f(x_0 + h) - f(x_0) - hf'(x_0)| \le Kh^2.$$

This condition holds at almost every point of E and this is exactly as we were required to prove in the lemma.

The theorem now follows since, if f satisfies the condition of the theorem at every point of E, then, by Lemma 4.103 and Lemma 4.104, it satisfies the condition of Lemma 4.101 and hence of Lemma 4.102 at almost every point of E.

5

Monotonicity

5.1 Introduction

A function with a positive derivative is increasing; a function with a positive
second derivative is convex. These facts from the elementary calculus have been
generalized countless times. This chapter contains an account of some monotonicity and convexity theorems within the context of the first and second symmetric derivatives.

The basic monotonicity theorem for the symmetric derivative is discussed in
a number of variants in the first few sections. Briefly and roughly it asserts that
a function with a positive symmetric derivative is increasing off of a countable
set. In Section 5.3.3 a proof of the Charzyński theorem is given based on the
monotonicity results.

The story of the monotonicity theorem for the approximate symmetric
derivative is rather more curious. In the literature can be found several articles each claiming a proof of some such monotonicity theorem. It was well
known in real analysis circles by the 1980's that these proofs were incorrect
and many people tried to find a correct proof. Interesting contributions were
made by Foran and Larson [107] and Matousek [195]. The theorem was finally
established by Freiling and Rinne [112]. This is presented in Section 5.6.

Finally a number of convexity theorems are discussed in Section 5.7. Convexity theorems are a natural companion for monotonicity theorems in general;
here their discussion is particularly natural since they can be obtained from
assertions about the second symmetric derivative just as our monotonicity theorems derive from assertions about the first symmetric derivative.

In an appendix to this chapter (Section 5.8.1) there is a brief rundown of
the history of monotonicity theorems for the symmetric derivative. It has taken

some time for the real nature of these theorems to emerge and the reader may be interested to see the development.

5.2 Some Basic Monotonicity Theorems

Our starting point for a discussion of monotonicity theorems for the symmetric derivative is a simple and elegant result that by itself illustrates the nature of such theorems and anticipates the later versions. It is a direct application of the covering Theorem 3.13. Here the assertion is that a function that has everywhere a positive lower symmetric derivative is mostly increasing. We shall prove more refined versions later but these serve as the models.

Theorem 5.1 *Let the function f have a lower symmetric derivate that is positive everywhere. Then f is increasing off some splattered set. More precisely there is an increasing function g such that the sets*

$$\{x \,:\, f(x) > g(x)\} \quad and \quad \{x \,:\, f(x) < g(x)\}$$

are, respectively, right scattered and left scattered.

Proof. Under this assumption the collection

$$V = \{(x, t) \,:\, f(x + t) - f(x - t) > 0, \ t > 0\}$$

is a transitive, full symmetric cover of the real line; if an interval $[a, b]$ belongs to V then certainly $f(a) < f(b)$. It is easy to see, then, that the theorem follows immediately from the covering theorem just cited.

There are a number of further corollaries that can be drawn here. If the function f is continuous and everywhere $\underline{SD} f(x)$ is positive then it must be increasing everywhere; even if not continuous it is increasing on its set of points of continuity.

Corollary 5.2 *Let the function f have a positive lower symmetric derivate everywhere. Then f is increasing on the set of its points of continuity.*

By adding a condition on the upper symmetric derivate to the hypothesis of Theorem 5.1 we can sharpen somewhat the assertion of that theorem. The exceptional set is now, not just splattered, but scattered. This follows immediately from the theorem together with an appeal to the Theorem of Charzyński; under the conditions here the function f is continuous off of a scattered set. This has been stated before in Section 2.5.1.

Corollary 5.3 *Let the function f satisfy*

$$0 < \underline{SD}\, f(x) \le \overline{SD}\, f(x) < +\infty$$

everywhere. Then f is increasing off some scattered set.

The Uher covering theorems (Theorems 3.25 and 3.26) permit similar monotonicity theorems to be derived.

Theorem 5.4 *Let the function f have a lower symmetric derivate that is positive everywhere on a set E that has the Baire property. Then there is an open set G so that $E \setminus G$ is first category and f is increasing on each component of G.*

Proof. As before

$$V = \{(x,t) : f(x+t) - f(x-t) > 0, \ t > 0\}$$

is a transitive, full symmetric cover E. By the transitivity $V = V^5$. By Theorem 3.25 there is an open set G so that $E \setminus G$ is first category and for every interval $[x-t, x+t] \subset G$, $(x,t) \in V$ so that $f(x-t) < f(x+t)$. It follows that f is increasing on each component of G.

Theorem 5.5 *Let the function f have a lower symmetric derivate that is positive everywhere on a measurable set E of positive measure. Then f is locally increasing at almost every point $x \in E$.*

Proof. Once again

$$V = \{(x,t) : f(x+t) - f(x-t) > 0, \ t > 0\}$$

is a transitive, full symmetric cover of E. By Theorem 3.26, for almost every $x \in E$, there is an open interval G_x containing x so that for every interval $[y, x]$ or $[x, z] \subset G_x$, $f(y) < f(x) < f(z)$. By definition then f is locally increasing at each such point x.

5.3 Splattered and Scattered Versions

If we have a lower symmetric derivate everywhere nonnegative (rather than positive) then the best we can conclude from a direct and elementary appeal to the covering lemmas is that f is nondecreasing off some countable set.

The argument would go like this. For each n the collection

$$V = \{(x,t) : (f(x+t) - f(x-t))/2t > -n^{-1}\}$$

produces a countable set C_n so that $f(x) + x/n$ is increasing off C_n; thus f is nondecreasing off $C = \bigcup_{n=1}^{\infty} C_n$. Thus we have proved: *if the function f has a nonnegative lower symmetric derivate everywhere then f is nondecreasing off some countable set.* We seem to have lost the scattered-splattered nature of the countable set by taking the union $C = \bigcup_{n=1}^{\infty} C_n$. Fortunately the result can be salvaged.

We prove the scattered and splattered variants here. They follow too from more general material in Sections 5.3.1 and 5.3.2 below but the presentation here is more transparent and can be used to introduce those ideas.

Theorem 5.6 (Frieling) *Let the function f have a lower symmetric derivate that is nonnegative everywhere. Then f is nondecreasing off some splattered set.*

We know from the arguments just sketched above that f is nondecreasing off some countable set and it remains to show that that set is in fact splattered. This will follow from the following lemma (reproduced from [117]) which is of independent interest.

Lemma 5.7 *Let f be a real function and N a countable set. Suppose that f is increasing [nondecreasing] on $\mathbb{R} \setminus N$ and that*

$$-\infty < \underline{SD} f(x)$$

at every point $x \in \mathbb{R} \setminus N$. Then f is increasing [nondecreasing] off a splattered set.

Proof. We give just the "nondecreasing" proof; the "increasing" proof is similar. Let Z denote the set of points of the form $(x + y)/2$ for x, $y \in N$ and write $Z \cup N = \{z_1, z_2, z_3, \ldots\}$. To start our Baire category argument write E_n for the set

$$\{z_1, z_2, z_3, \ldots z_n\} \cup \{x \notin Z \cup N : -nt < f(x+t) - f(x-t) \text{ if } 0 < t < n^{-1}\}.$$

These sets cover the real line and are expanding so the scattered Baire theorem (Theorem A.8) may be applied. Let G_η denote the open sets and let R and L denote the set of right and left endpoints from that application. We claim that f is nondecreasing off of the splattered set $R \cup L$.

To show this it is enough to show that for any $x_1 < x < x_2$ with x_1, $x_2 \notin N$ and $x \notin R \cup L$ the inequality $f(x_1) \leq f(x) \leq f(x_2)$ must hold. This is already true if $x \notin N$ so we may suppose that $x \in N$. There is a first ordinal so that $x \in G_{\eta+1}$ but $x \notin G_\eta$. Let E_i be dense in $(a_\eta, b_\eta) \setminus G_\eta$.

Recall that x is not in either R or L; let us use first the fact that it is not in L. Then for any $0 < \delta$ the set $(x, x + \delta) \setminus G_\eta$ must be infinite and so also E_i is infinite in $(x, x + \delta)$.

Choose a point z from $E_i \cap (x, x + \frac{1}{2}(x_2 - x))$ with $x < z < x + \delta$ so that $z \neq z_j$ for any $j \leq i$. As $x \in N$ and $z \neq z_j$ it follows that $2z - x$ cannot be in N so $f(2z - x) \leq f(x_2)$ and, as $z \in E_i$, hence $-i(z - x) < f(2z - x) - f(x)$. This proves that $f(x) < f(x_2) + i\delta$. If we let $\delta \to 0$ we see that $f(x) \leq f(x_2)$.

Similarly, using the fact that x is not in R we can obtain that $f(x_1) \leq f(x)$ and the lemma is proved.

A semi-scattered version (analogous to Corollary 5.3) can be proved in much the same way. Alternatively, of course, we might simply appeal to the Charzyński theorem, but that theorem seems rather deeper and less elementary than this. In fact as we shall see in Section 5.3.3 the two are very closely related and will lead to a "monotonicity" proof for the Charzyński theorem.

Theorem 5.8 (Frieling) *Let the function f satisfy*

$$0 \leq \underline{SD} f(x) \leq \overline{SD} f(x) < +\infty$$

everywhere. Then f is nondecreasing off some scattered set.

We already know that f is nondecreasing off some countable set; thus the proof is complete if we show that an analogue of Lemma 5.7 holds.

Lemma 5.9 *Let f be a real function and N a countable set. Suppose that f is increasing [nondecreasing] on $\mathbb{R} \setminus N$ and that*

$$-\infty < \underline{SD} f(x) \leq \overline{SD} f(x) < +\infty$$

at every point $x \in \mathbb{R} \setminus N$. Then f is increasing [nondecreasing] off a scattered set.

Proof. Again we give just the "nondecreasing" proof. Let Z denote the set of points of the form $(x + y)/2$ for $x, y \in N$ and write $Z \cup N = \{z_1, z_2, z_3, \ldots\}$. To start our Baire category argument write E_n for the set

$$\{z_1, z_2, z_3, \ldots z_n\} \cup \{x \notin Z \cup N : -nt < f(x+t) - f(x-t) < nt \text{ if } 0 < t < n^{-1}\}.$$

These sets cover the real line and are expanding so the scattered Baire theorem (Theorem A.8) may be applied. Let G_n denote the open sets and let R and L denote the set of right and left endpoints from that application. We claim that f is nondecreasing off of the scattered set $R \cap L$.

To show this it is enough to show that for any $x_1 < x < x_2$ with $x_1, x_2 \notin N$ and $x \notin R \cap L$ the inequality $f(x_1) \leq f(x) \leq f(x_2)$ must hold. This is already true if $x \notin N$ so we may suppose that $x \in N$. There is a first ordinal so that $x \in G_{\eta+1}$ but $x \notin G_\eta$. Let E_i be dense $(a_\eta, b_\eta) \setminus G_\eta$. As x is not in both R and

L let us suppose that it is not in L. Then for any $0 < \delta$ the set $(x, x + \delta) \setminus G_\eta$ must be infinite and so also E_i is infinite in $(x, x + \delta)$.

Choose a point z from $E_i \cap (x, x + \frac{1}{2}(x_2 - x))$ with $x < z < x + \delta$ so that $z \neq z_j$ for any $j \leq i$. As $x \in N$ and $z \neq z_j$ it follows that $2z - x$ cannot be in N. So $f(2z - x) \leq f(x_2)$ and, as $z \in E_i$, hence $-i(z - x) < f(2z - x) - f(x)$. This proves that $f(x) < f(x_2) + i\delta$. If we let $\delta \to 0$ we see that $f(x) \leq f(x_2)$.

Select $x_1' \in (x_1, x) \setminus N$ and, for any $\delta > 0$, choose a point z from $E_i \cap (x, x+\delta)$ so that $z \neq z_j$ for any $j \leq i$. Again, as $x \in N$ and $z \neq z_j$ it follows that $2z - x$ cannot be in N so $f(x_1) \leq f(x_1') \leq f(2z - x)$. Also, since $z \in E_i$, $f(2z-x)-f(x) < i(z-x) < i\delta$. Letting $\delta \to 0$ shows that $f(x_1) \leq f(x_1') \leq f(x)$.

Summarizing these two inequalities we have that, for any $x_1 < x < x_2$ with $x_1, x_2 \notin N$ and $x \notin R \cap L$, the inequality $f(x_1) \leq f(x) \leq f(x_2)$ must hold. It follows that f is nondecreasing off of the set $R \cap L$ and the lemma is proved.

5.3.1 Freiling Semi-Scattered Theorem

The monotonicity theorems of the preceding section can be generalized further still. As is usual in ordinary monotonicity theorems the condition $\underline{SD}\, f(x) \geq 0$ holding at every point can be relaxed to an "almost everywhere" statement if it is assumed that $\underline{SD}\, f(x) > -\infty$ holds everywhere. The theorem can also be developed within the setting of interval functions thus allowing a greater applicability.

An interval function F is simply a real-valued function assigning a number $F(I)$ to every compact, nonempty interval $I = [a, b]$. We think of it too as a function of two variables, $F(a, b)$. It may be defined on the upper half-plane or for all pairs of real numbers. In the latter case though only the values $F(a, b)$ for $a < b$ have any bearing in the sequel but it is convenient to allow $F(a, b)$ to have meaning more generally.

The symmetric derivates of an interval function F are defined in the obvious way using the differential quotient $F(x - h, x + h)/(2h)$. Symmetric continuity at a point x means $F(x - h, x + h) \to 0$ as $h \to 0$ and symmetric semi-continuity is defined similarly.

Theorem 5.10 (Freiling) *Let F be an interval function such that*

1. *If $a < b < c$ then $F(a, b) + F(b, c) \leq F(a, c)$,*

2. *$\underline{SD}\, F(x) \geq 0$ almost everywhere,*

3. *$\underline{SD}\, F(x) > -\infty$ everywhere except at the points of a countable set N,*

4. *F is symmetrically upper semi-continuous at every point of N .*

Then there is a right scattered set R and there is a left scattered set L such that $F(a, b) \geq 0$ for every $a \notin R$, $b \notin L$, $a < b$.

The proof is obtained in two steps; we first show that under these hypotheses $F(a,b) \geq 0$ for every pair a, b not in some countable set. Then the refinement to semiscattered sets is accomplished as in the preceding section.

Lemma 5.11 *Under the hypotheses of the theorem there is a countable set N such that $F(a,b) \geq 0$ for every pair a, $b \notin N$ with $a < b$.*

Proof. The proof uses a standard covering argument (cf. [281]). Let A_1 be the set of points at which $\underline{SD}\, F(x) \geq 0$. Let A_2 be the (measure zero) set of points at which $0 > \underline{SD}\, F(x) > -\infty$. And let A_3 be the (countable) set of points at which $\underline{SD}\, F(x) = -\infty$ but at which F is symmetrically upper semi-continuous.
  Let $\epsilon > 0$. The collection

$$V_1 = \{(x,t) \,:\, x \in A_1,\, t > 0,\, F(x - t, x + t) > -\epsilon t\}$$

is a full symmetric cover of A_1. For each integer m choose open sets G_m with $G_m \supset A_2$ and $|G_m| < \epsilon m^{-1} 2^{-m}$ and define

$$V_{2m} = \{(x,t) \,:\, x \in A_2,\, t > 0,\, F(x - t, x + t) > -mt,\, (x - t, x + t) \subset G_m\}.$$

The union of the V_{2m} is a symmetric cover of A_2. Finally if $\{c_1, c_2, c_3, \ldots\}$ is an enumeration of A_3 let

$$V_{3k} = \left\{(c_k, t) \,:\, t > 0,\, F(c_k - t, c_k + t) > -\epsilon 2^{-k}\right\}.$$

The union of the collections V_{3k} is a symmetric cover of A_3. Let V be the union of all the V_1, V_{2m} and V_{3k}. Then V is a symmetric cover for the real line. Thus by the covering Theorem 3.13 there is a countable set $N(\epsilon)$ so that V partitions $[a, b]$ if a and b do not belong to $N(\epsilon)$, $a < b$. If we write the elements of the partition as $a = x_0 < x_1 < x_2 < \ldots < x_n = b$ then, by the superadditivity of F,

$$F(a,b) \geq \sum_{k=1}^{n} F(x_{k-1}, x_k).$$

We simply split this sum into the parts corresponding to pairs

$$((x_k + x_{k-1})/2, (x_k - x_{k-1})/2)$$

that belong to some collection V_1, V_{2m} or V_{3k}. The sum corresponding to V_1 must exceed $-\epsilon(b - a)$; the sum corresponding to any one V_{2m} must exceed $-\epsilon 2^{-m}$ and the sum corresponding to any one V_{3k} must exceed $-\epsilon 2^{-k}$. In total then we must have

$$F(a,b) > -\epsilon(b - a + 2)$$

if a and b do not belong to $N(\epsilon)$, $a < b$. Finally if we write $N = \bigcup_{n=1}^{\infty} N(n^{-1})$ then we have a countable set with the desired property.

Now the semi-scattered part of the proof is obtained exactly as in the proof of Lemma 5.7.

Lemma 5.12 *Suppose that F is superadditive and that there is a countable set N such that $F(a,b) \geq 0$ for every pair a, $b \notin N$ with $a < b$. If $\underline{SD}F(x) > -\infty$ at every point with perhaps countably many exceptions then there is a right scattered set R and and there is a left scattered set L such that $F(a,b) \geq 0$ for every $a \notin R$, $b \notin L$, $a < b$.*

As an immediate corollary we can apply this to improve the monotonicity assertions of Section 5.2.

Corollary 5.13 *Let the function f have a nonnegative lower symmetric derivate almost everywhere and have $-\infty < \underline{SD}f(x)$ everywhere except possibly in a countable set at each point of which f is upper symmetrically semi-continuous. Then f is nondecreasing off some splattered set. More precisely there is an nondecreasing function g such that $\{x : f(x) \geq g(x)\}$ is right scattered and $\{x : f(x) \leq g(x)\}$ is left scattered.*

Proof. Apply the theorem to the interval function $F(x,y) = f(y) - f(x)$.

5.3.2 Freiling Scattered Theorem

The main result of the last section (Theorem 5.10) has a scattered version as well. Note that we assume in the statement of this theorem that for any triple of real numbers, a, b, c in any order the inequality $F(a,b) + F(b,c) \leq F(a,c)$ holds. Thus we assuming that $F(x,y)$ is defined for all pairs.

Theorem 5.14 (Freiling) *Let F be an interval function such that*

1. *For any a, b, c, the inequality $F(a,b) + F(b,c) \leq F(a,c)$ holds,*

2. *$\underline{SD}F(x) \geq 0$ almost everywhere,*

3. *$-\infty < \underline{SD}F(x) \leq \overline{SD}F(x) < +\infty$ everywhere except at the points of a countable set N,*

4. *F is symmetrically upper semi-continuous at every point of N .*

Then there is a scattered set S and such that $F(a,b) \geq 0$ for every a, $b \notin S$, $a < b$.

The proof is sufficiently close to the variant in Lemma 5.9 and the proof of Theorem 5.10 already given that it may be omitted. Again we have an immediate corollary that extends our earlier monotonicity theorems.

Corollary 5.15 *Let the function f have a nonnegative lower symmetric derivate almost everywhere and have*

$$-\infty < \underline{SD}\,f(x) \le \overline{SD}\,f(x) < +\infty$$

everywhere except possibly in a countable set at each point of which f is upper symmetrically semi-continuous. Then f is nondecreasing off some scattered set.

Proof. Apply the theorem to the interval function $F(x,y) = f(y) - f(x)$.

5.3.3 Freiling Proof of Charzyński Theorem

Using the scattered monotonicity theorem (Theorem 5.14) Freiling [111] has given an elegant proof of Charzyński's Theorem (Theorem 2.9). The proof uses that monotonicity theorem, the Pesin–Preiss Theorem (Theorem 2.26) and the fundamental Khintchine Theorem in a clever and simple way. Note that the statement of the theorem is slightly more general than the original as it allows a countable exceptional set.

Theorem 5.16 *Let f be a function that satisfies the condition*

$$\limsup_{h\to 0}\left|\frac{f(x+h)-f(x-h)}{2h}\right| < +\infty \tag{5.1}$$

at every point x with countably many exceptions at each of which f is symmetrically continuous. Then f is continuous at every point with the exception only of a scattered set.

Proof. The proof follows from the general monotonicity Theorem of Freiling (Theorem 5.14). Our version is different in some details from Freiling's original proof so that it more closely parallels the proof of Theorem 5.17 below which is the even analogue of the Charzyński Theorem and which permits a nearly identical proof.

We suppose that f satisfies the condition (5.1) at every point x except possibly for x in a countable set C and that f is symmetrically continuous at every point in C. Thus f is symmetrically continuous everywhere. For any $a < b$ define $F(a,b)$ as the supremum of all numbers $-\epsilon$ with the property that there exists a $\delta > 0$ such that

$$|f(b+h)-f(a-h)-f(b)+f(a)| < \epsilon \tag{5.2}$$

whenever $|h| < \delta$. The function F provides the proof of the continuity properties of the function f by virtue of the fact that should $F(a,b) = 0$ then f is continuous at a if and only if f is continuous at b. Conversely if f is continuous at both points a and then b then certainly $F(a,b) = 0$.

We check first the additivity properties of F. For any a, b, c (in any order) we show that

$$F(a,c) \geq F(a,b) + F(b,c). \tag{5.3}$$

If $F(a,b) = -\infty$ or $F(b,c) = -\infty$ then (5.3) holds trivially. Otherwise let $-\epsilon_1 < F(a,b)$, $-\epsilon_2 < F(b,c)$ and $\epsilon_3 > 0$. Then there is a positive δ for which $|h| < \delta$ implies

$$|f(b+h) - f(a-h) - f(b) + f(a)| < \epsilon_1,$$

$$|f(c+h) - f(b-h) - f(c) + f(b)| < \epsilon_2,$$

$$|f(b+h) - f(b-h)| < \epsilon_3,$$

and hence that

$$|f(c+h) - f(a-h) - f(c) + f(a)| < \epsilon_1 + \epsilon_2 + \epsilon_3. \tag{5.4}$$

From (5.4) we may conclude the relation (5.3). Thus F satisfies the first hypothesis of Freiling's Theorem (Theorem 5.14).

Now let us show that F satisfies the condition (3) in the statement of Freiling's Theorem. Note that $-\infty \leq F(a,b) \leq 0$. At any point x not in the exceptional set C we may use (5.1) to determine positive numbers δ and K so that

$$|f(x+s) - f(x-s)| < Ks$$

if $0 < s < \delta$. Then if $0 < t < \delta/2$ and $|h| < t$ we have from this inequality that

$$|f(x+t+h) - f(x-t-h)| < K(t+h)$$

and

$$|f(x+t) - f(x-t)| < Kt$$

and hence

$$|f(x+t+h) - f(x-t-h) - f(x+t) + f(x-t)| < 3Kt.$$

This gives

$$0 \geq F(x-t, x+t) > -3Kt$$

for all $0 < t < \delta/2$ and consequently

$$\limsup_{t \to 0+} |F(x-t, x+t)/t| < +\infty$$

as we require at these points. Condition (4) is similarly proved.

Finally, in order to apply the Freiling Theorem it remains to verify the property (2). Since f is symmetrically continuous everywhere we know from Theorem 2.26 that f is continuous almost everywhere. Fix a point c at which f is continuous; then at any other point x at which f is also continuous we know

$F(c, x) = 0$. Thus we can define the function $g(x) = F(c, x)$. This function vanishes almost everywhere and so is measurable.

Because of (5.3) we have at any point x that

$$F(x - t, x + t) \leq F(c, x + t) - F(c, x - t) = g(x + t) - g(x - t).$$

Thus

$$\frac{g(x + t) - g(x - t)}{t} \geq \frac{F(x - t, x + t)}{t}$$

and so, from what we have already proved, the lower symmetric derivate of g is always greater than $-\infty$ outside of C. By the Theorem of Khintchine (Theorem 1.23) then g is almost everywhere differentiable. At any point at which $g'(x)$ exists it is easy to see that $g'(x) = 0$.

Suppose that x is now a point of continuity of f and that $g(x) = g'(x) = 0$; almost every point x has this property. We apply (5.3) three times more to obtain

$$F(x, x + t) \geq F(x, c) + F(c, x + t) = F(c, x + t),$$

$$F(x, x - t) \geq F(x, c) + F(c, x - t) = F(c, x - t)$$

and

$$F(x - t, x + t) \geq F(x - t, x) + F(x, x + t).$$

Together these give

$$F(x - t, x + t) \geq g(x - t) + g(x + t).$$

Thus

$$\frac{F(x - t, x + t)}{t} \geq \frac{g(x - t)}{t} + \frac{g(x + t)}{t}.$$

Since each of these expressions on the right of the inequality tends to $g'(x) = 0$ as $t \to 0$ we must have

$$\liminf_{t \to 0+} F(x - t, x + t)/t \geq 0$$

at any such point x and hence almost everywhere. This is condition (2) of the Freiling theorem.

We now apply the Freiling Theorem and the proof of the theorem is complete. There is a scattered set S so that $F(a, b) = 0$ for every $a, b \notin S$, $a < b$. Take any point $a \notin S$ at which f is continuous and we see that f is continuous at any point $b \notin S$.

5.3.4 The Even Analogue of Charzyński Theorem

In Section 2.3.5 an even analogue of the Charzyński Theorem was already obtained as a corollary of the odd version. The theorem that we now state will follow in a similar way from Theorem 5.16 by applying the reduction theorem (Theorem 2.5). As an application of the Freiling monotonicity Theorem we present an independent proof here (reproduced from [285]).

Theorem 5.17 *Let f be a measurable function that satisfies the condition*

$$\limsup_{h \to 0} \left| \frac{f(x+h) + f(x-h) - 2f(x)}{2h} \right| < +\infty$$

at every point x with countably many exceptions at each of which f is symmetric. Then f is continuous at every point with the exception only of a scattered set.

Proof. Under these hypotheses f is symmetric everywhere and hence (Theorem 2.3) its set of discontinuities has measure zero. We wish to show that this set of discontinuity points is scattered.

The proof follows from the general monotonicity Theorem of Freiling. For any pair of real numbers a, b define $F(a,b)$ as the supremum of all numbers $t < 0$ with the property that there exists a $\delta > 0$ such that

$$|f(b+h) + f(a-h) - f(b) - f(a)| < -t$$

whenever $|h| < \delta$. The function F provides the proof of the continuity properties of the function f by virtue of the fact that should $F(a,b) = 0$ then f is continuous at a if and only if f is continuous at b. Conversely if f is continuous at both points a and then b then certainly $F(a,b) = 0$.

We check first the additivity properties of F. For any a, b, c (in any order) we show that

$$F(a,c) \geq F(a,b) + F(b,c). \tag{5.5}$$

If $F(a,b) = -\infty$ or $F(b,c) = -\infty$ then (5.5) holds trivially. Otherwise let ϵ_1, ϵ_2 and ϵ_3 be arbitrary positive numbers such that

$$-\epsilon_1 < F(a,b), \quad -\epsilon_2 < F(b,c) \text{ and } \epsilon_3 > 0.$$

Then there is a positive δ for which $|h| < \delta$ implies

$$|f(b+h) + f(a-h) - f(b) - f(a)| < \epsilon_1,$$

$$|f(c+h) + f(b-h) - f(c) - f(b)| < \epsilon_2,$$

$$|f(b+h) + f(b-h) - 2f(b)| < \epsilon_3,$$

and hence that

$$|f(c+h) + f(a-h) - f(c) - f(a)| < \epsilon_1 + \epsilon_2 + \epsilon_3.$$

From this we may conclude the relation (5.5). Both the superadditivity and the quasi-subadditivity of F follow now since we have not assumed any order requirements on the points a, b and c.

Now let us show that F satisfies the condition (3) in the statement of Freiling's Theorem. Note that $-\infty \leq F(a,b) \leq 0$. At any point x at which the condition (4.42) holds (i.e. at every point not in the countable set C) we may determine positive numbers δ and K so that

$$|f(x+s) + f(x-s) - 2f(x)| < Ks$$

if $0 < s < \delta$. Then if $0 < t < \delta/2$ and $|h| < t$ we have from this inequality that

$$|f(x+t+h) + f(x-t-h) - 2f(x)| < K(t+h)$$

and

$$|f(x+t) + f(x-t) - 2f(x)| < Kt$$

and hence

$$|f(x+t+h) + f(x-t-h) - f(x+t) - f(x-t)| < 3Kt.$$

This gives

$$0 \geq F(x-t, x+t) > -3Kt$$

for all $0 < t < \delta/2$ and consequently

$$\limsup_{t \to 0+} |F(x-t, x+t)/t| < +\infty$$

as we require.

A nearly identical proof shows that property (4) also holds because f is symmetric at every point and so at every point in C.

Finally, in order to apply the Freiling Theorem it remains to verify the property (2). We recall that f is continuous almost everywhere. Fix a point c at which f is continuous; then at any other point x at which f is also continuous we know $F(c,x) = 0$. Thus we can define the function $g(x) = F(c,x)$. This function vanishes almost everywhere and so is measurable.

Because of (5.5) we have at any point x that

$$F(x-t, x+t) \leq F(c, x+t) - F(c, x-t) = g(x+t) - g(x-t).$$

Thus

$$\frac{g(x+t) - g(x-t)}{t} \geq \frac{F(x-t, x+t)}{t}$$

and so, from what we have already proved, the lower symmetric derivate of g is always greater than $-\infty$ outside of C. By the Khintchine Theorem then g is

almost everywhere differentiable. At any point at which $g'(x)$ exists it is easy
to see that $g'(x) = 0$.

Suppose that x is now a point of continuity of f and that $g(x) = g'(x) = 0$;
almost every point x has this property. We apply (5.5) three times more to
obtain

$$F(x, x + t) \geq F(c, x) + F(c, x + t) = F(c, x + t),$$

$$F(x, x - t) \geq F(c, x) + F(c, x - t) = F(c, x - t)$$

and

$$F(x - t, x + t) \geq F(x - t, x) + F(x, x + t).$$

Together these give

$$F(x - t, x + t) \geq g(x - t) + g(x + t).$$

Thus

$$\frac{F(x - t, x + t)}{t} \geq \frac{g(x - t)}{t} + \frac{g(x + t)}{t}.$$

Since each of these expressions on the right of the inequality tends to $g'(x) = 0$
as $t \to 0$ we must have

$$\liminf_{t \to 0+} F(x - t, x + t)/t \geq 0$$

at any such point x and hence almost everywhere. This is condition (2) of the
Freiling theorem.

We now apply the Freiling Theorem and the proof of the theorem is com-
plete. There is a scattered set S so that $F(a, b) = 0$ for every $a, b \notin S$, $a < b$.
Take any point $a \notin S$ at which f is continuous and we see that f is continuous
at any point $b \notin S$.

5.4 Evans–Larson Theorem

Evans and Larson [96] show how some symmetric properties of functions can be
used to extend an elementary monotonicity theorem of Zygmund. That theorem
(see [251, pp.203–204]) asserts that a function f is nondecreasing on an interval
if at every point x of that interval both of the conditions $\overline{D}^+ f(x) \geq 0$ and

$$\limsup_{y \to x-} f(y) \leq f(x)$$

hold.

Theorem 5.18 (Evans–Larson) *Let f be measurable and smooth. If $\overline{D}^+ f(x) \geq$
0 at every point then f is nondecreasing.*

Theorem 5.23 (Larson) *Let the function f have everywhere a finite symmetric derivative. Then if a < b are points of continuity of f there are two sets of
type G_δ and of positive measure C, D ⊂ (a, b) so that*

$$SD\, f(c) \le \frac{f(b) - f(a)}{b - a} \le SD\, f(d)$$

for all c ∈ C and d ∈ D.

As a corollary Larson shows that there is a kind of Darboux property available for the symmetric derivative itself.

Corollary 5.24 (Larson) *Let the function f have everywhere a finite symmetric derivative g. Then for all x*

$$\liminf_{h \to 0} \frac{g(x+h) + g(x-h)}{2} \le g(x) \le \limsup_{h \to 0} \frac{g(x+h) + g(x-h)}{2}.$$

The next two mean value theorems are due to Freiling [111]. We give two
variants.

Theorem 5.25 (Freiling) *Let f have finite symmetric derivates everywhere
on an interval I. Then there is a scattered set S ⊂ I so that for all a, b ∈ I \ S,
a < b and all ε > 0 there is a sequence of numbers*

$$a = x_0, x_1, \ldots, x_n = b$$

such that

$$\sum_{i=1}^{n} \underline{SD}\, f\left(\frac{x_i + x_{i-1}}{2}\right)(x_i - x_{i-1}) - \epsilon$$
$$\le\; f(b) - f(a) \le \sum_{i=1}^{n} \overline{SD}\, f\left(\frac{x_i + x_{i-1}}{2}\right)(x_i - x_{i-1}) + \epsilon.$$

In the second variant the sequence of points $a = x_0, x_1, \ldots, x_n = b$ is chosen
increasing but then the exceptional set can be claimed only to be splattered.

Theorem 5.26 (Freiling) *Let f have finite symmetric derivates everywhere
on an interval I. Then there is a splattered set S ⊂ I so that for all a, b ∈ I \ S,
a < b and all ε > 0 there is a sequence of numbers*

$$a = x_0 < x_1 < \ldots < x_n = b$$

such that

$$\sum_{i=1}^{n} \underline{SD} f \left(\frac{x_i + x_{i-1}}{2} \right) (x_i - x_{i-1}) - \epsilon$$

$$\leq f(b) - f(a) \leq \sum_{i=1}^{n} \overline{SD} f \left(\frac{x_i + x_{i-1}}{2} \right) (x_i - x_{i-1}) + \epsilon.$$

5.6 Freiling–Rinne Theorem

We ask now for monotonicity theorems that use the approximate symmetric derivative. From the outset it is clear that different arguments will be needed. While measurability concerns play no role in the monotonicity theorems for the symmetric derivates this is not so for the approximate symmetric derivative. As we shall discover in Section 6.2.2 there is (under CH) a nonmeasurable function that has a zero approximate symmetric derivative everywhere. Thus any theorems that we might hope to prove that are approximate analogues of those for the ordinary symmetric derivative should lie considerably deeper and require measurability assumptions.

The basic monotonicity theorem to be expected is that a measurable function with a nonnegative approximate symmetric derivative is equivalent to a nondecreasing function. While such a problem must have been known since the 1920's (especially in view of the obvious applications to trigonometric series) there are no early accounts of such a monotonicity theorem. With the publication of the important report of Goffman and Neugebauer [124] on the approximate derivative a number of authors were inspired to push these methods over to the approximate symmetric case. Thus several articles appeared each claiming a proof of such a monotonicity theorem, using essentially the ideas of Goffman and Neugebauer: for example Mukhopadhyay [206] in 1966, Kubota [160] (see the review of H. Burkill MR47#2010) and Kundu [164] both in 1973.

It was well known in real analysis circles by the 1980's that these proofs could not be salvaged and many people tried to find a correct proof. It was not until the work of Freiling and Rinne [112] that the theorem was established.

Theorem 5.27 (Freiling–Rinne) *Suppose that f is measurable and*

$$\underline{ASD} f(x) \geq 0$$

everywhere. Then f is a.e. equal to a monotonic nondecreasing function.

We shall prove a variant of this (in the same spirit as Theorem 5.10). This is from [232, p. 546]; the proof is an application of the covering lemma of Section 3.6. (Recall, however, that the covering lemma itself is equivalent to the monotonicity theorem.)

Theorem 5.28 *Let f be measurable and have a lower approximate symmetric derivate that satisfies the following conditions:*

(i) $\underline{ASD}\,f(x) \geq 0$ *almost everywhere,*

(ii) $\underline{ASD}\,f(x) > -\infty$ *everywhere except at the points of a countable set N,*

(iii) f *is approximately symmetrically upper semi-continuous at every point of N.*

Then f is a.e. equal to a monotonic nondecreasing function.

Proof. The proof employs the same covering argument that we have seen in Theorem 5.10. Let A_1 be the set of points at which $\underline{ASD}\,f(x) \geq 0$. Let A_2 be the (measure zero) set of points at which $0 > \underline{ASD}\,f(x) > -\infty$. Let A_3 be the (countable) set of points at which $\underline{ASD}\,f(x) = -\infty$ but at which f is approximately symmetrically upper semi-continuous. It can be shown that each of these sets is measurable; only A_1 is a problem (see Section 7.4.2).

Let $\epsilon > 0$. The collection

$$V_1 = \{(x,t) \,:\, x \in A_1,\, t > 0,\, f(x+t) - f(x-t) > -\epsilon t\}$$

is a measurable approximate symmetric cover of A_1. For each integer m choose open sets G_m with $G_m \supset A_2$ and $|G_m| < \epsilon m^{-1}2^{-m}$ and define V_{2m} as the set

$$\{(x,t) \,:\, x \in A_2,\, t > 0,\, f(x+t) - f(x-t) > -mt,\, (x-t, x+t) \subset G_m\}.$$

The union of the V_{2m} is a measurable approximate symmetric cover of A_2. Finally if $\{c_1, c_2, c_3, \ldots\}$ is an enumeration of A_3 let

$$V_{3k} = \left\{(c_k, t) \,:\, t > 0,\, f(c_k + t) - f(c_k - t) > -\epsilon 2^{-k}\right\}.$$

The union of the collections V_{3k} is a measurable approximate symmetric cover of A_3. Let V be the union of all the V_1, V_{2m} and V_{3k}. Then V is a measurable approximate symmetric cover for the real line. Thus by Theorem 3.33 there is a set $N(\epsilon)$ of measure zero so that V contains any interval $[a, b]$ for which a and b do not belong to $N(\epsilon)$, $a < b$. If we write the transitions as

$$a = x_0 < x_1 < x_2 < \ldots < x_n = b$$

then

$$f(b) - f(a) = \sum_{k=1}^{n} f(x_k) - f(x_{k-1}).$$

We simply split this sum into the parts corresponding to pairs

$$((x_k + x_{k-1})/2, (x_k - x_{k-1})/2)$$

that belong to some collection V_1, V_{2m} or V_{3k}. The sum corresponding to V_1 must exceed $-\epsilon(b-a)$; the sum corresponding to any one V_{2m} must exceed $-\epsilon 2^{-m}$ and the sum corresponding to any one V_{3k} must exceed $-\epsilon 2^{-k}$. In total then we must have

$$f(b) - f(a) > -\epsilon(b-a+2)$$

if a and b do not belong to $N(\epsilon)$, $a < b$. Finally if we write $E = \bigcup_{n=1}^{\infty} N(n^{-1})$ then we have a measure zero set and f is nondecreasing off of E.

5.7 Convexity Theorems

The second symmetric derivative was introduced in Chapter 1 as the limit of the expression

$$\frac{f(x+h) + f(x-h) - 2f(x)}{h^2}$$

as $h \to 0$ and has been denoted by the symbol $SD_2 f(x)$. As we have already indicated in Section 1.4, the existence of $f''(x)$ requires the existence of the second symmetric derivative and both have the same value. We may also define extreme derivates of this type:

$$\underline{SD_2}\, f(x) = \liminf_{h \to 0} \frac{f(x+h) + f(x-h) - 2f(x)}{h^2}$$

and

$$\overline{SD_2}\, f(x) = \limsup_{h \to 0} \frac{f(x+h) + f(x-h) - 2f(x)}{h^2}.$$

The most immediate and important application of these derivates in the study of real functions is the observation that these notions characterize convex functions. Recall that a function f is convex on an interval (a, b) if

$$f\left(\lambda x + (1-\lambda)y\right) \le \lambda f(x) + (1-\lambda)f(y)$$

for all x, $y \in (a, b)$ and all $0 < \lambda < 1$. In the elementary calculus the usual test for convexity involves checking the sign of the second derivative f''. One can show that a function f convex on an open interval (a, b) is continuous there and must have a derivative f' that exists and is continuous on $(a, b) \setminus N$ for some denumerable subset N of (a, b). This set N however might be dense in (a, b) and this would preclude any use of the second derivative as a test in general for convexity.

It is in this situation that the second symmetric derivative has one of its most interesting, if elementary, applications. We reproduce a proof from [313, p. 23].

Theorem 5.29 *A necessary and sufficient condition for a continuous function f to be convex in an interval (a, b) is that*

$$\overline{SD_2}\, f(x) \geq 0$$

for every $x \in (a, b)$.

Proof. The necessity is obvious; indeed if f is convex it is easy to see that $\underline{SD_2}\, f(x) \geq 0$ at every point. To prove the sufficiency consider the function $g_n(x) = f(x) + x^2/n$. Since a limit of a sequence of convex functions is also convex it will be enough to show that each g_n is convex.

At each point x we have

$$\overline{SD_2}\, g_n(x) = \overline{SD_2}\, f(x) + 2/n > 0. \tag{5.6}$$

But a function with a positive upper second symmetric derivate must be convex. To see this assume not: then if g_n is not convex there is some choice of numbers m and c so that $g(x) = g_n(x) + mx + c$ has a maximum at a point z inside the interval (a, b). That means that $g(z + h) + g(z - h) - 2g(z) \leq 0$ for sufficiently small h. As a consequence we would have

$$\overline{SD_2}\, g_n(z) = \overline{SD_2}\, g(z) \leq 0$$

and this contradicts (5.6). Thus each function g_n is convex and so, letting $n \to \infty$, we obtain that f is convex too.

In the statement of Theorem 5.29 we cannot replace continuity by anything much weaker. For example there is a Darboux, Baire 1 function f which is not convex and yet for which $\overline{SD_2}\, f(x) \geq 0$ everywhere. As pointed out in [18, p. 197] one takes the Croft function g (for a construction see [18, Ex. 2.2, p. 12]) which is a nonnegative, nonconstant Darboux, Baire 1 function which vanishes almost everywhere. Then with $f = -g$ we have $f(x+h) + f(x-h) - 2f(x) = 2g(x) \geq 0$ for all x and almost all h; one sees immediately that $\overline{SD_2}\, f(x) \geq 0$ everywhere and that f is not convex.

There are a number of useful refinements of this theorem and the rest of this section is devoted to exploring these. We present them as corollaries even though some further work is needed to carry some of the proofs over. The first corollary allows us to neglect a countable set in checking the conditions on the derivates.

Corollary 5.30 *A sufficient condition for a continuous function f to be convex in an interval (a, b) is that*

$$\overline{SD_2}\, f(x) \geq 0$$

for all $x \in (a, b) \setminus N$ where N is denumerable and f is smooth at each point of N.

Proof. (cf. Bary [11, vol. II, pp. 344–346]) As in the preceding corollary we may reduce the proof to the case where $\overline{SD_2}\, f(x) > 0$ at every point outside of N. We refine the argument there. If f is not convex there is some choice of numbers m_0 and c so that $g(x) = f(x) + m_0 x + c$ has a maximum at some point $z = z(m_0)$ inside the interval (a, b). For sufficiently small ϵ every number $m \in (m_0 - \epsilon, m_0 + \epsilon)$ has again the same property. For each choice of $m \in (m_0 - \epsilon, m_0 + \epsilon)$ we may make the choice of $z = z(m)$ specific by choosing the largest of all the possible maxima.

As before $\overline{SD}_2 f(z(m)) \leq 0$ but this is no longer a contradiction; it says only that the point $z(m)$ must belong to N. The contradiction is obtained by showing that there are uncountably many distinct possibilities for $z(m)$ which violates the assumption that N is countable. Indeed we shall show that $f'(z(m)) = m$ for each permissible value of m and the contradiction is evident.

Each point $z(m)$ belongs to N and so is a point of smoothness for the function f and hence for the function g. But at a point of maximum of g at which g is also smooth the derivative must necessarily vanish. Thus for each choice of m we have $g'(z(m)) = 0$ and accordingly $f'(z(m)) = m$ as stated.

We can refine this further still.

Corollary 5.31 (Burkill) *Let f be continuous. Suppose that*

$$\overline{SD}_2 f(x) \geq 0$$

almost everywhere in the interval (a, b) and

$$\overline{SD}_2 f(x) > -\infty$$

for all $x \in (a, b) \setminus N$ where N is denumerable and f is smooth at each point of N. Then f is convex in (a, b).

Proof. (cf. Burkill [35, Theorem 1.1, p. 47]) For the refinement we shall, given a set $E \subset (a, b)$ of measure zero and a natural number n, construct a function g_n that is convex and smooth in (a, b) and such that $SD_2 g_n(x) = +\infty$ everywhere in E but with $0 = g_n(a) \leq g_n(x) \leq g_n(b) < 1/n$. If we grant the existence of these functions then, with E chosen as the set of points where $\overline{SD}_2 f(x) < 0$, the function $f + g_n$ satisfies the hypotheses of the preceding corollary. Thus each function $f + g_n$ is convex and so too then is the limit function f by allowing $n \to \infty$.

Let $\epsilon = 1/(n(b - a))$ and $k = 1, 2, 3, \ldots$. Choose an open set $G_k \supset E$ with measure smaller than $\epsilon 4^{-k}$. Define the functions $h_k(x) = 2^k$ for $x \in G_k$ and $h_k(x) = 0$ otherwise. Set $H_k(x) = \int_a^x h_k(t)\, dt$. Then H_k is continuous and nondecreasing with $H_k(a) = 0$, $H_k(b) < \epsilon 2^{-k}$. Write finally

$$H(x) = \sum_{k=1}^{\infty} H_k(x) \text{ and } g_n(x) = \int_a^x H(t)\, dt.$$

It is easy to check that g_n is continuous, smooth, convex and that $SD_2\, g_n(x) = +\infty$ everywhere in E. We easily check that $g_n(b) < \epsilon(b - a) = 1/n$. This completes the proof of the corollary.

As an obvious additional corollary we have a criterion for a function to be linear in an interval. The condition is clearly both necessary and sufficient.

Corollary 5.32 *If f is continuous in an interval (a, b) and*

$$SD_2\, f(x) = 0$$

for every $x \in (a, b)$ then f is linear in this interval.

This too can be refined further (see Bary [11, vol. II, pp. 346–348] for a proof.)

Corollary 5.33 *Let f be Lebesgue integrable in an interval (a, b), F continuous there and suppose that*

$$-\infty < \underline{SD}_2\, F(x) \leq f(x) \leq \overline{SD}_2\, F(x) < +\infty$$

at every point $x \in (a, b)$ excepting possibly in a denumerable set E at each point of which F is smooth. Then

$$F(x) = mx + c + \int_a^x \int_a^t f(\xi)\, d\xi$$

everywhere in (a, b) where m and c are suitable constants.

Note in particular that this Corollary allows to recover a function (up to a linear function) if we are given its second symmetric derivative and that derivative is known to be integrable. This should be compared with the situation in Chapter 9 where the same problem will be addressed without any information on the integrability of the derivative (it may not be Lebesgue integrable).

Finally as a further corollary we have the usual calculus test for convexity.

Corollary 5.34 *If f'' exists everywhere on (a, b) then f is convex there if and only if $f''(x) \geq 0$ on that interval.*

5.7.1 Further Convexity Theorems

The convexity theorems of the preceding section can be considered as the most basic ones. In this section we shall report on several others that depart in some way from those but still stated within the setting of the second symmetric derivative. The main theme here is the relaxation of the continuity hypothesis which was maintained throughout the last section. Since the hypothesis of continuity is removed some strong regularity condition is needed. (Of course continuity does not disappear really, for convex functions are continuous in any case.)

The first of these is from Verblunsky [300, pp. 255–256]. He shows that a semicontinuity condition can be used.

Theorem 5.35 (Verblunsky) *Let f be upper semicontinuous in an interval* (a, b). *If*

$$\overline{SD_2}\, f(x) \geq 0$$

for all $x \in (a, b)$ *then* f *is convex.*

A refinement of this is due to Taylor [280, pp. 255–256]. It is needed for his development of a trigonometric integral.

Theorem 5.36 (Taylor) *Let f be approximately continuous in an interval* (a, b) *and have the following property: for every perfect set* $P \subset (a, b)$ *there is a nonempty portion* $P \cap (c, d)$ *on which* f *is upper semicontinuous. If*

$$\overline{SD_2}\, f(x) \geq 0$$

for all $x \in (a, b)$ *then* f *is convex.*

Larson [175, p. 94] gives a version of these theorems that assumes the existence everywhere of the second symmetric derivative. Note that under the assumption that $0 \leq SD_2\, f(x) < +\infty$ everywhere the function f is continuous off a scattered set and so convex in every open interval contiguous to that set. If $S \neq \emptyset$ is the set of points of discontinuity of f then there is an isolated point $x_0 \in S$ and f is convex on intervals (s, x_0) and (x_0, t). But then the existence of $SD_2\, f(x_0)$ and the Darboux property show that f is continuous at x_0 giving a contradiction.

Theorem 5.37 (Larson) *Let f be a function that is measurable and has the Darboux property. and such that* $0 \leq SD_2\, f(x) < +\infty$ *everywhere. Then* f *is convex.*

In a similar direction, but going deeper, Weil [304] also drops continuity from the hypotheses. The proof in [304] depends on the function belonging to the class of Darboux, Baire 1 functions. Buczolich [21] has analyzed the proof and shown how the assumption that the function is Baire 1 can be reduced; the Darboux property is replaced by the assumption that the function is the sum of a continuous function and a function that has the Darboux property. Of course any Darboux function has this property trivially. The existence of non-measurable midpoint linear functions (Section 4.2) shows that something is needed. There are even non-measurable midpoint linear functions that have the Darboux property as pointed out in [304] and Section A.6.

Theorem 5.38 (Weil–Buczolich) *Let f be a measurable function on an interval (a, b) and suppose that f can be expressed as the sum of a continuous function and a function that has the Darboux property. If $\underline{SD}_2 f(x) \geq 0$ everywhere in (a, b) then f is convex there.*

The original proofs in [304] and [21] are quite complicated. A clearer and simpler analysis has been given by Fejzić [100]. He asks for the structure of a measurable function which has a nonnegative lower second symmetric derivate everywhere. An easy example shows what might happen. Let G be any open dense set and define a function f so that $f(x) = x^2$ for $x \in G$ and $f(x) = 0$ elsewhere. Then f is convex on each component of G but not convex on any set containing points outside G and $\underline{SD}_2 f(x) \geq 0$ at every point x. The theorem asserts that the general behaviour is no worse than this. Our proof follows [100] closely.

Theorem 5.39 (Fejzić) *Let f be a measurable function on an open interval (a, b) and suppose that $\underline{SD}_2 f(x) \geq 0$ everywhere in (a, b). Then there is an open dense set $G \subset (a, b)$ so that f is convex in each component of G.*

Proof. It is enough in the statement of the theorem for us to assume that $\underline{SD}_2 f(x) > 0$ everywhere in (a, b). For if $\underline{SD}_2 f(x) \geq 0$ everywhere then we may apply the theorem to the function $f(x) + x^2$: this function has a strictly positive lower second symmetric derivative and so will be convex on an open dense set. But then it is continuous there and so f is continuous there too. The classical Schwarz theorem shows immediately that f itself is convex on the open dense set in any case.

We suppose that $\underline{SD}_2 f(x) > 0$ everywhere in (a, b) and write

$$1/\delta(x) = \min\{n \in \mathbb{N} : f(x + h) + f(x - h) - 2f(x) > 0 \text{ for all } 0 < h < 1/n\}$$

and

$$A_n = \{x \in (a, b) : \delta(x) > 1/n \text{ and } f(x) \leq n\}.$$

The sets $\{A_n\}$ cover the interval.

Let G be the union of all open subintervals of (a, b) on which f is convex. We shall prove that G is dense. If not then there is an interval I_0 on each subinterval of which f fails to be convex. We shall obtain a contradiction.

We shall require a boundedness observation from [21]. Under the assumptions on f here suppose that $f(x) \leq B$ for all x in a set dense in an interval (c, d): then $f(x) \leq B$ for all $x \in (c, d)$.

We prove this (following [100]). Let $E = \{x \in (c, d) : f(x) \leq B\}$. We wish to show that $E = (c, d)$. If we know at least that E has full measure in (c, d)

then this is easy. There must be for any $x \in (c, d)$ a sufficiently small number h with both $x + h$ and $x - h$ in E and $f(x + h) + f(x - h) - 2f(x) > 0$. It follows that $f(x) \leq B$.

Suppose, in order to obtain a contradiction, that E does not have full measure. Then for some integer m the set

$$C_m = \{x \in (c, d) : B + m < f(x) \leq B + 2m\}$$

has positive measure. So also does $C_m \cap A_n$ for some n. Consider a subinterval $I \subset (c, d)$ with $|I| < 1/n$ and where the relative outer measure of $C_m \cap A_n$ exceeds $5/6$. Let I_0 denote the middle third of this interval. Then $D = I_0 \cap C_m \cap A_n$ has relative outer measure bigger than $1/2$ in I_0. Choose a point $x \in E$ before I_0 but sufficiently close. Then the set $2D - x$ will have relative outer measure greater than $1/3$ in I. But the values of f on $2D - x$ exceed $B + 2m$ and the values of f on C_m are less than $B + 2m$. Accordingly the sets

$$f^{-1}(B + 2m, \infty) \quad \text{and} \quad f^{-1}(-\infty, B + 2m)$$

are disjoint, measurable and have relative outer measure greater than $1/3$ and $5/6$ respectively in I. As this is a contradiction our boundedness claim is proved.

By the Baire category theroem we may choose a subinterval $J \subset I_0$ and a set A_n dense in J and with $|J| < 1/n$. As f fails to be convex on any subinterval of J there is a further subinterval (c, d) so that the graph of f does not lie below the line segment connecting the points $(c, f(c))$ and $(d, f(d))$. Write $g(x)$ for the function

$$g(x) = f(x) - \frac{x - c}{d - c} f(d) - \frac{d - x}{d - c} f(c)$$

so that $g(c) = g(d) = 0$ and $g(x_0) > 0$ at some point $x_0 \in (c, d)$. Note that

$$g(x + h) + g(x - h) - 2g(x) > 0$$

for all x, $x + h$, $x - h \in (c, d)$ with $x \in A_n$.

Let $s = \sup\{g(t) : t \in (c, d)\}$. Since $s \geq g(x_0) > 0$ we see that s is positive. Since f (and hence g) is bounded on a dense subset of (c, d) we see (by the boundedness observation above) that s is finite. From this we will obtain a contradiction.

There is a sequence $\{y_k\}$ of points in $A_n \cap (c, d)$ so that $g(y_k) \to s$. For if not then

$$\sup\{g(y) : y \in A_n \cap (c, d)\} < s' < s.$$

Pick $x \in (c, d)$ so that $g(x) > \frac{1}{2}(s' + s)$. Since A_n is dense in (c, d) there is a point

$$y \in A_n \cap (c, d) \cap (x - \delta(x), x + \delta(x))$$

with $2x - y \in (c, d)$. For this

$$g(x) < \frac{1}{2}(g(y) + g(2x - y)) < \frac{1}{2}(s' + s)$$

which contradicts our choice of x.

As the sequence $\{y_k\}$ is bounded we may, by passing to a subsequence, assume that it is convergent to a point $y \in [c, d]$. We claim that $g(y) \neq s$. For if $g(y) = s$ then y is in the open interval (c, d). Choose z in $(c, d) \cap (y - \delta(y), y + \delta(y))$ so that $2y - z \in (c, d)$. Then

$$g(z) + g(2y - z) - 2g(y) > 0$$

so that one of $g(z)$ or $g(2y - z)$ must exceed s which is not possible. But it is also impossible for $g(y) < s$. For $g(y_k) \to s$ and $g(2y_k - y) \leq s$ and $g(2y_k - y) + g(y) - 2g(y_k) > 0$ for large k. Finally we have obtained our contradiction and the theorem is proved.

From this theorem Fejzić [100] derives the following theorem which is an improvement of the Buczolich–Weil theorem.

Theorem 5.40 (Fejzić) *Let f be a measurable function on an open interval (a, b). Suppose that f has no finite or infinite jumps on either side and that $\underline{SD_2}\, f(x) \geq 0$ everywhere in (a, b). Then f is convex in (a, b).*

Here the "no jumps" condition asserts that if the right (left) limit of f exists at a point x_0 then it must be finite and assume the value $f(x_0)$. See [100] for details.

For another variant on this same theme we have the following theorem of Bruckner [18, p. 197]. Recall that a VBG_* function is measurable, in fact is Baire 1 (since it has only countably many discontinuities).

Theorem 5.41 (Bruckner) *Let f be Darboux and VBG_* with $\overline{SD}_2\, f(x) \geq 0$ everywhere. Then f is convex.*

O'Malley [215, p. 80] points out that with much the usual proof one can obtain an approximate version of some of these convexity theorems. The key here is that approximately continuous functions have approximate extrema; see Section 4.8.7 for other applications of this idea to approximate smoothness.

Theorem 5.42 (O'Malley) *Let f be a function that is approximately continuous and such that*

$$\text{ap} \limsup_{h \to 0} \frac{f(x + h) + f(x - h) - 2f(x)}{h^2} \geq 0$$

at every point $x \in (a, b) \setminus E$ where E is countable. If f is approximately smooth at each point of E then f is convex in (a, b).

5.8 Monotonicity Appendix

5.8.1 A Brief History of Symmetric Monotonicity Theorems

In order to give some idea of the earlier development of the subject we shall present here a short rundown of some of the monotonicity theorems established over the years. Most of these are easy and direct corollaries of the material in the preceding sections. To begin with the earliest, we have a result of Khintchine already proved directly as Theorem 1.4. This is an obvious consequence of Theorem 5.6.

(Khintchine) *A continuous function is necessarily nondecreasing (increasing) in any interval in which its lower symmetric derivative is nonnegative (positive).*

The basic monotonicity theorem for the ordinary derivative on which so many generalizations have been based is the following: if a function f satisfies $\underline{D}\,f(x) \geq 0$ almost everywhere and $\underline{D}\,f(x) > -\infty$ everywhere then f is monotonic nondecreasing. The symmetric analogue that one expects to be true should be the statement: if a function f satisfies $\underline{SD}\,f(x) \geq 0$ almost everywhere and $\underline{SD}\,f(x) > -\infty$ everywhere then (with some kind of continuity hypotheses) f is monotonic nondecreasing. A number of theorems of this type have appeared in the literature. Each of these can be obtained directly from Corollary 5.13; in the literature more frequently a modification of the "ordinary" proof was employed.

Perhaps the earliest theorem of this type was given by Verblunsky. It follows easily from our monotonicity results. Of course we now know that the assumption in advance that f is measurable is not needed.

(Verblunsky) *Suppose that f is measurable, that $\underline{SD}\,f(x) > -\infty$ everywhere and that $\overline{SD}\,f(x) \geq 0$ a.e.. Then f is equivalent to a nondecreasing function.*

This appears to be one of the earliest efforts to carry the usual analogue over to the symmetric derivates. A later variant was stated by Mukhopadhyay [207] (but see the correction in Pu and Pu [239]). (The statement we present is slightly more general.)

(Mukhopadhyay) *Let the function f have a nonnegative lower symmetric derivate almost everywhere and have $\underline{SD}\,f(x) > -\infty$ everywhere. If*

$$\limsup_{y \to x-} f(y) \leq f(x) \leq \limsup_{y \to x+} f(y)$$

at each point x then f is nondecreasing.

As an improvement to this Evans [83] (see also Weil [304]) relaxes the semicontinuity conditions. In the original there is an addded measurability hypothesis; as we shall see (in Theorem 6.51 given later on) a function f for which $\underline{SD}\,f(x) > -\infty$ everywhere is measurable in any case. Our proof does not require this, however, and measurability plays no role.

(Evans) *Let f be a function that has a nonnegative lower symmetric derivate almost everywhere and satisfies $\underline{SD}\,f(x) > -\infty$ everywhere. If*

$$\liminf_{y \to x} f(y) \leq f(x) \leq \limsup_{y \to x} f(y)$$

at each point x then f is nondecreasing.

Proof. It follows from Corollary 5.13 that f is nondecreasing off a splattered set E; we can take E to be maximal in the sense that no point can be omitted and let g be the equivalent nondecreasing function. So $f = g$ outside of E.

We have only to show that the semicontinuity conditions do not permit E to have any points. If E is nonempty then, since it is splattered, there is a point $x_0 \in E$ isolated on one side at least. Let us suppose that it is isolated on the left; the other case is similarly proved since the hypotheses are symmetrical. Evidently by the semicontinuity conditions $f(x_0) \geq g(x_0-)$; as E is maximal this requires $f(x_0) > g(x_0+)$ otherwise x_0 could be eliminated.

Choose $x_1 > x_0$, $x_1 \notin E$, $g(x_1) < f(x_0)$. For any number $\alpha \in (g(x_1), f(x_0))$ write

$$z_\alpha = \sup \{x \in (x_0, x_1) : f(x) > \alpha\}.$$

The semicontinuity conditions require $f(z_\alpha) \geq \alpha$ and do not allow $f(z_\alpha) > \alpha$ since that would contradict the definition of z_α. Thus $f(z_\alpha) = \alpha$. Since $\alpha > f(x_1)$ and $z_\alpha < x_1$ we have $z_\alpha \in E$. But this means there are continuum many points in E and that contradicts that E is splattered and hence countable. From this contradiction the theorem follows.

A theorem of Larson gives yet another variant on the same theme and illustrates the role of the semi-continuity assumptions in the above two versions. It too follows from Corollary 5.13. Later on (in Section 6.4.1) we shall see that, under the hypotheses here, f must be measurable anyway.

(Larson) *Let f be a measurable function having a nonnegative lower symmetric derivate almost everywhere and having $\underline{SD} f(x) > -\infty$ everywhere. Then f is nondecreasing on the set of its continuity points.*

A slight departure from this theme appears in Kundu [165] where the upper derivates are used. (This is likely inspired by theorems such as [251, Theorem 72., p. 204]). This theorem improves, formally at least, the result of Mukhopadhyay given above by replacing the lower symmetric derivate by the upper. Note however that, while it appears to be a monotonicity theorem that employs the upper derivate in an essential manner, we know from Khintchine's classical theorem that if f is measurable and $\underline{SD} f(x) > -\infty$ almost everywhere then f is differentiable a.e. in any case and so the condition on the upper derivate immediately applies to the lower derivate.

(Kundu) *Let the function f have a nonnegative upper symmetric derivate almost everywhere and have $\underline{SD} f(x) > -\infty$ everywhere except possibly in a countable set. If*

$$\limsup_{y \to x-} f(y) \leq f(x) \leq \limsup_{y \to x+} f(y)$$

at each point x then f is nondecreasing.

Weil [304] offers another variant that places stronger hypotheses on the regularity of the function.

(**Weil**) *If f is Darboux Baire 1 and $\underline{SD}\, f(x) \geq 0$ then f is nondecreasing.*

Again continuing this theme, but with a twist, Larson [174] gives a symmetric monotonicity theorem that might be considered an analogue of the Goldowski-Tonelli Theorem ([251, p. 206]). The proof can be obtained by showing that, under the conditions stated, $SD\, f(x) = -\infty$ cannot happen; then the theorem reduces to one of the statements above.

(**Larson**) *Let f be a measurable function having everywhere a symmetric derivative (possibly infinite) such that $SD\, f(x) \geq 0$ almost everywhere and such that the symmetric derivative itself has the Darboux property. Then f is nondecreasing on the set of its continuity points.*

Finally in a different direction Pu and Pu [239] obtain the following result. This should be compared with the classical monotonicity theorem for the Dini derivatives due to Zygmund (see Saks [251, Theorem 7.1, p. 203]); a proof can be fashioned by studying the classical methods.

(**Pu and Pu**) *Let the function f satisfy the condition*

$$\limsup_{y \to x-} f(y) \leq f(x) \leq \limsup_{y \to x+} f(y)$$

at each point x and suppose that the image under f of the set

$$\{x : \underline{SD}\, f(x) \leq 0\}$$

contains no nondegenerate interval. Then f is nondecreasing.

6
Odd Properties

6.1 Introduction

In this chapter we complete a systematic study of the odd properties of real functions. We recall that the study of the odd properties of a function reveals its even symmetric structure. At the simplest level a function is even if and only if its odd part vanishes. Similarly a greater or lesser degree of control on the behavior of the odd part of a function expresses some characteristic of its even symmetric structure.

The symmetry properties that we propose now to study all concern the nature of a function f that satisfies some requirement on the size of the expression

$$f(x + h) - f(x - h) \quad (\text{as } h \to 0).$$

For example the most severe requirement is to ask that at each point x there is a neighbourhood in which the symmetry $f(x + h) = f(x - h)$ holds. We shall say a function f is exactly locally symmetric at x if this holds. One expects that this vanishing odd part, giving a local even symmetry at each point, should arise only for a function that is symmetric about every vertical line (i.e. a constant function). This is very nearly the case and leads to some interesting results.

A much weaker requirement might be to ask only that

$$(f(x + h) - f(x - h))/2h$$

remain bounded as $h \to 0$; a function with a symmetric derivative would have this property. We have already encountered some results of this type in the first two chapters and we have seen a large collection of analogous problems stated in terms of even properties in Chapter 4. While there are parallels in the behavior of the odd and even properties our study reveals a broad range of differences and introduces different techniques.

Properties relating more directly to the symmetric derivatives themselves will be studied in Chapter 7.

6.2 Symmetry

We begin our study with exact symmetry conditions. A set E would be exactly symmetric at a point x if the condition $x + h \in E$ is equivalent to $x - h \in E$. A function would be exactly symmetric at a point x if there is an identity $f(x+h) = f(x-h)$. Naively this is our starting point and we carry investigations on this theme as far as they can go turning to local versions later on.

A function cannot be everywhere exactly symmetric unless it is constant; a set cannot be everywhere exactly symmetric unless it is empty or contains all reals. The remaining questions are not so easy or predictable.

6.2.1 Exact Symmetry

In this section we determine the nature of the set of points at which a function (set) may be exactly symmetric. It is easy to see that a function can have many points of symmetry. The function $f(x) = \cos \pi x$ is symmetric at the origin and at 1; since f is periodic it is easy to see that f is also symmetric at every integer point. The points at which it is symmetric form an additive group related to the periodic structure of f. This example goes a long way towards suggesting what the general picture must look like.

Definition 6.1 A real number x_0 is a point of *exact symmetry* of a function f if $f(x_0 + h) = f(x_0 - h)$ for all $h \in \mathbb{R}$. A real number x_0 is a point of *exact symmetry* of a set S if $x_0 + h \in S$ if and only if $x_0 - h \in S$ ($h \in \mathbb{R}$).

Definition 6.2 The set of all points at which a function f is exactly symmetric will be denoted $\mathcal{E}_f$. The set of points at which a set S is exactly symmetric will be denoted $\mathcal{E}(S)$.

Definition 6.3 A function f is *periodic* and p is a *period* of f if

$$f(x + p) = f(x) \quad \text{for all } x \in \mathbb{R}.$$

Definition 6.4 A set $H \subset \mathbb{R}$ is *translation homogeneous* if $x + y - z \in H$ whenever $x, y, z \in H$.

It is easy to prove (see [201, p. 244]) that any translation homogeneous set H is of the form $x_0 + G$ for $x_0 \in H$ and some additive subgroup $G \subset \mathbb{R}$ and that this characterizes the subsets of $\mathbb{R}$ that are translation homogeneous.

The next three lemmas show how the problem of characterizing the sets $\mathcal{E}_f$ can be reduced. The second lemma shows that the study of the sets $\mathcal{E}_f$ reduces to that of characterizing the translation homogeneous sets and, hence, merely to a discussion of additive subgroups of $\mathbb{R}$. There is a large literature on additive subgroups of the reals. Nonmeasurable subgroups are easily obtained by the use of Hamel bases (eg. [132]). Any subgroup G for which $\mathbb{R}/G$ is countable must be nonmeasurable; there are subgroups G for which $\mathbb{R}/G$ has cardinality c and which are universally nonmeasurable (a simple proof appears in Stromberg [278]).

Lemma 6.5 *Let* $H = x_0 + G$ *be a translation homogeneous set where* G *is a nontrivial additive subgroup of* $\mathbb{R}$. *Then, with* $E = x_0 + 2G$, *the function* $f = \chi_E$ *is periodic,* $2G$ *is its set of periods and* $\mathcal{E}_f = H$.

Proof. It is easy to see that the function $f = \chi_E$ is periodic and that $2G$ is its set of periods; let us prove that $\mathcal{E}_f = H$. Suppose that $g \in G$. Then for any h, $x_0 + g + h \in x_0 + 2G$ would imply that $g + h = 2g_1$ (some $g_1 \in G$) so that $g - h = 2(g - g_1)$ and hence

$$x_0 + g - h \in x_0 + 2G.$$

Thus $x_0 + g \in \mathcal{E}_f$ for any $g \in G$. Conversely if $y \notin G$ then choose some $x_0 + h \in x_0 + 2G$. The symmetric point about $x_0 + y$ is

$$x_0 + y - (x_0 + h - x_0 - y) = x_0 + 2y - h$$

which cannot belong to $x_0 + 2G$. Consequently f is not symmetric about $x_0 + y$ and so $x_0 + y \notin \mathcal{E}_f$ for any $y \notin G$.

Note, in particular, that if the set G is an additive subgroup of $\mathbb{R}$ then its set of points of exact symmetry is precisely the group $\mathcal{E}(G) = \frac{1}{2}G$.

Lemma 6.6 *Let* f *be an arbitrary function. Then* $\mathcal{E}_f$ *is translation homogeneous. Conversely if* E *is translation homogeneous then* $\mathcal{E}_f = E$ *for some function* f.

Proof. Suppose that $x, y, z \in \mathcal{E}_f$. We show that $x + y - z \in \mathcal{E}_f$ and the lemma follows. Simply write, for any $h \in \mathbb{R}$,

$$f(x + y - z + h) = f(x - y + z - h) = f(z + x - y - h)$$
$$= f(z - x + y + h) = f(y + z - x + h) = f(y - z + x - h)$$

so $f([x + y - z] + h) = f([x + y - z] - h)$ and we are done. The converse is already contained in the preceding lemma.

The third lemma shows that functions with more than one point of exact symmetry are periodic; this is just the simple geometric observation that two reflections produce a translation.

Lemma 6.7 *Let f be an arbitrary function and suppose that $\mathcal{E}_f$ contains at least two points. Then f is periodic and $2(\mathcal{E}_f - \mathcal{E}_f)$ is the set of its periods.*

Proof. Let $a, b \in \mathcal{E}_f$ and write $p = (b - a)$. Then, for any $x \in \mathbb{R}$,

$$f(x+p) = f(b-a+x) = f(b+a-x) = f(a+b-x) = f(a-b+x) = f(x-p).$$

This shows that f is $2p$ periodic and the lemma follows.

Having related the study of points of exact symmetry to the study of periodic functions and subgroups of real numbers we may now appeal to the literature of that subject to obtain some insight into the structure we are studying. The first two theorems are due to Burstin (see [201, pp. 236–237]). The condition that the set of points of exact symmetry $\mathcal{E}_f$ is dense forces the function f to be periodic with a dense set of periods; this is equivalent to the statement that f has arbitrarily small periods (microperiodic) or that the group of periods of f is not generated by a single element (multiperiodic).

Theorem 6.8 (Burstin) *Let f be measurable and suppose that $\mathcal{E}_f$ is dense. Then f is constant off of a set of measure zero.*

Theorem 6.9 (Burstin) *Let f have the Baire property and suppose that $\mathcal{E}_f$ is dense. Then f is constant off of a set of first category.*

It would be tempting to take these two theorems together and come to the conclusion that a function f having both properties (measurable and the Baire property) must be constant off a set that is both measure zero and first category. This is not so. Burstin provides a simple example (see [201, pp. 238]) of a set A of measure zero, whose complement $\mathbb{R} \setminus A$ has first category, so that χ_A is symmetric about every rational number.

These two theorems, applied to sets, have the following corollaries.

Corollary 6.10 *Let the set S be measurable and suppose that $\mathcal{E}(S)$ is dense. Then either S or $\mathbb{R} \setminus S$ has measure zero.*

Corollary 6.11 *Let the set S have the Baire property and suppose that $\mathcal{E}(S)$ is dense. Then either S or $\mathbb{R} \setminus S$ is first category.*

Continuing this theme we may ask just how complicated may be the set $\mathcal{E}(S)$ of points of exact symmetry of a set S. The following results illustrate the situation. Some are extracted from [201, pp. 239–241]. The two Erdös theorems are proved in [13]; the Foran result is from [105].

Theorem 6.12 (Burstin) *There are 2^c many sets S that are measurable [have the Baire property] and that are exactly symmetric about each rational number.*

Theorem 6.13 (Erdös) *(ZFC+CH). There is a measurable subgroup $S \subset \mathbb{R}$ so that $\frac{1}{2}S = \mathcal{E}(S)$ is measure zero and second category.*

Theorem 6.14 (Erdös) *(ZFC+CH). There is a subgroup $S \subset \mathbb{R}$ with the Baire property so that $\frac{1}{2}S = \mathcal{E}(S)$ is first category and has positive measure.*

Theorem 6.15 (Foran) *Let $0 \leq \alpha \leq 1$. Then there is a subgroup $S \subset \mathbb{R}$ so that $\frac{1}{2}S = \mathcal{E}(S)$ has Hausdorff dimension exactly α.*

Theorem 6.16 (Smítal–Erdös) *(ZFC+CH). There is an uncountable subgroup $S \subset \mathbb{R}$ so that $\frac{1}{2}S = \mathcal{E}(S)$ has only countably many points in common with any set that has measure zero [is first category].*

We conclude with a related construction from Uher [294, p. 429] where it is attributed to Preiss. There are other similar constructions in the literature; see Ponomarev [225] for measure and Belna [12] for category. The construction in Theorem 6.17 handles the measure and the category together.

Theorem 6.17 *There is a subgroup H of the irrationals so that every uncountable Borel subset of $\mathbb{R}$ meets H.*

Proof. We let Q denote the rationals and $Q[Z]$ the linear span of the set $Z \subset \mathbb{R}$ over the rationals. Let $\{F_\alpha : \alpha < 2^\omega\}$ be the family of all uncountable Borel sets indexed by ordinal numbers α. We define a sequence of points $\{x_\alpha : \alpha < 2^\omega\}$ inductively in such a way that

$$Q \cap Q[\{x_\alpha : \alpha < \beta\}] = \emptyset \tag{6.1}$$

for every ordinal $\beta < 2^\omega$. Choose an arbitrary point $x_1 \in F_1 \backslash Q$. If $\{x_\alpha : \alpha < \beta\}$ has been chosen so that (6.1) holds then it is possible to choose $x_\beta \in F_\beta$ so that

$$Q \cap Q[\{x_\alpha : \alpha \leq \beta\}] = \emptyset \tag{6.2}$$

too. To see this an elementary cardinality argument suffices. Since $\beta < 2^\omega$

$$\mathrm{card}\,(Q[Q \cup \{x_\alpha : \alpha < \beta\}]) \leq \max\{\omega, \beta\} < \mathrm{card} F_\beta. \tag{6.3}$$

Thus a point $x_\beta \in F_\beta$ can be chosen with

$$x_\beta \notin Q[Q \cup \{x_\alpha : \alpha < \beta\}].$$

This then gives (6.2) as required.

Now we set $H = Q[\{x_\alpha : \alpha < 2^\omega\}]$. Every uncountable Borel set clearly meets H, H contains only irrationals and H is an additive group.

As a corollary we express the theorem in terms of the characteristic function of the set there. This provides an example of a highly symmetric function which is nonetheless very irregular. The characteristic function χ_H for the set H of this theorem provides an example of a function which is nonmeasurable and nowhere continuous and yet satisfies at every point of H a severe symmetry condition: it is exactly symmetric and so also symmetrically continuous and symmetrically differentiable with zero derivative at every point of $\frac{1}{2}H$. This will provide in the sequel many counterexamples showing that some kind of measurability assumption is needed in a variety of theorems.

Corollary 6.18 *There is a nonmeasurable and nowhere continuous function f such that $\mathcal{E}_f$ is full in the sense of outer measure and the sense of category.*

6.2.2 Essential Symmetry

Except for the trivial cases $E = \emptyset$ and $E = \mathbb{R}$ a set cannot be symmetric about every point. A remarkable thing happens if we relax this. Let us say that E is *essentially symmetric* at a point x if $x + y \in E$ if and only if $x - y \in E$ with at most countably many exceptions. Sierpiński [258] has shown, under the continuum hypothesis, the existence of sets that are everywhere essentially symmetric.

Theorem 6.19 (Sierpiński) *(ZFC + CH). There is a nonmeasurable set E such that the set of points*

$$\{t : x - t \in E \text{ if and only if } x + t \in E\}$$

is co-countable for each $x \in \mathbb{R}$.

Taking $f = \chi_E$ in the theorem gives a similar statement for "essentially symmetric" functions.

Theorem 6.20 (Sierpiński) *(ZFC + CH). There is a nonmeasurable function f such that the set of points*

$$\{t : f(x - t) \neq f(x + t)\}$$

is countable for each $x \in \mathbb{R}$.

Theorems 6.19 and 6.20 follow from the more precise formulation we now give. Again this is due to Sierpiński; he gave as well a partial proof using Hamel bases which he ascribed to Ruziewicz. We employ a simpler argument due to Freiling [110]. Note that in this version CH is not assumed (i.e. the theorem is valid in ZFC) but that both Theorems 6.19 and 6.20 are then immediate.

Theorem 6.21 (Sierpiński) *There is a nonmeasurable set Y of full outer measure so that the set*

$$Y \setminus (2x - Y)$$

has cardinality $< 2^{\aleph_0}$ for each $x \in \mathbb{R}$.

Proof. Let H be a Hamel basis with full outer measure. Well order H by a relation $\prec$ so that each element of H has fewer than $2^{\aleph_0}$ predecessors. To each $r \in \mathbb{R}$ there is a unique element $h(r)$ in H defined as that element of the basis which appears in the expansion of r and is maximal in the order $\prec$. Note that the set

$$\{t \in \mathbb{R} : h(t) \prec h(r) \text{ or } h(t) = h(r)\} \tag{6.4}$$

has cardinality $< 2^{\aleph_0}$.

Define Y as the set of real numbers t so that the coefficient corresponding to $h(t)$ is $+1$ or -1. Then $Y \supset H$ so Y has full outer measure and also Y is disjoint from $2H$ so Y cannot be measurable.

Fix $x \in \mathbb{R}$ and consider any $t \in \mathbb{R}$ with $h(x) \prec h(t)$. It is clear that $h(2x - t) = h(t)$ so that $t \in Y$ if and only if $2x - t \in Y$. This means that for each $x \in \mathbb{R}$ the set $Y \setminus (2x - Y)$ is contained in the set (6.4) and hence has cardinality $< 2^{\aleph_0}$ as we were required to show.

To present yet another perspective on these theorems we give the following variant from Preiss and Thomson [232] that includes the Sierpiński construction. Here there is an asymmetry at certain points as well as an essential symmetry condition.

Theorem 6.22 *(ZFC + CH). The following statement holds for an arbitrary linear subspace E of $\mathbb{R}$ over the field of rational numbers. There is a set $Y \subset \mathbb{R}$ such that both the sets Y and $\mathbb{R} \setminus Y$ are of full outer measure and such that for every $x \in E$ the set $Y \setminus (2x - Y)$ is countable, and for every $x \in \mathbb{R} \setminus E$ the set $Y \setminus (2x - Y)$ is of full outer measure.*

Proof. To simplify the notation, let $Q[Z]$ denote the linear span of the set $Z \subset \mathbb{R}$ over the field of rational numbers.

Under the continuum hypothesis we may arrange the set of all pairs (x, C), where $x \in \mathbb{R}$ and C is an uncountable compact subset of $\mathbb{R}$, into a sequence (x_τ, C_τ) indexed by countable ordinals τ. By transfinite induction we choose points y_τ and z_τ, and sets Y_τ and Z_τ as follows. (As usual, to include the first step of the construction in the general description, we interpret the union over an empty family of indices as the empty set.)

For y_τ choose an arbitrary point of

$$C_\tau \setminus Q \left[\{x_\tau\} \cup \bigcup_{\sigma < \tau} Z_\sigma \right].$$

Define

$$Y_\tau = Q \left[\{y_\tau\} \cup \{x_\sigma; \; \sigma \leq \tau, \; x_\sigma \in E\} \right] \setminus \bigcup_{\sigma < \tau} Z_\sigma,$$

choose

$$z_\tau \in C_\tau \setminus Q \left[\{x_\tau, y_\tau\} \cup \bigcup_{\sigma < \tau} Z_\sigma \right],$$

and put

$$Z_\tau = Q \left[\{x_\tau, y_\tau, z_\tau\} \cup \bigcup_{\sigma < \tau} Z_\sigma \right].$$

Let $Y = \bigcup_{\sigma < \omega_1} Y_\sigma$. Observing that $Y_\tau \subset Z_\tau = Q[Z_\tau] \subset Z_\sigma$ and $Y_\sigma \cap Z_\tau = \emptyset$ whenever $\tau < \sigma$, we easily see that $y_\tau \in Y \cap C_\tau$ and $z_\tau \in (\mathbb{R} \setminus Y) \cap C_\tau$ for each countable ordinal τ. This implies that both the sets $\mathbb{R} \setminus Y$ and Y have full outer measure.

If $x \in E$, we find τ such that $x = x_\tau$ and we obtain the fact that $Y \setminus (2x - Y)$ is countable by establishing the inclusion

$$Y \setminus (2x - Y) \subset \bigcup_{\sigma \leq \tau} Y_\sigma.$$

To prove this, assume that $y \in Y_\nu$, $2x_\tau - y \notin Y$, and $\nu > \tau$. Then y and x_τ belong to

$$Q \left[\{y_\nu\} \cup \{x_\sigma; \; \sigma \leq \nu, \; x_\sigma \in E\} \right].$$

Hence

$$2x_\tau - y \in Q \left[\{y_\nu\} \cup \{x_\sigma; \; \sigma \leq \nu, \; x_\sigma \in E\} \right].$$

However, since $2x_\tau - y$ does not belong to Y_ν, this can happen only if

$$2x_\tau - y \in \bigcup_{\eta < \nu} Z_\eta.$$

Let $\tau \leq \eta < \nu$ be such that $2x_\tau - y \in Z_\eta$. Then, since $x_\tau \in Z_\eta$ and since $Z_\eta = Q[Z_\eta]$, we infer that $y \in Z_\eta$. But this contradicts $y \in Y_\nu$.

We finish the proof by showing that for every $x \in \mathbb{R} \setminus E$ the set $Y \setminus (2x - Y)$ intersects every uncountable compact subset C of $\mathbb{R}$. To prove this, we find a countable ordinal τ such that $(x, C) = (x_\tau, C_\tau)$. Since $y_\tau \in C \cap Y$, we just need to prove that $2x_\tau - y_\tau \notin Y$. But this is almost obvious, since $2x_\tau - y_\tau$ belongs to Z_τ but not to $\bigcup_{\sigma < \tau} Z_\sigma$ otherwise

$$y_\tau \in Q \left[\{x_\tau\} \cup \bigcup_{\sigma < \tau} Z_\sigma \right].$$

This means that $2x_\tau - y_\tau$ can be in Y only if it is in Y_τ. However, since

$$y_\tau \notin Q \left[\{x_\tau\} \cup \bigcup_{\sigma < \tau} Z_\sigma \right],$$

$2x_\tau - y_\tau \in Y_\tau$ implies that

$$x = x_\tau \in Q \left[\{x_\sigma; \, \sigma < \tau, \, x_\sigma \in E\} \right] \subset E.$$

We should explore in a little more detail the asymmetry part of Theorem 6.22. One might ask, in general, to what extent is it possible to have a "large" set Y so that everywhere $Y \cap (2x - Y)$ is "small", that is so that Y is not particularly symmetric? Note that for a measurable set Y having x as a point of density this intersection could not be very small. Freiling has supplied to this end the following theorem, utilizing the same methods as his proof of Theorem 6.21. (See also the discussion in Section 2.8 on antisymmetric sets.)

Theorem 6.23 (Freiling) *There is a nonmeasurable set Y of full outer measure so that $Y \cap (-Y) = \emptyset$, $Y \cup (-Y) = \mathbb{R} \setminus \{0\}$ and the set*

$$Y \cap (2x - Y)$$

has cardinality $< 2^{\aleph_0}$ for each $x \in \mathbb{R}$.

Proof. Let H, $\prec$, and $h(r)$ be as in the proof of Theorem 6.21. This time write Y as the set of points y so that the coefficient corresponding to the element $h(y)$ in the Hamel expansion of y is positive. Clearly

$$Y \cap (-Y) = \emptyset \text{ and } Y \cup (-Y) = \mathbb{R} \setminus \{0\}.$$

Then $Y \supset H$ so Y has full outer measure and also Y is disjoint from $-H$ so Y cannot be measurable.

Fix $x \in \mathbb{R}$ and consider any $t \in \mathbb{R}$ with $h(x) \prec h(t)$. It is clear that $h(2x - t) = h(t)$ so that $t \in Y$ if and only if $2x - t \in (-Y)$. This means that for each $x \in \mathbb{R}$ the set $Y \cap (2x - Y)$ is contained in the set (6.4) and hence has cardinality $< 2^{\aleph_0}$ as we were required to show.

The notion of essential symmetry can also be continued by investigating other classes of exceptional sets than the countable sets. Let $\mathcal{N}$ denote any σ-ideal of subsets of $\mathbb{R}$. The classes of countable sets, first category sets and measure zero sets are the ones that come most immediately to mind. We can say that a function f is $\mathcal{N}$-*essentially symmetric* at a point x_0 if there is a set $N \in \mathcal{N}$ so that

$$f(x_0 + t) = f(x_0 - t) \quad (t \in \mathbb{R},\ t \notin N).$$

We see, because of Theorem 6.22, that even if N is the class of countable sets there is a function f that is everywhere N-essentially symmetric but f is not close to being constant. With some measurability assumptions this cannot happen.

The following two theorems are the analogues of Theorems 6.8 and 6.9. They can be proved in much the same way by relating the set of points at which a function is N-essentially symmetric to its "essential" periodic structure. For details see [201, pp. 243–244].

Theorem 6.24 *Let f be measurable, let N denote the class of sets of measure zero and suppose that f is N-essentially symmetric at each point of a dense set. Then f is constant off of a set of measure zero.*

Theorem 6.25 *Let f have the Baire property, let N denote the class of sets of first category and suppose that f is N-essentially symmetric at each point of a dense set. Then f is constant off of a set of first category.*

6.2.3 Exact Local Symmetry Everywhere

We turn now to local considerations. A function f is *exactly locally symmetric* at a point x if there is a neighbourhood in which the symmetry $f(x + h) = f(x - h)$ holds. Similarly a set E is *(exactly) locally symmetric* at a point x if $x + h \in E$ precisely when $x - h \in E$ for sufficiently small h; this is equivalent to asking that the characteristic function χ_E be exactly locally symmetric at a point x. It is not difficult to see that this is also equivalent to asking that χ_E be symmetrically continuous at x.

As we have reported in Chapter 2, there was a query posted in an early issue of the Real Analysis Exchange ([78]) whether a set or function that is everywhere locally symmetric need be measurable. In view of Charzyński's Theorem (Theorem 2.9) this property, which is more severe than the property of that theorem, would require that such a function be in fact continuous everywhere excepting a scattered set. As Davies [60] and Ruzsa [250] pointed out in the next issue of the Exchange a sharper result than this is possible. We have previously proved this result (Theorem 2.27) as an illustration of the basic methods of Charzyński. We repeat this result here but we now supply a proof that flows more directly from covering arguments.

Theorem 6.26 *Let f be a function having the property that, at each point x, there is a positive number $\delta(x)$ so that*

$$0 < t < \delta(x) \implies f(x + t) - f(x - t) = 0.$$

Then f is constant off of a closed, countable set.

Proof. Define the covering relation

$$V = \{(x,t) : f(x - t) = f(x + t)\}.$$

Notice that V is transitive. From the exact local symmetry of f we must have a positive function $\delta(x)$ so that

$$0 < |t| < \delta(x) \Longrightarrow (x,t) \in V.$$

Theorem 3.8 induces a partition $\{E_n\}$ of the real line so that

$$x \in E_n, \ 0 < |t| < 1/n \Longrightarrow (x,t) \in V,$$

and the Baire category Theorem then supplies an interval (a,b) and a set E_n dense in (a,b) with $b - a < 1/n$. By Theorem 3.11 we have, for all $c, d \in (a,b)$, that the interval $[c,d]$ belongs to $V^8 = V$ and, consequently, $f(c) = f(d)$. This proves that f is constant on the interval (a,b).

Now that we have at least one interval of constancy of f we can apply Ruzsa's argument from Theorem 2.27 on page 48. Alternatively the covering Lemma 3.14 provides another argument. If $c \in (a,b)$ then there is a countable closed set $C \subset (c, \infty)$ so that $(c, (x - c)) \in V$ for every $x \in (c, \infty) \setminus C$, and hence then $f(2c - x) = f(x)$.

In fact the chain in V that connects $2c - x$ and x can be arranged to be of the form

$$2c - x = x_{-n}, \ \ldots, \ x_{-1}, x_1, \ldots, x_n = x$$

with both x_{-1} and x_1 inside the interval (a,b). (We need only make $\delta(x)$ smaller.) This would give $f(2c - x) = f(x_{-1}) = f(x_1) = f(x)$ for all such x and so f is constant off of the countable closed set $C \cup 2c - C$.

As a corollary we can characterize sets that are locally symmetric at every point of the real line.

Corollary 6.27 *Let E be a set that is locally symmetric at every point of the real line. Then either E or its complement has a countable closure.*

Notice that we cannot say more; if a set E is locally symmetric at every point of the real line then the same assertion holds for the set $E \setminus \{c\}$ for any point c. This result is also sharp in the sense that, for any countable closed set E, there is a countable closed set $F \supset E$ that is everywhere locally symmetric. Both Ruzsa [250, p. 86]) and Freiling [111] state this without proof. When pressed by the author the latter supplied the following reasoning.

Theorem 6.28 (Freiling) *Let E be any countable closed set. Then there is a countable closed set $F \supset E$ that is everywhere locally symmetric.*

Proof. We can assume that E is compact since any countable closed set could be broken into compact fragments and each handled separately. Associate with each point its usual Cantor-Bendixson rank. Every compact set C can be assigned a rank, defined as the maximum of the rank of its elements. Since C is compact there must exist a point with maximum rank: for any sequence $x_i \in C$ with increasing ranks some subsequence would converge to a point $x \in C$ and the rank of x would have to be higher since it is not isolated from the points in the sequence.

We apply induction on the rank of C. Consider the statement that for every compact set C of rank α and every interval $(a, b) \supset C$ there is a countable compact set $(a, b) \supset D \supset C$ that is everywhere locally symmetric. This statement is trivially true for sets of the first rank since such a set is finite and so already locally symmetric.

Assume the statement true for all ordinals $\alpha < \beta$ and suppose that C is a compact set of rank β and $J = (z_1, z_2)$ any containing interval. Since C has rank β there are finitely many points of exactly this rank in C. (If there were an infinite sequence, then some subsequence would converge to a point of higher rank.)

Let these points be arranged $a < b < c \ldots$. We argue inside each interval (z_1, a), (a, b), (b, c) ... as follows. Take a sequence of points $\ldots < x_{-2} < x_{-1} < x_0 < x_1 < x_2 < \ldots$ in the complement of C with x_n as $n \to \infty$ tending to the right hand endpoint and with x_n as $n \to -\infty$ tending to the left hand endpoint. Inside each $C \cap (x_{i-1}, x_i)$ we have a compact set of rank less than β and so, by the induction hypothesis, each may be enlarged to a countable compact set that is locally symmetric. Let C_1 be the set C so enlarged by applying this process to each of the intervals in turn.

Now C_1 is countable, compact and is locally symmetric except possibly at the finite collection of points a, b, c ... of maximum rank. An easy adjustment can be made at each point. Say at a we can choose an open interval $I \subset J$ centered at a, containing no other point of maximum rank and with endpoints in the complement of C_1. The set $(C_1 \cup 2a - C_1) \cap I$ is now locally symmetric at a and remains locally symmetric at every other point of I. If we apply this process to each of the points a, b, c ... of maximum rank we can construct a countable compact set $D \subset J$ that contains C and is everywhere locally symmetric. The theorem now follows by induction.

Let us return to the argument in the proof of Theorem 6.26: this could be replaced by an appeal to the covering Theorems of Uher (Theorem 3.25 or 3.26) rather than Theorem 3.11. In particular the following versions would follow by these arguments.

Theorem 6.29 *Let f be a function having the property that at each point x in a set E with the Baire property there is a positive number $\delta(x)$ so that*

$$0 < t < \delta(x) \implies f(x+t) - f(x-t) = 0.$$

Then there is an open set G so that $E \setminus G$ is first category and f is constant on each component of G.

Theorem 6.30 *Let f be a function having the property that at each point x in a measurable set E with positive measure there is a positive number $\delta(x)$ so that*

$$0 < t < \delta(x) \implies f(x+t) - f(x-t) = 0.$$

Then there is an open set G containing almost every point of E so that f is constant on each component of G.

Kostryko, Neubrunn, Smítal and Šalát [158] study other problems related to the structure of locally symmetric functions. They point out that from Theorem 6.26 one can deduce that such functions belong to the first Baire class. They show that the class of locally symmetric functions is not closed under uniform limits.

6.2.4 Points of Local Symmetry of a Set

A set S is *locally symmetric* at a point x provided for some $\delta = \delta(x) > 0$ that $x + h \in S \implies x - h \in S$ for all $0 < |h| < \delta$. The set of all points at which a set S is locally symmetric will be denoted S^*. In this section we consider some problems relating to the structure of the set S^*.

We illustrate the utility of this concept by proving a simple lemma from Chlebík [43, p. 34] that will be useful later.

Lemma 6.31 (Chlebík) *Let A be a set of real numbers and f a function that is everywhere symmetrically continuous. Suppose that $\lim_{y \to x} f(y) = 0$ for every $x \in \mathbb{R} \setminus A^*$. Then the product $f\chi_A$ is symmetrically continuous.*

Proof. It is almost immediate that $f\chi_A$ is symmetrically continuous at each point x at which $\lim_{y \to x} f(y) = 0$. If $x \in A^*$ then there is a $\delta > 0$ so that $\chi_A(x+h) = \chi_A(x-h)$ for all $0 < |h| < \delta$. Writing $g(x) = f(x)\chi_A(x)$ we easily see that

$$g(x+h) - g(x+h) = (f(x+h) - f(x-h))\chi_A(x+h)$$

and so g is symmetrically continuous at any such point x. This completes the proof.

We now investigate the structure of the set S^*. We have already seen (Corollary 6.27) that $S^* = \mathbb{R}$ only if S or its complement has countable closure. We will show later (Theorem 6.37) that S^* is measurable whenever S is. We have seen (Theorem 2.44) that S^* is a relatively simple Borel set if S is either a G_δ or an F_σ. On the other hand as we now find, in Theorem 6.32, there is a nonmeasurable set S for which S^* is nonmeasurable. This can also be found in Jaskula and Szkopińska [140, p. 32] where they give a nonmeasurable set E with $E = E^*$. Erdös [80] also constructs an example but under the continuum hypothesis.

Theorem 6.32 *There is a set E such that E^* is nonmeasurable and does not have the Baire property.*

Proof. Take $E = H$ where H is the set of Theorem 6.17. Certainly $E^* \supset E$. We can directly show that E^* is nonmeasurable. This follows also if we grant some later material. The function $f = \chi_E$ is nonmeasurable and symmetrically continuous at every point in E^*. The set E^* cannot be measurable since, if it were, then f is symmetrically continuous almost everywhere and, as we shall see later, such a function must be measurable. The category version is similar.

In the positive direction we prove that the set S^* is at least co-analytic when S is a Borel set. (See Theorem 2.44 for another proof.)

Lemma 6.33 *If S is Borel then S^* is co-analytic.*

Proof. Let S_n denote the set of all points x with the property that $x + h \in S$ if and only if $x - h \in S$ for all $0 < |h| < 1/n$. Then $S^* = \cup_{n=1}^\infty S_n$ and so it is enough to show that S_n is co-analytic for any n.

Fix n and let

$$A = \{(x,y) : x \in \mathbb{R}, \; y \in S\}$$

and

$$B = \{(x,y) : x,y \in \mathbb{R}, \; 0 < |x - y| < 1/n\}.$$

Let $E = A \cap B$ and $E' = g(E)$ where g is the continuous one to one function from $\mathbb{R}^2 \to \mathbb{R}^2$ defined by $g(x,y) = (x, 2x - y)$. Since A is Borel and B is open the sets E and E' are Borel too.

We show that $\mathbb{R} \setminus S_n$ is the projection onto the first coordinate of the Borel set $E \setminus E'$. This shows that $\mathbb{R} \setminus S_n$ is analytic and so S_n is co-analytic. If $x \notin S_n$ then there must exist a point $x + h \in S$ so that $x - h \notin S$ and $0 < |h| < 1/n$. If we let $y = x + h$ then $(x,y) \in A$ and, since $|x - y| = |h| < 1/n$, $(x,y) \in B$ also. Since $2x - y = x - h \notin S$ it follows that $(x,y) \notin E'$. Thus $(x,y) \in E \setminus E'$. We have shown that the projection of $E \setminus E'$ contains $\mathbb{R} \setminus S_n$.

Conversely if x is a member of that projection then there is a $y \neq x$ with $(x,y) \in E \setminus E'$. This means that $y \in S$, that $2x - y \notin S$ and that $|x - y| < 1/n$. By definition $x \notin S_n$ as we wished to show.

This lemma cannot be improved. Hammill [128] showed that there is a Borel set A such that A^* is not Borel. As her proof has not to date been published we shall reproduce it here. The arguments exploit the fact that there is a Hamel basis that contains a perfect set and use some clever methods developed by Laczkovich [170]. The Borel class of the set A in the proof can be reduced further than here but cannot be reduced too far: because of Theorem 2.44, if A is a G_δ or an F_σ then A^* is a $G_{\delta\sigma\delta}$.

Theorem 6.34 (Hammill) *There is a Borel set A such that A^* is not Borel.*

Proof. Let $P \subset [0,1]$ be a perfect set that is linearly independent over the rational numbers (see Section A.6 in the Appendix). Let P_1 and P_2 be disjoint perfect subsets of P and let $K \subset P_1$ be an analytic non Borel set. Then there is a G_δ set $U \subset P_1 \times P_2$ such that K is the projection onto P_1 of U. We now construct a set A of type $F_{\sigma\delta\sigma}$ such that $P_1 \setminus A^* = K$. Since K is not Borel it follows that A^* cannot be Borel or analytic.

For each natural number n define A_n as:

$$A_1 = \{\pm(x + \frac{1}{k}y) : (x,y) \in U, k \in \mathbb{N}\}$$

$$A_n = \{2(a_n x_n + \cdots + a_2 x_2) + a_1(x_1 + \tfrac{1}{k}y) : (x_1,y) \in U, k \in \mathbb{N},$$
$$x_1, \ldots, x_n \in P_1, x_i \neq x_j \, (i \neq j), a_1 = \pm 1, |a_i| \in \mathbb{N} \, (i = 2, \ldots, n)\}.$$

Then let $A = \bigcup_{n=1}^\infty A_n$. We will show that A is of type $F_{\sigma\delta\sigma}$. For each natural number n define:

$$B_n = \{(x_n, \ldots, x_1, y) : x_n > \cdots > x_2, x_i \neq x_1 (i = 2, \ldots, n)\}$$

and

$$C_n = B_n \cap \underbrace{P_1 \times \cdots \times P_1}_{n-1} \times U.$$

Since U is a G_δ, P_1 is closed and B_n is open, C_n is a $F_{\sigma\delta}$ subset of $\mathbb{R}^{n+1}$. For each n let

$$\mathcal{A}_n = \{(a_n, \ldots, a_1, k) : a_1 = \pm 1, |a_j| \in \mathbb{N}(j = 2, \ldots, n), k \in \mathbb{N}\}$$

and for each $a \in \mathcal{A}_n$ define $g_a : \mathbb{R}^{n+1} \to \mathbb{R}$ by :

$$g_a(x_n, \ldots, x_1, y) = 2(a_n x_n + \cdots + a_2 x_2) + a_1(x_1 + \tfrac{1}{k}y).$$

Then g_a is continuous and one to one on $B_n \cap P_1 \times \cdots \times P_1 \times (P_1 \times P_2)$. To see this let $(x_n, \ldots, x_1, u)$ and $(y_n, \ldots, y_1, w)$ be in $B_n \cap P_1 \times \cdots \times P_1 \times (P_1 \times P_2)$ and

$$2(a_n x_n + \cdots + a_2 x_2) + a_1(x_1 + \frac{1}{k}u) = 2(a_n y_n + \cdots + a_2 y_2) + a_1(y_1 + \frac{1}{k}w).$$

Then by the rational linear independence of P, $u = w, x_1 = y_1$ and since $x_n > \cdots > x_2$ and $y_n > \cdots > y_2$ we have $(x_n, \ldots, x_1, u) = (y_n, \ldots, y_1, w)$.

Thus, since g_a is continuous and one to one and C_n is a $F_{\sigma\delta}$, $g_a(C_n)$ is a $F_{\sigma\delta}$. Then since

$$A = \bigcup_{n=1}^{\infty} \bigcup_{a \in A_n} g_a(C_n)$$

A is an $F_{\sigma\delta\sigma}$.

Now we show that $P_1 \setminus A^* = K$. Let $x \in P_1 \setminus K$ and let $h \in \mathbb{R}$ be given. Suppose first that $x + h \in A_1$. Then $x + h = a_1(x_1 + \frac{1}{k}y)$ with $a_1 = \pm 1, k \in \mathbb{N}$ and $(x_1, y) \in U$. Since $x \notin K, x \neq x_1$. So

$$x - h = 2x - a_1(x_1 + \frac{1}{k}y) \in A_2.$$

Now suppose $x + h \in A_n, n \geq 2$. Then

$$x + h = 2(a_n x_n + \cdots + a_2 x_2) + a_1(x_1 + \tfrac{1}{k}y)$$

with $a \in A_n$, $x_i \neq x_j$, $(i \neq j)$, $x_j \in P_1$ and $(x_1, y) \in U$. Then $x \neq x_1$ and

$$x - h = 2(x - a_n x_n - \cdots - a_2 x_2) - a_1(x_1 + \tfrac{1}{k}y).$$

If $x = x_j$ for some $j \in \{2, \ldots, n\}$ then either (i) $x = a_j x_j$ in which case $x - h \in A_{n-1}$ or (ii) $x \neq a_j x_j$ in which case $x - h \in A_n$. If $x \neq x_j$ then $x - h \in A_{n+1}$. In any case $x - h \in A$. Thus $x \in A^*$.

Now we turn to $x \in K$. We wish to show that x is not a point of local symmetry. As $x \in K$ there is a point $y \in P_2$ such that $(x, y) \in U$. We show that the sequence $x + \frac{1}{k}y \searrow x$ belongs to A but that the sequence $x - \frac{1}{k}y \nearrow x$ does not.

By the definition of A, $x + \frac{1}{k}y \in A$ for each k. If also $x - \frac{1}{k}y \in A$ then

$$x - \frac{1}{k}y = 2(a_n x_n + \cdots + a_2 x_2) + a_1(x_1 + \frac{1}{j}u)$$

with $x_n, \ldots, x_1 \in P_1$, $(x_1, u) \in U$, $x_i \neq x_j$, $(i \neq j)$, $n \in \mathbb{N}$ and $a \in A_n$. Then by the rational linear independence of P, $a_1 = -1, j = k$ and $u - y$. Then

$$x = 2(a_n x_n + \cdots + a_2 x_2) - x_1.$$

Since the x_i are distinct this is impossible and we conclude that $x - \frac{1}{k}y \notin A$.

Thus $x \notin A^*$ as we wished to show.

6.2.5 Points of Local Symmetry of a Function

The example in Theorem 6.22 shows that a function can be exactly symmetric (not just locally) at each point of a large set E and yet meet no minimal regularity conditions. Also the set of points where a function is exactly locally symmetric need not be measurable nor have the Baire property. We can, using the covering Lemmas 3.25 and 3.26, however prove some positive results. In this section we examine further the structure the set of points at which an arbitrary function may be locally symmetric mainly by comparing this set with the set of points of constancy.

We let LS_f denote the set of points at which f is exactly locally symmetric and we let L_f denote the set of points at which f is constant (a function is *constant* at a point x if it is constant in a neighbourhood of the point x). Evidently the set L_f of points of constancy of f is open and an elementary compactness argument shows that f is constant on each component of that set.

Our theorem is the first in a series of results all of which follow from the basic covering theorems and all of which have as their theme a comparison of conditions that hold symmetrically with conditions that hold in a more ordinary sense. Thus here we compare a form of "symmetric constancy" with ordinary "constancy"; later we compare symmetric continuity with ordinary continuity, symmetric differentiability with ordinary differentiability, symmetric monotonicity with ordinary monotonicity and so on. In each case the arguments and statements of results are closely related and, as far as possible, we try to preserve the same form of expression and proof in order to show the unity. A single general theorem encompassing all the variants is possible but not very informative.

The theorem asserts that the set $LS_f \setminus L_f$ is very small except in what might be considered pathological cases; the example in Theorem 6.18 shows that these extreme cases can occur. Note in the statement of the theorem that a set that contains no measurable set of positive measure has inner measure zero and a set that contains no second category set with the Baire property is residual in no open interval; the theorem could have been expressed in this language instead. The definitions of qq-continuity and qa-continuity are given in Sections A.7 and A.8 in the Appendix; recall however that qa-continuity is a.e. equivalent to approximate continuity so that the only application of that part of the theorem would be to measurable functions.

Theorem 6.35 *Let f be an arbitrary function, let LS_f denote the set of points at which f is exactly locally symmetric and let L_f denote the set of points at which f is constant. Then the set*

$$LS_f \setminus L_f$$

contains no measurable set of positive measure and no second category set having the Baire property. If, furthermore qaC_f and qqC_f denote the sets of points at *which f is quasi-approximately continuous and quasi-qualitatively continuous respectively, then*

$$qaC_f \cap (LS_f \setminus L_f)$$

has measure zero and

$$qqC_f \cap (LS_f \setminus L_f)$$

is first category.

Proof. It is enough to indicate the measure-theoretic version as the translation to the category version is easily carried out. The proofs of the two versions are nearly identical. Recall that we write $d(E)$ for the set of density points of a set E.

Let E be a subset of $LS_f \setminus L_f$ and consider the covering relation

$$V = \{(x,t) : x \in E, \ t > 0, \ f(x - t) = f(x + t)\}.$$

By the assumptions in the theorem, at each point $x \in E$, there is a positive number $\delta(x)$ so that

$$0 < t < \delta(x) \implies (x,t) \in V.$$

Note that V is transitive.

For the first part of the theorem we assume that E is measurable. We may apply the covering Theorem 3.26. For almost every point $x \in E$ there is a neighbourhood U_x of x so that if $y \in U_x$ then the interval $[x,y]$ (or $[y,x]$) is in $V^5 = V$. Thus f is constant in each set U_x. Consequently almost every point in E is a point of constancy of f and from this it follows that E itself must have measure zero as required.

We turn now to the second part of the theorem, we drop the assumption that E is measurable and take $E = LS_f \setminus L_f$. We apply the covering Lemma 3.8 to obtain a partition $\{E_n\}$ of E so that

$$x \in E_n, \ 0 < |t| < 1/n \implies (x,t) \in V.$$

Let z be a point in both $d(E_n)$ and qaC_f. We have the conditions to apply the covering Lemma 3.25; therefore there is a positive number δ so that, for every point $0 < |z - x| < \delta$ there is a set A_x having z as a point of density and for all $a \in A_x$, $z < a < x \implies [a,x]$ belongs to V^2 and $z > a > x \implies [x,a]$ belongs to V^2.

Since $V^2 = V$ this gives $f(a) = f(x)$ for such points. But z is a point of qa-continuity of f so that, since A_x has density 1 at z some sequence of points a_n in A_x can be found with $f(a_n) \to f(z)$. From this we conclude that $f(z) = f(x)$ for all $|z - x| < \delta$ so that, again, z is a point of constancy of f. Thus $qaC_f \cap d(E_n) \subset L_f$ and so $qaC_f \cap E \cap d(E_n) = \emptyset$ for each n. As before then we conclude that $E \cap qaC_f$ itself must have measure zero as required.

As corollaries we have immediately the following observations.

Corollary 6.36 *Let f be an arbitrary measurable function. Then the set LS_f of points at which f is exactly locally symmetric is measurable.*

Corollary 6.37 *Let S be an arbitrary measurable set and let S^* denote the set of points at which S is locally symmetric. Then S^* is measurable.*

Corollary 6.38 *Let f be a function with the Baire property. Then the set LS_f of points at which f is exactly locally symmetric has the Baire property.*

Corollary 6.39 *Let f be a function that is exactly locally symmetric at each point of a measurable set E. Then there is an open set G containing almost every point of E so that f is constant on each component of G.*

Corollary 6.40 *Let f be a function that is exactly locally symmetric at each point of a set E that has the Baire property. Then there is an open set G containing all but a first category subset of E so that f is constant on each component of G.*

Corollary 6.41 *Let f be a function that is exactly locally symmetric at each point of a set residual in $\mathbb{R}$. Then f is constant on each component of a dense open set G.*

6.3 Symmetric Monotonicity

We say that a point x is a point of *symmetric increase* for a function f or that f is *symmetrically increasing* at x if there is some positive number δ so that

$$0 < t < \delta \Longrightarrow f(x - t) < f(x + t).$$

Similarly one defines symmetrically decreasing, symmetrically nondecreasing and symmetrically nonincreasing. Certainly if the lower symmetric derivate of f at x is positive then x must be a point of symmetric increase but not conversely. The fact that x is a point of symmetric monotonicity need not say too much about the behavior of the function at that point: the function $f(x) = x + x^{-2}$ is symmetrically increasing at 0 but not even bounded in any neighbourhood of 0.

We shall investigate first the nature of functions that are symmetrically monotonic at every point and then we turn to the study of the set of points of symmetric increase for arbitrary functions.

6.3.1 Everywhere Symmetric Increase

The first elementary result follows directly from our covering lemmas of Chapter 4. If a function f satisfies a symmetric monotonicity requirement at every point then f is monotonic in the normal sense off of some splattered set S.

Theorem 6.42 *If f is symmetrically increasing [nondecreasing] at every point then f is increasing [nondecreasing] off some splattered set.*

Proof. Let $V = \{(x,t) : f(x-t) < f(x+t)\}$. Note that V is a transitive, full symmetric covering relation. By Lemma 3.13 there is a splattered set S so that V partitions $[a,b]$ for every $a < b$, $a, b \in \mathbb{R} \setminus S$. This immediately gives $f(a) < f(b)$ as required. The proof for the nondecreasing case (rather than increasing) is identical but merely uses the collection $V = \{(x,t) : f(x-t) \leq f(x+t)\}$.

6.3.2 Points of Symmetric Increase

We turn now to an investigation of Belna, Evans and Humke [13] on the structure of the set of points of symmetric monotonicity. Let SI_f denote the set of points at which the function f is symmetrically nondecreasing and let I_f denote the set of points at which f is nondecreasing in the ordinary sense. That is at points $x \in SI_f$ the relation $f(x-t) \leq f(x+t)$ must hold for all sufficiently small t while at points $x \in I_f$ the relation $f(x') \leq f(x) \leq f(x'')$ holds for $x' < x < x''$ sufficiently close to x. The Borel structure of I_f is easy to obtain; it is of type $G_{\delta\sigma}$ (see for example [282, p. 116]). Belna, Evans and Humke proposed to study the nature of the set $SI_f \setminus I_f$. We obtain the following theorem in our usual fashion from the covering Lemmas 3.25 and 3.26. As the proof follows the same lines as that for Theorem 6.35 it may be omitted.

Theorem 6.43 *Let f be an arbitrary function, let SI_f denote the set of points at which f is symmetrically nondecreasing and let I_f denote the set of points at which f is nondecreasing in the ordinary sense. Then the set*

$$SI_f \setminus I_f$$

contains no measurable set of positive measure and no second category set having the Baire property. If, furthermore qaC_f and qqC_f denote the sets of points at which f is quasi-approximately continuous and quasi-qualitatively continuous respectively, then

$$qaC_f \cap (SI_f \setminus I_f)$$

has measure zero and

$$qqC_f \cap (SI_f \setminus I_f)$$

is first category.

From this we immediately obtain the following corollaries. The first is a theorem of Belna, Evans and Humke (as cited above) and the second is a category analogue; their category version uses a different assumption on the function.

Theorem 6.44 (Belna, Evans and Humke) *If f is measurable then SI_f is measurable and $SI_f \setminus I_f$ has measure zero.*

Theorem 6.45 *If f has the Baire property then SI_f has the Baire property and $SI_f \setminus I_f$ is first category.*

Thus for well behaved functions the set $SI_f \setminus I_f$ is small both in the sense of measure and of category. Some examples illustrate that this cannot be much improved. First note that because of Theorem 6.17 there is a function f that is symmetrically nondecreasing (actually symmetrically constant) at the points of a set that is large in the sense of measure and category and yet nowhere nondecreasing; this function is nonmeasurable and has no points of continuity so that the set $SI_f \setminus I_f$ is nonmeasurable, and does not have the Baire property in any interval.

There is a measurable function f so that the set $SI_f \setminus I_f$ is large in the sense of category.

Theorem 6.46 *(ZFC+CH). There is a measurable function f for which the set $SI_f \setminus I_f$ is second category.*

Proof. Let $f = \chi_G$ where G is the measure zero, second category additive group of Theorem 6.13. Since G has measure zero it follows that f is measurable. The set G in the construction is dense (in fact it contains all the rationals) so that $I_f = \emptyset$ and because it is a group $SI_f \supset G$ so that SI_f is second category.

Finally let us give an example of Belna, Evans and Humke from [13] showing that the set $SI_f \setminus I_f$, while small, need not be countable; they give an example of a *continuous* function for which this set is uncountable.

Theorem 6.47 (Belna, Evans and Humke) *There is a continuous function f for which the set $SI_f \setminus I_f$ is uncountable.*

Proof. Let C be a symmetric perfect set in the interval $[0,1]$ formed by removing at each stage the middle $\frac{3}{5}$. In each of the contiguous intervals (a_n, b_n) to C in $[0,1]$ we shall construct a continuous, piecewise linear function with the desired property. Let g be defined on $[0,1]$ so as to be continuous and pass through each of the points

$$(0,0), \ (\tfrac{1}{6}, \tfrac{1}{6}), \ (\tfrac{1}{3}, 0), \ (\tfrac{2}{3}, 0), \ (\tfrac{5}{6}, -\tfrac{1}{6}) \text{ and } (1,0)$$

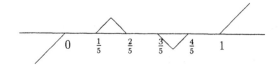

Fig. 6.1. First steps in the construction of f in Theorem 6.47.

and be linear in each of these segments. Now we just place scaled down copies of g in each of the intervals (a_n, b_n). Specifically let $f(x) = 0$ for $x \in C$, let $f(x) = x$ for $x \leq 0$, let $f(x) = x - 1$ for $x \geq 1$ and if $a_n < x < b_n$ then let

$$f(x) = (b_n - a_n) g \left(\frac{x - a_n}{b_n - a_n} \right).$$

The figure illustrates the first step in the construction. In each of the intervals $[0, \frac{1}{5}]$ and $[\frac{4}{5}, 1]$ the saw-tooth jags are repeated at a smaller scale.

As defined it is easy to check that f is continuous. Let

$$S = \{x \in C : f(x - t) \leq f(x + t), 0 < t < 1\}.$$

Since f is continuous the set S must be closed and it certainly is contained in SI_f; moreover, since no point of C can belong to I_f, we have that $S \subset SI_f \setminus I_f$. The proof follows by showing that S is dense in C so that, since S is closed, S must be all of C and accordingly is uncountable.

We have only to check that every point a_n belongs to S since these points are dense in C. By the way in which f was constructed, $|f(x)|$ cannot exceed the distance from x to C. Then it is a simple matter to check that $f(a_n - t) \leq f(a_n + t)$ for each $0 < t < 1$ and we have $a_n \in S$ as stated.

6.3.3 Porosity Properties

We have seen that the set $SI_f \setminus I_f$ of points where a function f is symmetrically nondecreasing but not nondecreasing in the ordinary sense is small in the sense of both measure and category, at least for very regular functions. A closer study of this set was undertaken by Belna, Evans and Humke [13] who showed that, under additional assumptions on the function f, this set was smaller still: it must be σ-porous. For material on the σ-ideal of σ-porous sets one can consult the appendix in [282] for example or the survey of Zajicek [310]. Briefly the porosity of a set E at a point x is computed as

$$\limsup_{r \to 0+} \frac{\ell(E, x, r)}{r}$$

where $\ell(E, x, r)$ is the length of the largest open subinterval of $(x-r, x+r)$ that contains no point of E. A set is porous if it has positive porosity at each of its points; a set is σ-porous if it can be expressed as a countable union of porous sets. It is evident that at a point of porosity of a set E there are "substantial" gaps, so that at such a point E can neither be dense nor have density 1. It follows that porous sets are nowhere dense and have measure zero and that σ-porous sets are measure zero and first category.

The theorem asserts that the set $SI_f \setminus I_f$, under special assumptions on f, is not only measure zero and first category but also σ-porous, which is a strictly stronger statement.

The role of the porosity in the proof rests on the following elementary lemma from [13].

Lemma 6.48 *Let x be a point in a set E and suppose that $\ell(E, x, r) < r/2$ for all $0 < r < \eta$. Then for any open interval $(a, b) \subset (x-\eta, x)$ [resp. $(a, b) \subset (x, x+\eta)$] there is a finite sequence of right [resp. left] reflections through points in E that carries (a, b) to an interval (a', b') that contains x.*

Proof. If I is an interval and $e \in E$ lies to the right of I then by a right reflection of I we mean the interval $2e - I$. Thus we are required to find a sequence of points $e_1, e_1, \ldots e_{k-1}$ so that I_k contains x where $I_1 = I$ and $I_{i+1} = 2e_i - I_i$ each being a right reflection about e_i.

If $b = x$ then choose any $e_1 \in (x, 2x - a)$ and if $b < x$ choose $e_1 \in (b, (b + x)/2)$; by the porosity requirements each of these intervals must contain a point $e_1 \in E$. If the interval $I_1 = (a, b)$ has length larger than (b, x) then we are done for then $I_2 = 2e_1 - I_1$ must contain x; if not then repeat this process. Since each I_i has the same length we must reach some stage where $|I_{k-1}| > d(I_{k-1}, x)$ and at the next reflections I_k or I_{k+1} must contain x.

This proves the lemma for the case of right reflections; the left hand reflection argument is identical.

Theorem 6.49 (Belna, Evans and Humke) *If f has a dense set of points of continuity then $SI_f \setminus I_f$ is σ-porous.*

Proof. For each integer n define the set of points

$$S_n = \{x : f(x - h) \le f(x + h),\ 0 < h < 1/n\}$$

and then SI_f is the union of the sets $\{S_n\}$. We shall show first that if a point $x \in S_n$ is both a point of continuity of f and a point at which the porosity of S_n is less than $1/2$ then $x \in I_f$. Then we show that there are only countably many points $x \in S_n$ that are points of discontinuity of f and such that the porosity of S_n at x is less than $1/2$. From these two statements the theorem evidently follows.

For the first of these statements suppose that x_0 is a point at which $\ell(S_n, x_0, r) < r/2$ for all $0 < r < \eta < 1/n$ and that f is continuous at x_0. Then $f(y) \leq f(x_0)$ for all $x_0 - \eta < y < x_0$. If not then there is such a point y with $f(y) > f(x_0)$. As in the proof of Lemma 6.48 we may choose a sequence of points $e_1 < e_2 < e_3 \ldots$ in S_n so that

$$f(x_0) < f(y) \leq f(y_2) \leq f(y_3) \leq \ldots f(y_k) \to f(x_0)$$

where $y = y_1$ and $y_{k+1} = 2e_k - y_k$, and this is a contradiction. In an identical way we can also show that $f(y) \geq f(x_0)$ for all $x_0 < y < x_0 + \eta$. Consequently $x \in I_f$ as we wished to prove.

Let us now show that there are only countably many points $x \in S_n$ that are points of discontinuity of f and such that the porosity of S_n at x is less than $1/2$. For rationals α, β and integers m write $A_{\alpha, \beta, m}$ for the set of points x where

$$\sup_{0 < h < 1/m,\, x - h \in C_f} f(x - h) \leq \alpha < \beta \leq \inf_{0 < h < 1/m,\, x + h \in C_f} f(x + h).$$

Since C_f (the set of points where f is continuous) is dense each such set is countable. Indeed an obvious argument shows that if $x \in A_{\alpha, \beta, m}$ then

$$A_{\alpha, \beta, m} \cap \left(x - \frac{1}{m}, x + \frac{1}{m} \right) = \{x\}.$$

To establish that there are only countably many points $x \in S_n$ that are points of discontinuity of f and such that the porosity of S_n at x is less than $1/2$ it is therefore enough to show that for nearly every such point there are rationals α, β and an integer m so that $x \in A_{\alpha, \beta, m}$.

Suppose that $x_0 \in S_n$ is a point at which $\ell(S_n, x_0, r) < r/2$ for all $0 < r < \eta < 1/n$ and that f is not continuous at x_0. For nearly every such point x_0 rational numbers α, β may be chosen such that

$$\liminf_{x \to x_0} f(x) < \alpha < \beta < \limsup_{x \to x_0} f(x).$$

(This just excludes the countable number of points at which f has a removable discontinuity.) Now we have only to verify that

$$\sup_{0 < h < \eta,\, x_0 - h \in C_f} f(x_0 - h) \leq \alpha < \beta \leq \inf_{0 < h < \eta,\, x_0 + h \in C_f} f(x_0 + h)$$

and it will follow that x_0 is in some set $A_{\alpha, m}$.

Suppose, in order to obtain a contradiction, that

$$\sup_{0 < h < \eta,\, x_0 - h \in C_f} f(x_0 - h) > \alpha$$

Then there is an open interval (a, b) contained in in $(x_0 - \eta, x_0)$ on which f exceeds α. By Lemma 6.48 there must then be an interval (a', b') containing x_0 with this property: for every $z \in (a', b')$ there is a sequence

$$z = s_1 > s_2 > \ldots > s_k \in (a, b),$$

each $s_2, s_3, \ldots s_{k-1}$ belongs to S_n and $s_{i+1} = 2s_i - s_{i-1}$ for $i = 2, 3 \ldots, k - 1$. Our contradiction is at hand: from $s_k \in (a, b)$ we have $f(s_k) > \alpha$ and since each $s_2, s_3, \ldots s_{k-1}$ belongs to S_n we have $f(s_{i-1}) > f(s_i)$ for $i = 2, 3 \ldots, k - 1$ and consequently $f(z) > \alpha$. This cannot hold for every z in a neighbourhood of x_0 since that conflicts with $\liminf_{x \to x_0} f(x) < \alpha$ and thus the proof is complete.

6.4 Symmetric Continuity

We turn now to the problem of determining the continuity properties of functions that are symmetrically continuous. Recall that the first result of this type in the literature is the Theorem of Fried asserting that an everywhere symmetrically continuous function is continuous on a residual set. The measure-theoretic version is given by the Theorem of Zygmund and Stein (Theorem 2.1 on page 25) showing that if a measurable function is symmetrically continuous on a measurable set then it is almost everywhere continuous in that set. See Chapter 2 for more background.

The complete analysis of the situation was provided by Uher [294] in 1986 who showed that both measure-theoretic and category versions follow from the same geometric arguments. His focus, as is ours here, is on the more natural assumption of symmetric semicontinuity.

Definition 6.50 A function f is said to be *upper symmetrically semicontinuous* at a point x provided

$$\limsup_{h \to 0+} (f(x + h) - f(x - h)) \leq 0$$

and *lower symmetrically semicontinuous* there provided

$$\liminf_{h \to 0+} (f(x + h) - f(x - h)) \geq 0.$$

A function is *symmetrically continuous* at a point if it is both upper and lower symmetrically semicontinuous there. A function which has at a point a finite symmetric derivative or two finite symmetric extreme derivates is certainly symmetrically continuous there. If the function has at a point an infinite symmetric derivative or just one finite symmetric extreme derivate then it may be only symmetrically semicontinuous there. This last observation is one of the reasons for studying the semicontinuity properties rather than focusing just on the continuity properties.

6.4.1 Theorem of Uher

Our main theorem in this section is due to Uher [294] and it combines a measure-theoretic version with a category version by appealing to the two covering theorems, Theorem 3.25 and Theorem 3.26. Note that it is directly analogous to Theorem 6.35 on page 229 both in statement and in proof. In contrast to the positive statements of this theorem we have already seen an example (Theorem 6.18) showing that certain hypotheses may not be relaxed: directly from the construction in Theorem 6.17 there is a set E large in the sense of measure and in the sense of category and a function f that is symmetrically continuous at every point of E and yet f is nonmeasurable and is nowhere continuous. This example together with the theorem we now prove gives a rather full picture of the situation for symmetrically continuous or symmetrically semicontinuous functions.

Theorem 6.51 (Uher) *Let f be an arbitrary function, let SS_f denote the set of points at which f is symmetrically semicontinuous and let C_f denote the set of points at which f is continuous in the ordinary sense. Then the set*

$$SS_f \setminus C_f$$

contains no measurable set of positive measure and no second category set having the Baire property. If, furthermore qaC_f and qqC_f denote the sets of points at which f is quasi-approximately continuous and quasi-qualitatively continuous respectively, then

$$qaC_f \cap (SS_f \setminus C_f)$$

has measure zero and

$$qqC_f \cap (SS_f \setminus C_f)$$

is first category.

Proof. As before, it is enough to indicate the measure-theoretic version as the translation to the category version is easily carried out.

Let E be a subset of $SS_f \setminus C_f$, and consider the covering relation

$$V(\epsilon) = \{(x, t) : x \in E, \, f(x + t) - f(x - t) < \epsilon\}.$$

Note that $V(\epsilon)$ is not transitive but that $V^2(\epsilon) \subset V(2\epsilon)$ and $V^5(\epsilon) \subset V(5\epsilon)$; the latter plays a role in this part of the proof and the former a role in the second part of the proof.

Let E^+ denote the set of points in E at which f is upper semicontinuous, and let E^- denote the set of points there at which f is lower semicontinuous. By the assumptions in the theorem at each point $x \in E$ there is a positive number $\delta(x)$ so that $V(\epsilon)$ has the properties

$$x \in E^+, 0 < t < \delta(x) \implies (x, t) \in V(\epsilon)$$

and

$$x \in E^-, 0 > t > -\delta(x) \implies (x, t) \in V(\epsilon).$$

Thus $V(\epsilon)$ satisfies the hypotheses of Theorem 3.25 (in the category case with E having the Baire property) or Theorem 3.26 (in the measure case with E measurable).

For the first part of the theorem, if E is measurable, then we apply Theorem 3.26 to obtain a neighbourhood U_x of almost every point $x \in E$ such that for all $x \pm t$ in U_x either

(i) $(x + t/2, t/2)$ and $(x - t/2, t/2)$ are in $V^5(\epsilon)$ if $t > 0$ or

(ii) $(x + t/2, -t/2)$ and $(x - t/2, -t/2)$ are in $V^5(\epsilon)$ if $t < 0$.

This gives that at almost every point x one or other of the two alternatives holds: either

$$\liminf_{z \to x-} f(z) \geq f(x) - 5\epsilon \quad \text{and} \quad \limsup_{z \to x+} f(z) \leq f(x) + 5\epsilon$$

or else the two opposite inequalities

$$\liminf_{z \to x+} f(z) \geq f(x) - 5\epsilon \quad \text{and} \quad \limsup_{z \to x-} f(z) \leq f(x) + 5\epsilon.$$

It follows now that one of these two statements holds for infinitely many $\epsilon = 1/n$ ($n = 1, 2, 3, \ldots$) at almost every $x \in E$. Thus at these points f is lower semicontinuous from one side and upper semicontinuous from the other. By a well known Theorem of Young (see [251, Theorem 1.1, p. 261]) this means f is continuous at every such point with at most countably many exceptions. Finally then f is continuous at almost every point of E.

We turn now to the second part of the theorem, we drop the assumption that E is measurable and take $E = SS_f \setminus C_f$. We apply the covering Lemma 3.8 to the set E^+ and the collections $V(1/m)$ to obtain a partition $\{E^+_{nm}\}$ of that set with the properties that

$$x \in E^+_{nm}, 0 < t < 1/n \implies (x, t) \in V(1/m).$$

Let z be a point in both $d(E^+_{mn})$ and in $\mathrm{qa}C_f$.

We have the conditions to apply the covering Theorem 3.26; therefore there is a positive number δ so that for every point $0 < |z - x| < \delta$ there is a set A_x having z as a point of density and for all $a, b \in A_x$, $z < a < x \implies [a, x]$ belongs to $V^2(1/m)$ and $z > b > x \implies [x, b]$ belongs to $V^2(1/m)$.

This gives $f(x) - f(a) < 2/m$ and $f(b) - f(x) < 2/m$ for all such points. But z is a point of quasi-approximate continuity of f so that, since A_x has density 1 at z, some sequences of points $a_n > z > b_n$ in A_x can be found with $f(a_n) \to f(z)$ and $f(b_n) \to f(z)$. Thus we obtain

$$\liminf_{x \to z-} f(x) \geq f(z)$$

and

$$\limsup_{y \to z+} f(y) \leq f(z)$$

for all

$$z \in \bigcap_{m=1}^{\infty} \bigcup_{n=1}^{\infty} d(E_{mn}^+).$$

Again we argue using the set E^- in the obvious manner, appeal to the Theorem of Young again and we can conclude that the set $E \cap \mathrm{qaC}_f$ itself must have measure zero as required.

This theorem can be restated in somewhat different language; here is a version from Uher [294, Theorem 1, p. 425] which follows from Theorem 6.51 above both in the measure-theoretic version (where $d(E)$ denotes the set of density points of E) and in the category version (where $c(E)$ denotes the set of category density points of E).

Corollary 6.52 (Uher) *Let SSD_f denote the set of points at which a function f is not symmetrically semicontinuous. Then f is continuous at almost every point of the set*

$$\mathbb{R} \setminus d\,(SSD_f).$$

Corollary 6.53 (Uher) *Let SSD_f denote the set of points at which a function f is not symmetrically semicontinuous. Then f is continuous at all but a first category subset of the set*

$$\mathbb{R} \setminus c\,(SSD_f).$$

We may draw a number of further corollaries from Theorem 6.51.

Corollary 6.54 *Let the function f be symmetrically semicontinuous at every point. Then f is continuous at every point excepting a set of measure zero and first category. In particular f is measurable and has the Baire property.*

Corollary 6.55 *Let the function f be measurable; then f is continuous at almost every point at which it is symmetrically semicontinuous.*

Corollary 6.56 *Let f be continuous relative to some residual set; then f is continuous at all but a first category subset of the set of points at which it is symmetrically semicontinuous .*

6.4.2 Symmetrically Continuous Functions

Most of the properties of functions that are symmetrically continuous are obtainable as corollaries of the Theorem of Uher. For convenient reference we restate these theorems as statements about symmetric continuity. The form is similar to that used in Belna [12, pp. 99–100] and imitates the statements of Theorems 6.35 and 6.51 given above; indeed it is just a corollary of the latter. Recall that a set that contains no measurable set of positive measure has inner measure zero and a set that contains no second category set with the Baire property is residual in no open interval; the theorem could have been expressed in this language instead. The example of Theorem 6.18 on page 218 shows that in this theorem, while the set C_f has a well known simple Borel structure, the set SC_f need not even be measurable.

Theorem 6.57 *Let f be an arbitrary function and let SC_f and C_f denote the sets of points at which f is symmetrically continuous and continuous respectively. Then the set $SC_f \setminus C_f$ contains no measurable set of positive measure and no second category set with the Baire property. If, furthermore qaC_f and qqC_f denote the sets of points at which f is quasi-approximately continuous and quasi-qualitatively continuous respectively, then*

$$qaC_f \cap (SC_f \setminus C_f)$$

has measure zero and

$$qqC_f \cap (SC_f \setminus C_f)$$

is first category.

We have immediately as corollaries the following theorems of Belna [12] and Fried [119].

Corollary 6.58 (Belna) *Let the function f be symmetrically continuous at almost every point of a measurable set E. Then f is continuous almost everywhere in E.*

Corollary 6.59 (Fried) *Let the function f be symmetrically continuous on a set residual in $\mathbb{R}$. Then f is continuous at every point excepting a set of the first category.*

If f is measurable then Theorem 6.57 shows that SC_f must be measurable since SC_f differs from the Borel set C_f by a set of measure zero.

Corollary 6.60 *Let f be a measurable function. Then the set SC_f of points at which f is symmetrically continuous is measurable.*

We repeat here, without proof, Theorem 2.46 that contains some related ideas. This shows that, while we have been able to make some measurability and category assertions for the set SC_f, we cannot make any useful topological assertions except in simple cases. The methods are closely related to those just used in the proof of Theorem 6.34 .

Theorem 6.61 (Darji) *Let M be any zero-dimensional co-analytic set of reals. Then there is a Baire 2 function f such that the set SC_f of points where f is symmetrically continuous is homeomorphic to M.*

6.4.3 Discontinuities of Symmetrically Continuous Functions

A function that is everywhere symmetrically continuous can have only a small set of discontinuities in the ordinary sense; this set is, by the above theorems, both measure zero and first category. Apparently this is the best that is so far known. An example from Preiss [230] shows that this set may however be uncountable. For greater detail see Section 2.7 where this theorem is proved and the general problem of characterizing the discontinuity points of symmetrically continuous functions is discussed.

Theorem 6.62 (Preiss) *There is a function that is everywhere symmetrically continuous and yet discontinuous at the points of an uncountable set.*

6.4.4 Weak Symmetric Continuity

A function f is said to be *weakly continuous* at a point x if there are sequences $a_n \nearrow x$ and $b_n \searrow x$ so that

$$\lim_{n \to \infty} f(a_n) = \lim_{n \to \infty} f(b_n) = f(x).$$

While this does not demand very much of the function at a given point, one might suspect that it is possible to find a function that possesses this kind of continuity property at no point. A remarkable theorem from the theory of cluster sets asserts that for an arbitrary function f the set of points where f fails to be weakly continuous is at most countable (see [282, Chap. 2] for an account).

The symmetric version is not so easy. Let us say that a function f is *weakly symmetrically continuous* at a point x if there is a sequence $h_n \to 0$ so that

$$\lim_{n \to \infty} (f(x + h_n) - f(x - h_n)) = 0.$$

If f is weakly symmetrically continuous at a point it need not be weakly continuous there, nor conversely. The question naturally arises whether any symmetric version of the theorem above is available.

If f is a function that is nowhere weakly symmetrically continuous then f must be quite irregular. A simple density argument at points of approximate continuity shows that any measurable function must be weakly symmetrically continuous almost everywhere. The category analogue is immediate too: a function with the Baire property must be weakly symmetrically continuous at all points outside some first category set. The existence then of a function (necessarily neither measurable nor possessing the Baire property) that is nowhere weakly symmetrically continuous may seem doubtful.

The problem of the existence of such functions was posed by Evans and Larson at the 1984 Real Analysis Summer Symposium and, independently, by Kostyrko [156]. The existence has been established by Ciesielski and Larson [44] who call such functions *uniformly antisymmetric*. The proof depends on a clever argument using a Hamel basis.

Theorem 6.63 (Ciesielski–Larson) *There exists a function $f : \mathbb{R} \to \mathbb{N}$ that is nowhere weakly symmetrically continuous.*

Proof. Let B be a Hamel basis for the reals. Consider $\mathbb{Q}$ with the discrete topology and $\mathbb{Q}^B$ with the product topology. Then, by a result from general topology, $\mathbb{Q}^B$ is separable, being a product of continuum many separable spaces. Let $D = \{d_n : n \in \mathbb{N}\}$ be a dense subset of $\mathbb{Q}^B$.

Define an embedding $E : \mathbb{R} \to \mathbb{Q}^B$ by putting $E(x)(b) = q_b$, where $x = \sum_{b \in B} q_b b$ is the unique representation of x in the Hamel basis B. Notice that for every $x \in \mathbb{R}$ the set $\mathrm{supp}(x) = \{b \in B : E(x)(b) \neq 0\}$ is finite.

For $x \in \mathbb{R}$ let

$$[x] = \left\{ g \in \mathbb{Q}^B : g(b) = E(x)(b) \text{ for all } b \in \mathrm{supp}(x) \right\}.$$

and note that $[x]$ is a basic open set in $\mathbb{Q}^B$. Define

$$f(x) = \min \left\{ n : d_n \in [x] \right\}.$$

Let $P_n = f^{-1}(n)$ for every $n \in \mathbb{N}$. We claim that $\{P_n : n \in \mathbb{N}\}$ is a partition of the real numbers such that for every $x \in \mathbb{R}$ the set

$$S_x = \bigcup_{n \in \mathbb{N}} \{h > 0 : x - h, x + h \in P_n\} \tag{6.5}$$

is finite. It follows from this that f is not weakly symmetrically continuous at any point, since for every $x \in \mathbb{R}$ the set

$$S_x = \bigcup_{n \in \mathbb{N}} \{h > 0 : x - h, x + h \in P_n\} = \{h > 0 : f(x - h) = f(x + h)\}$$

is finite and $|f(x - h) - f(x + h)| \geq 1$ for every $h > 0$ such that $h \notin S_x$.

First notice that, in order to prove that (6.5) is finite for each $x \in \mathbb{R}$, it is enough to show that there exists only a finite number of $h > 0$ such that

$$[x - h] \cap [x + h] \neq \emptyset.$$

Now to prove this, choose $x, h \in \mathbb{R}$, $h > 0$, such that $[x - h] \cap [x + h] \neq \emptyset$. Notice that

$$E(x - h)(b) + E(x + h)(b) = 2 E(x)(b) \quad \text{for all } b \in B$$

and, since $[x - h] \cap [x + h] \neq \emptyset$,

$$E(x - h)(b) = E(x + h)(b) = E(x)(b)$$

for all $b \in \operatorname{supp}(x - h) \cap \operatorname{supp}(x + h)$. Hence,

(i) $\operatorname{supp}(x - h) \cup \operatorname{supp}(x + h) = \operatorname{supp}(x)$

and for all $b \in \operatorname{supp}(x)$ exactly one of the following holds:

(ii) $b \in \operatorname{supp}(x - h) \cap \operatorname{supp}(x + h)$ and $E(x - h)(b) = E(x + h)(b) = E(x)(b)$,

(iii) $b \in \operatorname{supp}(x - h) \setminus \operatorname{supp}(x + h)$ and $E(x - h)(b) = 2E(x)(b)$, $E(x + h)(b) = 0$, or

(iv) $b \in \operatorname{supp}(x + h) \setminus \operatorname{supp}(x - h)$ and $E(x + h)(b) = 2E(x)(b)$, $E(x - h)(b) = 0$.

It is clear that there are at most 3^n numbers $h > 0$ satisfying (i)–(iv), where n is the cardinality of $\operatorname{supp}(x)$. This completes the proof.

Some of the considerations of this section were already raised in Section 2.8 in a brief mention of antisymmetric sets. The fact that no antisymmetric set exists is equivalent to the observation that there does not exist a function that is nowhere weakly symmetrically continuous and having a range that consists of just two values. The range of the function in Theorem 6.63 is countable. One can ask then whether such a function can be found with a finite range (more than two values of course). Komjáth and Shelah [152] show, by employing a combinatorial theorem of Erdös and Radó, that it is not possible to amend the construction there in this way. In fact they show that if $f : \mathbb{R} \to \{1, 2, 3, \ldots, n\}$ is an arbitrary function then, for some $x \in \mathbb{R}$, the set

$$S_x = \{h > 0 : f(x + h) = f(x - h)\}$$

is infinite. (This is does not rule out some other method whereby a nowhere weakly symmetrically continuous with finite range might be found; the condition that each S_x is finite is stronger.) In [152] they show too that the continuum hypothesis is equivalent to the existence of a function $f : \mathbb{R} \to \mathbb{N}$ for which each set S_x has no more than one element.

The articles [44] and [152] should be consulted for further details about these and related ideas.

6.5 Chlebík Theorem

How irregular can a symmetrically continuous function be? We know that such a function is at least measurable. A number of authors have asked whether such a function need be Borel measurable. This was answered negatively by Chlebík [43, p. 34]. His argument takes the example of Preiss from Theorem 6.62 and builds on it by constructing 2^c different symmetrically continuous functions where c is the power of the continuum. Since the cardinality of the class of Borel measurable functions is smaller than this we see, as a corollary, that there must exist symmetrically continuous functions that are not Borel measurable.

Theorem 6.64 (Chlebík) *The power of the set of symmetrically continuous functions is 2^c.*

Corollary 6.65 *There exist symmetrically continuous functions that are not Borel measurable.*

The strategy of the proof is to use Theorem 2.33 (especially Lemma 2.36) to construct a symmetrically continuous function f with a large set E of discontinuities. Lemma 6.31 is then the elementary key to constructing, on the basis of this function, a large class (2^c many) of symmetrically continuous functions. A set $G \subset E$ with power c is chosen so that for each distinct subset $H \subset G$ the function $f\chi_{2E+H}$ is a distinct, symmetrically continuous function. The details of the construction are clever and rather intricate. Since they depend very heavily on the arithmetic of the set E constructed we shall refer the reader to the original source [43].

6.6 Boundedness

While continuous functions are bounded in every compact interval the assumption of symmetric continuity has no such consequence; indeed the function $f(x) = x^{-2}$ is symmetrically continuous (even symmetrically differentiable) but unbounded at the origin. Even so the points of unboundedness of a symmetrically continuous functions cannot be too large as the following theorem shows. Larson [172] has some closely related ideas.

Theorem 6.66 *Let f be an everywhere symmetrically continuous function. Then the set*

$$E = \{x : \limsup_{y \to x} |f(y)| = \infty\}$$

is closed, everywhere locally symmetric and countable.

Proof. Clearly the set E is closed. We show that it is exactly locally symmetric at any point x. Choose $\delta > 0$ so that

$$|f(x + t) - f(x - t)| < 1 \quad (0 < t < \delta.)$$

If $y \in E$ and $|x - y| < \delta/2$ then there is a sequence of points $y_n \to y$ with $|f(y_n)| \to \infty$. But

$$|f(y_n) - f(2x - y_n)| < 1$$

for large enough n and so $|f(2x - y_n)| \to \infty$ as well. Thus $2x - y \in E$ and we see that E is exactly locally symmetric at point x.

From Corollary 6.27 now either E or its complement must be countable. Theorem 6.57 shows that it must be E itself that is countable as required.

We can ask too whether a weaker condition such as

$$|f(x + h) - f(x - h)| = O(1) \quad (h \to 0)$$

requires f to be bounded at any points. By our usual methods the following theorem may be proved. In the theorem we let BS_f denote the set of points at which f is symmetrically bounded, i.e. points x at which

$$\limsup_{h \to 0} |f(x + h) - f(x - h)| < +\infty$$

and we let B_f denote the set of points at which f is bounded (a function is bounded at a point x if it is bounded in a neighbourhood of the point x). Evidently the set B_f of points of boundedness of f is open and an elementary compactness argument shows that f is bounded on each compact subset of that set. For a proof one need only adapt the arguments in the proof of Theorem 6.35 to this situation.

Theorem 6.67 *Let f be an arbitrary function, let BS_f denote the set of points at which f is symmetrically bounded and let B_f denote the set of points at which f is bounded. Then the set*

$$BS_f \setminus B_f$$

contains no measurable set of positive measure and no second category set having the Baire property. If, furthermore qaC_f and qqC_f denote the sets of points at which f is quasi-approximately continuous and quasi-qualitatively continuous respectively, then

$$qaC_f \cap (BS_f \setminus B_f)$$

has measure zero and

$$qqC_f \cap (BS_f \setminus B_f)$$

is first category.

As corollaries we have immediately the following observations.

Corollary 6.68 *Let f be a measurable function. Then the set BS_f of points at which f is symmetrically bounded is measurable.*

Corollary 6.69 *Let f be a measurable function. Then f is bounded at almost every point at which it is symmetrically bounded.*

Corollary 6.70 *Let f be a function possessing the Baire property. Then f is bounded at every point at which it is symmetrically bounded except possibly for a set of the first category.*

6.7 Symmetric Lipschitz Conditions

In this section we investigate the condition

$$f(x + h) - f(x - h) = O(h)$$

as $h \to 0$. Note that this is a more severe condition than mere symmetric continuity, and so one can expect that more can be said about the points of continuity of f. If this condition holds everywhere, then the points of discontinuity (by Theorem 6.51) must be a set of measure zero and first category. As we have already seen it is considerably smaller still; this is the Theorem of Charzyński (Theorem 2.9 on page 32) which we repeat here for reference. The original proof has appeared in Section 2.3 and Freiling's proof in Section 5.3.3.

Theorem 6.71 (Charzyński) *Let the function f satisfy at every point x the condition*

$$\limsup_{h \to 0} \left| \frac{f(x + h) - f(x - h)}{2h} \right| < +\infty. \tag{6.6}$$

Then the set of points of discontinuity of f is scattered.

If there is a uniform bound on the symmetric difference quotient (6.6) then the function f is very nearly Lipschitz. The following theorem, due also to Charzyński, we have seen before (Theorem 2.19); the set of points where f and g disagree is scattered.

Theorem 6.72 (Charzyński) *Let the function f satisfy at every point x of an interval $[a, b]$ the condition*

$$\limsup_{h \to 0} \left| \frac{f(x + h) - f(x - h)}{2h} \right| \leq M.$$

Then there exists a function g satisfying a Lipschitz condition

$$|g(x) - g(y)| \leq M|x - y|$$

such that f and g agree on the set of points of continuity of f in $[a, b]$.

See Ponomarev [229] for sufficient conditions for a continuous function f, satisfying (6.6) everywhere on an interval, to be an ACG_*-function or to satisfy Lusin's condition (N).

7

The Symmetric Derivative

7.1 Introduction

A great deal is known now about the behavior of ordinary derivatives of real functions. As research has progressed to place this analysis in other settings or employ generalized versions of ordinary differentiation it has been natural to seek this same behavior in the more generalized settings. The program that has been frequently followed is to investigate the extent to which the various generalized derivatives share or do not share in the host of properties possessed by ordinary derivatives. Such a program, while it has its more mundane aspects and has inspired not a few tedious reports, is frequently brightened by startling differences or puzzling problems that require a different technical approach.

In this chapter we follow this program as it applies to the symmetric and the approximate symmetric derivatives. One theme that appears often is that of the measurability properties of symmetric derivates. There is now a fairly complete understanding of the exact Baire classification of most of these kinds of derivates; Sections 7.2.4, 7.3.1 and 7.4.1 cover this material. The success of the Denjoy-Young-Saks theorem in classifying the relations that hold among the four Dini derivatives has prompted a similar study in almost every other setting where the ideas make sense; in Sections 7.2.1, 7.2.7 and 7.4.4 presents this material for the various symmetric derivates.

There are a number of other directions that we have not discussed. Ponomarev [226] shows that a complex function defined in a region of the complex plane and everywhere there symmetrically differentiable (in the obvious sense) must be analytic. We shall make no other mention of complex functions.

There are as well other modifications of symmetric derivatives that could be introduced. The notion of symmetric differentiability may be interpreted in a uniform sense. We have not reported on these ideas as they require no new methods. The interested reader might consult Mukhopadhyay [204] and [208], Humke and Šalát [137] and others.

Pokorny [222] has studied a symmetric version of the selective derivative of O'Malley. (A brief account of selective derivatives with references can be found in [282].) This is a genuine symmetric derivative but rather specialized and of narrow interest.

One can define ordinary derivatives that ignore sets in some σ–ideal. See [282] for the notion of *negligent derivative* and the special case of *qualitative derivatives* when the σ–ideal is taken as the class of first category sets. Evans and Larson [95] should be consulted for an extensive account of the qualitative derivative including symmetric variants. See too Evans and Vallin [97]. The methods for the symmetric side are mostly a blend of the ordinary negligent methods that can be found in [282] and symmetric methods such as we encounter in this work. Related ideas can be found in some of the articles of Starcev cited in the Bibliography.

There are too some derivatives employing parameters that have been labeled as "symmetric". Since they do not really evolve from the even/odd structure of functions they have no place here. See Larson's survey [173] for references.

Finally higher order symmetric derivatives do not enter our discussion beyond the brief mention in Chapter 1; although they are of considerable interest and do continue themes developed here we have had to limit our discussion mainly to first order theories so this, already bloated work, does not grow out of hand. The interested reader can construct a course of study of this topic by consulting Anghelutza [2], Ash [4], Bullen and Mukhopadhyay [30], Butzer [37], Butzer and Kozakiewicz [38], Corominas [45], Cross [55], De Sarkar and Das [62], Dutta [75], Gorlich and Nessel [125], Kassimatis [149], Lee [179], Marcinkiewicz and Zygmund [186], Riesz [244], Verblunsky [299] and, of course, many others.

7.2 Extreme Symmetric Derivatives

We recall the definitions of the extreme symmetric derivates.

Definition 7.1 For an arbitrary function f the extreme symmetric derivates are defined as

$$\overline{SD}\, f(x) = \limsup_{h \to 0} \frac{f(x+h) - f(x-h)}{2h}$$

and

$$\underline{SD}\, f(x) = \liminf_{h \to 0} \frac{f(x+h) - f(x-h)}{2h}.$$

In the next few sections we shall investigate properties of these derivates, motivated mostly by the program of comparing their properties to those of ordinary extreme derivates.

7.2.1 Relations Among the Derivates

There is by now an extensive literature devoted to the study of the relations that hold among the various generalized derivatives. The first such theorem traces back to Beppo Levi who showed that, while a function can easily have different one sided derivatives at a given point, there can be a disagreement between the two one sided derivatives only on a countable set. For a review of some of this literature as it applies to Dini derivatives, approximate Dini derivatives, qualitative derivatives and others see [282, Chapters VI and VII].

We have already seen a number of results that can be considered to establish relations between the ordinary and symmetric derivates. For example Theorem 4.69 asserts that if a function f is smooth at a point x then the derivates $\overline{SD}\, f(x)$ and $\overline{D}\, f(x)$ necessarily agree. Under the weaker assumption that f is everywhere symmetric Theorem 4.54 shows that these derivates mostly agree. We repeat the statement of the theorem here for convenience.

Theorem 7.2 (Neugebauer) *Let the function f be measurable and symmetric. Then the set*

$$\left\{ x\ :\ \overline{D}^+ f(x) \neq \overline{D}^- f(x)\ or\ \underline{D}^+ f(x) \neq \underline{D}^- f(x) \right\}$$

is of the first category.

For other similar relations between the symmetric derivates and the sharp (or strong) derivates see Belna, Evans, and Humke [14] who prove that if f has the Baire property then for all but a first category set of points the upper symmetric and upper sharp derivates of f agree. They deduce that if f is measurable and has a symmetric derivative everywhere then the sharp derivative exists at all but a first category set of points. Early similar, but weaker, results are given in Esser and Shisha [81] and Mukhopadhyay [208]. For more details about the sharp derivative and its relations to the other derivates see [282] which presents a setting in which these results should be viewed.

The most important relation between the symmetric derivates and the ordinary derivates of measurable functions was first established by Khintchine (see Lemma 1.25 on page 21 for a version of this). This relationship was completed by Uher [294, Lemma 2, p. 426]. The following is a version of that theorem with a few refinements. This theorem is expressed in a form we have seen repeated often in Chapter 6.

Theorem 7.3 *Let f be an arbitrary function. Then the set*

$$\left\{x \ : \ \overline{SD}\, f(x) \neq \overline{D}\, f(x)\right\}$$

contains no measurable set of positive measure and no second category set having the Baire property. If, furthermore qaC_f and qqC_f denote the sets of points at which f is quasi-approximately continuous and quasi-qualitatively continuous respectively, then

$$qaC_f \cap \left\{x \ : \ \overline{SD}\, f(x) \neq \overline{D}\, f(x)\right\}$$

has measure zero and

$$qqC_f \cap \left\{x \ : \ \overline{SD}\, f(x) \neq \overline{D}\, f(x)\right\}$$

is first category.

Notice that the relation $\overline{SD}\, f(x) \leq \overline{D}\, f(x)$ must hold at every point; consequently the set of points where $\overline{SD}\, f(x) \neq \overline{D}\, f(x)$ can be written as a countable union

$$\bigcup_{r \in Q} \left\{x \ : \ \overline{SD}\, f(x) < r < \overline{D}\, f(x)\right\}$$

where the union is over all rational numbers r. Therefore the theorem is proved by showing that each set of this form has the required property. This leads us to the following lemma which, when proved, provides a proof of the theorem.

Lemma 7.4 *Let f be an arbitrary function and K a real number. Then the set*

$$\left\{x \ : \ \overline{SD}\, f(x) < K\right\} \setminus \left\{x : \overline{D} f(x) \leq K\right\}$$

contains no measurable set of positive measure and no second category set having the Baire property. If, furthermore qaC_f and qqC_f denote the sets of points at which f is quasi-approximately continuous and quasi-qualitatively continuous respectively, then

$$qaC_f \cap \left\{x \ : \ \overline{SD} f(x) < K\right\} \setminus \left\{x : \overline{D} f(x) \leq K\right\}$$

has measure zero and

$$qqC_f \cap \left\{x \ : \ \overline{SD} f(x) < K\right\} \setminus \left\{x : \overline{D} f(x) \leq K\right\}$$

is first category.

Proof. We indicate only the measure-theoretic version with $d(E)$ denoting the set of density points. Let E be a subset of

$$\left\{x \ : \ \overline{SD}\, f(x) < K\right\} \setminus \left\{x : \overline{D}\, f(x) \leq K\right\}$$

and consider the covering relation

$$V = \{(x,t) : x \in E, \ t > 0, \ f(x+t) - f(x-t) < 2Kt\}.$$

V is transitive. By the assumptions in the lemma there is a positive number $\delta(x)$ at each point $x \in E$ so that V has the properties

$$x \in E, 0 < t < \delta(x) \Longrightarrow (x,t) \in V.$$

For the first part of the lemma we assume that E is measurable. We may apply Theorem 3.26 to obtain a neighbourhood U_x of almost every point $x \in E$ such that for all y in U_x the interval $[x,y]$ (or $[y,x]$ if $y < x$) belongs to $V^5 = V$. But that means that $f(z) - f(x) < K(z - x)$ and $f(x) - f(y) < K(x - y)$ for all $y < x < z$ in U_x. Consequently

$$\overline{D} f(x) \leq K$$

at each such point. This proves the first part of the lemma.

We turn now to the second part of the lemma. We drop the assumption that E is measurable and take E as the set

$$\left\{x : \ \overline{SD} f(x) < K\right\} \setminus \left\{x : \overline{D} f(x) < K\right\}.$$

We apply the elementary covering Lemma 3.8 to the set E and the relation V to obtain a partition $\{E_n\}$ of that set with the properties that

$$x \in E_n, 0 < t < 1/n \Longrightarrow (x,t) \in V.$$

Let z be a point in both $d(E_n)$ and in qaC$_f$. We have the conditions to apply the covering Theorem 3.24; therefore there is a positive number δ so that for every point $0 < |z - x| < \delta$ there is a set A_x having z as a point of density and for all $a, b \in A_x$,

$$a < x < z \Longrightarrow [a, x] \text{ belongs to } V^2$$

and

$$z < b < y \Longrightarrow [b, y] \text{ belongs to } V^2.$$

Once again this gives $f(x) - f(a) < K(x - a)$ and $f(y) - f(b) < K(y - b)$ for such points. But z is a point of quasi-approximate continuity of f so that, since A_x has density 1 at z, some sequences of points $a_n > z > b_n$ in A_x can be found with $f(a_n) \to f(z)$ and $f(b_n) \to f(z)$. Thus, in almost the same manner as before, we obtain

$$\overline{D} f(z) \leq K$$

at each such point. We conclude, in the usual fashion, that the set $E \cap$ qaC$_f$ itself must have measure zero as required. This completes the proof of the lemma and the theorem then evidently follows.

As corollaries to Theorem 7.3 we have immediately the following assertions.

Corollary 7.5 (Filipczak) *Let the function f be measurable. Then the relations*

$$\overline{SD}f = \overline{D}f(x) \quad and \quad \underline{SD}f(x) = \underline{D}f(x)$$

between the extreme symmetric derivates and the ordinary derivates hold almost everywhere.

Corollary 7.6 *Let the function f have the Baire property. Then the relations between the extreme symmetric derivates and the ordinary derivates*

$$\overline{SD}f = \overline{D}f(x) \quad and \quad \underline{SD}f(x) = \underline{D}f(x)$$

hold off of a set of the first category.

Since the ordinary derivates of a measurable function are measurable and agree almost everywhere with the symmetric derivates these too must be measurable. This provides a further corollary of Theorem 7.3.

Corollary 7.7 (Filipczak) *Let the function f be measurable. Then the extreme symmetric derivates $\overline{SD}f$ and $\underline{SD}f$ are measurable too.*

Finally we should mention that these results are clearly related to the Khintchine Theorem. It is well known (see, for example, Saks [251, Theorem 7.2, p. 230 and Theorem 10.1, p. 234]) that a function f is differentiable at almost every point at which the inequality $\overline{D}f(x) < \infty$ holds. The estimates just obtained show that (for measurable functions) the inequality $\overline{D}f(x) < \infty$ is almost everywhere equivalent to the inequality $\overline{SD}f(x) < \infty$ and so the Khintchine Theorem follows too as a corollary.

Corollary 7.8 (Khintchine) *Let f be a measurable function. Then f is differentiable at almost every point x at which $\overline{SD}f(x) < \infty$.*

It is worth mentioning that, while this condition ensures the existence of the derivative off of a set of measure zero, that set might be any set of measure zero; for example it need not be first category or σ-porous. Take an arbitrary set Z of measure zero; then there is a measurable set E whose metric density fails to exist at any point in Z (see Theorem A.18 given in Section A.10 of the Appendix). The measurable function

$$f(x) = \int_0^x \chi_E(t)\,dt$$

satisfies

$$-1 \le \underline{SD}\,f(x) \le \overline{SD}\,f(x) \le 1$$

everywhere and f' fails to exist at every point of Z.

7.2.2 Measurability of Derivates

It is essential in many investigations of derivates to know that the functions we are working with are measurable. It has long been known that the ordinary extreme derivates $\overline{D} f(x)$ and $\underline{D} f(x)$ of a measurable function f are measurable. It was Banach who first showed that these derivates are measurable even if f is not. Hájek completed this line of research in 1957 by showing, surprisingly, that these derivates are always in the second Baire class.

The situation for the symmetric derivates differs somewhat. If we assume that f is measurable then these derivates are too. Filipczak [102] and [103] established that the extreme symmetric derivates of a measurable function f are always measurable. The next section considers the case where f is not assumed to be measurable.

Theorem 7.9 (Filipczak) *If f is measurable then the extreme symmetric derivates $\overline{SD} f(x)$ and $\underline{SD} f(x)$ are measurable too.*

Proof. We have already obtained this as a corollary to Theorem 7.3.

7.2.3 Nonmeasurable Derivates

In comparison with the situation for the ordinary derivates the results stated in the preceding section might seem terribly weak. We know the symmetric derivates of a function f are measurable but we needed there to assume that f itself is measurable. Recall that the situation for the ordinary derivates is much better: for a completely arbitrary function f the derivates $\overline{D} f(x)$ and $\underline{D} f(x)$ are in fact Borel measurable. In light of this the next theorem presents an interesting contrast between the properties of ordinary and symmetric derivates.

Theorem 7.10 *There is a nonmeasurable function f with nonmeasurable extreme symmetric derivates $\overline{SD} f(x)$ and $\underline{SD} f(x)$.*

Proof. Let f be the characteristic function of the nonmeasurable set H in Theorem 6.18. This has a symmetric derivative equal to 0 at every point at which H is symmetric and it has infinite symmetric derivates elsewhere. Clearly then both extreme symmetric derivates are nonmeasurable.

The example in the proof of the theorem employs a function with extreme symmetric derivates that are infinite on a set of positive measure. It is easy to see that this is essential. If f has one or both extreme symmetric derivates finite almost everywhere then, by Theorem 6.51, f must be measurable. It would follow then, from Theorem 7.9, that the derivates themselves are measurable too.

7.2.4 Baire Class of Symmetric Derivates

We might still wish to know in what cases the symmetric derivates are Borel measurable. The parallel situation for the ordinary derivates is very pleasant: for a completely arbitrary function f the extreme ordinary derivates $\overline{D} f(x)$ and $\underline{D} f(x)$ are Borel measurable, belonging to the second Baire class (see [18, p. 57]). The articles of Filipczak [102] and [103], just cited in Section 7.2.2, established that the symmetric derivates $\overline{SD} f(x)$ and $\underline{SD} f(x)$ are Baire two provided f is approximately continuous (hence itself Baire one). This was continued by Wajch [303] in 1986 under the weaker hypothesis that f is merely Baire one. In particular then, if we start with a Baire 1 function, the extreme symmetric derivates are Borel measurable. The next section will provide a sharp contrast to this theorem.

Theorem 7.11 (Wajch) *If f is in the first class of Baire then the derivates $\overline{SD} f(x)$ and $\underline{SD} f(x)$ are in the third Baire class.*

We indicate the structure of the proof in [303] by stating two lemmas without proof; the proofs are straightforward and can be found in the article cited.

Lemma 7.12 *Let f be upper semicontinuous, let g be lower semicontinuous and suppose that $0 < a < b$. Then the function*

$$s(x) = \sup_{a \leq h \leq b} \frac{f(x+h) - g(x-h)}{2h}$$

is upper semicontinuous.

Lemma 7.13 *Let f_n be a nondecreasing sequence of functions converging point- wise to f and let g_n be a nonincreasing sequence of functions converging point- wise to g. Then the sequence of functions*

$$s_n(x) = \sup_{a \leq h \leq b} \frac{f_n(x+h) - g_n(x-h)}{2h}$$

is nondecreasing and converges pointwise to the function

$$s(x) = \sup_{a \leq h \leq b} \frac{f(x+h) - g(x-h)}{2h}.$$

From these two lemmas the proof of Theorem 7.11 now easily follows. Since f is Baire 1 there is a a nondecreasing sequence f_n of upper semicontinuous functions converging pointwise to f and there is a a nonincreasing sequence g_n of lower semicontinuous functions also converging pointwise to f. Define

$$s_{i,j,n}(x) = \sup_{1/i \leq h \leq 1/j} \frac{f_n(x+h) - g_n(x-h)}{2h}$$

for natural numbers i, j and k (with $i > k$).

By Lemma 7.12 these functions are upper semicontinuous. By Lemma 7.13

$$\lim_{n \to \infty} s_{i,j,n}(x) = \sup_{1/i \leq h \leq 1/j} \frac{f(x+h) - f(x-h)}{2h}.$$

Consequently

$$\overline{SD}\, f(x) = \lim_{j \to \infty} \lim_{i \to \infty} \lim_{n \to \infty} s_{i,j,n}(x)$$

exhibits the derivate as belonging to the third Baire class. (In fact this proof places this function in one of the Young classes; see [303] for details.) Similarly we can prove that $\underline{SD}\, f(x)$ belongs to the third Baire class.

7.2.5 Borel Measurability of Derivates

Because of Theorem 7.10 we know that the hypothesis of measurability in Theorem 7.9 cannot be relaxed. It occurs to us to ask whether, in Theorem 7.11 just proved, the true picture really parallels that of Banach's Theorem for the Dini derivatives (see [18, p. 56]). If f is in the class Baire α then are the derivates $\overline{SD}\, f(x)$ and $\underline{SD}\, f(x)$ in the class Baire $\alpha + 2$? Wajch posed the question in her paper and it was answered negatively by Laczkovich [170]. The derivates may fail to be Borel measurable although, because of Theorem 7.9, they are certainly Lebesgue measurable. This places a startling limitation on the theorems of the preceding section.

Theorem 7.14 (Laczkovich) *There is a function f of Baire class two that has an upper symmetric derivate $\overline{SD}\, f(x)$ that is not Borel measurable.*

Proof. The proof is obtained from the following statement: there is a set A of type G_δ and a set B of type F_σ so that the characteristic function $f = \chi_{A \cup B}$ has a non Borel measurable upper symmetric derivate $\overline{SD}\, f(x)$. This function f is evidently in the second Baire class and the theorem follows.

As the methods and the theorem itself are closely related to material we have discussed in Section 6.2.4 we shall not supply further details.

7.2.6 Porosity Relations for Symmetric Derivates

The relations between the extreme ordinary and symmetric derivates that we have established in Section 7.2.1 show that, for a function that is both measurable and has the Baire property, the set of points

$$\left\{x \ : \ \overline{SD}\,f(x) \neq \overline{D}\,f(x)\right\}$$

has measure zero and is first category. One naturally asks if it is smaller still. Belna, Evans and Humke [13, Theorem 2, p. 265] showed that under additional hypotheses this set is σ–porous.

Theorem 7.15 (Belna, Evans, Humke) *Let f be a function possessing a dense set of points of continuity. Then the sets*

$$\left\{x \ : \ \overline{SD}\,f(x) \neq \overline{D}\,f(x)\right\}$$

and

$$\left\{x \ : \ \underline{SD}\,f(x) \neq \underline{D}\,f(x)\right\}$$

are σ–porous.

Proof. It is enough to consider the latter set. Thus we shall show that the set of points

$$\{x \ : \ \underline{SD}\,f(x) \neq \underline{D}\,f(x)\}$$

is σ–porous by expressing it as the union of the sequence of sets

$$E_r = \{x \ : \ \underline{SD}\,f(x) > r > \underline{D}\,f(x)\}$$

taken over all rational numbers r and showing that each set E_r is σ–porous. Let $g(x) = f(x) - rx$. Then $\underline{SD}\,g(x) > 0$ and $0 > \underline{D}\,g(x)$ at each point $x \in E_r$. In the language of Theorem 6.49 this means that each point $x \in E_r$ belongs to the set $SI_g \setminus I_g$. Since f has a dense set of points of continuity so too has g. and Theorem 6.49 shows that E_r is σ–porous exactly as required to prove the theorem.

7.2.7 Denjoy Relations for Symmetric Derivates

A simple version of the classical Denjoy-Young-Saks Theorem asserts that for an arbitrary function f it is the case that, at almost every point x, either f has a finite ordinary derivative or else both assertions

$$\overline{D}\,f(x) = +\infty \quad \text{and} \quad \underline{D}\,f(x) = -\infty$$

hold. As pointed out in Ezzel and Nymann [99] the Khintchine Theorem gives exactly the same relations for the symmetric derivates of measurable functions.

A number of authors (eg. Uher [295], Pu and Pu [241]) have considered a version of the Denjoy-Young-Saks theorem for symmetric derivates of general functions. It is not immediately clear what the situation is for *nonmeasurable* functions; for example, a problem posed in [295] asks if there is a function f for which the set

$$\left\{x \; : \; -\infty < \underline{SD}\, f(x) < \overline{SD}\, f(x) < +\infty\right\}$$

has positive outer measure? It has zero inner measure because of Theorem 7.3. The complete analysis was carried out by Freiling and Rinne [114]. Part of the proof continues a suggestion of Uher [295, p. 37] by making clever use of Hamel bases to construct the positive possibilities.

Theorem 7.16 (Freiling–Rinne) *For an arbitrary real-valued function f define the sets:*

1. $S_0 = \left\{x \; : \; -\infty = \underline{SD}\, f(x) < \overline{SD}\, f(x) = +\infty\right\}$,

2. $S_1 = \left\{x \; : \; -\infty < \underline{SD}\, f(x) = \overline{SD}\, f(x) < +\infty\right\}$,

3. $S_2 = \left\{x \; : \; -\infty < \underline{SD}\, f(x) < \overline{SD}\, f(x) = +\infty\right\}$,

4. $S_3 = \left\{x \; : \; -\infty = \underline{SD}\, f(x) < \overline{SD}\, f(x) < +\infty\right\}$,

5. $S_4 = \left\{x \; : \; -\infty < \underline{SD}\, f(x) < \overline{SD}\, f(x) < +\infty\right\}$,

6. $S_5 = \left\{x \; : \; -\infty < \underline{SD}\, f(x) = \overline{SD}\, f(x) = +\infty\right\}$.

7. $S_6 = \left\{x \; : \; -\infty = \underline{SD}\, f(x) = \overline{SD}\, f(x) < +\infty\right\}$.

Any of the pairs $\{S_0, S_i\}$ ($i = 1, 2 \ldots, 6$) can simultaneously have full outer measure, as can either of the triples $\{S_0, S_2, S_5\}$ or $\{S_0, S_3, S_6\}$. If, however, S is one of the sets S_1, $S_2 \cup S_5$, $S_3 \cup S_6$ or S_4 then almost every outer density point of S belongs to $S \cup S_0$.

Proof. We construct first the examples. Let H be a Hamel basis for the reals that has full outer measure (see Section A.6 in the Appendix). Fix an element h_1 of H. Write $Q(A)$ as the linear rational span of any $A \subset H$. Let J and K be subsets of H, each of full outer measure. Write $G = Q(H \setminus \{h_1\})$ and

$$U = \bigcup_{h \in H}^{\infty} Q((-\infty, h] \cap H) \cap [h, +\infty).$$

Then, expressed in words, G is the set of reals x whose coefficient corresponding to the element h_1 in the Hamel expansion of x is zero; U is the set of reals x whose Hamel expansion uses no elements of H that exceed x. For any $x \in \mathbb{R}$ and $A \subset H$ we write $n_A(x)$ for the number of elements in A used in writing the Hamel expansion of x and $k_A(x)$ for the number of elements in A that exceed x and are used in writing the Hamel expansion of x. The examples follow.

$\{S_0, S_1\}$: $f(x) = \chi_G(x)$, $S_1 = G$, $S_0 = \mathbb{R} \setminus G$.

$\{S_0, S_2\}$: $f(x) = \chi_U(x)$, $S_2 = U$, $S_0 = \mathbb{R} \setminus U$.

$\{S_0, S_3\}$: $f(x) = -\chi_U(x)$, $S_3 = U$, $S_0 = \mathbb{R} \setminus U$.

$\{S_0, S_4\}$: $f(x) = x\chi_G(x)$, $S_4 = G$, $S_0 = \mathbb{R} \setminus G$.

$\{S_0, S_5\}$: $f(x) = n_H(x) - 2k_H(x)$.

$\{S_0, S_6\}$: $f(x) = 2k_H(x) - n_H(x)$.

$\{S_0, S_2, S_5\}$: $f(x) = n_J(x) - 2k_J(x)$, $S_0 = \mathbb{R} \setminus (H \cup \{0\})$, $S_2 = K \cup \{0\}$, $S_5 = J$.

$\{S_0, S_3, S_6\}$: $f(x) = 2k_J(x) - n_J(x)$.

These can each be verified. Note that in the case $\{S_0, S_5\}$ the symmetric derivative is infinite on H since, if $h \in H$ is the center of an interval (a, b), then
$f(b) - f(a) \geq 1$.

   To prove the final assertion of the theorem we require the following lemma
from [114].

Lemma 7.17 *Let f be an arbitrary function, $\epsilon > 0$ and*

$$A = \{x : f(x + h) - f(x - h) < 0, \quad \text{for all } 0 < h < \epsilon\}.$$

*If x is a point of outer density of A and f has a symmetric derived number
$d > 0$ at x then f has also a symmetric derived number $d_1 \leq -3d$ at x.*

   We prove the lemma for $d < \infty$ but a similar argument shows the case
$d = \infty$ too. Choose a sequence $h_k \to 0+$ so that $h_k < \epsilon$,

$$f(x + h_k) - f(x - h_k)/(2h_k) \to d$$

and

$$|A \cap (x - h_k, x + h_k)| > 2h_k(1 - (k + 1)^{-1}).$$

Then there are points a_k, b_k with $k > 2$ and

$$a_k \in A \cap (x - \tfrac{1}{3}h_k - \tfrac{2}{k+1}h_k, x - \tfrac{1}{3}h_k)$$

and

$$b_k \in A \cap (x + \tfrac{1}{3}h_k, x + \tfrac{2}{k+1}h_k + \tfrac{1}{3}h_k).$$

Reflect $x - h_k$ about the points a_k and b_k to obtain

$$c_k = 2b_k - (2a_k - (x - h_k))$$

and reflect $x + h_k$ about the points b_k and a_k to obtain

$$d_k = 2a_k - (2b_k - (x + h_k)).$$

Since a_k, $b_k \in A$, we have

$$f(x + h_k) - f(2a_k - (x + h_k)) < 0 \tag{7.1}$$

$$f(2a_k - (x + h_k)) - f(d_k) < 0 \tag{7.2}$$

$$f(2a_k - (x - h_k)) - f(x - h_k) < 0 \tag{7.3}$$

and

$$f(c_k) - f(2a_k - (x - h_k)) < 0 \tag{7.4}$$

From (7.1), (7.2), (7.3) and (7.4) we obtain

$$f(c_k) - f(d_k) < - (f(x + h_k) - f(x - h_k)). \tag{7.5}$$

Also

$$c_k - d_k = 4b_k - 4a_k - 2h_k < 4(\tfrac{2}{3}h_k + \tfrac{4}{k+1}h_k) - 2h_k = (\tfrac{2}{3} + \tfrac{16}{k+1})h_k.$$

Thus (7.5) provides

$$\frac{f(c_k) - f(d_k)}{c_k - d_k} < - \frac{f(x + h_k) - f(x - h_k)}{2h_k \left(\tfrac{1}{3} + \tfrac{8}{k+1}\right)}. \tag{7.6}$$

Note that $d_k < x < c_k$, $\tfrac{1}{2}(c_k + d_k) = x$ and the right hand side of the inequality (7.6) tends to $-3d$ as $k \to \infty$. Thus there must be at least one symmetric derived number of f at x that is less than or equal to $-3d$. This proves the lemma.

Note that from Lemma 7.17 we can derive the following fact: for almost every outer density point x_0 of the set $\{x : \overline{SD}\, f(x) < r\}$ and any $d > 0$

$$\overline{SD}\, f(x_0) > r + d \implies \underline{SD}\, f(x_0) < r - 3d. \tag{7.7}$$

Just consider the sets

$$A_n = \{x : f(x + h) - f(x - h) < 2hr,\ 0 < h < 1/n\}$$

and apply the lemma to the function $f(x) - rx$. Since the A_n increase to the set $\{x : \overline{SD}\, f(x) < r\}$ almost every outer density point x_0 of the latter set is an outer density point of some A_n. A similar assertion is, of course, available for almost every outer density point of the set $\{x : \underline{SD}\, f(x) > s\}$.

We may now return to the proof of Theorem 7.16. The final statement of our theorem now follows from Lemma 7.17. For example consider the set S_1. We need to show that almost every outer density point of S_1 is in $S_0 \cup S_1$. Write for $r \in \mathbb{Q}$, $n \in \mathbb{N}$

$$A_{rn} = \{x : r < \underline{SD}\, f(x) = \overline{SD}\, f(x) < r + 1/n\}$$

and note that $S_1 = \bigcap_{n=1}^{\infty} \bigcup_{r \in \mathbb{Q}} A_{rn}$. For each n almost every outer density point of S_1 is an outer density point of a set A_{rn} for some r.

By using (7.7) we see that almost every point x_0 that is an outer density point of a set A_{rn} fits one of the following cases:

1. If $\overline{SD} f(x_0) = +\infty$ then $\underline{SD} f(x_0) = -\infty$.

2. If $\underline{SD} f(x_0) = -\infty$ then $\overline{SD} f(x_0) = +\infty$.

3. If $\underline{SD} f(x_0) > r$ and $\overline{SD} f(x_0) > r + 1/n$ then

$$\underline{SD} f(x_0) < r + 1/n - 3\left(\overline{SD} f(x_0) - (r + 1/n)\right)$$

and so $\overline{SD} f(x_0) - \underline{SD} f(x_0) < \frac{4}{3n}$.

4. If $\underline{SD} f(x_0) < r$ and $\overline{SD} f(x_0) < r + 1/n$ then

$$\overline{SD} f(x_0) > r + 3\left(r - \underline{SD} f(x_0)\right)$$

and so $\overline{SD} f(x_0) - \underline{SD} f(x_0) < \frac{4}{3n}$.

But almost every $x_0 \in d(S_1)$ satisfies one of these for all integers n. In the first two cases we see that such a point x_0 belongs to S_0 (since the derivates are infinite) and in the latter case that such a point x_0 belongs to S_1 itself (since the derivates are finite and equal).

This completes the proof for the set S_1. Each of the remaining cases of the theorem is similarly handled.

7.3 Symmetric Derivatives

7.3.1 Baire Class of Symmetric Derivatives

If f is everywhere differentiable (finitely) then its derivative f' may be expressed, trivially, as the pointwise limit of a sequence of continuous functions

$$f'(x) = \lim_{n \to \infty} n(f(x + 1/n) - f(x)).$$

Thus ordinary derivatives are necessarily of the first Baire class; this fact, while elementary, plays a key role in investigations of derivatives. As we pass to generalized derivatives the hope usually remains that these more general derivatives will have such a simple structure. Perhaps surprisingly, most of the important derivation processes produce derivatives in the first Baire class. Even if f is not continuous and f' is allowed infinite values this is the case; this is true too for approximate derivatives.

For the symmetric derivative Filipczak [103] in 1976 showed that an exact symmetric derivative SD $f(x)$ is in the first Baire class provided the primitive f is assumed to be everywhere approximately continuous. Since approximately continuous functions are themselves in the first Baire class this is a considerable restriction. The theorem was completed by Larson [172] who removed the hypothesis on the primitive and allowed infinite values for the derivative. Note that the primitive function must be measurable if it is everywhere symmetrically differentiable, since it is continuous almost everywhere by Theorem 6.51. The proof, however, does not take advantage of this.

Theorem 7.18 (Larson) *Let the function f be everywhere symmetrically differentiable (allowing infinite values). Then the derivative* SD $f(x)$ *is in the first Baire class.*

Proof. The proof is an application of the general criterion in Theorem A.17 for Baire 1 functions. If $g(x) = $ SD $f(x)$ is not in the first Baire class then there is a perfect set P and real numbers $\alpha < \beta < \delta < \gamma$ such that, if we define the sets

$$A = \{x \in P \ : \ g(x) \leq \alpha\}$$

$$B = \{x \in P \ : \ g(x) > \beta\}$$

$$C = \{x \in P \ : \ g(x) < \gamma\}$$

$$D = \{x \in P \ : \ g(x) \geq \delta\},$$

then both A and D are dense in P. We shall use the covering theorem of Charzyński (Theorem 3.10) to obtain a contradiction.

Evidently $P = B \cup C$ so that, as P is perfect, one at least of these two sets is second category in P. Let us suppose that it is B which has this property; identical arguments would handle the other situation in any case. For each point $x \in B$ there is a $\delta(x) > 0$ so that

$$0 < t < \delta(x) \Longrightarrow f(x+t) - f(x-t) > 2\beta t$$

and then, applying the elementary covering Theorem 3.8, there is a partition $\{B_n\}$ of B so that

$$0 < t < 1/n, x \in B_n \Longrightarrow f(x+t) - f(x-t) > 2\beta t. \tag{7.8}$$

By our category assumptions on B there must now exist an interval (c, d) and a set B_m so that B_m is dense in $P \cap (c, d)$; we may assume that $d - c < 1/m$.

By our assumption the set A is also dense here thus we may select a point $x \in A \cap P \cap (c, d)$ which we may assume is not isolated in P on one side at least, say the right. Because $x \in A$ there is an $h_0 < 1/m$ so that

$$0 < t < h_0 \Longrightarrow f(x+t) - f(x-t) < 2\beta t. \tag{7.9}$$

In order to contradict this inequality we choose a point $x_2 \in A \cap (c, d)$ with $0 < x_2 - x < h_0$ and then a point $x_1 \in B_m \cap (x, x_2)$ sufficiently close to x so that

$$0 < t < x_1 - x \Longrightarrow f(x_2 + t) - f(x_2 - t) < 2\beta t. \tag{7.10}$$

Now we apply the covering Theorem 3.10 with the points as labeled here and with $h = x_2 - x_1 < h_0$, $h_2 = x_1 - x$ and $h_1 = h + h_2$. Note that

$$x_2 - h_2 = x + h, \quad x - h = x_1 - h_1 \text{ and } x_1 + h_1 = x_2 + h_2. \tag{7.11}$$

Since $x_1 \in B_m$ we have, by (7.8), (7.10) and (7.11), that

$$\begin{aligned} f(x + h) - f(x - h) &= f(x_1 + h_1) - f(x_1 - h_1) - (f(x_2 + h_2) - f(x_2 - h_2)) \\ &\geq 2\beta h_1 - 2\beta h_2 = 2\beta h. \end{aligned}$$

As this is in contradiction with (7.9) the theorem is proved.

7.3.2 Symmetric Differentiability

It has been known since Khintchine's fundamental paper that a measurable function must have an ordinary derivative at almost every point at which it has a symmetric derivative. This relation between symmetric differentiability and ordinary differentiability can be examined in more detail. The theorems of the previous sections provide the necessary estimates connecting the symmetric derivates to the ordinary derivates. If we wish to extend these ideas to arbitrary functions then we should recall that Theorem 6.18 provides a (nonmeasurable) function f that is nowhere differentiable and yet has a zero symmetric derivative at the points of a set that is large in the senses of both measure and category. Apparently the best assertion that can be made assumes the following form.

Theorem 7.19 *Let f be an arbitrary function, let DS_f denote the set of points at which f has a finite or infinite symmetric derivative and let D_f denote the set of points at which f has a finite or infinite ordinary derivative. Then the set*

$$DS_f \setminus D_f$$

contains no measurable set of positive measure and no second category set having the Baire property. If, furthermore qaC_f and qqC_f denote the sets of points at which f is quasi-approximately continuous and quasi-qualitatively continuous respectively, then $qaC_f \cap DS_f \setminus D_f$ has measure zero and $qqC_f \cap DS_f \setminus D_f$ is first category.

Proof. This follows directly from Theorem 7.3.

As a corollary we have almost immediately the following assertions connecting the ordinary derivative with the symmetric derivative.

Corollary 7.20 (Uher) *Let f be an everywhere symmetrically differentiable function (infinite values allowed). Then f is differentiable everywhere excepting at the points of a set that has measure zero and is of the first category.*

Corollary 7.21 *Let f be a measurable function. Then f is differentiable at almost every point at which it has a symmetric derivative (infinite values allowed).*

Corollary 7.22 *Let f have the Baire property. Then f is differentiable at every point at which it has a symmetric derivative (infinite values allowed) except possibly at the points of a set of first category.*

7.3.3 Theorem of Belna, Evans and Humke

As a consequence of Theorem 7.19 we know that a symmetrically differentiable function has an ordinary derivative at every point saving only the points of some set of measure zero and first category. In fact this set must be σ-porous. (The original proof in [13] had to assume that f was measurable since it was unknown at the time whether a symmetrically differentiable function needed to be measurable.) The theorem follows from material in Section 7.2.6.

Theorem 7.23 (Belna–Evans–Humke) *Let f have a finite or infinite symmetric derivative at every point. Then f is differentiable in the ordinary sense at all but a σ-porous set.*

Proof. A function f possessing a finite or infinite symmetric derivative everywhere must be symmetrically semicontinuous and so continuous almost everywhere by Theorem 6.51. Accordingly f satisfies the hypotheses of Theorem 7.15 and the ordinary derivates agree with the symmetric derivative at all but a σ-porous set.

Pu and Pu [240] point out that this theorem combines with other results to show that if f is everywhere symmetrically differentiable, allowing infinite derivatives, then f itself must be continuous off of a σ-porous set. (Naturally one might ask whether this set must be countable or scattered.) On the basis of this observation Pu and Pu give a direct proof of the measurability of such a function f without invoking the Uher theorem.

There are more refined versions of porosity available. Even in the original work of Denjoy, where the notion first arose, a symmetric version of porosity was used. Zajíček [311], Evans [88] and [89], Evans, Humke and Saxe [92] and Repický [243] have studied this symmetric porosity. It does appear to play a role in the study of the symmetric derivatives as the following theorem of Zajíček [311] shows.

Theorem 7.24 (Zajíček) *Let f be a continuous function. Then the set of points where f is symmetrically differentiable but not differentiable in the ordinary sense is σ-r-symmetrically porous for any $0 < r < 1$.*

For definitions of the notion of a σ–r–symmetrically porous set see any of the works just cited. Evans [88] and [89] proves a more refined version of the Zajíček Theorem. He also shows that this symmetric porosity does not enter into the statements of many of the ordinary differentiation results where σ–porosity plays a role. That the two notions of porosity are distinct is proved in [92] and [243]. It is not clear at this stage of the study of such porosity notions whether this refinement will prove very useful and we shall not report further on these ideas.

7.3.4 Example of Foran–Ponomarev

To balance the results of Sections 7.3.2 and 7.3.3 we should show that the exceptional set

$$\mathrm{DS}_f \setminus \mathrm{D}_f$$

of points at which a symmetric derivative exists but an ordinary one does not need not be countable. Such an example was constructed by Foran [106] and, later but independently, by Ponomarev [228]. The latter supplied moreover an example of an exact symmetric derivative that is not Denjoy-Perron integrable. Incidentally, then, one knows immediately that it must fail to be a derivative on some uncountable set.

We shall state this without proof. Note that if we do not insist on *finite* derivatives then an easier example comes to mind. Take the classical Cantor function; it has a symmetric derivative $+\infty$ at each point of the Cantor set itself and has a zero derivative elsewhere.

Theorem 7.25 (Foran–Ponomarev) *There exists a continuous function f that has a finite symmetric derivative $SD f(x)$ everywhere on an interval (a, b) but which is not differentiable at the points of an uncountable set and for which $SD f(x)$ fails to be Denjoy-Perron integrable on $[a, b]$.*

A characterization of the exceptional set where f' fails to exist is not available. For example it is not known whether, given a closed, σ-porous set $E \subset (0, 1)$, there is a symmetrically differentiable function f that fails to have a derivative in the ordinary sense at each point of E. In Foran's example the Hausdorff dimension of the set where f' fails to exist is easily checked. His example has Hausdorff dimension zero but it seems unlikely that this would be a necessary requirement of any such example.

7.3.5 Points of Non Symmetric Differentiability

Need a continuous function have a symmetric derivative anywhere? Of course it has been known for a long time that there are continuous, nowhere differentiable functions. Naturally one expects that there should be continuous functions that do not possess derivatives in any reasonable sense. In fact since the typical continuous function (typical in the sense of category) does not possess a derivative one naturally asks whether the typical continuous function also lacks a symmetric derivative.

This result was obtained by Kostyrko [153].

Theorem 7.26 (Kostyrko) *The typical function f in $C[0,1]$ has the property that*

$$\limsup_{h \to 0} \frac{f(x+h) - f(x-h)}{2h} = +\infty$$

and

$$\liminf_{h \to 0} \frac{f(x+h) - f(x-h)}{2h} = -\infty$$

for all $0 < x < 1$.

One can construct specific examples of continuous functions that are nowhere symmetrically differentiable. Savotin [253] investigates the symmetric differentiability of a function f given by a series

$$f(x) = \sum_{n=0}^{\infty} a_n \varphi(b_n x + c_n),$$

where $\{a_n\}$, $\{b_n\}$ and $\{c_n\}$ are sequences of real numbers, $\sum_{n=0}^{\infty} |a_n| < \infty$, and the function φ is continuous, periodic and not constant. He gives some conditions on the function φ and the sequences $\{a_n\}$, $\{b_n\}$ and $\{c_n\}$ so that the function f: (i) is everywhere nondifferentiable (in the ordinary sense), (ii) is everywhere nondifferentiable and, at the same time, has a finite symmetric derivative in a dense countable set of points, or (iii) is everywhere symmetrically nondifferentiable.

Let us mention also in this context a theorem of Filipczak [102, Theorem 21, p. 47]. He characterizes the set of points at which a continuous function may fail to have a symmetric derivative. This is a result inspired by and similar to a theorem of Zahorski for the ordinary derivative ([18, Theorem 3.1, p. 228]). As the methods are very specialized and have no other bearing on the study of symmetric derivatives we do not reproduce the proof. Pu, Pu and Teng [237] also address this problem for a wider class of functions (measurable, symmetric functions), obtaining the same classification.

Theorem 7.27 (Filipczak) *A necessary and sufficient condition for E to be the set of points at which a continuous function fails to have a symmetric derivative is that $E = A \cup B$ where A is a Borel set of type G_δ and B is a measure zero set of type $G_{\delta\sigma}$.*

7.3.6 Infinite Symmetric Derivatives

The set of points x at which an ordinary derivative $f'(x)$ may assume infinite values is of measure zero. This observation is due to Denjoy and dates back to 1916 (see Saks [251, p. 236]). In fact this is the best that can be said about the set $\{x : f'(x) = \pm\infty\}$ for if E is a measure zero set of type G_δ then there is a differentiable, absolutely continuous function f with $f'(x) = +\infty$ precisely on the set E and elsewhere finite (see, for example, [18, Theorem 3.2, p. 229]).

An easy example illustrates a natural situation with infinite symmetric derivatives. Uher [295] points out that the Cantor function, which is a continuous monotonic function f on $[0,1]$ that is constant on each interval contiguous to the Cantor set, has a zero derivative off the Cantor set and a symmetric derivative equal to $+\infty$ at every point of the Cantor set.

Here the set of points where the function has an infinite symmetric derivative has measure zero. In fact from the Khintchine Theorem we can conclude, in general, that a measurable function f can have an infinite symmetric derivative only on a set of measure zero: if $E = \{x : \mathrm{SD}\, f(x) = +\infty\}$ and if f is measurable then E is measurable and that theorem assures us that $f'(x)$ exists for almost every point $x \in E$ which is only possible if E has measure zero.

It remains to describe the situation for nonmeasurable functions. Larson [172, Theorem 2.2, p. 591] showed that for a completely arbitrary function f the set
$$\{x : |\mathrm{SD}\, f(x)| = \infty\}$$
can contain no interval. However should such a set contain an interval (a, b) then f would be symmetrically semicontinuous everywhere on that interval; we have seen in Section 6.4.1 that such a function must be measurable on that interval and so the Khintchine Theorem again applies.

For arbitrary functions we can prove the following. This follows from material in Section 7.2.1.

Theorem 7.28 *For an arbitrary function f the set*
$$\{x : |SD f(x)| = \infty\} \setminus \{x : |f'(x)| = \infty\}$$
contains no measurable set of positive measure and no set of second category with the Baire property.

Proof. Again this is a corollary of Theorem 7.3.

There are nonmeasurable functions for which the set
$$\{x : |\mathrm{SD}\, f(x)| = \infty\}$$
fails to have measure zero, indeed has full outer measure. This one of the cases of Theorem 7.16 stated earlier.

7.3.7 Steep Infinite Derivatives

The Cantor function F has the curious property that at every point it possesses a symmetric derivative, either zero or infinite. Buczolich and Laczkovich [23] have shown that the symmetric difference quotients are growing at a certain geometric rate: there is a $b > 1$ so that at each point x in the Cantor set not only does F have an infinite symmetric derivative but in fact

$$\limsup_{h \to 0+} \frac{F(x+bh) - F(x-bh)}{F(x+h) - F(x-h)} < b. \tag{7.12}$$

This can be written also as

$$\limsup_{h \to 0+} \frac{(F(x+bh) - F(x-bh))/(2bh)}{(F(x+h) - F(x-h))/(2h)} < 1 \tag{7.13}$$

which shows that, for small enough h, there is a factor larger than one by which the slope of the chord joining $x \pm h$ exceeds the slope of the chord joining $x \pm bh$. This is a stronger statement than that the symmetric derivative at these points must be $+\infty$.

Lemma 7.29 *Let F be a nondecreasing function and suppose that x is a point at which F is not locally constant. If F satisfies (7.13) for some $b > 1$ then $SD\,F(x) = +\infty$.*

Proof. There are r_0 and $h_)$ with $0 < r_0 < 1$ and $h_0 > 0$ so that

$$\frac{F(x+bh) - F(x-bh)}{2bh} < r_0 \frac{F(x+h) - F(x-h)}{2h} \tag{7.14}$$

for all $0 < h < h_0$. Write $\delta_0 = (F(x+h_0) - F(x-h_0))/(2h_0)$. Since F is not locally constant at x, $\delta_0 > 0$. For any $h \in [h_0, bh_0]$, since $b > 1$ and F is nondecreasing,

$$\frac{F(x+h) - F(x-h)}{2h} \geq \frac{F(x+h_0) - F(x-h_0)}{2bh_0} = \delta_0/b. \tag{7.15}$$

For every $k \in \mathbb{N}$ and $h \in [h_0 b^{-k}, h_0 b^{-k+1}]$ we see, from (7.14) and (7.15), that

$$\frac{F(x+h) - F(x-h)}{2h} \geq \frac{\delta_0}{br_0^k}. \tag{7.16}$$

Since $r_0^k \to 0$ as $k \to \infty$ it follows immediately from (7.16) that $SD\,F(x) = +\infty$.

The conditions in (7.12) and (7.13) allow a porosity computation that we will need. This, like all the material in this section, is from [23]. Recall that, for any set E and any interval (a, b), $\lambda(E, a, b)$ denotes the length of the longest component of $(a, b) \setminus E$.

Lemma 7.30 *Let f be a continuous, nondecreasing function, let $b > 1$, $h_0 > 0$ and $0 < r < 1$ and let H denote the set of points x at which f is not locally constant and for which the inequality*

$$\frac{(f(x + bh) - f(x - bh))/(2bh)}{(f(x + h) - f(x - h))/(2h)} \leq r \qquad (0 < h < h_0) \qquad (7.17)$$

holds. Then there is a $p > 0$ so that

$$\liminf_{h \to 0+} \frac{\lambda(H, x, x + h)}{h} \geq p \quad and \quad \liminf_{h \to 0+} \frac{\lambda(H, x - h, x)}{h} \geq p \qquad (7.18)$$

holds for all $x \in H$.

Proof. Choose $N \in \mathbb{N}$ so large that

$$\frac{\log 8N}{\log N} + \frac{\log b}{\log N} < \frac{\log b}{\log rb}$$

and take $M \in \mathbb{N}$ so that

$$\frac{\log 8N}{\log b} < M \leq \frac{\log 8N}{\log b} + 1.$$

In particular we note, after some computations, that

$$b^M > 8N \quad and \quad (rb)^M < N. \qquad (7.19)$$

We prove that the inequalities in (7.18) are satisfied with $p = 1/(4N)$. Let $x \in H$ and $0 < h < h_0$; we prove the first of these inequalities by finding an subinterval I of $(x, x + h)$ with $I \cap H = \emptyset$ and $|I| \geq h/(4N) = hp$.

We can assume that

$$A = (f(x + h) - f(x))/(2h) > 0$$

since otherwise, as f is nondecreasing, f is constant in this interval and we may merely take $I = (x, x + h)$.

Consider the intervals

$$I_i = \left(x + \frac{i - 1}{4N} h, x + \frac{i}{4N} h \right) \qquad (i = 1, 2, \dots 2N).$$

It is enough if we can find an i for which $I_i \cap H = \emptyset$ for then take $I = I_i$ and $|I| \geq h/(4N)$ as we wished. It suffices then to obtain a contradiction from the assumption that there is a point y_i in $I_i \cap H$ for each $i = 1, 2, \dots, 2N$.

As $y_i \in H$ and $0 < h < h_0$ we have

$$A = \frac{f(x+h) - f(x)}{2h}$$
$$\leq \frac{f(y_i + h) - f(y_i - h)}{2h} \leq r^m \frac{f(y_i + b^{-m}h) - f(y_i - b^{-m}h)}{2b^{-m}h} \quad (7.20)$$

for each $m = 1, 2, \ldots$.

By our choice of M we have from (7.19) that $b^{-M}h < h/(8N)$. Thus in this case the intervals

$$\left(y_{2j} - b^{-M}h, y_{2j} + b^{-M}h\right) \quad (j = 1, 2, \ldots N)$$

are pairwise disjoint subintervals of $(x, x+h)$. It follows, using the fact that f is nondecreasing and (7.20), that

$$2hA = f(x+h) - f(x)$$
$$\geq \sum_{j=1}^{N} \left(f(y_{2j} + b^{-M}h) - f(y_{2j} - b^{-M}h)\right) \geq N \frac{2b^{-M}hA}{r^M}$$

and hence that $(rb)^M \geq N$. But this contradicts (7.19) and the proof is complete.

It has become fashionable of late to refer to the classical Cantor function as "the Devil's staircase". As this terminology has invaded both the popular and the serious literature it seems likely that it will survive. In the same spirit we can make the following definition in order to promote the study of functions having the property discussed above.

Definition 7.31 Let f be a continuous, nondecreasing function. If there is a $b > 1$ so that at each point x either f is locally constant at x or else

$$\limsup_{h \to 0+} \frac{f(x+bh) - f(x-bh)}{f(x+h) - f(x-h)} < b \quad (7.21)$$

then f shall be called a *symmetric devil*.

If b is a real number for which (7.21) holds we shall say that f is *b-concentrated* at x. We have the following theorem describing the properties of symmetric devils. Recall that the set of points where a function f is locally constant is open. Its complement is a closed set called the *support* of f.

Theorem 7.32 (Buczolich–Laczkovich) *Let f be a symmetric devil. Then f is continuous, nondecreasing, and singular and everywhere possesses a symmetric derivative. The support of f is a perfect, σ-porous set, $SD f(x) = +\infty$ for each x in the support of f and $SD f(x) = 0$ elsewhere.*

Proof. We already know from Lemma 7.29 that SD $f(x) = +\infty$ for each x at which (7.21) holds. By definition this holds at every point in the support of f; at all other points f is locally constant so, of course, SD $f(x) = 0$ there.

The only part of the theorem then remaining to be proved are the statements about the support. The support can contain no isolated points because of this property of the symmetric derivative. That the support is σ-porous follows directly from Lemma 7.30 in an obvious way. Any σ-porous set is measure zero and consequently f is locally constant off a measure zero set and so is singular.

The following theorems from [23] provide concrete examples of symmetric devils.

Theorem 7.33 (Buczolich–Laczkovich) *The classical Cantor function is a symmetric devil that is b–concentrated for all b sufficiently large but is not b–concentrated for b = 4.*

In fact the Cantor function is b-concentrated for all $b \geq 81$. It is not known precisely for what values of b the statement may be made. We refer the reader to the proof in [23, p. 354] for the computations.

Theorem 7.34 (Buczolich–Laczkovich) *There is a symmetric devil that is b–concentrated for all b > 2.*

7.3.8 Zero Symmetric Derivative

A function with a zero ordinary derivative is constant. In this section we address the analogous question for the symmetric derivative. This repeats material we have already seen in Chapter 1. The following result is best possible.

Theorem 7.35 *Let the function f have a zero symmetric derivative at every point. Then f is constant off some scattered set.*

Proof. By Charzyński's Theorem f is continuous off some scattered set. By using the covering Theorem 3.3 it is easy to show that there is a countable set C so that for every $\epsilon > 0$ the functions $f(x) - \epsilon x$ and $\epsilon x - f(x)$ are nondecreasing off of C. Thus f is constant on its set of points of continuity.

That Theorem 7.35 is sharp was first established by Jurek and then later and independently by Szpilrajn [279]; this has been proved in Section 2.6.

Theorem 7.36 (Jurek–Szpilrajn) *Let E be a scattered set of reals. Then there is a nonnegative function f that has a zero symmetric derivative at every point and $E = \{x : f(x) > 0\}$.*

7.3.9 Stationary Sets for Symmetric Derivatives

The structure of a family of functions can be partially revealed by determining which sets are stationary sets. If $\mathcal{F}$ is a family of functions then we say that E is a *stationary set* for $\mathcal{F}$ if

$$f \in \mathcal{F}, \ f \text{ constant on } E \Longrightarrow f \text{ constant.}$$

The stationary sets for ordinary derivatives (possibly infinite) of continuous functions has a measure-theoretic character: E is a stationary set for the class of such functions if and only if E is the complement of a set of inner Lebesgue measure zero. This is due to Boboc and Marcus (see [18, p. 201]). Corollary 7.41, given later, shows that a similar situation holds for finite symmetric derivatives; this was first observed in Kulbacka [162]. For possibly infinite symmetric derivatives the characterization has a metric character; this is due to Uher [295] and, again, to Kulbacka [162].

Theorem 7.37 (Kulbacka–Uher) *A set E is a stationary set for the family of (possibly infinite) symmetric derivatives of continuous functions if and only if E meets every perfect set.*

7.3.10 Larson's Primitive

If $F' = f$ everywhere then F is determined uniquely up to an addditive constant. A symmetric primitive is much more ambiguous. Let S be any scattered set; then, by Theorem 2.23, there is a function G with a symmetric derivative zero everywhere and $S = \{x : G(x) \neq 0\}$. Consequently if $SD\,F(x) = f(x)$ everywhere then $F(x) + aG(x) + b$ is also a symmetric primitive for f for any $a, b \in \mathbb{R}$. From this abundant supply of symmetric primitives for f it seems unlikely that any canonical choice might be made. Larson [172] shows how.

We reproduce some of his ideas here. The full treatment allows infinite derivatives in some statements and so is more general than our account. We simplify the details somewhat by treating only the problem of finding a reasonable primitive of an everywhere finite symmetric derivative. Of course there is no natural choice of a primitive for symmetric or ordinary derivatives if infinite values are allowed.

Let F be an everywhere symmetrically continuous function. Then we recall, by Theorem 6.66, that the set

$$M_F = \{x : \limsup_{y \to x} |F(y)| = \infty\}$$

is closed, everywhere locally symmetric and countable. For such a function we define

$$\mu_F(x) = \begin{cases} F(x) & \text{if } x \in M_F \\ \limsup_{t \to x,\, t \in C(F)} F(t) & \text{if } x \in \mathbb{R} \setminus M_F \end{cases}$$

where $C(F)$ is the set of points of continuity of F.

For a symmetrically differentiable function F the function μ_F serves as an alternative primitive for SD $F(x)$ that has certain semicontinuity properties outside of the closed set M_F.

Theorem 7.38 (Larson) *Suppose that* $SD\,F(x)$ *exists, finitely, everywhere. Then*

(i) μ_F *is continuous at every point at which F is continuous and* $\mu_F(x) = F(x)$ *at such points.*

(ii) *For each* $x \in \mathbb{R} \setminus M_F$,

$$\mu_F(x) = \limsup_{t \to x+} \mu_F(t) = \limsup_{t \to x-} \mu_F(t).$$

(iii) $SD\,F(x) = SD\,\mu_F(x)$ *everywhere.*

Proof. The first property (i) is obvious. At every point $x \notin M_F$, it follows directly from the definition that

$$\mu_F(x) = \limsup_{t \to x} \mu_F(t). \tag{7.22}$$

To verify (ii) then we need to show that the limit in (7.22) is two-sided. Let $x \in \mathbb{R} \setminus M_F$. The set $C(F)$ of continuity points of F is residual, in fact all but a scattered set. There must be, then, a sequence of intervals $I_n \subset C(F)$ with $I_n \to x$ and $F(I_n) \to \mu_F(x)$. Select $x_n \in I_n$ so that $2x - x_n \in C(F)$. By symmetric continuity $F(x_n) - F(2x - x_n) \to 0$ and so both $\mu_F(x_n) = F(x_n)$ and $\mu_F(2x - x_n) = F(2x - x_n)$ converge to $\mu_F(x)$. Thus the limit in (7.22) is valid on both sides of x.

It remains to check the derivative in statement (iii). Consider the statement

$$SD\,F(x) = \lim_{h \to 0,\, x \pm h \in M_F} \frac{\mu_F(x+h) - \mu_F(x-h)}{2h}. \tag{7.23}$$

This is trivially true because, by definition, F and μ_F agree on M_F. Since M_F is locally symmetric at every point it is sufficient in order to prove statement (iii) to show that

$$SD\,F(x) = \lim_{h \to 0,\, x \pm h \notin M_F} \frac{\mu_F(x+h) - \mu_F(x-h)}{2h}. \tag{7.24}$$

Let $x \in \mathbb{R}$, $h > 0$ with $x + h \notin M_F$ and $x - h \notin M_F$. As before there must be a sequence of intervals $I_n \subset C(F) \cap (x - 2h, x)$ with $I_n \to x - h$ and

$F(I_n) \rightarrow \mu_F(x-h)$. Select $x - h_n \in I_n$ so that $x + h_n \in C(F)$. Note that $F(x - h_n) \rightarrow \mu_F(x-h)$ and $\limsup_{n\to\infty} F(x + h_n) \leq \mu_F(x+h)$ and so

$$\limsup_{n\to\infty} \frac{F(x+h_n) - F(x-h_n)}{2h_n} \leq \frac{\mu_F(x+h) - \mu_F(x-h)}{2h}. \qquad (7.25)$$

Thus, for every such $h > 0$, we may use (7.25) to select a $0 < h' < 2h$ so that

$$\frac{F(x+h') - F(x-h')}{2h'} \leq \frac{\mu_F(x+h) - \mu_F(x-h)}{2h} + h. \qquad (7.26)$$

From (7.26) we obtain

$$\mathrm{SD}\, F(x) \leq \liminf_{h\to 0,\, x\pm h\notin M_F} \frac{\mu_F(x+h) - \mu_F(x-h)}{2h}. \qquad (7.27)$$

Identical arguments produce the opposite inequality for the limit superior and (7.24) follows as required to prove (*iii*).

As corollaries we obtain the following, also from [172]. These can be readily obtained by arguing from our monotonicity theorems of Chapter 5.

Corollary 7.39 (Larson) *Let F be everywhere symmetrically differentiable with $SD\, F(x) \geq 0$ almost everywhere. Then μ_F is continuous and nondecreasing.*

Corollary 7.40 (Larson) *Let F be everywhere symmetrically differentiable with $SD\, F(x)$ bounded. Then μ_F is continuous.*

Corollary 7.41 (Larson) *Let F and G be everywhere symmetrically differentiable with $SD\, F(x) = SD\, G(x)$ almost everywhere. Then $SD\, F(x) = SD\, G(x)$ everywhere.*

In a later paper Larson [174] gives a symmetric monotonicity theorem that continues these same themes. The proof can be obtained from our usual monotonicity results of Chapter 5 by showing that, under the conditions stated, $SD\, f(x) = -\infty$ cannot happen. The negative of the Cantor function shows that the assumption that the derivative has the Darboux property is needed in the statement of the theorem. That function has a symmetric derivative that is zero off the Cantor set and $-\infty$ on it.

Theorem 7.42 (Larson) *Let F have everywhere a symmetric derivative (possibly infinite) such that $SD\, F(x) \geq 0$ almost everywhere and such that the symmetric derivative itself has the Darboux property. Then μ_f is nondecreasing.*

There are further applications of the primitive μ_f. Kostyrko [155] has employed the Larson primitive to answer a question posed by Larson elsewhere

[Real Anal. Exchange 9 (1983/84), no. 1, p. 295, Query 172]. The problem is to provide a characterization of the situation under which finite symmetric derivatives have the Darboux property. He defines a class, MVT, of functions "fulfilling the mean value theorem" and shows that a locally bounded symmetric derivative has the Darboux property if and only if it is the symmetric derivative of a function in MVT.

7.3.11 The Range of Symmetric Derivatives

The range of ordinary derivatives is easy enough to sort out. If f is continuous and has a derivative everywhere, even allowing infinite values, then f' has the Darboux property. Thus the range of f' must be an interval or a single point.

For symmetric derivatives these questions are rather more delicate. For example the continuous function $f(x) = |x|$ is everywhere symmetrically differentiable and its symmetric derivative assumes just the three values 0, 1 and -1. The Cantor function is also continuous and everywhere symmetrically differentiable and its symmetric derivative assumes just the two values 0 and $+\infty$. Buczolich and Laczkovich [23, Theorem 5.1, p. 359] show that there is no possibility of two *finite* values.

The proof we give here (reproduced from [288]) is based almost entirely on the elementary fact that a symmetric derivative of a continuous function is in the first Baire class.

Theorem 7.43 (Buczolich–Laczkovich) *There is no symmetrically differentiable function whose symmetric derivative assumes just two finite values.*

Proof. Our first observation is that the theorem can be reduced to showing that there is no *continuous* function with this property. This exploits the "nice" primitive notion of Larson [172] (see Section 7.3.10): if a function g exists with a bounded, symmetric derivative everywhere then there is a continuous function f for which SD $f(x) =$ SD $g(x)$ everywhere.

We assume then, contrary to the theorem, that there is a continuous, symmetrically differentiable function f whose symmetric derivative assumes only the two distinct values α and β, $\alpha < \beta$. From the fact that $\alpha \leq$ SD $f(x) \leq \beta$ our basic monotonicity theorem shows that both $f(x) - \alpha x$ and $\beta x - f(x)$ are nondecreasing.

Since SD $f(x)$ is Baire 1 there are points of continuity of SD $f(x)$ in every interval. But at a point of continuity there must be an interval in which SD $f(x)$ assumes only the value α or the value β. In such an interval the monotonicity theorem, applied once again, shows that f is linear with slope α or β. Thus there is a maximal open set G so that in every component of G the function f is linear with slope α or β.

Let P denote the complement of G. P can have no isolated points. For if $b \in P$ and (a, b), $(b, c) \subset G$ then f is linear with slope α or β in each interval $[a, b]$, $[b, c]$. If the slope is the same in the two intervals then f is linear on $[a, c]$ which contradicts the maximality of G. If the slope is different in the two intervals then SD $f(b) = \frac{1}{2}(\alpha + \beta)$ and this value is not allowed for the symmetric derivative.

In fact P must be empty. If not then P is perfect and, again using the fact that SD $f(x)$ is Baire 1, there is a point of continuity of SD $f(x)$ relative to P. Thus there must be a nonempty portion $P \cap (a, b)$ so that either SD $f(x) = \alpha$ for all $x \in P \cap (a, b)$ or SD $f(x) = \beta$ for all $x \in P \cap (a, b)$.

Let us suppose the latter case; the argument for the former is similar. Consider some interval $[c, d]$ contiguous to P in (a, b). In the interval $[c, d]$ the function f is linear with slope α or β. Since SD $f(c) = \beta$ and $f'_+(c)$ is either α or β it follows that $f'_-(c)$ exists too. But, since $f(x) - \alpha x$ and $\beta x - f(x)$ are nondecreasing, $\alpha \leq f'_-(c) \leq \beta$. This shows that

$$f'_+(c) = 2\,\mathrm{SD}\,f(c) - f'_-(c) \geq 2\beta - \beta = \beta$$

and so f cannot have slope α in $[c, d]$. Thus in this case in every interval contiguous to P in (a, b) the function f is linear with slope β. This means that SD $f(x) = \beta$ for all $x \in (a, b)$ and hence f is linear in (a, b) which contradicts the fact that the portion $P \cap (a, b)$ is nonempty.

We can conclude that P must be empty and so we see that f can only be linear. This contradicts the fact that its symmetric derivative assumes two values and the conclusion of the theorem follows.

A symmetric derivative may, as already stated, assume three distinct finite values. Indeed let α, $\beta \in \mathbb{R}$ with $\alpha \neq \beta$. Then there is a continuous, symmetrically differentiable function f such that its symmetric derivative assumes just the three finite values α, β and $\frac{1}{2}(\alpha + \beta)$. (Simply bend the example $f(x) = |x|$ into the right shape.) Using the arguments of Theorem 7.43, we can show that no other configuration is possible. This is reproduced from [288].

Theorem 7.44 *Let α, β, $\gamma \in \mathbb{R}$ with $\alpha < \gamma < \beta$ and $\gamma \neq \frac{1}{2}(\alpha + \beta)$. Then there is no symmetrically differentiable function whose symmetric derivative assumes just the three values α, β and γ.*

Proof. As in the preceding proof we need only show that there is no continuous function f with this property. If there is then, as before, both $f(x) - \alpha x$ and $\beta x - f(x)$ are nondecreasing.

We show that this cannot happen. Since SD $f(x)$ is Baire 1 there are points of continuity of SD $f(x)$ in every interval. But at a point of continuity there must be an interval in which SD $f(x)$ assumes only the value α, β or γ; in such an interval f is linear with slope α, β or γ. Thus there is a maximal open set G so that in every component of G the function f is linear with slope α, β or γ.

Let P denote the complement of G. Exactly as before P can have no isolated points. If P is not empty then P is perfect and, yet again using the fact that SD $f(x)$ is Baire 1, there is a point of continuity of SD $f(x)$ relative to P. Thus there must be a nonempty portion $P \cap (a, b)$ so that SD $f(x)$ assumes just one of the three values α, β or γ for all $x \in P \cap (a, b)$.

Let us suppose the value assumed is α. Consider some interval $[c, d]$ contiguous to P in (a, b). In the interval $[c, d]$ the function f is linear with slope α, β or γ. But, exactly as argued in the proof of Theorem 7.43, it cannot have slope β. This means that in the entire interval (a, b) the symmetric derivative assumes only the two values α or γ. But by Theorem 7.43 itself no function can exist with just two values for its symmetric derivative in an interval. Thus this case cannot occur.

In the same way we may suppose that the value assumed is β and again obtain a contradiction.

Thus we arrive now at the case that SD $f(x)$ assumes just the value γ for all $x \in P \cap (a, b)$. We may suppose, without loss of generality that $\gamma > \frac{1}{2}(\alpha + \beta)$. Consider some interval $[c, d]$ contiguous to P in (a, b). In the interval $[c, d]$ the function f is linear with slope α, β or γ.

Since SD $f(c) = \gamma$ and $f'_+(c)$ is either α, β or γ it follows that $f'_-(c)$ exists too. But, since $f(x) - \alpha x$ and $\beta x - f(x)$ are nondecreasing, $\alpha \leq f'_-(c) \leq \beta$. This shows that

$$f'_+(c) = 2 \operatorname{SD} f(c) - f'_-(c) \geq 2\gamma - \beta > \alpha$$

and so f cannot have slope α in $[c, d]$. Thus in this case in every interval contiguous to P in (a, b) the function f is linear with slope β or γ. This means that in the entire interval (a, b) the symmetric derivative assumes only the two values β or γ. Again by Theorem 7.43 no function can exist with just two values for its symmetric derivative in an interval. Thus this case cannot occur.

As we have eliminated all possible cases we see that, as before, P must be empty so that f can only be linear; this contradicts the fact that its symmetric derivative assumes three distinct values.

7.4 Approximate Symmetric Derivative

7.4.1 Baire Class of Approximate Symmetric Derivative

We have now seen a complete picture of the measurability properties of symmetric derivatives and extreme symmetric derivates. We turn now to a similar problem for the approximate symmetric derivative. As for the ordinary symmetric derivative (Theorem 7.18) the Baire classification of approximate symmetric derivative was carried out by Larson [173]. The proof we present here follows closely the argument of Theorem 7.18 and just takes advantage of the measurability of f to replace the covering argument in the earlier proof.

Theorem 7.45 (Larson) *If f is measurable and everywhere approximately symmetrically differentiable (allowing infinite values) then its approximate symmetric derivative, ASD $f(x)$, is in the first Baire class.*

Proof. The proof we give follows that for Theorem 7.18 with some changes, and so once again is an application of the general criterion in Theorem A.17 for Baire 1 functions. If $g(x) = \text{ASD}\, f(x)$ is not in the first Baire class then there is a perfect set P and real numbers $\alpha < \beta < \delta < \gamma$ such that, if we define the sets

$$A = \{x \in P : g(x) \leq \alpha\}$$
$$B = \{x \in P : g(x) > \beta\}$$
$$C = \{x \in P : g(x) < \gamma\}$$
$$D = \{x \in P : g(x) \geq \delta\},$$

then both A and D are dense in P.

We use these sets to obtain a contradiction. Evidently $P = B \cup C$ so that, as P is perfect, one at least of these two sets is second category in P. Let us suppose that it is B which has this property; identical arguments would handle the other situation in any case. For any $0 < \kappa < 1$ and for each point $x \in B$ there is a $\delta(x) > 0$ so that

$$|\{t : f(x+t) - f(x-t) > 2\beta t\} \cap (0, h)| > \kappa h \qquad (7.28)$$

if $0 < h < \delta(x)$. We need a number κ close enough to 1 so that the sets in the intersection (7.29) below have a point in common; $\kappa = .99$ is more than enough. Then, applying the covering Lemma 3.8, there is a partition $\{B_n\}$ of B so that (7.28) holds if $0 < h < 1/n$ and $x \in B_n$. By our category assumptions on B there must now exist an interval (c, d) and a set B_m so that B_m is dense in $P \cap (c, d)$; we may assume that $d - c < 1/m$.

By our assumption the set A is also dense here thus we may select a point $x \in A \cap P \cap (c, d)$ which we may assume is not isolated in P on one side at least, say the right. Because $x \in A$ there is an h_0 so that $(x - h_0, x + h_0) \subset (c, d)$ and so that the set

$$Q = \left\{t : f(x+t) - f(x-t) < \frac{1}{2}(\alpha + \beta)2t\right\} \cap (0, h_0)$$

has $|Q| > \kappa h_0$. In the interval $(x - h_0, x + h_0)$ choose a compact subset P_0 so that

$$|(x - h_0, x + h_0) \setminus P_0| < (1 - \kappa)2h_0.$$

and so that f is continuous on P_0; this just uses Lusin's Theorem which is allowed since f is measurable. Now we may choose a number $0 < \eta < h_0/2$ so that whenever $y, z \in P_0$ with $|y - z| < \eta$ then $|f(y) - f(z)| < (\beta - \alpha)h_0/2$.

In order to obtain a contradiction we choose a point $x_1 \in B_m \cap (x, x + \eta)$ and a number h in the intersection

$$Q \cap (P_0 - x) \cap (x - P_0) \cap (P_0 - x_1) \cap (x_1 - P_0) \cap (h_0/2, h_0). \qquad (7.29)$$

Each of these sets is measurable and has large relative measure in this interval and so such a number may be chosen. Now each of the points $x + h$, $x - h$, $x_1 + h$ and $x_1 - h$ are in P_0 and distances between the corresponding pairs of points is smaller than η so that we must have

$$|f(x + h) - f(x_1 + h)| < (\beta - \alpha)h_0/2$$

and

$$|f(x - h) - f(x_1 - h)| < (\beta - \alpha)h_0/2.$$

This gives

$$\begin{aligned} f(x + h) - f(x - h) &> f(x_1 + h) - f(x_1 - h) - 2(\beta - \alpha)h_0/2 \\ &> 2\beta h - (\beta - \alpha)h = (\alpha + \beta)h. \end{aligned}$$

But this contradicts the fact that $h \in Q$ and thus completes the proof.

7.4.2 Baire Class of Approximate Symmetric Derivates

In this section we prove a theorem of Larson [171] showing that the extreme approximate symmetric derivates of a measurable a.e. finite function are at least in the third Baire class. It seems not to be known if this can be reduced to the second Baire class; it is known that a nonmeasurable function can have nonmeasurable extreme approximate symmetric derivates. We show later on that an exact approximate symmetric derivative of a measurable or a nonmeasurable function must be measurable.

The earliest result classifying the properties of approximate symmetric derivates appears to be that of Kundu [164] who showed that for continuous functions the extreme approximate symmetric derivates are measurable. Pu, Pu and Teng [237] also study the Baire class of symmetric and approximate symmetric derivates but for measurable, symmetric functions; since these latter are themselves in the first Baire class this places a strong restriction on the study. Their main result is that the upper and lower symmetric derivates of any measurable, symmetric real-valued function are of Baire class 2 and that its approximate symmetric derivates are of Baire class 4. Larson's theorem takes these results much further.

Theorem 7.46 (Larson) *Let f be a real-valued measurable function defined almost everywhere and let $-\infty \le \beta \le +\infty$. Then the sets*

$$\left\{x \ : \ \overline{ASD}\, f(x) > \beta\right\} \text{ and } \{x \ : \ \underline{ASD}\, f(x) < \beta\}$$

are of type $G_{\delta\sigma}$ and the sets

$$\left\{x \ : \ \overline{ASD}\, f(x) < \beta\right\} \text{ and } \{x \ : \ \underline{ASD}\, f(x) > \beta\}$$

are of type $F_{\sigma\delta\sigma}$.

The proof is given in a series of lemmas. We write

$$Q(x,t) = \frac{f(x+t) - f(x-t)}{2t} \tag{7.30}$$

and note that at every point x the approximate upper and lower limits of $Q(x,t)$ as $t \to 0$ are these derivates. The first lemma is a parallel for Lemma 4.43 after which the proof is modeled; see also [171, pp. 127–130].

Lemma 7.47 *Let M be a bounded measurable set and let $\epsilon > 0$ and $\delta > 0$. Then there is a number $r > 0$ so that*

$$|\{t \in M \ : \ |Q(x,t) - Q(y,t)| < \delta\}| > |M| - \epsilon$$

if $|x - y| < r$.

Proof. Since f is measurable and M is bounded and measurable there is a compact set $P \subset M$ so that $0 \notin P$, so that f is continuous relative to P and $|P| > |M| - \epsilon/2$. Let $m = \inf\{|p| \ : \ p \in P\}$. Choose $r_1 > 0$ so that $|f(x') - f(x'')| < m\delta$ if $x', x'' \in P$ and $|x' - x''| < r_1$.

Choose $r_2 > 0$ so that

$$|(P - x) \cap (P - y) \cap P| > |M| - \epsilon/2 \tag{7.31}$$

if $|x - y| < r_2$. This can be justified much like we justified the similar statement (4.39) in the proof of Lemma 4.43.

Now with $r = \min\{r_1, r_2\}$ we can verify that

$$\left|\left\{t \in M \ : \ \left|\frac{f(x+t) - f(y+t)}{2t}\right| < \delta/2\right\}\right| > |M| - \epsilon/2.$$

Indeed this is clear from the fact that this set includes the set $(P - x) \cap (P - y) \cap P$: if t is a member of this latter set then $x + t$, $y + t$ are in P so that $|f(x+t) - f(y+t)| < m\delta$. But t is also in P so that $|t| \ge m$ and hence

$$\left|\frac{f(x+t) - f(y+t)}{2t}\right| < \delta/2$$

as claimed.

A similar argument provides

$$\left| \left\{ t \in M : \left| \frac{f(x-t) - f(y-t)}{2t} \right| < \delta/2 \right\} \right| > |M| - \epsilon/2$$

and together these assertions prove the lemma.

The next lemma parallels Lemma 4.44; indeed the same proof works here and so may be omitted.

Lemma 7.48 *Let $t > 0$, $0 < \rho < 1$ and c a real number. Then the set*

$$G(t, \rho, c) = \{x : |\{\tau \in (0, t) : Q(x, \tau) > c\}| > \rho t\}$$

is open.

Lemma 7.49 *Let $g(x)$ denote the upper approximate symmetric derivate of f at x. Then*

$$\{x : g(x) > c\} = \bigcup_{n=2}^{\infty} \bigcap_{k=1}^{\infty} \bigcup_{0 < a < 1/k} G\left(a, \frac{1}{n}, c\right). \tag{7.32}$$

Proof. Let us first show that any point x that belongs to the set on the right of the identity (7.32) must have $g(x) > c$. Any such point belongs to

$$\bigcap_{k=1}^{\infty} \bigcup_{0 < a < 1/k} G\left(a, \frac{1}{n}, c\right).$$

for some n. Then for each k there is a number $0 < a_k < 1/k$ so that the set

$$\{\tau \in (0, a_k) : Q(x, \tau) > c\}$$

has measure exceeding a_k/n. This requires $g(x) > c$ as we wished to prove.

In the other direction suppose $g(x) > c$; then it is possible to choose an integer n_0 large enough so that

$$\limsup_{a \to 0} \frac{|\{t \in (0, a) : Q(x, t) > c\}|}{a} > 1/n_0.$$

Thus for every k there is a point $a_k \in (0, 1/k)$ such that $x \in G\left(a_k, \frac{1}{n_0}, c\right)$. Again this means that

$$x \in \bigcap_{k=1}^{\infty} \bigcup_{0 < a < 1/k} G\left(a, \frac{1}{n_0}, c\right)$$

for some n_0 depending on x. The point x now must belong to the set on the right of identity (7.32) and so the proof is complete.

Now the proof of the theorem is clear. From Lemmas 7.48 and 7.49 we see that for every real c the set $\{x : g(x) > c\}$ is a Borel set of type $G_{\delta\sigma}$. It is also clear that this means too that the $\{x : g(x) > -\infty\}$ is a $G_{\delta\sigma}$. The same arguments (obtained by replacing f by $-f$) would handle the lower approximate symmetric derivative g_1 showing that $\{x : g_1(x) < c\}$ is also of type $G_{\delta\sigma}$. Finally to see that $\{x : g(x) < c\}$ is a Borel set of type $F_{\sigma\delta\sigma}$ we need only see that it is the complement of the set $\{x : g(x) \geq c\}$ which in turn is the union of the sets $\{x : g(x) > c - 1/n\}$ for $n = 1, 2, 3 \ldots$ and so is itself a $G_{\delta\sigma\delta}$.

7.4.3 Measurability of the Approximate Symmetric Derivative

It is possible for a nonmeasurable function to have everywhere an approximate symmetric derivative. We have seen that the approximate symmetric derivative of a measurable function is Baire 1. Here we show that this derivative is measurable even if the primitive function is not. The theorem and proof are reproduced from Preiss and Thomson [232].

Theorem 7.50 *If an almost everywhere defined function is almost everywhere approximately symmetrically differentiable, then its approximate symmetric derivative is a measurable function.*

The statement follows immediately from the following assertion if we use the characterization of measurability described in Section A.11 of the Appendix.

Lemma 7.51 *Let f be an arbitrary function defined on a measurable set S and let $a < b$ be real numbers. Then the sets $\{x \in S; \overline{ASD} f(x) < a\}$ and $\{x \in S; \underline{ASD} f(x) > b\}$ can be separated by a measurable set.*

Proof. Let us assume, to the contrary, that the sets

$$U = \{x \in S; \overline{ASD} f(x) < a\}$$

and

$$V = \{x \in S; \underline{ASD} f(x) > b\}$$

cannot be separated by a measurable set. For each $x \in U$ we choose $h(x) > 0$ such that for every $0 < h < h(x)$ the set

$$\left\{t \in (0, h); \frac{f(x+t) - f(x-t)}{2t} \geq a\right\}$$

has outer measure at most $h/72$. Similarly, for each $x \in V$ we choose $h(x) > 0$ such that for every $0 < h < h(x)$ the set

$$\left\{t \in (0, h); \frac{f(x+t) - f(x-t)}{2t} \leq b\right\}$$

has outer measure at most $h/72$.

Since U and V cannot be separated by a measurable set, we may use A.20 from the Appendix to find a measurable set P of positive measure such that the sets

$$U_0 = \{x \in U \cap P; \, h(x) > \epsilon\}$$

and

$$V_0 = \{x \in V \cap P; \, h(x) > \epsilon\}$$

are both of full outer measure in P.

Let $x \in U_0$ be a density point of P and let $y \in U_0 \cap (x, x + \epsilon/2)$ be such that the set $(x, y) \setminus P$ has measure at most $(y - x)/5$. Then we can find a point z belonging to the set

$$(3x/4 + y/4, (x + y)/2) \cap V_0.$$

From the definition of the function $h(x)$ we infer that there is a measurable subset T of the interval $(0, 2(y - x))$ with measure at least

$$2(y - x) - 3[2(y - x)]/72 = 2(y - x) - (y - x)/12$$

such that for each $t \in T$ the inequalities $f(x + t) - f(x - t) < 2at$, $f(y + t) - f(y - t) < 2at$, and $f(z + t) - f(z - t) > 2bt$ hold. Let N be the set of all points $u \in (y, 2y - x)$ such that $u - y \in T$, $2y - u - x \in T$, $z - 2x + 2y - u \in T$, $u - z \in T$, $x - 2z + u \in T$, and $y - 2x + 2z - u \in T$. Observing that for every $u \in (y, 2y - x)$ all the points $u - y$, $2y - u - x$, $z - 2x + 2y + u$, $u - z$, $x - 2z + u$, and $y - 2x + 2y - u$ belong to $(0, 2(y - x))$, we easily infer that N has measure at least $(y - x) - 6[(y - x)/12] > 0$. We conclude that $N \neq \emptyset$. On the other hand, using the above inequalities, we easily see that every $u \in N$ fulfills

$$\begin{aligned} f(u) \; &< \; f(2y - u) + 2a(u - y) < f(2x - 2y + u) + 2a(y - x) \\ &< \; f(2z - 2x + 2y - u) + 2a(y - x) - 2b(z - 2x + 2y - u) \end{aligned}$$

as well as

$$\begin{aligned} f(u) \; &> \; f(2z - u) + 2b(u - z) \\ &> \; f(2x - 2z + u) + 2b(z - u) - 2a(x - 2z + u) \\ &> \; f(2y - 2x + 2z - u) + 2b(u - z) - 2a(y - x). \end{aligned}$$

Thus

$$2b(u - z) - 2a(y - x) < 2a(y - x) - 2b(z - 2x + 2y - u),$$

which is $4b(y - x) < 4a(y - x)$. This finishes the proof, since we know that $a < b$ and $y - x > 0$. (These computations are closely related to those in [172, p. 590]).

This theorem should be balanced by the following assertion from Preiss and Thomson [232]. While Theorem 7.50 asserts that an exact approximate symmetric derivative must be measurable even if the primitive function itself need not be, the extreme derivates of a nonmeasurable function do not have to be measurable.

Theorem 7.52 *(Under CH.) Let E be an arbitrary linear subspace of the real numbers over the field of rational numbers. Then there is a nonmeasurable function f such that for every $x \in E$ there is a countable set S_x such that $f(x + h) - f(x - h) = 0$ for every $h \in \mathbb{R} \setminus S_x$, and such that $\overline{ASD}\, f(x) = +\infty$ and $\underline{ASD}\, f(x) = -\infty$ for every $x \in \mathbb{R} \setminus E$.*

Proof. The proof is a continuation of the proof of Theorem 6.22. Using the notation there we let f be the characteristic function of the set Y from the proof of Theorem 6.22. For each $x \in E$ we put

$$S_x = [(Y \setminus (2x - Y)) - x] \cup [x - (Y \setminus (2x - Y))].$$

Then clearly each of the sets S_x is countable and the first statement follows from the observation that $f(x + h) - f(x - h) \neq 0$ implies $h \in S_x$. If $x \notin E$, we observe that the sets

$$U = \{t \in (0, \infty); \; x + t \in Y, \; x - t \notin Y\}$$

and

$$V = \{t \in (0, \infty); \; x + t \notin Y, \; x - t \in Y\}$$

have full outer measure in $(0, \infty)$, and that

$$\lim_{t \searrow 0, \, t \in U} (f(x + t) - f(x - t))/2t = +\infty$$

and

$$\lim_{t \searrow 0, \, t \in V} (f(x + t) - f(x - t))/2t = -\infty$$

which completes the proof.

Corollary 7.53 *(Under CH.) There is a nonmeasurable function f whose extreme approximate symmetric derivates $\overline{ASD}\, f(x)$ and $\underline{ASD}\, f(x)$ are not measurable.*

7.4.4 Approximate Version of Khintchine Theorem

We present in this section a theorem of Freiling and Rinne [115] which can be considered as the approximate version of the classical Khintchine Theorem. This theorem was stated first in Russo and Valenti [248] but there is an oversight in the proof. Russo [247] also studies the same problem.

The assumption in the theorem that f is measurable is essential. Recall the Sierpiński example (Theorem 6.19) of a nonmeasurable function that is everywhere nearly symmetric. This shows that the measurability assumption may not be dropped; such a function would have a zero approximate symmetric derivative everywhere and yet have a finite approximate derivative nowhere.

Theorem 7.54 (Freiling–Rinne) *Let f be measurable. Then f has a finite approximate derivative at almost every point x where $\overline{ASD} f(x) < +\infty$.*

As a consequence the Denjoy relations for the approximate symmetric derivates may be stated as follows: for any measurable function f it is the case that at almost every point x either there is a finite approximate symmetric derivative or else both

$$\overline{ASD} f(x) = +\infty \quad \text{and} \quad \underline{ASD} f(x) = -\infty.$$

Theorem 7.54 can be used in conjunction with Theorem 1.19 to obtain a theorem of Lusin in a transparent way.

Theorem 7.55 (Lusin) *Suppose that the series*

$$a_0/2 + \sum_{n=1}^{\infty} (a_n \cos nx + b_n \sin nx) \tag{7.33}$$

converges to a finite sum $s(x)$ at every point of a set E with $|E| > 0$. Let L denote the Lebesgue function for the trigonometric series (7.33) Then $AD\,L(x) = s(x)$ for almost every $x \in E$.

Proof. Since the series (7.33) converges on a set of positive measure it follows from the Cantor-Lebesgue Lemma that a_n, $b_n \to 0$. Thus the Lebesgue function L is defined almost everywhere and is measurable. By Theorem 1.19 $ASD\,L(x) = s(x)$ for every $x \in E$. By Theorem 7.54 $AD\,L(x) = ASD\,L(x)$ almost everywhere in E.

7.4.5 Relations for Monotone Functions

If a function f is monotone then there is no distinction between approximate differentiability and ordinary differentiability. In fact the ordinary and approximate Dini derivatives agree at every point. This was first observed by Khintchine and extended by Mišik. The key is really that sets that have density 1 at a point are nonporous there. Then a geometrical lemma for monotone functions due to

Evans and Humke allows these conclusions to be drawn. For a full account see [282, pp. 155–157, p. 191].

Kostyrko [156] showed that, by very much the same methods, one obtains the same relations between the ordinary and approximate versions of the symmetric derivative of monotone functions.

Theorem 7.56 (Kostyrko) *Let f be monotone. Then, at every point x,*

$$\overline{ASD}\,f(x) = \overline{SD}\,f(x) \quad and \quad \underline{ASD}\,f(x) = \underline{SD}\,f(x).$$

A· corollary follows quite easily.

Corollary 7.57 (Kostyrko) *Let f be measurable and approximately symmetrically differentiable on an interval (a,b). Suppose that g is symmetrically differentiable on (a,b) and that*

$$ASD\,f(x) \le SD\,g(x) \quad (a < x < b).$$

Then there exists a function h that is symmetrically differentiable on (a,b) and $ASD\,f(x) = SD\,h(x)$ everywhere.

These statements can be exploited much like the original Khintchine–Mišik version to obtain properties of approximate symmetric derivatives from properties of symmetric derivatives. For example, from this result Kostyrko is able readily to deduce that the class of approximate symmetric derivatives is, like the class of symmetric derivatives, closed under uniform limits.

Another consequence of this property of monotone functions is pointed out in Evans and Humke [93, p. 611]. Their theorem on the equality of the right and left Dini derivatives of monotone functions extends then immediately to symmetric and approximate symmetric derivates.

Theorem 7.58 (Evans–Humke) *Let f be monotone. Then at every point x, excepting only a σ-porous set,*

$$\overline{D}^-\,f(x) = \overline{D}^+\,f(x) = \overline{D}\,f(x) = \overline{SD}\,f(x) = \overline{ASD}\,f(x).$$

7.4.6 Typical Continuous Functions

We have already noted in Section 7.3.5 that the typical continuous function is nowhere symmetrically differentiable. To this we can add the statement of Evans [86] regarding the existence of an approximate symmetric derivative. This is the approximate extension of Theorem 7.26. (The even analogue has been given as Theorem 4.88 in Chapter 4.)

Theorem 7.59 (Evans) *The typical continuous function f in $C[0,1]$ has the property that*

$$\text{ap} \limsup_{h \to 0} \frac{f(x+h) - f(x-h)}{2h} = +\infty$$

and

$$\text{ap} \liminf_{h \to 0} \frac{f(x+h) - f(x-h)}{2h} = -\infty$$

for all $0 < x < 1$.

We mention as well a related result of Matousek [194] extending these ideas. He shows that the typical continuous function in $C[0,1]$ has the property

$$\limsup_{h \to 0} |S_f(x, K) \cap (-h, h)| / (2h)$$
$$= \limsup_{h \to 0} |[\mathbb{R} \setminus S_f(x, -K)] \cap (-h, h)| / (2h) = 1$$

for all $x \in (0, 1)$ and for arbitrarily large K. Here,

$$S_f(x, K) = \{h : [f(x+h) - f(x-h)]/(2h) > K\}.$$

7.5 Borel Symmetric Derivative

The symmetric–Borel derivates of an integrable f at a point x are defined as

$$\overline{\text{SBD}} f(x) = \limsup_{h \to 0} \frac{1}{h} \left\{ \int_0^h \frac{f(x+t) - f(x-t)}{2t} \, dt \right\}$$

and

$$\underline{\text{SBD}} f(x) = \liminf_{h \to 0} \frac{1}{h} \left\{ \int_0^h \frac{f(x+t) - f(x-t)}{2t} \, dt \right\}.$$

If the two extreme derivates are equal and finite then f is said to have a symmetric Borel derivative at the point x and the common value is denoted as SBD $f(x)$.

Some comments are in order on the sense in which the integration is to be interpreted; in this we are following [186, pp. 35–37] and [270]. To keep things sufficiently general we assume that f is Denjoy-Perron integrable in a neighbourhood of the point x and then $\int_0^h$ is taken to mean $\lim_{\epsilon \to 0+} \int_\epsilon^h$. Thus if we write

$$\phi(f, x, h) = \lim_{\epsilon \to 0+} \int_\epsilon^h \frac{f(x+t) - f(x-t)}{2t} \, dt \qquad (7.34)$$

then, provided that this exists as a finite or infinite limit, $\overline{\text{SBD}} f(x)$ and $\underline{\text{SBD}} f(x)$ are defined as the upper and lower limits of $\phi(f, x, h)/h$. If the limit in (7.34) does not exist finitely or infinitely then we shall write $\overline{\text{SBD}} f(x) = +\infty$ and $\underline{\text{SBD}} f(x) = -\infty$.

There is a close connection between the symmetric Borel derivates of a function f and second symmetric derivates of an indefinite integral of f. We have already seen the following theorem of Khintchine in Section 1.7 of Chapter 1.

Theorem 7.60 (Khintchine) *Let f be integrable in a neighbourhood of a point x and let F be an indefinite integral of f. If the second symmetric derivative of F exists at a point x then the symmetric Borel derivative of f exists there and*

$$SD_2 F(x) = SBD f(x).$$

We wish now to elaborate on this theorem by showing the full extent of these relations. The investigation was carried out by Marcinkiewicz and Zygmund [186] with a refinement given by Skvorcov [270]. (A related set of inequalities obtained by the same methods can be found in Bullen and Mukhopadhyay [29, p. 171].) The theorem we now prove includes the preceding theorem as a special case and illustrates the real character of the relationship.

Theorem 7.61 (Marcinkiewicz–Zygmund) *Let f be integrable in a neighbourhood of a point x and let F be an indefinite integral of f. We suppose either that the limit in (7.34) exists or else that F is smooth at the point x.*
Then

(*i*) $\underline{SD_2}\, F(x) \leq \underline{SBD}\, f(x) \leq \overline{SBD}\, f(x) \leq \overline{SD_2}\, F(x).$

(*ii*) *The second symmetric derivative of F exists at a point x if and only if the symmetric Borel derivative of f exists there and then*

$$SD_2 F(x) = SBD f(x).$$

(*iii*) *A necessary and sufficient condition for*

$$\limsup_{h \to 0} \left| \frac{1}{h} \int_0^h \frac{f(x+t) - f(x-t)}{2t}\, dt \right| < +\infty$$

is that $|\underline{SD_2}\, F(x)| + |\overline{SD_2}\, F(x)| < +\infty.$

Proof. We begin the proof by showing that information about the Borel symmetric derivates of f leads to the stated relations for the second symmetric derivates of F. To simplify the notation let us introduce the two functions ϕ and ψ by writing

$$\phi(h) = \int_0^h \frac{f(x+t) - f(x-t)}{2t}\, dt$$

and

$$\psi(h) = \int_0^h \left(f(x+t) - f(x-t) \right) dt$$

so that $\psi(h) = F(x+h) + F(x-h) - 2F(x)$. The proof is obtained by relating these two expressions.

If the limit in (7.34) exists as a finite number then the integral defining ϕ exists and

$$\psi(h) = 2 \left[\phi(h)h - \int_0^h \phi(t)\, dt \right]. \qquad (7.35)$$

We obtain this from the identity

$$\psi(h) - \psi(k) = \int_k^h \left(f(x+t) - f(x-t) \right) dt = 2 \int_k^h \phi'(t) t\, dt \qquad (7.36)$$

by integrating by parts and letting $k \to 0$.

From (7.35) part of the theorem is now evident. If $\mathrm{SBD}\, f(x) = c$ exists then, by definition, $\phi(h)/h \to c$. Because of (7.35) this means that $\psi(h)/h^2 \to c$ and this means that the second symmetric derivative of F exists and has the value c. This proves half of the assertion (ii) of the theorem.

For the same reasons if $\phi(h) = O(h)$ then by (7.35) this means that $\psi(h) = O(h^2)$ This proves half of the assertion (iii) of the theorem.

In this direction there remains only assertion (i). Suppose that $\overline{\mathrm{SBD}}\, f(x) > \alpha$. We show that this requires $\overline{\mathrm{SD}}_2\, F(x) \geq \alpha$. Without loss of generality we take $\alpha = 0$. Thus $\overline{\mathrm{SBD}}\, f(x)$ is positive and so ϕ must assume some positive values in every neighbourhood of 0. Since $\phi(h') \to 0$ as $h' \to 0$ it is possible to select a sequence of points $h_k \searrow 0$ so that $\phi(h_k) \geq \phi(h)$ for all $0 < h' \leq h_k$; in particular

$$h_k \phi(h_k) - \int_0^{h_k} \phi(t)\, dt \geq 0.$$

From (7.35) then it follows that $\psi(h_k) \geq 0$ and consequently that $\overline{\mathrm{SD}}_2\, F(x) \geq 0$ as required. In general then $\overline{\mathrm{SBD}}\, f(x) \leq \overline{\mathrm{SD}}_2\, F(x)$ at every point; in a similar way it may be proved that $\underline{\mathrm{SD}}_2\, F(x) \leq \underline{\mathrm{SBD}}\, f(x)$ and so assertion (i) is established in this particular case.

If the limit in (7.34) exists as $+\infty$ or $-\infty$ then the integral defining ϕ and the above relations cannot be used. Suppose that the limit in (7.34) is $+\infty$ so that $\overline{\mathrm{SBD}}\, f(x) = \underline{\mathrm{SBD}}\, f(x) = +\infty$. We wish to prove that $\overline{\mathrm{SD}}_2\, F(x) = +\infty$. Suppose not; then, by subtracting a linear function, we can suppose that $\overline{\mathrm{SD}}_2\, F(x) < 0$. Put

$$\phi^*(u) = \int_u^h \frac{f(x+t) - f(x-t)}{2t}\, dt.$$

With some modifications the argument that gave (7.35) now produces the identity

$$\psi(h) = -2 \left[\phi^*(h)h - \int_0^h \phi^*(t)\, dt \right]. \qquad (7.37)$$

This is less transparent though. We use the identity

$$\psi(h) - \psi(k) = \int_0^h \left(f(x+t) - f(x-t) \right) dt = -2 \int_k^h \phi^{*\prime}(t) t \, dt \qquad (7.38)$$

in place of (7.36). Once again we integrate by parts and let $k \to 0$. To make the argument work we have to use the fact that $u\phi^*(u) \to 0$ with u. To see this we use the fact that $\overline{SD}_2 F(x) < 0$ to find a $\delta > 0$ so that $\psi(t) \le -\delta t^2$ for small t. We then compute

$$\begin{aligned}
\phi^*(u) &= \int_u^h \frac{f(x+t) - f(x-t)}{2t} \, dt \\
&= \int_u^h \frac{\psi'(t)}{2t} \, dt \\
&= \frac{1}{2} \left[\frac{\psi(h)}{h} - \frac{\psi(u)}{u} + \int_u^h \frac{\psi(t)}{t^2} \, dt \right] \\
&\le \frac{1}{2} \left[\frac{\psi(h)}{h} - \frac{\psi(u)}{u} - \delta(h-u) \right].
\end{aligned}$$

Now we see that $\limsup_{u \to 0+} u\phi^*(u) \le 0$; since $\lim_{u \to 0+} \phi^*(u) = +\infty$ the limit $\lim_{u \to 0+} u\phi^*(u) = 0$ must follow as we have stated. Thus we have established the identity (7.37).

Since $\phi^*(h') \to +\infty$ as $h' \to 0$ it is possible to select a sequence of points $h_k \searrow 0$ so that $\phi(h_k) \le \phi(h')$ for all $0 < h' \le h_k$; in particular

$$h_k \phi(h_k) - \int_0^{h_k} \phi(t) \, dt \le 0.$$

From (7.37) then it follows that $\psi(h_k) \ge 0$ and consequently that $\overline{SD}_2 F(x) \ge 0$. Since this contradicts our assumption the assertion (i) is proved in this case too.

Finally to prove assertion (i) in the only case that remains suppose that the limit in (7.34) does not exist either finitely or as $\pm\infty$. Then, by the assumptions in the statement of the theorem, F is smooth at x. By definition in this situation $\overline{SBD} f(x) = +\infty$ and $\underline{SBD} f(x) = -\infty$. We use the fact that

$$(F(x+t) + F(x-t) - 2F(x))' = f(x+t) - f(x-t)$$

for a.e. value of t in some neighbourhood of 0 to obtain for any $0 < \epsilon < h$ that

$$\begin{aligned}
\int_\epsilon^h \frac{f(x+t) - f(x-t)}{2t} \, dt &= \frac{F(x+h) + F(x-h) - 2F(x)}{2h} \\
&- \frac{F(x+\epsilon) + F(x-\epsilon) - 2F(x)}{2\epsilon} + \frac{1}{2} \int_\epsilon^h \frac{F(x+t) + F(x-t) - 2F(x)}{t^2} \, dt.
\end{aligned}$$

The argument is now clear. The first two expressions on the right hand side of the identity approach 0 with h because F is smooth at x. In order for the limit in (7.34) to fail to exist finitely or as $\pm\infty$ the expression

$$\frac{F(x+t) + F(x-t) - 2F(x))}{t^2}$$

must oscillate infinitely as $t \to 0$. Thus $\underline{SD}_2 F(x) = -\infty$ and $\overline{SD}_2 F(x) = +\infty$ and since $\overline{SBD} f(x) = +\infty$ and $\underline{SBD} f(x) = -\infty$ also this establishes the inequalities in (i) for this case.

We must now complete the proof by showing that the stated information in (ii) and (iii) about the Borel symmetric derivates of f can be derived from knowledge of the second symmetric derivates of F. We may assume that the latter are finite. We shall need the analogue of (7.35) which is

$$\phi(h) = \frac{1}{2}\left[\frac{\psi(h)}{h} + \int_0^h \frac{\psi(t)}{t^2}\, dt.\right] \tag{7.39}$$

As before we obtain this from the identity

$$\phi(h) - \phi(k) = \int_k^h \frac{f(x+t) - f(x-t)}{2t}\, dt = \int_k^h \frac{\psi'(t)}{2t}\, dt$$

by integrating by parts and letting $k \to 0$.

From (7.39) the remainder of the theorem is now evident. If $SD_2 F(x) = c$ exists then, by definition, $\psi(h)/h^2 \to c$. Because of (7.39) this means that $\phi(h)/h \to c$ and this means that the Borel symmetric derivative of f exists and has the value c. This complete the proof of the assertion (ii) of the theorem. For the same reasons if $\psi(h) = O(h^2)$ then by (7.39) this means that $\phi(h) = O(h)$ and this proves the remainder of the assertion (iii) of the theorem.

We complete this section by applying the theorem just proved to a monotonicity result for the Borel symmetric derivative.

Theorem 7.62 *Let f be integrable in an interval $[a, b]$ and let N be a denumerable subset of $[a, b]$ containing only Lebesgue points for the function f. Suppose that $\overline{SBD} f(x) \geq 0$ a.e. in $[a, b]$ and that $\overline{SBD} f(x) > -\infty$ everywhere in $[a, b] \setminus N$. Then f is equivalent in that interval to a nondecreasing function.*

Proof. The proof is obtained from properties of an indefinite integral F of f. The conditions stated in the hypotheses can be translated into assertions about F by using Theorem 7.61. We must have that $\overline{SD}_2 F(x) \geq 0$ a.e. in $[a, b]$ and $\overline{SD}_2 F(x) > -\infty$ everywhere in $[a, b] \setminus N$. At each point of the countable set N the function F is smooth since each such point is a Lebesgue point for f. Consequently, by Theorem 5.31, F must be convex on this interval. Its derivative F' then exists nearly everywhere and is nondecreasing. Since $F' = f$ a.e. it is easy now to see that f is equivalent to some nondecreasing function.

8

Symmetric Variation

8.1 Introduction

In the study of real functions, particularly with attention to differentiation properties, the notion of the variation of a function plays a key role. In elementary courses we simply define the concept of bounded variation and obtain differentiation results from that. Later we see how to generate a measure from a function of bounded variation, a measure which carries the variational information about f.

More directly and more generally it is possible to define a measure μ_f for an arbitrary function f that immediately carries this information. Let $E \subset \mathbb{R}$ and let δ be a positive function on E. Then we write

$$V_\delta(f, E) = \sup \sum_{i=1}^{n} |f(b_i) - f(a_i)|$$

where the supremum is with regard to all sequences $\{[a_i, b_i]\}$ of nonoverlapping intervals containing some point $x_i \in E$ and with $b_i - a_i < \delta(x_i)$. We write then

$$\mu_f(E) = \inf V_\delta(f, E)$$

where the infimum is taken over all positive functions δ on E.

For an arbitrary function f the set function μ_f is an outer measure for which all Borel sets are measurable. If f is continuous and has locally bounded variation then μ_f is the usual measure associated with f as its "total variation measure". There is a close connection between the differentiation properties of f and the properties of μ_f. See [283], [282] and [286] for a full account of this theory.

These same variational ideas may be extended to various generalized derivatives. In this chapter we wish to apply these notions to a study of the first and second order symmetric derivatives. The material here for the first order case is mostly reproduced from Thomson [287]. The second order variation has not appeared elsewhere; it is based on ideas developed in Freiling, Rinne and Thomson [116].

One could develop as well a variational theory associated with the approximate symmetric derivative by modeling the theory after that presented here. We shall not go into this beyond the brief sketch in Section 9.3.5 in Chapter 9 where the ideas are introduced to provide a characterization of an approximate symmetric integral.

8.2 First Order Symmetric Variation

For an arbitrary function f the symmetric variation VS_f is defined much as the measure μ_f was defined in the introduction to this chapter. The notion was probably first suggested in [283]. E. Freund [118] investigated the structure of functions having zero symmetric variation on an interval and [287] extended his study to the finite and σ–finite case.

Definition 8.1 Let $E \subset \mathbb{R}$ and let δ be a positive function on E. Then we write

$$S_\delta(f, E) = \sup \sum_{i=1}^{n} |f(x_i + h_i) - f(x_i - h_i)|$$

where the supremum is with regard to all sequences $\{[x_i - h_i, x_i + h_i]\}$ of nonoverlapping intervals with centers $x_i \in E$ and with $h_i < \delta(x_i)$. We write then

$$VS_f(E) = \inf S_\delta(f, E)$$

where the infimum is taken over all positive functions δ on E.

It is common to call a positive function δ on a set E a *gauge*. We can say that an interval $[x - h, x + h]$ is δ–fine at its center if $0 < h < \delta(x)$. We recall that some of our partitioning theorems of Chapter 3 provide partitions that are δ–fine at the centers.

This expression is called the *symmetric variation* of f on E and the set function VS_f is called the *symmetric variational measure* associated with f. It is not difficult to see that VS_f is an outer measure.

Lemma 8.2 *The set function VS_f is an outer measure on the real line for which all Borel sets are measurable.*

Proof. Clearly VS_f is monotone and $\mathrm{VS}_f(\emptyset) = 0$. Suppose that $E \subset \bigcup_{n=1}^{\infty} E_n$ where the $\{E_n\}$ are disjoint.

Let $\epsilon > 0$. For each interger n we may choose a gauge δ_n on E_n so that

$$S_{\delta_n}(f, E_n) \le \mathrm{VS}_f(E_n) + \epsilon/2^n.$$

Take any gauge δ with $\delta(x) < \delta_n(x)$ for $x \in E_n$. Then an easy argument shows that

$$\mathrm{VS}_f(E) \ \le \ S_\delta(f, E) \le \sum_{n=1}^{\infty} S_{\delta_n}(f, E_n)$$

$$\le \ \sum_{i=1}^{\infty} \left\{ \mathrm{VS}_f(E_i) + \epsilon/2^i \right\}.$$

Since ϵ is arbitrary we can conclude that $\mathrm{VS}_f(E) \le \sum_{i=1}^{\infty} \mathrm{VS}_f(E_i)$. Thus we have established that VS_f is an outer measure.

To see that all Borel sets are VS_f–measurable it suffices, by the standard theory (see, for example [246, pp. 30–33]), to show that

$$\mathrm{VS}_f(E_1 \cup E_2) = \mathrm{VS}_f(E_1) + \mathrm{VS}_f(E_2)$$

if $\mathrm{dist}(E_1, E_2) > 0$. Let δ be any gauge with $\delta(x) < \frac{1}{2}\mathrm{dist}(E_1, E_2)$. Then

$$S_\delta(f, E_1 \cup E_2) = S_\delta(f, E_1) + S_\delta(f, E_2)$$

since any intervals used in computing the two expressions on the right cannot overlap. From this we can argue that $\mathrm{VS}_f(E_1 \cup E_2) = \mathrm{VS}_f(E_1) + \mathrm{VS}_f(E_2)$ and the lemma is proved.

8.2.1 Zero Variation

A function with zero variation should be constant. It is easy to see that a function with zero symmetric variation need not be precisely constant. In this section we characterize such functions. Our result is from Freund [118, Theorem 2.3]; note that it is almost an immediate consequence of the covering theorems of Section 3.3.3.

Theorem 8.3 (Freund) *If $\mathrm{VS}_f((a, b)) = 0$ then there is a constant function g so that*

$$\{x \in (a, b) : \ f(x) \ne g(x)\} \tag{8.1}$$

is countable and so that each set

$$\{x \in (a, b) : \ |f(x) - g(x)| > \epsilon\} \qquad (\epsilon > 0)$$

has countable closure.

Proof. For each $n = 1, 2, 3 \ldots$ choose a gauge δ_n so that $S_{\delta_n}(f, (a, b)) < 1/n$. By Theorem 3.13, there is a countable set C_n so that there is a partition of any subinterval of (a, b) with endpoints not in C_n, a partition that is δ_n–fine at its centers. It follows that, if x, $y \in (a, b) \setminus C_n$, then

$$|f(y) - f(x)| \leq S_{\delta_n}(f, (a, b)) < 1/n.$$

Set $C_0 = \bigcup_{n=1}^{\infty} C_n$ and we have $f(x) = f(y)$ for all x, $y \in (a, b) \setminus C_0$.

To simplify the notation we may suppose that f vanishes outside of the countable set C_0 so that g may be taken as identically zero. Consider the set of points x where $|f(x)| > \epsilon$. Choose a gauge δ so that $S_\delta(f, (a, b)) < \epsilon$. Take any $x_0 \in (a, b)$ and choose a point c with $a < 2c - x_0 < c < x_0$ so that c does not belong to the countable set

$$S = \{(x + y)/2 : f(x) \neq 0, f(y) \neq 0\}.$$

We show that the set

$$\{x : |f(x)| > \epsilon\} \cap (c, x_0) \tag{8.2}$$

has countable closure.

By Lemma 3.14 there is a set $D \subset (c, x_0)$ with countable closure so that there is a partition of any interval $[2c - x, x]$ with $x \in (c, x_0) \setminus D$, a partition that is δ–fine at its centers. Let $x \in (c, x_0)$ with $|f(x)| > \epsilon$. We claim that $x \in D$. If x is not in D then, since $f(x) \neq 0$ and $c \notin S$, it follows that $f(2c - x) = 0$; as x is not in D there is a δ–fine partition of $[2c - x, x]$ and so

$$|f(x)| = |f(2c - x) - f(x)| \leq S_\delta(f, (a, b)) < \epsilon$$

which is a contradiction.

Consequently the set (8.2) is contained in D and so has countable closure. Since this holds for every point x_0 it follows that the set $\{x : |f(x)| > \epsilon\}$ has countable closure. This is the final assertion of the theorem.

The set in (8.1) cannot be further described: any countable set may appear as such for some constant function g and some f with $VS_f((a, b)) = 0$. To see this let $C \subset (a, b)$ be countable and write its elements as $c_1, c_2, c_3, \ldots$. Define $f(c_i) = 2^{-i}$ and $f(x) = 0$ if $x \in (a, b) \setminus C$. Such a function is symmetrically continuous and so $VS_f(\{c\}) = 0$ for each $c \in C$. It follows then that $VS_f(C) = 0$. Choose a positive function δ so that if $x \in (a, b) \setminus C$ then $(x - \delta(x), x + \delta(x))$ contains no point $c_1, c_2, \ldots c_{N+1}$. Evidently then

$$VS_f((a, b) \setminus C) \leq S_\delta(f, (a, b) \setminus C) \leq 2^{-N}$$

and, since N is arbitrary, $VS_f((a, b) \setminus C) = 0$. It follows that $VS_f((a, b)) = 0$.

Freund [118, Theorem 2.4] points out that it follows from Theorem 8.3 that functions with zero symmetric variation are Baire class one. This extends an easier observation that locally symmetric functions (see Section 6.2.5) are Baire class one.

8.2.2 Finite Variation

A function with finite variation in the ordinary sense is said to have bounded variation. A function with σ–finite variation in the ordinary sense (i.e. with the measure μ_f, defined in the introduction, σ–finite) is usually said to be VBG$_*$ (generalized bounded variation in the restricted sense). These two conditions play a strong role in the study of differentiation and integration in the ordinary sense.

In this section we consider the analogous concepts for the symmetric variation. This material is from [287]. We find that the condition $\mathrm{VS}_f((a,b)) < +\infty$ can be handled by the same methods as in the preceding section and the assertion is similar.

Theorem 8.4 *Suppose that* $\mathrm{VS}_f((a,b)) < +\infty$. *Then there is a function g of bounded variation on (a,b) so that*

$$\{x \in (a,b) : f(x) \neq g(x)\}$$

is scattered.

Proof. Choose a gauge δ so that $S_\delta(f,(a,b)) < +\infty$. By Theorem 3.13, there is a splattered set C so that there is a δ–fine partition of any subinterval of (a,b) with endpoints not in C. It follows that, for any sequence of nonoverlapping intervals $\{[x_i, y_i]\}$ with $x_i, y_i \in (a,b) \setminus C$,

$$\sum |f(y_i) - f(x_i)| \leq S_\delta(f,(a,b)) < +\infty.$$

Thus f has bounded variation relative to the set $(a,b) \setminus C$ for a splattered set C. With a little more work (which might amuse the reader) one can push this to show that C can be taken as scattered. The function g thus is just f on $(a,b) \setminus C$ extended to all of (a,b).

This theorem has a parallel with Theorem 8.3. We should recall a similar but older result of Charzyński [42] already presented as Theorem 2.19:

Suppose that

$$\limsup_{h \to 0+} h^{-1}|f(x+h) - f(x-h)| \leq M$$

at every point $x \in (a,b)$. Then there is a Lipschitz function g with Lipschitz constant no more than M so that

$$\{x \in (a,b) : f(x) \neq g(x)\}$$

is scattered.

The analogy cannot be pushed too far. One might hope that the condition

$$\limsup_{h \to 0+} h^{-1}|f(x+h) - f(x-h)| < +\infty$$

at every point $x \in (a, b)$ would force f to agree with a reasonable function g except on a small set. The example $f(x) = \cos x^{-1}$ shows that f cannot agree even with a continuous function outside of anything fairly small.

Notice that for the function $f(x) = \cos x^{-1}$ the measure VS_f is not finite but it is σ-finite. One asks then what properties this will impose on f. An answer is available by applying some of the covering arguments of Section 3.3.2.

Theorem 8.5 *Suppose that VS_f is σ-finite on an interval (a, b). Then there is a dense set of subintervals of (a, b) on each of which f has bounded variation.*

Proof. Note first that f is symmetrically continuous at each point of (a, b) excepting only the points of a countable set C. This follows since $VS_f(\{x_0\})$ can only be positive at a countable number of points in (a, b) if VS_f is σ-finite and, if $VS_f(\{x_0\}) = 0$, then f is symmetrically continuous at x_0.

Let $\{A_m\}$ be a sequence of disjoint sets covering (a, b) such that each $VS_f(A_m)$ is finite. Choose a positive function δ on (a, b) so that $S_\delta(f, A_m) < +\infty$ for each m and write

$$E_{nm} = \{x \in A_m \ : \ \delta(x) > 1/n\}.$$

The countable collection $\{E_{nm}\}$ covers (a, b) and so, by the Baire theorem, there must be an interval $(c, d) \subset (a, b)$ and a set E_{NM} that is second category in each subinterval of (c, d). We can assume that the length of (c, d) is less than $1/N$.

Suppose that $\{[x_i, y_i] \ : \ i = 1, 2, \ldots, P\}$ are nonoverlapping intervals contained in (c, d). We shall show that

$$\sum_{i=1}^{P} |f(y_i) - f(x_i)| \leq 6S_\delta(f, A_M) + 2 \tag{8.3}$$

and it will follow that f has bounded variation on (c, d). Since the collection of intervals with this property is dense in (a, b) the proof is then complete.

Take $[x, y]$ as any one of these intervals $[x_i, y_i]$. We shall construct points

$$x = x_1, x_2, x_3, x_4, x_5, x_6, x_7, x_8, x_9 = y$$

from inside the interval $[x, y]$ and write I_j as the interval with endpoints x_j and x_{j+1}. Write

$$f(I_j) = |f(x_{j+1}) - f(x_j)|$$

for $j = 1, 2, \ldots, 8$ and then we can employ the estimate

$$|f(y) - f(x)| \leq \sum_{j=1}^{8} f(I_j).$$

These points are supplied by the covering theorem, Theorem 3.11.

For convenience we reproduce the details of the construction from [287] although the reader should by now see how the proof can be completed.

Choose a point $z \in E_{NM}$ close to the midpoint of $[x, y]$; since E_{NM} is dense in (c, d) we can choose such a point, say in the middle 1/10th of $[x, y]$. Since E_{NM} is uncountable in this middle 1/10th of $[x, y]$ we can arrange that the points $(x + z)/2$ and $(z + y)/2$ do not belong to the countable set C, i.e. so that f is symmetrically continuous at both of these points.

Write $x = z - 2s$, $y = z + 2t$ and choose s', $t' > 0$ so that $z + t' \in E_{NM}$, $z - s' \in E_{NM}$, $z + t'$ is sufficiently close to $z + t/2$ so that

$$|f(z + 2t - 2t') - f(z + 2t')| < 1/P \qquad (8.4)$$

and $z - s'$ is sufficiently close to $z - s/2$ so that

$$|f(z - 2s') - f(z - 2s + 2s')| < 1/P. \qquad (8.5)$$

The inequalities (8.4) and (8.5) just employ the symmetric continuity of f at the points $z + t$ and $z - s$ by making $|t - 2t'|$ and $|s - 2s'|$ small. Now write

$$
\begin{aligned}
x_1 &= x \ (= z - 2s), \\
x_2 &= z - 2s' + 2s, \\
x_3 &= z - 2s + 2s', \\
x_4 &= z - 2s', \\
x_5 &= z, \\
x_6 &= z + 2t', \\
x_7 &= z + 2t - 2t', \\
x_8 &= z + 2t' - 2t, \\
x_9 &= y \ (= z + 2t).
\end{aligned}
$$

As long as $|t - 2t'|$ and $|s - 2s'|$ are not too big each of these points is inside the interval $[x, y]$. Thus we have produced 8 subintervals of $[x, y]$ and (because of (8.4) and (8.5)) we have arranged for $f(I_3) < 1/P$ and for $f(I_6) < 1/P$; notice that the centers of the remaining intervals I_1, I_2 I_4, I_5 I_7, and I_8 are in the set E_{NM}. Evidently we can do this same construction for each interval in $\{[x_i, y_i] : i = 1, 2, \ldots, P\}$ and, using an obvious notation, we produce intervals

$$\{I_{ij} : i = 1, 2, \ldots, P, \ j = 1, 2, \ldots 8\}$$

with $f(I_{i3}) < 1/P$ and $f(I_{i6}) < 1/P$. All the remaining intervals have midpoints in E_{NM} and they can be split into 6 subcollections of nonoverlapping intervals each of length less than $1/N$. Thus

$$\sum_{i=1}^{P} |f(y_i) - f(x_i)| \le \sum_{i=1}^{P} \sum_{j=1}^{8} f(I_{ij}) \le 6S_\delta(f, A_M) + P(1/P + 1/P)$$

which supplies the inequality (8.3) that we wished to prove.

We point out that the set complementary to the intervals in Theorem 8.5 where f has bounded variation may have positive Lebesgue measure. There exists an everywhere differentiable function whose set of points of non bounded variation has positive measure. The construction of such a function is given in [18, p. 73]

8.2.3 Variation and Symmetric Derivates

We now give the differential structure of a function f satisfying the condition $VS_f(E) = 0$ or $VS_f(E) < +\infty$ for a measurable set E. The first two lemmas carry most of the information needed to relate the measure VS_f to the symmetric derivates of f.

Lemma 8.6 *Suppose, at each point x of a set E, that the inequalities*

$$-\alpha < \underline{SD}\, f(x) \le \overline{SD}\, f(x) < \alpha$$

hold. Then $VS_f(E) \le \alpha|E|$.

Proof. Let $\beta > |E|$ and choose an open set $G \supset E$ so that $|G| < \beta$. There is a gauge δ so that for every $x \in E$ the interval $(x - \delta(x), x + \delta(x))$ is contained in G and so that

$$|f(x + h) - f(x - h)| < 2\alpha h$$

for all $0 < h < \delta(x)$.

If $\{[x_i - h_i, x_i + h_i]\}$ is any sequence of nonoverlapping intervals with centers $x_i \in E$ and with $h_i < \delta(x_i)$ then

$$\sum |f(x_i + h_i) - f(x_i - h_i)| < \sum 2\alpha h_i < \alpha|G| < \alpha\beta$$

and so $VS_f(E) \le S_\delta(f, A_m) \le \alpha\beta$. From this inequality the lemma now follows on letting $\beta \searrow |E|$.

Lemma 8.7 *Suppose, at each point x of a set E, that one of the two inequalities*

$$-\alpha > \underline{SD}\, f(x) \quad or \quad \overline{SD}\, f(x) > \alpha$$

is true. Then $VS_f(E) \ge \alpha|E|$.

Proof. We may assume that E is bounded. Let δ be any positive function on E and let $\mathcal{C}$ denote the class of all intervals $[x, y]$ for which

$$|f(y) - f(x)| > \alpha(y - x), \quad y - x < \delta((x+y)/2), \quad (x+y)/2 \in E.$$

If either $-\alpha > \underline{SD}\, f(x)$ or $\overline{SD}\, f(x) > \alpha$ is true at each point $x \in E$ then $\mathcal{C}$ is a Vitali cover of E. For any $\beta < |E|$ choose a nonoverlapping collection $\{[x_i - h_i, x_i + h_i]\} \subset \mathcal{C}$ so that

$$\beta < \sum_{i=1}^{n} |[x_i - h_i, x_i + h_i]|.$$

Then

$$\beta < \sum_{i=1}^{n} 2h_i = \alpha^{-1} \sum_{i=1}^{n} 2\alpha h_i$$

$$\leq \alpha^{-1} \sum_{i=1}^{n} |f(x_i + h_i) - f(x_i - h_i)| \leq \alpha^{-1} S_\delta(f, E).$$

From this it follows that $\alpha\beta \leq VS_f(E)$ and finally, on letting $\beta \nearrow |E|$, that $\alpha|E| \leq VS_f(E)$ as required.

Theorem 8.8 *Let f be a measurable function and let E be a measurable set. If $VS_f(E) < +\infty$ then the ordinary derivative $f'(x)$ exists for almost every $x \in E$. If $VS_f(E) = 0$ then $f'(x) = 0$ for almost every $x \in E$.*

Proof. For any natural number n write

$$A_n = \left\{ x \in E : \underline{SD}\, f(x) < -n \text{ or } \overline{SD}\, f(x) > n \right\}$$

and

$$B_n = \left\{ x \in E : -n^{-1} > \underline{SD}\, f(x) \text{ or } \overline{SD}\, f(x) > n^{-1} \right\}.$$

Note that

$$A = \left\{ x \in E : \underline{SD}\, f(x) = -\infty \text{ or } \overline{SD}\, f(x) = +\infty \right\} = \bigcap_{n=1}^{\infty} A_n$$

and

$$B = \{ x \in E : \underline{SD}\, f(x) \neq 0 \} = \bigcup_{n=1}^{\infty} B_n.$$

It follows directly from Lemma 8.7 that $n|A_n| \leq VS_f(E)$ and that $|B_n| \leq nVS_f(E)$. Thus in the case that $VS_f(E) < +\infty$ the set A has measure zero and it follows that the symmetric derivates of f are finite almost everywhere in E. By the Theorem of Khintchine $f'(x)$ exists for almost every $x \in E$. In the case that $VS_f(E) = 0$, B has measure zero and it follows that the symmetric derivative of f vanishes almost everywhere in E and so $f'(x) = 0$ for almost every $x \in E$ as required.

8.3 Second Order Variation

In the preceding sections we have studied the first order symmetric variation of a function. This is very much a natural generalization of the ordinary variation of a function. In order to do the same for the second order symmetric derivative a different approach is needed. This is suggested by Theorem 3.44 in Section 3.7.3 where a covering of 2–intervals is used to form partitions.

We wish a variation that somehow "adds" the increments

$$f(x_i + h_i) + f(x_i - h_i) - 2f(x_i)$$

for a function f in a manner similar to the definition of the ordinary variation and the first order symmetric variation and yet which plays a role in the study of the second symmetric derivative. Some reflection shows that nonoverlapping intervals play no role here. Indeed if

$$f(x + h) + f(x - h) - 2f(x) = O(h^2)$$

then we see that the pieces being added have magnitudes similar to squares of the length of the intervals and not to the lengths themselves. A sum over nonoverlapping intervals would be much too small.

This suggests that the problem needs to be lifted to two dimensions. For this the notion of a 2–interval, employed in Chapter 3, can be used. Since we wish to have access to a covering theorem we would prefer to consider a variation that adds the increments

$$f(x_i + h_i) + f(x_i - h_i) - f(x_i + k_i) - f(x_i - k_i)$$

over regular 2–partitions $\{([x_i - h_i, x_i + h_i], [x_i - k_i, x_i + k_i])\}$ of some 2–interval. In order to be able to use Theorem 3.44 we shall define a variation of the form

$$\inf \sup \sum_{i=1}^{n} \left(f(x_i + h_i) + f(x_i - h_i) - f(x_i + k_i) - f(x_i - k_i) \right)$$

where the sum is taken only for 2–intervals satisfying a regularity condition of the form $0 \le 3k_i \le h_i$.

This motivational exercise, necessarily vague, should prepare the reader for the definitions which lead up to the concept of the second symmetric variation.

8.3.1 Functions on Rectangles and 2–Intervals

There is a close connection between second symmetric derivatives and 2–intervals. We can first associate the 2–interval

$$([x - h, x + h], [x, x])$$

with the second symmetric difference

$$G(x + h) + G(x - h) - 2G(x)$$

of a function G. A connection with arbitrary r–regular 2–intervals will come about because of the following elementary computation: if $0 \leq k \leq rh$ then

$$h^2 + k^2 \leq (1 + r^2)h^2 = \frac{1 + r^2}{1 - r^2} \left(h^2 - r^2 h^2 \right) \leq \frac{1 + r^2}{1 - r^2} \left(h^2 - k^2 \right). \tag{8.6}$$

We will use the notation

$$C_r = \frac{1 + r^2}{1 - r^2}. \tag{8.7}$$

(Throughout the only regularity condition imposed is the use of $\frac{1}{3}$–regular 2–intervals and so $r = \frac{1}{3}$ and $C_{\frac{1}{3}} = 10/8$.)

Now the connection is easy to make. We convert the second symmetric derivative into a derivative taken over arbitrary r–regular 2–intervals. This goes back to Riemann although he did not express it in this language. By definition, the existence of the derivative $SD_2 G(x) = f(x)$ requires that for every $\epsilon > 0$ there is a $\delta > 0$ so that

$$\left| G(x + h) + G(x - h) - 2G(x) - f(x)h^2 \right| < \epsilon h^2$$

for $0 < h < \delta$. We see, using (8.6) and (8.7), that this translates into the requirement

$$\left| G(x + h) + G(x - h) - G(x + k) - G(x - k)) - f(x)(h^2 - k^2) \right|$$
$$< \epsilon C_{\frac{1}{3}}(h^2 - k^2) \tag{8.8}$$

for $0 \leq 3k \leq h < \delta$. We express this in the language of 2–interval functions. For any 2–interval $K = ([x - h, x + h], [x - k, x + k])$ and any real function G we define

$$\Delta_K G = G(x - h) + G(x + h) - G(x - k) - G(x + k). \tag{8.9}$$

Now by using (8.9) we see that the inequality (8.8) above is exactly the statement that

$$\left| \Delta_K G - f(x) \|K\| \right| < \epsilon C_{\frac{1}{3}} \|K\| \tag{8.10}$$

where $x = c(K)$ is the center of K and K a 2–interval that is $\frac{1}{3}$–regular with $|K| < \delta$.

Thus we can think of the second symmetric derivative as a limit of the difference quotient $\Delta_K G / \|K\|$. This is very suggestive of applications to integration and variation. The former will be covered in Chapter 9; here we continue with the variational ideas inspired by this formulation.

For further motivational discussion of the ideas of this section see Ash [6].

8.3.2 Symmetric Increments

In discussions of derivatives, integrals and variation the increment of a function F on an interval $[a, b]$ is frequently employed. In symmetric situations it is often more convenient to employ the expression

$$\lambda_F(a, b) = \lim_{h \to 0+} \{F(b - h) - F(a + h)\} \qquad (8.11)$$

if it exists. Note that if F is continuous on the right at a and on the left at b then

$$\lambda_F(a, b) = F(b) - F(a). \qquad (8.12)$$

If F is integrable with an indefinite integral G then $\lambda_F(a, b)$ can be directly obtained from G. For an arbitrary function G defined on an interval $[a, b]$ and a 2-interval $K_p = ([a, b], [a + p, b - p])$ we shall write

$$\Lambda_G(a, b) = \lim_{p \to 0+} \tfrac{1}{p} \Delta_{K_p} G = \lim_{p \to 0+} \frac{G(a) - G(a + p) - G(b - p) + G(b)}{p}. \qquad (8.13)$$

Note that if $G(x) = \int_a^x F(t)\,dt$ and $\lambda_F(a, b)$ exists then

$$\lambda_F(a, b) = \Lambda_G(a, b). \qquad (8.14)$$

Also if the one-sided derivatives $G'_+(a)$ and $G'_-(b)$ exist for a function G then

$$\Lambda_G(a, b) = G'_-(b) - G'_+(a). \qquad (8.15)$$

We shall need the following elementary lemma.

Lemma 8.9 *Let G be continuous and suppose that $\Lambda_G(a, b) = 0$ for every interval (a, b). Then G is linear.*

Proof. We show that $G'_+(x) = G'_-(y)$ for every pair x, y from which the lemma evidently follows. Take any interval $[x, y]$; we show $G'_+(x) = G'_-(y)$. We may suppose, by subtracting a linear function if necessary, that $G(x) = G(y)$. If G is constant in $[x, y]$ we are done. If not then it has a maximum or minimum there. Suppose that at $\xi \in (x, y)$ a maximum is assumed. Then from the fact that $\Lambda_G(x, \xi) = 0$ we can deduce that all derived numbers of G on the right at x are nonnegative; similarly from the fact that $\Lambda_G(\xi, y) = 0$ we can deduce that all derived numbers of G on the left at y are nonpositive. But from $\Lambda_G(x, y) = 0$ we see that all the derived numbers of G on the right at x are identical with the derived numbers of G on the left at y. The conclusion follows.

8.3.2.1 Additive rectangle functions. For any rectangle $R = [a, b] \times [c, d]$ and any function H of two variables we shall write the difference

$$H(R) = H(a, c) - H(a, d) - H(b, c) + H(b, d). \qquad (8.16)$$

Observe that, if the sequence $(R_1, R_2, \ldots, R_n)$ is a partition of R into subrectangles, then

$$H(R) = \sum_{i=1}^{n} H(R_i) \qquad (8.17)$$

so that H can be interpreted as an additive rectangle function.

Let G be a real function, let R be a rectangle and let $\pi(R)$ denote the 2–interval that is the projection of this rectangle (see equation 3.16 on page 99) under our usual mapping $\pi : (x, y) \rightarrow (x + y)$ interpreted as a projection from $\mathbb{R}^2$ to $\mathbb{R}$. If we write $H(x, y) = G(x + y)$ then notice that

$$H(R) = \Delta_{\pi(R)} G \qquad (8.18)$$

thus linking (8.16) and (8.9). As a result of this we see, using (8.17), that if the sequence $(K_1, K_2, \ldots, K_n)$ is a 2–partition of the 2–interval K then

$$\Delta_K G = \sum_{i=1}^{n} \Delta_{K_i} G. \qquad (8.19)$$

Thus the expression in (8.17) of H as an additive rectangle function translates in (8.19) to allow us to view $K \rightarrow \Delta_K G$ as an additive 2–interval function.

The following notation will be useful is going back and forth between two dimensional concepts and the projected one dimensional concepts. If f is any real function we write

$$\xi_f([x - h, x + h], [x - k, x + k]) = f(x + h) + f(x - h) - f(x - k) - f(x + k) \qquad (8.20)$$

for every 2–interval $K = ([x - h, x + h], [x - k, x + k])$ to obtain an associated 2–interval function. The connection with second symmetric differences is immediate:

$$\xi_f([x - h, x + h], [x - k, x + k]) = \Delta_s^2 f(x, h) - \Delta_s^2 f(x, k).$$

Now, if we denote by H_f the function $H_f(x, y) = f(x + y)$, then we have three convenient expressions for the same quantity:

$$H_f(R) = \xi_f(K) = \Delta_K f$$
$$= f(x + h) + f(x - h) - f(x - k) - f(x + k) \qquad (8.21)$$

where $K = ([x - h, x + h], [x - k, x + k])$ and R is a rectangle projecting to K.

8.3.3 Variational Definitions

The main definitions leading to the variational measure are presented in this section. The variation is here defined for an arbitrary 2–interval function.

Definition 8.10 Let ξ denote an arbitrary real function defined for all 2–intervals. Let K denote any 2–interval, let $E \subset \mathbb{R}$ and let δ be a gauge on $\mathbb{R}$. Then we write

$$V(\xi, K, E; \delta) = \sup \sum_{i=1}^{n} |\xi(K_i)|$$

where the supremum is taken over all 2–subpartitions $(K_1, K_2, \ldots, K_n)$ of K that are $\frac{1}{3}$–regular and δ–fine at centers $c(K_i)$ that belong to E.

Note that throughout the regularity r is taken as $\frac{1}{3}$ although other values are possible. For different values a different variation would result but with an identical theory as long as a corresponding covering theorem could be proved.

Definition 8.11 Let ξ denote an arbitrary real function defined for all 2–intervals. Let K denote any 2–interval and let $E \subset \mathbb{R}$. We write

$$V(\xi, K, E) = \inf_{\delta} V(\xi, K, E; \delta)$$

where the infimum is taken over all gauges δ on $\mathbb{R}$.

Lemma 8.12 *The set function* $E \to V(\xi, K, E)$ *is an outer measure on the real line for which all Borel sets are measurable.*

Proof. Let $\mu(E) = V(\xi, K, E)$ for K fixed. Clearly μ is monotone and $\mu(\emptyset) = 0$. Suppose that $E \subset \bigcup_{i=1}^{\infty} E_i$ where the $\{E_i\}$ are disjoint.
    Let $\epsilon > 0$ and choose gauges δ_i so that

$$V(\xi, K, E_i) + \epsilon/2^i \geq V(\xi, K, E_i; \delta_i).$$

Take any gauge δ with $\delta(x) < \delta_i(x)$ for $x \in E_i$. Then an easy argument shows that

$$V(\xi, K, E; \delta) \leq \sum_{i=1}^{\infty} V(\xi, K, E_i; \delta_i)$$

$$\leq \sum_{i=1}^{\infty} \left\{ V(\xi, K, E_i) + \epsilon/2^i \right\}.$$

We can conclude that $\mu(E) \leq \sum_{i=1}^{\infty} \mu(E_i)$ and hence that μ is an outer measure.
    To see that all Borel sets are measurable it suffices, by the standard theory, to show that $\mu(E_1 \cup E_2) = \mu(E_1) + \mu(E_2)$ if $\mathrm{dist}(E_1, E_2) > 0$. Let δ be any gauge with $\delta(x) < \frac{1}{2}\mathrm{dist}(E_1, E_2)$. Then

$$V(\xi, K, E_1 \cup E_2; \delta) = V(\xi, K, E_1; \delta) + V(\xi, K, E_2; \delta)$$

since any rectangles used in computing the two expressions on the right cannot overlap. It follows that $\mu(E_1 \cup E_2) = \mu(E_1) + \mu(E_2)$ and the lemma is proved.

Lemma 8.13 *Let ξ be an arbitrary real function defined for all 2–intervals, let $E \subset \mathbb{R}$ be a bounded set and let $K_p = ([a, b], [a+p, b-p])$ where $E \subset (a+p, b-p)$ and $p > 0$. Then*

$$\frac{1}{p} V(\xi, K_p, E)$$

is independent of the choice of a, b and p.

Proof. Write $K'_p = ([c, d], [c + p, d - p])$ where

$$E \subset (a + p, b - p) \subset (c + p, d - p).$$

Let δ be arbitrary but small enough, for $x \in E$, that

$$(x - \delta(x), x + \delta(x)) \subset (a + p, b - p).$$

Consider, for such a δ, any δ–fine subpartition of K'_p with centers in E. Certainly all of the 2–intervals of such a subpartition must be contained in $(a + p, b - p)$. Thus it is easy to see that

$$V(\xi, K_p, E; \delta) = V(\xi, K'_p, E; \delta)$$

In this way we see that $\frac{1}{p} V(\xi, K_p, E)$ is independent of the choice of a, b but may possibly depend on p.

Fix a and b and suppose that $(a + 2p, b - 2p) \supset E$. Write

$$K'_{2p} = ([a - p, b + p], [a + p, b - p]) \text{ and } K_{2p} = ([a, b], [a + 2p, b - 2p]).$$

We notice first that
$$V(\xi, K'_{2p}, E) = V(\xi, K_{2p}, E)$$

because of what we have just proved in the first paragraph of the proof.

Let us show first that

$$V(\xi, K'_{2p}, E) = 2V(\xi, K_p, E).$$

We can suppose that these are finite. Let R_1 and R_2 be two adjacent rectangles, one above the other, that project to the same 2–interval K_p. We can add two further squares S_1 and S_2 to the figure thus formed and obtain a rectangle R that projects to K'_{2p}. The rectangle R is made up of the two rectangles R_1 and R_2 and, at the corners, the two squares S_1 and S_2. We can argue that

$$V(\xi, \pi(R), E) = V(\xi, \pi(R_1), E) + V(\xi, \pi(R_2), E) = 2V(\xi, K_p, E) \quad (8.22)$$

by analyzing the subrectangles that would be used in estimating these expressions.

As the centers of these subrectangles must project to E there is no contribution to $V(\xi, \pi(R), E)$ from rectangles contained in S_1 and S_2 if we take the gauge δ small enough. Thus we can look just at R_1 and R_2. If ℓ denotes the line that is the common boundary of R_1 and R_2 note that for any $\epsilon > 0$ there must be a gauge δ sufficiently small so that the contribution in $V(\xi, \pi(R), E; \delta)$ from rectangles meeting ℓ cannot exceed ϵ; if this were not so then, by using many horizontal lines, we could show that $V(\xi, \pi(R), E)$ is unbounded. Thus $V(\xi, \pi(R), E; \delta)$ can be estimated separately from the contributions from subrectangles of R_1 and R_2. In this way (8.22) can be proved. Thus we obtain

$$V(\xi, K_{2p}, E) = V(\xi, K'_{2p}, E) = 2V(\xi, K_p, E).$$

For any integer n, by the same method, we can show that

$$V(\xi, K_{np}, E) = nV(\xi, K_p, E)$$

if $(a + np, b - np) \supset E$. From this we can deduce that if $0 < q < p$ and p/q is a rational, say m/n, then

$$nV(\xi, K_p, E) = mV(\xi, K_q, E)$$

and hence that

$$\frac{1}{p}V(\xi, K_p, E) = \frac{1}{q}V(\xi, K_q, E).$$

The general result now follows from the continuity of this expression.

Lemma 8.13 justifies our final definition of the expression we take for the variation of a 2–interval function.

Definition 8.14 Let ξ denote any real function defined for all 2–intervals and let $E \subset \mathbb{R}$ be a bounded set. We write

$$\xi^*(E) = \frac{1}{p}V(\xi, K_p, E)$$

where K_p denotes any 2–interval $K_p = ([a, b], [a + p, b - p])$ for which $E \subset (a + p, b - p)$. For arbitrary E (unbounded) we shall write

$$\xi^*(E) = \lim_{n \to \infty} \xi^*(E \cap (-n, n)).$$

Lemma 8.15 *The set function ξ^* is an outer measure on the real line for which all Borel sets are measurable.*

Proof. This follows from Lemma 8.12 and the definition of ξ^*.

8.3.4 Basic Properties

We now develop some of the simpler properties of the variation.

Lemma 8.16 *Let ξ_1, ξ_2 denote any real functions defined on 2–intervals, and let $E \subset \mathbb{R}$. Suppose that there is a gauge δ so that*

$$0 \leq \xi_1(K) \leq \xi_2(K)$$

holds for all $K = ([x-h, x+h], [x-k, x+k])$, $x \in E$ and $0 \leq 3k \leq h < \delta(x)$. Then $\xi_1^*(E) \leq \xi_2^*(E)$.

Proof. This is immediate.

Lemma 8.17 *Let ξ_1, ξ_2, ξ_3 denote any real functions defined for all 2–intervals, and let $E \subset \mathbb{R}$. Suppose that there is a gauge δ so that*

$$|\xi_1(K) - \xi_2(K)| \leq \xi_3(K)$$

holds for all $K = ([x-h, x+h], [x-k, x+k])$, $x \in E$ and $0 \leq 3k \leq h < \delta(x)$. Suppose that $\xi_3^*(E) = 0$. Then $\xi_1^*(E) = \xi_2^*(E)$.

Proof. This is immediate.

Lemma 8.18 *Let f be a convex function and (a, b) an interval. Then*

$$\xi_f^*((a, b)) = \Lambda_f(a, b).$$

Proof. We use the terminology in (8.20) and (8.21). To simplify the notation we suppose that f is continuous and linear outside of $[a, b]$, remaining convex, with $f'(a)$ and $f'(b)$ both existing (that is to say f could be replaced with such a function without changing any of the statements).

Let δ be arbitrary but at least sufficiently small that $\delta(x) < a - x$ for $x < a$, that $\delta(x) < x - b$ for $x > b$ and so that $(x - \delta(x), x + \delta(x)) \subset (a, b)$ for $x \in (a, b)$. Take $K_0 = ([c, d], [c+p, d-p])$ for any interval $(c+p, d-p) \supset [a, b]$. Let R_0 be a rectangle that projects to the 2–interval K_0.

We claim that

$$V(\xi_f, K_0, (a, b)) = \xi_f(K_0) = f(c) + f(d) - f(c+p) - f(d-p). \qquad (8.23)$$

Let $(K_1, K_2, \ldots, K_n)$ be any 2–subpartition of K_0 finer than δ and with centers in (a, b); let $(R_1, R_2, \ldots, R_n)$ be the associated subrectangles of R_0. Then

$$\sum_{i=1}^n \xi_f(K_i) = \sum_{i=1}^n H_f(R_i) = H_f(R_0) = \xi_f(K_0)$$

since H_f is nonnegative and additive as a rectangle function. It follows that

$$V(\xi_f, K_0, (a, b); \delta) \leq \xi_f(K_0).$$

This proves the inequality (8.23) in one direction.

Let $\epsilon > 0$. Choose a gauge δ as above and also sufficiently small at a and b that

$$V(\xi, K_0, \{a, b\}; \delta) < \epsilon/2. \tag{8.24}$$

The function f is smooth at a and b and this permits such a choice of δ. As before let $(K_1, K_2, \ldots, K_n)$ be any 2-subpartition of K_0 finer than δ and with centers in (a, b); let $(R_1, R_2, \ldots, R_n)$ be the associated subrectangles of R_0. By using Theorem 3.44 there is a countable set C so that for any gauge γ on $C \times \mathbb{N}$ we may add to this list a further set of rectangles making up a full partition of R_0 in the manner prescribed by the theorem. The additional rectangles fall in four classes. The first are still δ-fine with centers in (a, b) and so we can consider that they are already in the list $(R_1, R_2, \ldots, R_n)$. The second are rectangles which project outside of $[a, b]$ and hence contribute nothing to the sum since H_f vanishes on such rectangles. The third are δ-fine with centers that project to either a or b; by (8.24) these contribute no more than $\epsilon/2$ to the sum. The fourth and remaining rectangles can be controlled (by the gauge γ and the continuity of f) so as to contribute no more than $\epsilon/2$ to the sum. In this way we obtain

$$V(\xi, K_0, (a, b); \delta) > \xi(K_0) - \epsilon.$$

Now (8.23) follows.

We can deduce now, as required, that

$$\xi_f^*((a, b)) = \frac{1}{p}(f(c) + f(d) - f(c + p) - f(d - p)) = f'(b) - f'(a) = \Lambda_f(a, b)$$

since f is convex, linear outside of $[a, b]$ and $f'(a)$ and $f'(b)$ both exist.

Lemma 8.19 *Let ξ denote any real function defined for all 2-intervals, and let $E \subset \mathbb{R}$. Then if any one of the three cases hold it follows that $\xi^*(E) = 0$.*

(i) for all $x \in E$,

$$\lim_{h \to 0+, \, 0 \leq 3k \leq h} (h^2 - k^2)^{-1} \xi([x - h, x + h], [x - k, x + k]) = 0.$$

(ii) E has measure zero and for all $x \in E$,

$$\limsup_{h \to 0+, \, 0 \leq 3k \leq h} (h^2 - k^2)^{-1} |\xi([x - h, x + h], [x - k, x + k])| < +\infty.$$

(iii) E is countable and for all $x \in E$,

$$\lim_{h\to 0+,\,0\leq 3k\leq h}(h-k)^{-1}\xi([x-h,x+h],[x-k,x+k])=0.$$

Proof. It is enough to give the proof in the situation where E is bounded. Fix $K_p=([a,b],[a+p,b-p])$ where $E\subset(a+p,b-p)$. Suppose (i) holds. Then for any $\epsilon>0$ there is a gauge δ so that

$$|\xi([x-h,x+h],[x-k,x+k])|<\epsilon(h^2-k^2)\tag{8.25}$$

for $0\leq 3k\leq h<\delta(x)$. Write ξ_g for the 2-interval function associated with the function $g(x)=x^2$. Then the inequality (8.25) can be read as

$$\xi(K)\leq\tfrac{1}{2}\epsilon\xi_g(K)\tag{8.26}$$

for $K=([x-h,x+h],[x-k,x+k])$.

From Lemma 8.18 we know that $\xi_g^*((a,b))=\varLambda_g(a,b)=2(b-a)$. Thus from the inequality (8.26) we see that

$$p\xi^*(E)=V(\xi,K_p,E)\leq V(\epsilon\xi_g,K_p,E)\leq p\epsilon(b-a).$$

Since ϵ is arbitrary this measure vanishes as required.

Suppose (ii) holds. For each integer n let E_n denote the set of points in E at which there is a gauge δ so that

$$|\xi([x-h,x+h],[x-k,x+k])|<n(h^2-k^2)$$

for $0\leq 3k\leq h<\delta(x)$. Again write ξ_g for the 2-interval function associated with the function $g(x)=x^2$. Then this inequality can be read as

$$\xi(K)\leq n\xi_g(K)\tag{8.27}$$

for $K=([x-h,x+h],[x-k,x+k])$, $x\in E_n$ and $0\leq 3k\leq h<\delta(x)$.

We claim that each $\xi^*(E_n)=0$. Let $\epsilon>0$. Since E has measure zero we may choose open sets $O_n\supset E_n$ so that $|O_n|<\epsilon$. Let δ_1 be any gauge smaller than δ and also small enough so that if $x\in E_n$ then $\delta_1(x)$ is smaller than the distance from x to $\mathbb{R}\setminus O_n$.

Note that $V(\xi_g,K_p,E_n;\delta_1)<p\epsilon$. To see this let R_0 be a rectangle that projects to K_p and let G be the subset of that rectangle that projects to O_n. Observe that the two dimensional measure of G is less than $p\epsilon$ and that any rectangle K that projects to a 2-interval that is δ_1-fine and with center in E_n must be contained in G. For such a rectangle K recall that $\xi_g(K)$ is exactly twice the area of K. This provides the estimate.

Consequently, using (8.27), we obtain

$$V(\xi,K_p,E_n;\delta_1)\leq V(n\xi_g,K_p,E_n;\delta_1)<np\epsilon$$

and hence $\xi^*(E_n) = 0$ as claimed. Since $E = \bigcup_{n=1}^{\infty} E_n$ it follows that $\xi^*(E_n) = 0$ as required.

Suppose (iii) holds. We show that $\xi^*(\{x\}) = 0$ for any particular $x_0 \in E$. For any $\epsilon > 0$ there is a gauge δ so that

$$|\xi([x_0 - h, x_0 + h], [x_0 - k, x_0 + k])| < \epsilon(h - k)$$

for $0 \le 3k \le h < \delta(x_0)$.

Fix $K_p = ([a, b], [a + p, b - p])$ where $x_0 \in (a + p, b - p)$. To compute $V(\xi, K_p, \{x_0\}; \delta)$ we need to consider sums of the form

$$\sum_{i=1}^{n} |\xi(K_i)|$$

where $(K_1, K_2, \ldots, K_n)$ is a subpartition of K_p that is $\frac{1}{3}$-regular, δ-fine and whose centers $c(K_i) = x_0$. If we denote

$$K_i = ([x_0 - h_i, x_0 + h_i], [x_0 - k_i, x_0 + k_i])$$

then, by considering the geometry of the rectangles projecting to the K_i, we see that $\sum_{i=1}^{n} h_i - k_i < p$. Accordingly $V(\xi, K_p, \{x_0\}; \delta) < p\epsilon$ and it follows that $\xi^*(\{x_0\}) \le \epsilon$. Hence, since ϵ is arbitrary, this measure vanishes at each point of E. Finally, as E is countable, $\xi^*(E)$ vanishes too as required.

Lemma 8.20 *Let g be a measurable function and E a measurable set. Suppose that*

$$\lim_{h \to 0+, \, 0 \le 3k \le h} (h^2 - k^2)^{-1} \xi([x - h, x + h], [x - k, x + k]) = g(x).$$

Then

$$\xi^*(E) = \int_E |g(t)| \, dt.$$

Proof. To begin the proof we assume that g is a bounded function and E is a bounded set and establish the statement of the lemma in this case. We may assume that g vanishes off of E and, without loss of generality, we may take g as nonnegative throughout. Take any interval with $(a + p, b - p) \supset E$ and write

$$G_1(x) = \int_a^x g(t) \, dt \quad \text{and} \quad G(x) = \int_a^x G_1(t) \, dt$$

so that

$$G(a) + G(b) - G(a + p) - G(b - p) = p(G'(b) - G'(a)) = p \int_E g(t) \, dt.$$

There is a set N of measure zero so that $G_1'(x) = g(x)$ for each point $x \notin N$; consequently $\mathrm{SD}_2\, G(x) = |g(x)|$ also at such points. This means that $\xi_G - \xi$

satisfies the conditions of Lemma 8.19(i) on the set $\mathbb{R} \setminus N$ and it follows from Lemma 8.17 that $\xi_G^* = \xi^*$ on subsets of $\mathbb{R} \setminus N$. Also ξ satisfies the conditions of Lemma 8.19(ii) on the measure zero set N and so $\xi^*(N) = 0$. Since G_1 is Lipschitz it is also true that ξ_G satisfies the conditions of Lemma 8.19(ii) on the measure zero set N and so $\xi_G^*(N) = 0$ too. Since g vanishes off of E we also have $\xi^*(\mathbb{R} \setminus E) = 0$.

Putting these computations together we now have

$$\xi^*(E) = \xi^*((a, b)) = \xi^*((a, b) \setminus N) =$$
$$\xi_G^*((a, b) \setminus N) = \xi_G^*((a, b)) = \Lambda_G(a, b) = \int_E |g(t)|\, dt.$$

Thus the final statement of the lemma holds in case g is bounded.

As we have already observed ξ^* vanishes on all sets of Lebesgue measure zero. It follows then, from Lemma 8.15, that all Lebesgue measurable sets are also ξ^*–measurable. Write $E_n = \{x \in E : n - 1 \le |g(x)| < n\}$. Then, applying what we have just proved for each set E_n, we have

$$\xi^*(E) = \sum_{n=1}^{\infty} \xi^*(E_n) = \sum_{n=1}^{\infty} \int_{E_n} |g(t)|\, dt = \int_E |g(t)|\, dt$$

as required.

8.3.5 Variation of a Continuous Function

If f is a continuous function we adopt the notation
$$\mathrm{VS}_f^{(2)}(E) = \xi_f^*(E)$$
where ξ_f is, as usual, the 2–interval function
$$([x - h, x + h], [x - k, x + k]) \longrightarrow f(x + h) + f(x - h) - f(x + k) - f(x - k)$$
and the variation ξ_f^* is defined in Definition 8.14.

We refer to $\mathrm{VS}_f^{(2)}$ as the *second symmetric variational measure* of f. The first few results provide elementary properties of the variational measure $\mathrm{VS}_f^{(2)}$ associated with an arbitrary continuous function f.

Theorem 8.21 *The set function $\mathrm{VS}_f^{(2)}$ is an outer measure on the real line for which all Borel sets are measurable.*

Proof. This follows from Lemma 8.15.

Theorem 8.22 *Suppose that f is convex. Then for any interval (a, b)*
$$\mathrm{VS}_f^{(2)}((a, b)) = f'_-(b) - f'_+(a). \tag{8.28}$$

Proof. This follows from Lemma 8.18 and (8.15).

8.3.6 Zero Variation

One of the most frequent applications of a measure is in the study of situations where the measure vanishes. All of the results in this section concern sets on which a continuous function has zero variation.

Lemma 8.23 *Suppose that f is a continuous, real function and E a set of real numbers. Then, in any one of the three cases that follows, $VS_f^{(2)}(E) = 0$.*

(i) $f(x + h) + f(x - h) - 2f(x) = o(h^2)$ *as* $h \to 0+$ *for all* $x \in E$.

(ii) $f(x + h) + f(x - h) - 2f(x) = O(h^2)$ *as* $h \to 0+$ *for all* $x \in E$ *and* $|E| = 0$.

(iii) $f(x + h) + f(x - h) - 2f(x) = o(h)$ *as* $h \to 0+$ *for all* $x \in E$ *and* E *is countable.*

Proof. If (i) holds then there is a gauge δ so that for $0 \le t < \delta(x)$ the inequality $|f(x + t) + f(x - t) - 2f(x)| < \epsilon t^2$ holds. Then, using (8.6), with $0 \le 3k \le h < \delta(x)$, we have

$$|f(x + h) + f(x - h) - f(x + k) + f(x - k)| < \epsilon(h^2 + k^2) < \epsilon C_{\frac{1}{3}} \left(h^2 - k^2 \right).$$

Now the conclusion follows directly from Lemma 8.19(i). Similar computations supply the remaining statements.

Theorem 8.24 *A continuous function f is smooth at a point x_0 if and only if*

$$VS_f^{(2)}(\{x_0\}) = 0.$$

Proof. In one direction this follows directly from the definition of smoothness and Lemma 8.23. Suppose, conversely, that $VS_f^{(2)}(\{x_0\}) = 0$. Let $\epsilon > 0$ and take any $K_p = ([a, b], [a + p, b - p])$ with $a + p < x_0 < b - p$. There is a gauge δ so that

$$V(\xi_f, K_p, \{x_0\}; \delta) < p\epsilon/2.$$

We can specify that $\delta(x_0)$ is small compared to p.

Let R_0 be a rectangle that projects to K_p. Consider squares contained in R_0 that project to $([x_0 - h, x_0 + h], [x_0, x_0])$ and with $h < \delta(x_0)$. We can fit about p/h such nonoverlapping squares into R_0. In any case there are more than $\frac{1}{2}ph^{-1}$ such squares and so we can conclude that

$$\frac{p}{2h}|f(x_0 + h) + f(x_0 - h) - 2f(x_0)| \le V(\xi_f, K_p, \{x_0\}; \delta) < p\epsilon/2.$$

Hence $|f(x_0 + h) + f(x_0 - h) - 2f(x_0)| < \epsilon h$ for $h < \delta(x_0)$ and by definition then f is smooth at x_0.

Theorem 8.25 *Let f' be absolutely continuous and N a set of measure zero. Then* $VS_f^{(2)}(N) = 0$.

Proof. We can suppose that N is bounded. Let $\epsilon > 0$ and fix

$$K_p = ([a, b], [a + p, b - p])$$

where $N \subset (a + p, b - p)$. Let R_0 be a rectangle that projects to K_p. Since f'' is locally Lebesgue integrable the integral

$$H(R) = \int\int_R f''(x + y)\, dx\, dy$$

exists over every rectangle $R \subset R_0$. Write $F(x) = \int_a^x f''(t)\, dt$ and $G(x) = \int_a^x F(t)\, dt$. Then a direct computation shows that

$$\xi_f(\pi(R)) = \Delta_{\pi(R)} f = \Delta_{\pi(R)} G = H(R). \tag{8.29}$$

Choose an open set $O \supset N$ with measure sufficiently small so that

$$\sum |H(R_i)| < p\epsilon \tag{8.30}$$

for any sequence $\{R_i\}$ of subrectangles of R_0 with each $\pi(R_i) \subset O$. This just uses the absolute continuity of the integral.

We now define a gauge δ by setting $\delta(x)$ as the distance from x to $\mathbb{R} \setminus O$ if $x \in N$. Then in computing $V(\xi_f, K_p, N; \delta)$ we use only δ–fine 2–intervals $\{K_i\}$ with centers in N. Any subrectangles R_i of R_0 that project to such a K_i will satisfy (8.30). Consequently (8.29) and (8.30) show that

$$p\xi_f^*(N) = V(\xi_f, K_p, N) \leq V(\xi_f, K_p, N; \delta) < p\epsilon$$

and the theorem follows.

Theorem 8.26 *Let f be continuous. A necessary and sufficient condition that f be linear is that* $VS_f^{(2)}$ *vanishes.*

Proof. The necessity is obvious. For the converse we shall show that, if $VS_f^{(2)}$ vanishes, then $\Lambda_f(a, b) = 0$ for every interval (a, b). It follows then, from Lemma 8.9, that f is linear.

Let $\epsilon > 0$. Write

$$K_p = ([a, b], [a + p, b - p]) \quad \text{and} \quad K' = ([c, d], [c + p, d - p])$$

where $c + p < a < b < d - p$ and $p > 0$. As $VS_f^{(2)}$ vanishes we can choose a gauge δ so that

$$V(\xi, K', [a, b]; \delta) < p\epsilon.$$

We apply Theorem 3.44 now to this δ to obtain a countable set C so that for any gauge γ on $C \times \mathbb{N}$ there is a sequence $(I_1, I_2, \ldots, I_n)$ forming a 2–partition of K_p that is $\frac{1}{3}$–regular, such that each I_i is either δ–fine at the center or else I_i has an outer endpoint x in C with $|I_i| < \gamma(x, i)$. For this then

$$f(a) + f(b) - f(a + p) - f(b - p) = \xi_f(K_p) = \sum_{i=1}^{n} \xi_f(I_i).$$

We can use the continuity of f and the gauge γ to ensure that the contribution to this sum from any I_i having an outer endpoint in $x_j \in C$ does not exceed $p\epsilon$. Thus

$$|f(a) + f(b) - f(a + p) - f(b - p)| \leq \sum_{i=1}^{n} |\xi_f(I_i)| \leq \mathrm{V}(\xi, K', [a, b]; \delta) + p\epsilon < 2p\epsilon.$$

Since ϵ is arbitrary we can conclude that $\Lambda_f(a, b) = 0$ for every interval (a, b) as required.

8.3.7 Differentiation and Variation.

In this section we relate properties of the variational measure of a function with its derivation properties.

Theorem 8.27 *Let f be continuous and suppose that $SD_2 f(x) = g(x)$ for every x in a measurable set E. Then*

$$\mathrm{VS}_f^{(2)}(E) = \int_E |g(t)| \, dt.$$

Proof. This follows from Lemma 8.20.

Corollary 8.28 *Let f be continuous and suppose that $f''(x) = g(x)$ for every x in a measurable set E. Then*

$$\mathrm{VS}_f^{(2)}(E) = \int_E |g(t)| \, dt. \tag{8.31}$$

Proof. This follows directly from Theorem 8.27 since $SD_2 f(x) = g(x)$ at every point x at which $f''(x) = g(x)$.

Corollary 8.29 *Let g be Lebesgue integrable on an interval $[a, b]$ and suppose that*

$$f(x) = \int_a^x \int_a^t g(s) \, ds \, dt.$$

Then, for every measurable set $E \subset (a, b)$,

$$\mathrm{VS}_f^{(2)}(E) = \int_E |g(t)|\, dt. \tag{8.32}$$

Proof. This follows from Theorem 8.27 and Theorem 8.25.

The following is an extension of the computations in Theorem 8.27 where an exceptional set is allowed.

Theorem 8.30 *Let f be continuous. Suppose that f is smooth at each point of a measurable set E, that*

$$-\infty < \underline{SD}_2\, f(x) \leq \overline{SD}_2\, f(x) < +\infty$$

nearly everywhere in E and that $SD_2\, f(x) = g(x)$ almost everywhere in E. Then

$$\mathrm{VS}_f^{(2)}(E) = \int_E |g(t)|\, dt. \tag{8.33}$$

Proof. Let E_1 be the set of points x in E at which $SD_2\, f(x) = g(x)$. Let E_2 be the set of points x in $E \setminus E_1$ at which the derivates $\underline{SD}_2\, f(x)$ and $\overline{SD}_2\, f(x)$ are finite. Let $E_3 = E \setminus (E_1 \cup E_2)$.

By hypothesis E_3 is countable and so, since f is smooth, it follows from Theorem 8.24 that $\mathrm{VS}_f^{(2)}(E_3) = 0$. By hypothesis E_2 has measure zero thus it follows, from Theorem 8.19(ii), that $\mathrm{VS}_f^{(2)}(E_2) = 0$. Finally then

$$
\begin{aligned}
\mathrm{VS}_f^{(2)}(E) &= \mathrm{VS}_f^{(2)}(E_1) + \mathrm{VS}_f^{(2)}(E_2) + \mathrm{VS}_f^{(2)}(E_3) \\
&= \mathrm{VS}_f^{(2)}(E_1) = \int_{E_1} |g(t)|\, dt \\
&= \int_E |g(t)|\, dt
\end{aligned}
$$

proving (8.33).

The next two lemmas show just how intimately connected the variation is to estimates on the extreme derivates.

Lemma 8.31 *Suppose, at each point x of a measurable set E, that the inequalities*

$$-\alpha < \underline{SD}_2\, f(x) \leq \overline{SD}_2\, f(x) < \alpha$$

hold. Then $\mathrm{VS}_f^{(2)}(E) \leq \alpha C_{\frac{1}{3}} |E|$.

Proof. There is a gauge δ on E so that

$$|f(x + t) + f(x - t) - 2f(x)| < \alpha t^2$$

for $x \in E$ and $0 \le t < \delta(x)$. Now, by computations we have seen before, we have also

$$|f(x+h) + f(x-h) - f(x+k) - f(x-k)| < \alpha C_{\frac{1}{3}}(h^2 - k^2).$$

for $x \in E$ and $0 \le 3k \le h < \delta(x)$.

Write $g(x) = 0$ if $x \notin E$ and for $x \in E$ write $g(x) = \alpha C_{\frac{1}{3}}$. Then, from Lemma 8.16 and Lemma 8.20, we obtain

$$\mathrm{VS}_f^{(2)}(E) \le \int_E g(x)\, dx = \alpha C_{\frac{1}{3}}|E|$$

as required.

Lemma 8.32 *Suppose, at each point x of a set E, that one of the two inequalities*

$$-\alpha > \underline{SD}_2\, f(x) \quad or \quad \overline{SD}_2\, f(x) > \alpha$$

is true. Then $\mathrm{VS}_f^{(2)}(E) \ge \alpha|E|$.

Proof. We may assume that E is bounded and we choose a, b and $p > 0$ so that $E \subset (a+p, b-p)$. Fix $K_p = ([a,b],[a+p,b-p])$ and choose a rectangle R_0 that projects to K_p.

Let δ be any gauge and let C denote the class of all 2–intervals of the form $([x-h, x+h],[x,x])$ for which

$$|f(x+h)) + f(x-h) - 2f(x)| > \alpha h^2, \quad (0 < h < \delta(x),\ x \in E). \qquad (8.34)$$

Since either $-\alpha > \underline{SD}_2\, f(x)$ or $\overline{SD}_2\, f(x) > \alpha$ is true at each point $x \in E$ the collection C is similar to a Vitali cover of the set E; but we need to work in two dimensions.

Let E_0 denote the set of points in the rectangle R_0 that project into E, i.e.

$$E_0 = \{(x,y) \in R_0 : x + y \in E\}.$$

Note that the two dimensional Lebesgue measure of E_0 is $p|E|$.

Let R denote the collection of squares contained as subrectangles of R_0 and which project to members of the collection C. This family R is a Vitali cover of E_0 and so can be used to estimate the measure of E_0, and hence also the measure of E. Note that

$$\xi_f(\pi(R)) \ge \alpha\, \|\pi(R)\|$$

for each member R of R because of (8.34).

Let $\beta < |E|$ and choose from R a subpartition $(R_1, R_2, R_3, \ldots R_m)$ of R_0 so that

$$\sum_{j=1}^{m} \|R_i\| > p\beta.$$

Now

$$\sum_{j=1}^{m} \xi_f(\pi(R_j)) \geq \sum_{j=1}^{m} \alpha \|\pi(R_j)\| = \alpha \sum_{j=1}^{m} \|R_j\| \geq p\alpha\beta.$$

Thus we have

$$V(\xi, K_p, E; \delta) \geq p\alpha\beta$$

for any gauge δ. It follows that

$$V(\xi, K_p, E) \geq p\alpha\beta$$

and so

$$\xi^*(E) \geq \alpha\beta.$$

Finally then $\xi^*(E) \geq \alpha|E|$ follows on letting $\beta \nearrow |E|$.

Theorem 8.33 *Let f be continuous and suppose that the measure $\mathrm{VS}_f^{(2)}$ is σ-finite on a set E. Then, for almost every point $x \in E$,*

$$f(x+h) + f(x-h) - 2f(x) = O(h^2) \quad (h \to 0).$$

Proof. We may assume that $\mathrm{VS}_f^{(2)}(E)$ is finite. We show that for a.e. point $x \in E$ the upper derivate $\overline{\mathrm{SD}}_2 f(x) < +\infty$. Let E_0 denote the set of points $x \in E$ at which $\overline{\mathrm{SD}}_2 f(x) = +\infty$. Then, by Lemma 8.32,

$$k|E_0| \leq \mathrm{VS}_f^{(2)}(E_0) < +\infty$$

for every $k > 0$. It follows that $|E_0| = 0$ and so $\overline{\mathrm{SD}}_2 f(x) < +\infty$ for a.e. point $x \in E$. In a similar way it may be shown that $\underline{\mathrm{SD}}_2 f(x) > -\infty$ for a.e. point $x \in E$. The theorem now follows.

Theorem 8.34 *Let f be continuous and suppose that the measure $\mathrm{VS}_f^{(2)}$ is σ-finite on a set E. Then the derivative $f'(x)$ and the second order symmetric derivative $\mathrm{SD}_2 f(x)$ both exist at almost every point of E. Moreover*

$$AD f'(x) = SD_2 f(x)$$

a.e. in E.

Proof. The theorem follows from Theorem 8.33 by using the material in Section 4.9.3.

8.3.8 Finite Variation

We collect in this section analogues of some well known variational results in the theory of real functions. If f has bounded variation on an interval $[a, b]$ then f' exists a.e. and

$$\operatorname{Var}(f, [a, b]) \geq \int_a^b |f'(t)| \, dt. \tag{8.35}$$

If, moreover, f is absolutely continuous then (8.35) is an equality and f is an indefinite integral of its derivative. The second symmetric variation allows similar assertions.

Theorem 8.35 Let f be continuous and suppose that the measure $\mathrm{VS}_f^{(2)}$ is finite on (a, b). Then the second order symmetric derivative $SD_2 f(x) = g(x)$ exists at almost every point of (a, b), g is Lebesgue integrable there and

$$\mathrm{VS}_f^{(2)}((a, b)) \geq \int_a^b |g(t)| \, dt. \tag{8.36}$$

Proof. By Theorem 8.34 the derivative $SD_2 f(x) = g(x)$ exists at almost every point of (a, b), say at the points in the set E, where E has full measure in (a, b). Certainly g is measurable. Therefore, by Theorem 8.27,

$$\mathrm{VS}_f^{(2)}((a, b)) \geq \mathrm{VS}_f^{(2)}(E) = \int_E |g(t)| \, dt = \int_a^b |g(t)| \, dt$$

giving (8.36) as required. Integrability of g follows directly from this inequality.

Theorem 8.36 Let f be continuous and suppose that the measure $\mathrm{VS}_f^{(2)}$ is finite on (a, b) and vanishes on every subset of measure zero. Then the second order symmetric derivative $SD_2 f(x) = g(x)$ exists at almost every point of (a, b), g is Lebesgue integrable there,

$$\mathrm{VS}_f^{(2)}(E) = \int_E |g(t)| \, dt \tag{8.37}$$

for all measurable $E \subset (a, b)$ and

$$f(x) = cx + d + \int_a^x \int_a^t g(s) \, ds \, dt. \tag{8.38}$$

for some constants c and d.

Proof. The first assertions follow from Theorem 8.35. If $\mathrm{VS}_f^{(2)}$ vanishes on every subset of (a, b) of measure zero then (8.37) follows, for all measurable $E \subset (a, b)$, again from Theorem 8.35.

Since g is integrable we may write

$$G(x) = \int_a^x \int_a^t g(s)\, ds\, dt.$$

Then $G''(x) = g(x)$ and hence also $SD_2\, G(x) = g(x)$ at almost every point of (a, b). By Theorem 8.25 the variational measure $VS_G^{(2)}$ associated with G also vanishes on every subset of (a, b) of measure zero. Write $H = f - G$. Then H is continuous, $SD_2\, H(x) = 0$ at almost every point of (a, b) and the variational measure $VS_H^{(2)}$ associated with H vanishes on every subset of (a, b) of measure zero. It follows from Theorem 8.26 that H is linear, say that $H(x) = cx + d$. Hence $f = G + H$ yields (8.38).

Corollary 8.37 *A real-valued function f may be expressed as the difference of two convex functions if and only if f is continuous and the measure $VS_f^{(2)}$ is finite on each bounded set.*

Proof. One direction is elementary. The converse follows from Theorem 8.36 by expressing $g = g_1 - g_2$ in terms of its positive and negative parts. Then with

$$G_i(x) = \int_a^x \int_a^t g_i(s)\, ds\, dt \quad (i = 1, 2)$$

we can express f as $G_1 - G_2$ plus some linear function. It follows that f is the difference of two convex functions.

8.3.9 Generalization of Schwarz Theorem.

The classical Schwarz theorem asserts that a continuous function must be linear if it has everywhere a vanishing second symmetric derivative. One can allow exceptional sets by adding hypotheses. The following theorem is a generalization of the Schwarz theorem that exactly expresses the allowed exceptional sets. This allows us to extend the material of Section 1.5.

Theorem 8.38 *Let f be continuous. A necessary and sufficient condition for f to be linear is that $SD_2\, f(x) = 0$ for $VS_f^{(2)}$-almost every point x.*

Proof. The necessity is obvious. The sufficiency follows easily too since, if $SD_2\, f(x) = 0$ for $VS_f^{(2)}$-almost every point x in a set E, then certainly $VS_f^{(2)}(E) = 0$. Thus the theorem reduces merely to the assertion of Theorem 8.26.

As a corollary we can quickly express some simpler assertions that imply that $SD_2\, f(x) = 0$ for $VS_f^{(2)}$-almost every x and so imply linearity.

Corollary 8.39 *Let f be continuous. Any one of the following conditions implies that f is linear:*

(i) $SD_2 f(x) = 0$ *everywhere.*

(ii) f *is smooth and* $SD_2 f(x) = 0$ *nearly everywhere.*

(ii) f *is smooth,* $SD_2 f(x) = 0$ *almost everywhere and*

$$-\infty < \underline{SD_2} f(x) \le \overline{SD_2} f(x) < +\infty$$

nearly everywhere.

8.4 A Theorem of Denjoy

Denjoy [68, p. 257] obtained the following deep variational property of continuous functions that possess finite second symmetric derivates. The proof is lengthy and detailed; a simpler and more transparent proof would be welcome. The conclusion of the theorem may, perhaps, also follow if it is assumed that F has an absolutely continuous, σ–finite second symmetric variational measure $VS_f^{(2)}$ on (a,b).

Theorem 8.40 (Denjoy) *Let* F *be continuous on* $[a,b]$ *and suppose that*

$$-\infty < \underline{SD_2} F(x) \le \overline{SD_2} F(x) < +\infty$$

at every point of (a,b). *Then* F *is ACG on* $[a,b]$.

8.5 An Example of Skvorcov

We do not have a complete characterization of the class of continuous functions that possess second order symmetric derivatives. We know that such functions are continuous, smooth and ACG. We know, too, that they are almost everywhere differentiable. In particular if $SD_2 F(x) = f(x)$ everywhere then F' exists almost everywhere, AD $F'(x) = f(x)$ almost everywhere and

$$F(x) - F(a) = \int_a^x F'(x)\,dx$$

where the integral is interpreted in a quite broad sense, in the sense of the Denjoy–Khintchine integral. We cannot reasonably expect this to be a Lebesgue integral but it could be that the integral may be interpreted in the narrower Denjoy-Perron sense. Skvorcov [267] shows that this is not so.

Theorem 8.41 (Skvorcov) *There is a continuous function* F *on* $[0,1]$ *that has a finite second symmetric derivative everywhere on* $(0,1)$ *and* F *is not* ACG_* *on* $[0,1]$. *Hence* F *is not the indefinite integral of its derivative in the Denjoy-Perron sense.*

9
Symmetric Integrals

9.1 Introduction

In this chapter we shall discuss in detail a variety of integrals based on the first and second order symmetric derivatives and on the approximate symmetric derivative. To some degree this parallels the story of integrals based on the ordinary derivative, but with interesting differences. We would prefer of the reader a familiarity with the Perron and Denjoy integrals as they are presented, for example, in Saks [251] although the account here is self contained and covers this background briefly.

The original Perron integral, from 1911, provides a formal solution to the differential equation

$$\frac{dy}{dx} = f(x)$$

if it is given that f has an exact primitive F. This method "constructs" the values $F(x) - F(a)$ from the values of f and so defines an indefinite integral $\int_a^x f(t)\, dt$ that is more general than the Lebesgue integral. (For example the derivative of the function $x^2 \sin x^{-2}$ is not Lebesgue integrable.) The procedure usually demands little more than an appropriate monotonicity theorem in order to be applicable to more general kinds of derivatives.

In contrast to Perron's approach which merely describes the features of the integral rather than developing a process leading to the values of the integral, Denjoy produced at the same time an equivalent integral by a transfinite sequence of well-defined extensions of the Lebesgue integral. Over the years many characterizations of this integral have been presented. The most interesting of these has been the characterization of the Denjoy–Perron integral in the 1950's as a limit of Riemann sums due, independently, to Kurzweil and Henstock. These equivalent formulations of the Denjoy–Perron integral are well known. The survey article of Bullen [25] contains an account. In Section 9.1.1 we provide a short briefing.

One can ask whether there is a parallel story for the various generalized derivatives. For example the approximate derivative allows a completely analogous development; although the technical details may differ, the plot line is familiar and the results are nearly identical. For the symmetric derivatives (first, second and approximate) the story is not well known and results in the literature are isolated and occasionally limited. Also, while certain themes recur in this development, many details differ and a number of problems remain. In this chapter we shall present an account of a variety of symmetric integrals arising directly from these symmetric derivatives.

Within the broad topic of "symmetric integrals" there is a variety that we have not discussed here. The best known to date has been the James integral based on the second symmetric derivative and defined by a Perron method. This appears in Zygmund [313] as the M_2–integral and so is widely known. There is a collection of closely related integrals that should be mentioned in the same context.

These integrals have arisen directly from the problem of defining an integral so that the sum function $f(x)$ of a convergent (or summable) trigonometric series

$$f(x) = a_0/2 + \sum_{n=1}^{\infty} (a_n \cos nx + b_n \sin nx)$$

is integrable in some sense and so that the coefficients, a_n and b_n, can be written as Fourier coefficients of the function f. This same problem motivates the symmetric integrals which are our main study too but the integrals we shall discuss are mostly different from and even incompatible with these other symmetric integrals.

For this other side of the symmetric integral story, based mainly on the second symmetric derivative and its variants, the reader may wish to consult [34], [35], [56], [67], [68], [138], [186], [280], [301] and [313, Vol. II, pp. 86–91] among others. For a survey of the literature prior to 1955 see [139] and [142]. For studies of the properties of and the interrelations between the several integrals which have been constructed see [24], [27], [28], [29], [32], [33], [47], [48], [49], [50], [51], [52], [53], [54], [79], [160], [177], [178], [209], [259], [261], [263], [262], [264], [265], [267], [268], [269], and [270]. An approximate version of the James integral is given in [266] where it is also shown that Denjoy's second order totalization does not invert approximate, second symmetric derivatives.

We have to chosen to present, in Section 9.4, only one integral based on the second symmetric derivative. This has the advantage (perhaps) of being definable as a generalized Riemann integral. All of the other references just cited employ either a Perron definition or (in [56] only) a variational definition.

9.1.1 Integrals from Derivatives

If $F' = f$ everywhere then F is determined, up to an additive constant, by f. How, given f, may F be found?

This problem has motivated integration theory since Newton. Lebesgue explicitly announced that it was his motivation for producing his integral. If f is bounded then the Lebesgue integral solves the problem, i.e. $F(x)$ is $\int_a^x f(t)\,dt$ plus some constant. But it is easy to see that any integral of the Lebesgue type must fail in general to invert derivatives. The example $F(x) = x^2 \sin(1/x^2)$ illustrates. Here the total area between the curve $y = F'(x)$ and the x–axis is infinite and so $\int_0^1 |F'(t)|\,dt = +\infty$. Since the Lebesgue integral of a function f does not exist unless the Lebesgue integral of $|f|$ also exists there is no hope that F can be recovered from F' by an integration in this sense.

What is needed is a *nonabsolutely convergent* integral, that is an integral such that $\int f$ may exist when $\int |f|$ fails to exist. Consequently, too, then one needs an entirely different method of integration to solve the derivative problem.

There are by now several different approaches that solve this problem and we propose to sketch each of them briefly. In order to gain some insight into the problem of inverting the various symmetric derivatives it is obviously necessary to know first how to invert ordinary derivatives.

9.1.1.1 Newton's Integral. One can solve the derivative inversion problem in a completely tautological way. We simply declare, with Newton, that

$$(N) \int_a^b f(t)\,dt = F(b) - F(a)$$

provided there exists a continuous function F on $[a, b]$ for which $F'(x) = f(x)$ for each $a < x < b$. Indeed most calculus students of this generation hardly see the "integral" in any other light.

Such a definition is justified by the fact that for any two functions F_1 and F_2 with $F_1' = F_2' = f$ there is the identity $F_1(b) - F_1(a) = F_2(b) - F_2(b)$. This shows the value of the integral does not depend on the particular primitive chosen to express it. Any Newton type integral for more general derivatives must allow some similar technical statement.

While this concept is useful for discussions of the problem it completely avoids the problem itself. In many ways, though, it is no more cavalier than Perron's method discussed below. We shall introduce several symmetric versions of the Newton integral as a first step in our development of the various symmetric integrals.

9.1.1.2 Denjoy's Totalization. Denjoy's method, first announced in two notes in the 1912 Comptes Rendues [66] and [64], is to construct a sequence of extensions of the Lebesgue integral. There are just two procedures needed which then must be applied one after the other in turn.

The first operation is due to Cauchy and taught in every calculus class under the unfortunate heading "improper integrals". If f is integrable on every subinterval $[c, d]$ of $[a, b]$ $(a < c < d < b)$ but not on $[a, b]$ itself then one can extend the integral to the larger interval by writing

$$\int_a^b f(x)\, dx = \lim_{s,t\to 0+} \int_{a+s}^{b-t} f(x)\, dx$$

if this limit exists.

The second operation is due to Harnack and is less well known. If f is integrable on a perfect set $P \subset [a, b]$ and also on every interval $[c_i, d_i]$ contiguous to P in $[a, b]$ but is not integrable on $[a, b]$ itself then one can extend the integral to the larger interval by writing

$$\int_a^b f(x)\, dx = \int_P f(x)\, dx + \sum_{i=1}^\infty \int_{c_i}^{d_i} f(x)\, dx.$$

(Some assumptions must be made on the integrals so that this sum is absolutely convergent and produces a continuous integral.)

The two operations are applied one after another in a transfinite, but countable, sequence to produce an integral that Denjoy called a *total*. The integral obtained includes the Lebesgue integral (because it starts with that integral) and inverts derivatives. The latter fact follows from elementary properties of derivatives. Denjoy showed, too, that given any countable ordinal η and interval $[a, b]$ there is a derivative F' so that the construction of the integral $\int_a^b F'(x)\, dx$ requires at least η steps to achieve. The best exposition of Denjoy's totalization process is given by Saks [251].

Denjoy sketched as well two totalization processes based on symmetric derivatives. The first [68] presents a long and complicated series of steps, called by him the $(T_{2.s})_0$ totalization process, needed to invert second order symmetric derivatives.

This process can also be used to invert the symmetric derivative of a continuous function. If $SD\, F(x) = f(x)$ is given with F continuous then the second order symmetric derivative of the integral G of F is f. Accordingly a possible strategy for determining F is to determine instead G from f (up to a linear function of course) and then differentiate G to obtain F (up to a constant). Zygmund pointed this out to Denjoy and asked whether a simpler kind of totalization could be constructed for this special problem. The $(T_{2.s})_0$ totalization process requires nine steps; Denjoy's answer is the first order symmetric totalization process, T_s totalization, that requires five operations. The latter process is briefly sketched in [69] and [70] and there seems to have been no further attention paid to these ideas in the literature. In particular it is not known what relation such an integral would have to the other integrals discussed in this chapter.

We shall not report on these ideas in this work. The reader will have to consult the original works of Denjoy cited.

9.1.1.3 Perron's "Method".

Perhaps the most common and best known approach is the method of Perron [219]. In spite of the unrestrained contempt that Denjoy has frequently directed at this method it remains useful. The Perron integral originates in a monotonicity theorem: if $\underline{D} F(x) \geq 0$ almost everywhere and $\underline{D} F(x) > -\infty$ everywhere then F is nondecreasing. One then defines G to be an upper function for f if $\underline{D} G(x) \geq f(x)$ almost everywhere and $\underline{D} G(x) > -\infty$ everywhere; H is a lower function for f if $-H$ is an upper function for $-f$. The monotonicity theorem shows that

$$G(b) - G(a) \geq H(b) - H(a) \tag{9.1}$$

for any such pair (G, H) of upper and lower functions and for any interval $[a, b]$. The Perron integral $\int_a^b f$ is then taken as the common bound of the major and minor functions in the inequality (9.1). It is evident that if $F' = f$ everywhere this procedure recovers F from f at least formally. (For details beyond this brief sketch see [251, Chapter VI].)

Evidently any kind of derivation process would sponsor a Perron-type integral provided only that a monotonicity theorem of some kind is available. Integrals based on certain symmetric derivatives have been introduced by Marcinkiewicz and Zygmund [186], James [138], Burkill [35], Taylor [280], Skvorcov [266] and Ponomarev [229], all following a Perron scheme.

We shall give Perron type characterizations of some of our symmetric integrals.

9.1.1.4 Lusin's Characterization.

A function f is Lebesgue integrable with an indefinite integral F if and only if $F' = f$ almost everywhere and F is absolutely continuous. Lusin obtained a characterization of the Denjoy-Perron integral by modifying this statement. If we enlarge the class of functions to which F may belong from the class of absolutely continuous functions to the class ACG$_*$ then we have a similar descriptive characterization of the more general integral. The class of functions that are ACG$_*$ can be best expressed in the language of measure theory (although the original definitions are differently formulated). Primitives of Lebesgue integrable functions correspond to measures that are finite and absolutely continuous. Functions that are ACG$_*$ correspond to measures that are σ–finite and absolutely continuous.

We shall give Lusin type characterizations of some of our symmetric integrals. The variational measures of Chapter 8 will be used to express symmetric versions of these ideas.

9.1.1.5 The Generalized Riemann Integral. To invert a derivative one might take a completely naive approach and start with the definition of the derivative itself. If $F' = f$ everywhere then, at each point ξ and for every $\epsilon > 0$, there is a $\delta > 0$ so that

$$|F(x'') - F(x') - f(\xi)(x'' - x')| < \epsilon(x'' - x') \qquad (9.2)$$

for $x' \leq \xi \leq x''$ and $0 < x'' - x' < \delta$.

We shall attempt to recover $F(b) - F(a)$ as a limit of Riemann sums for f. Let $a = x_0 < x_1 < x_2 \ldots < x_n = b$ be a partition of $[a, b]$ and let $\xi_i \in [x_{i-1}, x_i]$. Then

$$F(b) - F(a) = \sum_{i=1}^{n} F(x_{i-1}) - F(x_i) = \sum_{i=1}^{n} f(\xi_i)(x_i - x_{i-1}) + R$$

where, of course,

$$R = \sum_{i=1}^{n} \Big(F(x_i) - F(x_{i-1}) - f(\xi_i)(x_i - x_{i-1}) \Big).$$

Thus $F(b) - F(a)$ has been given as a Riemann sum for f plus some error term R. But it appears now that, if the partition is finer than δ so that (9.2) may be used, we have

$$|R| \leq \sum_{i=1}^{n} \Big| F(x_i) - F(x_{i-1}) - f(\xi_i)(x_i - x_{i-1}) \Big|$$

$$< \sum_{i=1}^{n} \epsilon(x_i - x_{i-1}) = \epsilon(b - a).$$

Evidently then *if there are no mistakes here* we have just proved that f is Riemann integrable and that $\int_a^b f(t)\, dt = F(b) - F(a)$.

This is false of course. Even the Lebesgue integral does not invert all derivatives. The error, not very skillfully concealed, is that the choice of δ depends on the point ξ considered and so is not a constant. But instead of abandoning the argument one can change the definition of the Riemann integral to allow a variable δ. The definition then changes to look like this: a function f is *generalized Riemann integrable* on $[a, b]$ with value I if for every ϵ there is a positive function δ on $[a, b]$ so that

$$\left| \sum_{i-1}^{n} f(\xi_i)(x_i - x_{i-1}) - I \right| < \epsilon$$

whenever $a = x_0 < x_1 < x_2 < \ldots < x_n = b$ is a partition of $[a, b]$ with $\xi_i \in [x_{i-1}, x_i]$ and $0 < x_i - x_{i-1} < \delta(\xi_i)$.

Of course to justify the definition requires knowing that such partitions actually exist for any such gauge δ; this is supplied by the Cousin Lemma of Section 3.1.

These entirely elementary observations are the basis for a definition of a Riemann type integral that is equivalent to the Denjoy–Perron integral. This was discovered independently by Henstock and Kurzweil and these ideas have led to a number of other integration theories. There have been definitions of symmetric integrals as limits of Riemann sums. Preiss and Thomson [232] produce an integral based on the approximate symmetric derivative in this way that integrates all everywhere convergent trigonometric series. Kurzweil and Jarník [168] study a similar integral related to the ordinary symmetric derivative.

We present Riemann type definitions of all our symmetric integrals. In each case the technical problem that must be surmounted is to supply the existence of the partitions used in the definition in the same way that the Cousin Lemma supplies the existence of ordinary partitions for any ordinary gauge. Chapter 3 already contains all of our partitioning tools.

9.1.1.6 Variational Definitions. The Denjoy-Perron integral allows a characterization as follows: the relation $F(b) - F(a) = \int_a^b f(x)\,dx$ holds in that sense if and only if for every $\epsilon > 0$ there is a gauge δ and a nondecreasing function η on $[a, b]$ with $\eta(b) - \eta(a) < \epsilon$ so that

$$|F(x + t) - F(x) - f(x)t| \le |\eta(x + t) - \eta(x)|$$

for $a < x < b$ and $0 < |t| < \delta(x)$. Note the similarity between this statement and the definition of the derivative itself expressed in (9.2).

This is commonly called a variational definition and the ideas trace back to work of Ward (see Henstock [131]). We present variational characterizations of some of our symmetric integrals.

9.2 The Ordinary Symmetric Integral

9.2.1 Preliminaries

The first monotonicity theorem for the symmetric derivative is due to Khintchine [151] (Theorem 1.4). It is entirely elementary and limited: *if f is continuous and $\underline{SD}\,f(x) \ge 0$ everywhere then f is nondecreasing.* On the basis of this, though, a simple version of a symmetric Perron integral may be defined using continuous major and minor functions (for example see Ponomarev [229]).

In order to develop an adequate theory of symmetric integrals, however, one needs instead the covering theorem supplied in Section 3.3.3. As this is the basis for most of our work let us restate it here. As is common in many such theories we call any positive function δ defined on a set E, *a gauge* on E. A partition of an interval $[c, d]$ of the form

$$c = x_0 < x_1 < x_2 < \ldots < x_{n-1} < x_n = d$$

is said to be *symmetrically finer* than δ if

$$x_i - x_{i-1} < \delta\left(\frac{x_{i-1} + x_i}{2}\right)$$

for each $i = 1, 2, \ldots, n$. The smallness of the partition is measured by the gauge at the center of each interval of the partition.

Lemma 9.1 *Let δ be a gauge on an interval (a, b). Then there is a countable set $C \subset (a, b)$ so that for every interval $[c, d]$ with $c, d \in (a, b) \setminus C$ there is a partition of $[c, d]$ that is symmetrically finer than δ.*

9.2.2 A Symmetric Newton Integral

The ordinary Newton or calculus integral is defined by the expression

$$\int_a^b F'(x)\, dx = F(b) - F(a).$$

To make this work it is enough that $F'(x)$ exist everywhere in (a, b) and that F is continuous at a and b. This "integral" inverts derivatives in a completely tautological manner; we can do much the same for symmetric derivatives.

   While a symmetric Newton integral is not likely to serve many useful purposes it can illustrate very clearly the ideas involved in symmetric integrals and show how the theory must evolve in a different manner than for the usual integrals on the real line. For this reason we start our discussion with this limited but curious integral. In this we are following in spirit Jeffrey [142, p. 27] and Mařík [196] who develop a Newton integral based on the second order symmetric derivative which quite admirably helps introduce the rather arcane P^2–integral of James.

   The use of the increment $\lambda_F(a, b)$ in Definition 9.2, rather than something simpler such as $F(b) - F(a)$, is motivated by a desire to retain a kind of symmetry at the endpoints, as well as to overcome the ambiguity arising from the fact that two very different functions may have the same symmetric derivative.

Definition 9.2 A finite function f is said to be (N_s^1)–*integrable* on an interval $[a, b]$ if there is a function F on (a, b) so that SD $F(x) = f(x)$ everywhere on (a, b) and such that the expression

$$\lambda_F(a, b) = \lim_{h \to 0} F(b - h) - F(a + h) \tag{9.3}$$

exists. We write then

$$\lambda_F(a, b) = (N_s^1) \int_a^b f(x)\, dx.$$

If SD $F(x) = f(x)$ everywhere on an interval (a, b) then, while F is not quite unique, the expression for $\lambda_F(a, b)$ in (9.3), if it exists, does not depend on the choice of F. In fact if SD $F(x) =$ SD $G(x)$ for each $x \in (a, b)$ then there is a countable set C such that $F - G$ is constant on $(a, b) \setminus C$ (this follows immediately from Lemma 9.1). If both $\lambda_F(a, b)$ and $\lambda_G(a, b)$ exist they must clearly be equal. This argument shows that the integral in Definition 9.2 is well defined.

Here are some elementary properties of the (N_s^1)–integral. The proof for the first theorem is immediate.

Theorem 9.3 *Let f and g be (N_s^1)–integrable on an interval $[a, b]$. Then so too is any linear combination $sf + tg$ and*

$$\int_a^b (sf(x) + tg(x))\, dx = s \int_a^b f(x)\, dx + t \int_a^b g(x)\, dx$$

in the sense of this integral.

The integral does not quite have the usual interval additivity properties one expects to encounter in integration theories. In part this comes from the symmetric nature of the integral and in part from the fact that conditions ensuring integrability on two abutting intervals may break down at the common point. A simple example illustrates one aspect of the problem. Let

$$F(x) = \frac{1}{\sqrt{1 - x^2}} \quad (-1 < x < 1). \tag{9.4}$$

Then, while the integral $(N_s^1) \int_{-1}^1 F'(t)\, dt = 0$ can be seen to exist, the function F' is not integrable on $[-1, c]$ for any value of $c \in (-1, 1)$. Of course the integral does exist on $[c, d]$ for any $-1 < c < d < 1$. The example well illustrates how the (N_s^1)–integral, and all of the symmetric integrals we shall discuss, exploit the symmetry at the points -1 and 1.

This integral is rather more fragile than most integrals. The function $f(x) = x^{-3}$ $(x \neq 0)$, $f(0) = 0$ is everywhere an exact symmetric derivative and so, in particular, the integral

$$(N_s^1) \int_{-1}^1 f(t)\, dt$$

exists. However a product fg may not be integrable in this or any symmetric sense. Even the integral

$$(N_s^1) \int_{-1}^1 t f(t)\, dt$$

fails to exist. This same feature will be found in all of the symmetric integrals in the sequel.

It is possible for the integral to exist on $[a, b]$ and on $[b, c]$ but not exist on the interval $[a, c]$. This is a common defect in Newton-type integrals and an example is easy to construct. If we assume integrability on the larger interval then additivity is easy enough to see as the next theorem shows.

Theorem 9.4 *If f is (N_s^1)–integrable on $[a, b]$ and on $[a, c]$ $(a < b < c)$ then it is integrable on $[b, c]$ and*

$$\int_b^c f(x)\, dx = \int_a^c f(x)\, dx - \int_a^b f(x)\, dx \qquad (9.5)$$

in the sense of this integral.

Proof. Let F be a symmetric primitive for f on $[a, c]$. We may assume that both $\lambda_F(a, b)$ and $\lambda_F(a, c)$ exist. Note that F is symmetrically continuous at b. From this and the identity

$$
\begin{aligned}
F(c - h) - F(b + h) \;=\;& (F(c - h) - F(a + h)) - (F(b - h) - F(a + h)) \\
& - (F(b + h) - F(b - h))
\end{aligned}
$$

it readily follows that

$$\lambda_F(b, c) = \lambda_F(a, c) - \lambda_F(a, b). \qquad (9.6)$$

The identity (9.6) supplies (9.5) and the theorem is proved.

We have already noted (using example (9.4)) that (N_s^1)–integrability on an interval $[a, b]$ does not require integrability on all subintervals or even integrability on *any* interval $[a, c]$ for $a < c < b$. We do have following though.

Theorem 9.5 *Suppose that f is (N_s^1)–integrable on an interval $[a, b]$ with F any particular (N_s^1)–primitive of f there. Then F is continuous everywhere in $[a, b]$ outside of some scattered set and*

$$(N_s^1) \int_c^d f(t)\, dt = F(d) - F(c) \qquad (9.7)$$

at any points $c, d \in (a, b)$ at which F is continuous.

Proof. Note first that F, being everywhere symmetrically differentiable in (a, b), must satisfy the hypotheses of Charzyński's Theorem (Theorem 2.9). Hence F is continuous everywhere in $[a, b]$ outside of some scattered set. If F is continuous at c and d, then certainly $\lambda_F(c, d) = F(d) - F(c)$. This supplies the identity (9.7) and the theorem follows.

The (N_s^1)–integral is compatible with the Denjoy-Perron integral as the following result of Ponomarev [229, Theorem 3, p. 122] shows. The proof we shall give is rather less complicated than that in [229] although it retains some of the ideas.

Theorem 9.6 (Ponomarev) *Suppose that f is (N_s^1)–integrable on an interval $[a, b]$ and also Denjoy-Perron integrable there. If F is an (N_s^1)–primitive of f there then F agrees in $[a, b]$ with an ACG_* function G except on a scattered set and G is also an (N_s^1)–primitive of f there. In this case*

$$(N_s^1) \int_a^b f(t)\, dt = \int_a^b f(t)\, dt = G(b) - G(a).$$

Proof. We have SD $F(x) = f(x)$, $a < x < b$ and f is Denjoy-Perron integrable on $[a, b]$. Let $G(x) = \int_a^x f(t)\, dt$.

We may choose sequences of Perron major and minor functions U_n, L_n with $U_n(a) = L_n(a) = 0$, $U_n(b) - L_n(b) < n^{-1}$, $U_n \to G$ and $L_n \to G$ uniformly on $[a, b]$, and $\underline{D}\, U_n(x) \geq f(x) \geq \overline{D}\, L_n(x)$ for all $a < x < b$. It follows that

$$\mathrm{SD}\, (U_n(x) - F(x)) \geq \underline{D}\, U_n(x) - \mathrm{SD}\, F(x) \geq 0$$

and

$$\mathrm{SD}\, (F(x) - L_n(x)) \geq \mathrm{SD}\, F(x) - \overline{D}\, U_n(x) \geq 0$$

at each point of the interval (a, b). It follows readily from Lemma 9.1 that both functions $U_n - F$ and $F - L_n$ are nondecreasing off a countable set. From this we can conclude that $F = G$ off a countable set. By Theorem 9.5 the function F is continuous off a scattered set S. Since G is continuous we have the agreement of $F(x)$ and $G(x)$ everywhere off this same scattered set S.

It remains only to prove that SD $F(x) = $ SD $G(x)$ everywhere. Fix $x \in (a, b)$, $\epsilon > 0$ and choose $\delta > 0$ so that

$$|F(x + h) - F(x - h) - 2f(x)h| < \epsilon h/4 \tag{9.8}$$

for all $0 < h < \delta$. Fix any $h < \delta$. Choose $t > 0$ so that $x + h + t \notin S$, $x - h - t \notin S$, $h + t < \delta$, $2|f(x)|t < \epsilon h/4$,

$$|G(x + h) - F(x + h + t)| < \epsilon h/4 \tag{9.9}$$

and

$$|G(x - h) - F(x - h - t)| < \epsilon h/4. \tag{9.10}$$

Such a choice of t is possible since S is scattered (in particular nowhere dense), G is continuous and $F = G$ off S.

Together (9.8), (9.9) and (9.10) supply

$$|G(x + h) - G(x - h) - 2f(x)h| < \epsilon h$$

for all $0 < h < \delta$ and we have proved that $\operatorname{SD} F(x) = \operatorname{SD} G(x)$ everywhere as required.

In regards to this last theorem we should mention too another result of Ponomarev [228] showing that the (N_s^1)–integral is not included in the Denjoy-Perron integral even if the primitive function is continuous. We present this without proof.

Theorem 9.7 (Ponomarev) *There is a continuous function F with a symmetric derivative at every point of $[0, 1]$, such that the derivative $f(x) = \operatorname{SD} F(x)$ is not Denjoy-Perron integrable there. In particular f has a continuous (N_s^1)–primitive and $\int_0^1 f(t)\, dt$ exists in the (N_s^1)–sense but not in the Denjoy-Perron sense.*

Finally we present two theorems that arise often in the setting of nonabsolutely convergent integrals showing that such integrals are closely related to the Lebesgue integral. The first asserts that the integral is merely the Lebesgue integral when applied to nonnegative functions and the second that an integrable function must be Lebesgue integrable on a dense set of intervals.

Theorem 9.8 *Suppose that f is nonnegative and (N_s^1)–integrable on an interval $[a, b]$. Then f is Lebesgue integrable on $[a, b]$.*

Proof. If f has a continuous (N_s^1)–primitive F then an elementary monotonicity theorem shows that F is nondecreasing. By the Khintchine Theorem $F' = f$ almost everywhere and so f, as the a.e. derivative of a monotonic function, is Lebesgue integrable. Even if the primitive F is not continuous, it follows easily from Lemma 9.1 that it is equivalent to a nondecreasing function. Again f must be Lebesgue integrable since it is the a.e. derivative of an essentially monotonic function.

Theorem 9.9 *Suppose that f is (N_s^1)–integrable on an interval $[a, b]$. Then there is a set of intervals dense in $[a, b]$ on each of which f is Lebesgue integrable.*

Proof. If there is a *continuous* symmetrically differentiable function F on $[a, b]$ such that

$$\operatorname{SD} F(x) = f(x) \quad (a < x < b)$$

then obviously f is in the first Baire class and hence bounded on a set of intervals dense in $[a, b]$. On each of these intervals then f is clearly Lebesgue integrable. Thus this case is elementary. But in general it is known that *any* exact symmetric derivative is Baire 1 (Theorem 7.18) and so the same conclusion holds.

9.2.3 An Elementary Application to Trigonometric Series

We show that in very simple situations a symmetric integral may be used to solve the coefficient problem for trigonometric series when the conventional integrals fail to exist. Of course the full dress treatment of this problem requires a much more general integral; we shall return to this problem repeatedly in the sequel in Sections 9.2.3, 9.2.7, 9.3.3, 9.3.8 and 9.4.5. The observation in the final sentence of this theorem is due to Denjoy. Note as a special feature of this integral that, even though the function f has singularities at 0 and 2π, it is integrable over *any* period; it is this feature that the symmetry in Definition 9.2 is designed to produce.

Theorem 9.10 *Suppose that $\{b_n\}$ is a decreasing sequence of real numbers with $\lim_{n\to\infty} b_n = 0$. Then the trigonometric series*

$$\sum_{k=1}^{\infty} b_k \sin kx$$

converges everywhere to a finite value $f(x)$, f is (N_s^1)–integrable on any interval of length 2π and the series is the Fourier series for f in the sense of this integral, i.e. for each n,

$$(N_s^1) \int_0^{2\pi} f(t) \sin nt \, dt = \pi b_n \quad and \quad (N_s^1) \int_0^{2\pi} f(t) \cos nt \, dt = 0.$$

If $\sum_{n=1}^{\infty} b_n/n = \infty$ then f is not integrable in the senses of Riemann, Lebesgue or Denjoy-Perron.

Proof. Using elementary convergence theorems we see that the series for f converges everywhere, and converges uniformly on any closed interval that contains no point $0, \pm\pi, \pm2\pi, \ldots$. The formally integrated series converges at every point, for the same reason, except possibly at the points $0, \pm2\pi, \pm4\pi, \ldots$. If $\sum_{n=1}^{\infty} b_n/n = \infty$ then certainly there is divergence at these points.

If we let F denote the sum of the integrated series (perhaps not defined at the points $0, \pm2\pi, \pm4\pi, \ldots$) then $F' = f$ everywhere except at these exceptional points, but SD $F(x) = f(x)$ at every point without exception.

Since $F(2\pi - h) = F(-h)$ and F is symmetrically continuous at 0 we easily compute the expression

$$\lambda_F(0, 2\pi) = \lim_{h\to 0} F(2\pi - h) - F(h) = 0. \tag{9.11}$$

Consequently f has a vanishing (N_s^1)–integral on $[0, 2\pi]$. If $\sum_{n=1}^{\infty} b_n/n = \infty$ then F may be shown to be unbounded and cannot be also an indefinite integral of f in the senses of Riemann, Lebesgue or Perron. (The condition $\sum_{n=1}^{\infty} b_n/n < \infty$ is both necessary and sufficient in order that f be Lebesgue integrable; cf. [313, Vol. I, p. 185].)

To obtain a formula for the remaining coefficients is elementary. The function $f(x) \sin kx$ is continuous (see [313, Vol. I, p. 185]) and so its integral exists in any sense and can be easily computed. The function $f(x) \cos kx$ is odd and it has an even symmetric primitive G with $\lambda_G(0, 2\pi) = \lambda_G(-\pi, \pi) = 0$.

For a specific and familiar example of a trigonometric series that is a Fourier series in the (N_s^1)–sense but not in a more conventional sense (Lebesgue or Denjoy-Perron) take

$$f(x) = \sum_{n=1}^{\infty} \frac{\sin nx}{\log(n+1)}.$$

It was Fatou who first presented this example as an instance of an everywhere convergent trigonometric series whose sum function is not Lebesgue integrable. Of course in this case (and in the theorem in general) the series is a generalized Fourier sine series in the sense that the integrals $\int_0^{2\pi} f(t) \sin nt \, dt$ all exist as Lebesgue integrals even though f itself is not Lebesgue integrable.

9.2.4 A Symmetric Totalization

Recall that a finite function f is (N_s^1)–*integrable* on an interval $[a, b]$ if there is a function F on (a, b) so that SD $F(x) = f(x)$ everywhere on (a, b) and such that $\lambda_F(a, b) = \lim_{h \to 0} F(b - h) - F(a + h)$ exists. This number is then taken as the value of the integral.

The (N_s^1)–integral is *descriptive* in the sense that a situation is described for the value of the integral but no explicit procedure is given to determine that value. If f has a continuous symmetric primitive can we determine the integral in some constructive manner? One might give a mock process as follows. Suppose that f is (N_s^1)–integrable on an interval $[a, b]$. List all continuous functions $\{F_1, F_2, F_3, \ldots\}$ in a transfinite sequence. One at a time for each ordinal α determine if SD $F_\alpha(x) = f(x)$ everywhere on (a, b). At some ordinal the process stops and presumably the value $F_\alpha(b) - F_\alpha(a)$ is at hand. Unfortunately the transfinite sequence is rather long and the checking at each stage is also lengthy. The goal is to produce such a procedure with a *countable* transfinite process, at each stage of which a *countable* number of operations suffice to carry on the process.

We now develop such a process. The details are due to Freiling. Let there be given a function f about which we are assured that there is a continuous function F with SD $F(x) = f(x)$ everywhere on (a, b). Let $\epsilon > 0$. We shall determine transfinite sequences $\{G_\alpha\}$, $\{S_\alpha\}$, $\{R_\alpha\}$ such that $\{G_\alpha\}$ is an expanding sequence of open sets, each S_α is the (countable) set of endpoints of components of sets G_β for $\beta \leq \alpha$ and each R_α is the (countable) set of points which are at a rational distance from a point in S_α. For each ordinal α and for each pair of points c, d ($c < d$) with $c, d \in R_\alpha$, $(c, d) \subset G_\alpha$ a number $E(c, d)$ is determined satisfying the following property:

$$|F(d) - F(c) - E(c,d)| < \epsilon(d-c). \tag{9.12}$$

As the process will stop at some countable ordinal (by the Cantor-Baire stationary principle) we will have detemined an approximation to the increment $F(y) - F(x)$ on a dense subset of (a,b). This gives an estimate of $F(b) - F(a)$ with an error of less than $\epsilon(b-a)$. The precise value of $F(b) - F(a)$ can then be detemined by repeating the procedure for a sequence $\epsilon_n \searrow 0$.

Here is the construction, given inductively.

At the first stage: Set $G_1 = \emptyset$ and then there is nothing more to be determined.

At the limit stages: Suppose that α is a limit ordinal and that the process has been carried out successfully to this stage. Set $G_\alpha = \bigcup_{\lambda < \alpha} G_\lambda$. The sets S_α and R_α are then detemined.

Let c, d $(c < d)$ with $c, d \in R_\alpha$, $(c,d) \subset G_\alpha$. By a compactness argument we can determine points

$$c < x_0 < x_1 < \ldots < x_n < d$$

so that each interval $[x_i, x_{i+1}]$ is contained in some G_{α_i} with $\alpha_i < \alpha$, so x_i, x_{i+1} are in R_{α_i} and with x_0 and x_n arbitrarily close to the endpoints c and d.

We let $E(c,d)$ be any limit point of the sums $\sum_{i=1}^n E(x_i, x_{i+1})$ as above where we fix the x_i for $1 \le i \le n-1$ let $x_0 \to c$ and $x_n \to d$. As F is continuous at c and d and the previous approximations have

$$|F(x_{i+1}) - F(x_i) - E(x_i, x_{i+1})| < \epsilon(x_{i+1} - x_i)$$

we will have $|F(d) - F(c) - E(c,d)| < \epsilon(d-c)$ giving the required condition (9.12).

At the stage $\alpha+1$: If G_α is not dense in (a,b) then, since f is Baire 1, we may detemine a rational p and an interval A disjoint from G_α and with endpoints in R_α so that

$$p - \epsilon < f(x) < p + \epsilon$$

for every $x \in A$. The estimate $E(c,d) = p(d-c)$ will satisfy

$$|F(d) - F(c) - E(c,d)| < \epsilon(d-c)$$

for any $(c,d) \subset A$ and thus if we set $G_{\alpha+1} = G_\alpha \cup A$ we can easily determine $E(c,d)$ for those c, d required.

Suppose instead that G_α is dense in (a,b) but not equal to (a,b). Again since f is Baire 1, we may detemine a rational p and an interval A_1 so that

$$p - \epsilon/6 < f(x) < p + \epsilon/6 \tag{9.13}$$

for every $x \in A_1 \setminus G_\alpha$.

We pause from the constructive process to establish some properties of subintervals of A_1.

Let O be an arbitrary open set containing $A_1 \setminus G_\alpha$; we can choose this so that the measure $|O \cap G_\alpha|$ is arbitrarily small. Define a gauge δ on A_1 as follows. For $x \in G_\alpha$, $\delta(x) = \text{dist}(x, A_1 \setminus G_\alpha)$. For $x \in A_1 \setminus G_\alpha$ determine $\delta(x)$ so that $\delta(x) < \text{dist}(x, A_1 \setminus O)$ and such that if $0 < t < \delta(x)$ then

$$|F(x+t) - F(x-t) - 2f(x)t| < \epsilon t/6. \tag{9.14}$$

This latter just uses the fact that $SD\, F(x) = f(x)$ everywhere on (a, b). Note that, because of (9.13) and (9.14),

$$|F(x+t) - F(x-t) - 2pt| < \epsilon t/2 \tag{9.15}$$

for such points.

Consider an interval $[c, d] \subset A_1$ with endpoints in $G_\alpha \cap R_\alpha$. A partition $\mathcal{P}$ of $[c, d]$ into subintervals is said to be *admissible* if the endpoints of all intervals used are in $R_\alpha \cap G_\alpha$ and each interval $I \in \mathcal{P}$ is of one or other of the following types:

(*i*) $I \subset G_\alpha$, or

(*ii*) I is centered at an endpoint of a component of G_α and has rational radius.

Note that there are only countably many admissible partitions.

For an admissible partition $\mathcal{P}$ write $\mathcal{P}_1$ for the collection of intervals in $\mathcal{P}$ of type (*i*) and $\mathcal{P}_2$ for the collection of intervals of type (*ii*). Write also

$$S(\mathcal{P}) = \sum_{I \in \mathcal{P}_1} E(I) + \sum_{I \in \mathcal{P}_2} p|I|.$$

By a Baire category argument there must exist a subinterval $A_2 \subset A_1$ and a positive number κ so that $|A_2| < \kappa$ and the set of points

$$\{x \in A_1 \setminus G_\alpha : \delta(x) > \kappa\} \tag{9.16}$$

is dense in $A_2 \setminus G_\alpha$. (Note, however, that we cannot *construct* this interval because the gauge δ is not available to us in advance.) If $\mathcal{P}$ is an admissible partition of a subinterval J of A_2 then note that every type (*ii*) subinterval of $\mathcal{P}$ has a center c in $A_2 \setminus G_\alpha$ and is contained in the open set O.

Now for $I \in \mathcal{P}_1$ we have $I \subset G_\alpha$ and $|F(I) - E(I)| < \epsilon|I|$. Consider an interval $I \in \mathcal{P}_2$ which we may write as $I = [c - t, c + t]$ where $c \in A_2 \setminus G_\alpha$. There are points c' in the set (9.16) arbitrarily close to the center c of I at which $\delta(c') > |A_2|$; for these points, because of (9.15),

$$|F(c'+t) - F(c'-t) - 2pt| < \epsilon t/2 \tag{9.17}$$

But F is continuous and so, letting $c' \to c$ in (9.17), we have for each $I \in \mathcal{P}_2$ that

$$\left|F(I) - p|I|\right| < \tfrac{1}{4}\epsilon|I|. \tag{9.18}$$

Consequently

$$|S(\mathcal{P}) - F(J)| < \sum_{I \in \mathcal{P}_1} \epsilon|I| + \sum_{I \in \mathcal{P}_2} \tfrac{1}{4}\epsilon|I|.$$

In the last sum we write $|I| = |I \cap G_\alpha| + |I \setminus G_\alpha|$ and hence deduce that for such partitions

$$|S(\mathcal{P}) - F(J)| < \epsilon|J \cap G_\alpha| + \tfrac{1}{4}\epsilon|J \setminus G_\alpha|. \tag{9.19}$$

Consider now an arbitrary subinterval A of A_1 with endpoints in $R_\alpha \cap G_\alpha$. We wish to find an admissible partition $\mathcal{P}$ of A in such a way that (9.19) must hold. By an application of the symmetric covering theorem from Section 3.3.4 with the gauge δ there is a δ–fine partition $\mathcal{P}$ of A. By the way that δ has been defined each interval of the partition is either entirely in G_α or else has a midpoint in $A \setminus G_\alpha$, is δ–fine there at the center and is contained in O. Note that for the latter intervals then the estimate (9.15) can be used; that is if I is such an interval then $|F(I) - p|I|| < \tfrac{1}{4}\epsilon|I|$ as before. Since G_α is open and dense and F is continuous we may adjust the partition slightly (again as in Section 3.3.4) without affecting this inequality so that all the endpoints of the partition are in G_α.

We need to make a finite number of further adjustments to $\mathcal{P}$ to make it admissible. Suppose that $[u, v]$ is a member of $\mathcal{P}$ centered at a point c in $A \setminus G_\alpha$. Recall that $u, v \in G_\alpha$.

Arbitrarily close to c are points $c' \in S_\alpha$ and we can select a rational r so that $u < c' - r < c' + r < v$ with $[u, c' - r]$ and $[c' + r, v]$ contained in G_α. Since F is continuous at u and v we can choose c' and r in such a way that if $I = [c' - r, c' + r]$ then once again the inequality $|F(I) - p|I|| < \tfrac{1}{4}\epsilon|I|$ continues to hold. In this way the three intervals $[u, c' - r]$, $[c' - r, c' + r]$ and $[c' + r, v]$ replace $[u, v]$ and the partition $\mathcal{P}$ is amended. We do this with each such interval. All intervals now are of type (ii) and satisfy our inequality or else are subsets of G_α. These latter may not have endpoints in $G_\alpha \cap R_\alpha$ and so we make the final adjustment by moving the offending endpoints slightly so that these new intervals remain in G_α. This can be done without affecting the type (ii) intervals since each of these have endpoints in $G_\alpha \cap R_\alpha$ already. Thus by a finite number of adjustments to $\mathcal{P}$ we arrive at an admissible partition $\mathcal{Q}$ of A for which the type (ii) intervals satisfy the inequality (9.15). Consequently the analogue of (9.19) holds, namely that

$$|S(\mathcal{Q}) - F(A)| < \epsilon|A \cap G_\alpha| + \tfrac{1}{4}\epsilon|A \setminus G_\alpha|. \tag{9.20}$$

A fact we shall also need later is, for this admissible partition $\mathcal{Q}$ of A, that $I \cap G_\alpha$ is a subset of $O \cap G_\alpha$ for each $I \in \mathcal{Q}_2$; recall that $O \cap G_\alpha$ can be specified to have arbitrarily small measure.

We are now ready to define $G_{\alpha+1}$. Choose an interval $A'_2 \subset A_1$ with endpoints in $G_\alpha \cap R_\alpha$ in such a way that A'_2 meets the complement of G_α and so that the following condition is met: for every type (ii) subinterval J of A'_2 and every admissible partition $\mathcal{P}$ of J

$$\left| \sum_{I \in \mathcal{P}_1} E(I) - p|I| \right| < \tfrac{5}{4}\epsilon |J \cap G_\alpha| + \tfrac{1}{2}\epsilon |J \setminus G_\alpha|. \tag{9.21}$$

We then set $G_{\alpha+1} = G_\alpha \cup A'_2$.

The existence of such an interval A'_2 was guaranteed by our exhibition of the interval A_2 which evidently has this property. Indeed if $J \subset A_2$ is of type (ii) and $\mathcal{P}$ is an admissible partition of J then use (9.18) and (9.19) to obtain

$$\left| S(\mathcal{P}) - p|J| \right| \leq |S(\mathcal{P}) - F(J)| + \left| F(J) - p|J| \right|$$

$$< \epsilon |J \cap G_\alpha| + \tfrac{1}{4}\epsilon |J \setminus G_\alpha| + \tfrac{1}{4}\epsilon |J|.$$

and hence deduce that

$$\left| p|J| - S(\mathcal{P}) \right| < \tfrac{5}{4}\epsilon |J \cap G_\alpha| + \tfrac{1}{2}\epsilon |J \setminus G_\alpha|. \tag{9.22}$$

But $|J| = \sum_{I \in \mathcal{P}} |I|$ and $S(\mathcal{P}) = \sum_{I \in \mathcal{P}_1} E(I) + \sum_{I \in \mathcal{P}_2} p|I|$ and we see that (9.22) is equivalent to (9.21). Consequently the determination of the interval A'_2 can be made in a countable number of steps by testing (9.21).

It remains only to determine $E(J)$ for all appropriate subintervals J of A'_2. Let $\mathcal{P}$ be an admissible partition of J. Each interval $I \in \mathcal{P}_1$ has $|F(I) - E(I)| < \epsilon |I|$ with $I \subset G_\alpha$. For each interval $I \in \mathcal{P}_2$ we know that there exists a further admissible partition $\mathcal{Q}$ of I with $|F(I) - S(\mathcal{Q})|$ satisfying the inequality (9.20). But also $|p|I| - S(\mathcal{Q})|$ satisfies the inequality (9.22) by definition of our interval A'_2. This gives for $I \in \mathcal{P}_2$ that

$$\left| F(I) - p|I| \right| < \tfrac{9}{4}\epsilon |I \cap G_\alpha| + \tfrac{3}{4}\epsilon |I \setminus G_\alpha|.$$

Consequently summing over I in $\mathcal{P}_1$ and $\mathcal{P}_2$ separately we obtain

$$|S(\mathcal{P}) - F(J)| < \epsilon |J \cap G_\alpha| + \epsilon |J \setminus G_\alpha| + \tfrac{9}{4}\epsilon \sum_{I \in \mathcal{P}_2} |I \cap G_\alpha|. \tag{9.23}$$

Thus $S(\mathcal{P})$ is within $\epsilon |J|$ of $F(J)$ except for the extra error

$$\tfrac{9}{4}\epsilon \sum_{I \in \mathcal{P}_2} |I \cap G_\alpha|.$$

As remarked earilier in the proof we can choose our admissible partition $\mathcal{P}$ so that the type (ii) intervals cover as little in measure of G_α as we wish so that this sum becomes arbitrarily small. Thus by taking a limit point of a sequence of such partitions we can arrive at an estimate $E(J)$ that is within $\epsilon |J|$ of $F(d) - F(c)$.

These procedures now completely describe the process inductively and the construction is done.

9.2.5 A Symmetric Variational Integral

The (N_s^1)–integral is defined by the relation SD $F(x) = f(x)$ for $a < x < b$. This can be written as the requirement that for every $\epsilon > 0$ there is a gauge δ so that

$$|F(x + t) - F(x - t) - 2f(x)t| < 2t\epsilon \qquad (9.24)$$

for $a < x < b$ and $0 < t < \delta(x)$.

The classical Denjoy-Perron integral permits a variational characterization (see Section 9.1.1.6) that is reminiscent of this. The relation $F(b) - F(a) = \int_a^b f(x)\,dx$ holds in that sense if and only if for every $\epsilon > 0$ there is a gauge δ and a nondecreasing function η on $[a, b]$ with $\eta(b) - \eta(a) < \epsilon$ so that

$$|F(x + t) - F(x) - f(x)t| \leq |\eta(x + t) - \eta(x)| \qquad (9.25)$$

for $a < x < b$ and $0 < |t| < \delta(x)$. Putting (9.24) and (9.25) together in a natural manner leads to a variational integral that is immediately more general than both the Denjoy-Perron integral and the (N_s^1)–integral.

Definition 9.11 Let f be defined on an interval (a, b). We say that f is (V_s^1)–integrable in $[a, b]$ if there is a function F defined on (a, b) such that $\lambda_F(a, b)$ exists and with the property that for every $\epsilon > 0$ there is a gauge δ and a nondecreasing function τ on $[a, b]$ with $\tau(b) - \tau(a) < \epsilon$ and such that

$$|F(x + h) - F(x - h) - 2f(x)h| \leq \tau(x + h) - \tau(x - h) \qquad (9.26)$$

for all $a < x < b, 0 < h < \delta(x)$. In this case we write

$$(V_s^1) \int_a^b f(x)\,dx = \lambda_F(a, b).$$

We can drop the reference to (V_s^1) when it is apparent which integral is intended. We shall refer to any function F satisfying the properties here as a (V_s^1)–primitive for f. It is obvious, in view of the considerations in (9.24) and (9.25), that this integral extends both the (N_s^1) and Denjoy-Perron integrals.

The definition requires a short argument to justify that the value of the integral does not depend on the choice of function F; this is supplied by the following lemma.

Lemma 9.12 *Suppose that the functions F and G are defined on (a, b) and that for every $\epsilon > 0$ there is a gauge δ and a nondecreasing function τ on $[a, b]$ with $\tau(b) - \tau(a) < \epsilon$ and such that*

$$|F(x + h) - F(x - h) - (G(x + h) - G(x - h))| \leq \tau(x + h) - \tau(x - h) \quad (9.27)$$

for all $a < x < b$, $0 < h < \delta(x)$. Then, if both $\lambda_F(a,b)$ and $\lambda_G(a,b)$ exist,
$\lambda_F(a,b) = \lambda_G(a,b)$.

Proof. We apply (9.27), with $\epsilon = 1/n$ for each integer n, to obtain sequences τ_n and δ_n with $\tau_n(b) - \tau_n(a) < 1/n$. By Lemma 9.1 there is a countable set C so that for every c, $d \in (a,b) \setminus C$ and each n there is a partition

$$c = x_0 < x_1 < \ldots < x_m < x_{m+1} = d$$

with $x_{i+1} - x_i < \delta_n((x_i + x_{i+1})/2)$, $(i = 0, 1, \ldots m)$. From this it follows that

$$|F(d) - F(c) - (G(d) - G(c))| < 1/n$$

for each n and hence that $F(d) - F(c) = G(d) - G(c)$, for c and d outside of this countable set. It is now clear that, if both $\lambda_F(a,b)$ and $\lambda_G(a,b)$ exist, $\lambda_F(a,b) = \lambda_G(a,b)$.

This integral is as fragile as the (N_s^1) integral; see the discussion in Section 9.2.2. The example (9.4) there illustrates too that the (V_s^1)–integral lacks the usual additivity properties. As before take $F(x) = (1 - x^2)^{-\frac{1}{2}}$ $(-1 < x < 1)$ and extend the function F so as to be zero everywhere outside of $[-1, 1]$. The integral $(V_s^1) \int_{-1}^{1} F'(t)\, dt = 0$ exists, but F' is not integrable on $[-1, c]$ for any value of $c \in (-1, 1)$. Here too F' is (V_s^1)–integrable on each of the intervals $[-1, 1]$ and $[1, 2]$ but not integrable on $[-1, 2]$. Thus it is possible for the (V_s^1)–integral to exist on $[a, b]$ and on $[b, c]$ but not exist on the interval $[a, c]$.

If the conditions exist for a "symmetric" join then the additivity is available as the next theorem shows.

Theorem 9.13 *Let $a < b < c$ and suppose that f is (V_s^1)–integrable on $[a, b]$ with a primitive F and on $[b, c]$ with a primitive G. Then f is (V_s^1)–integrable on $[a, c]$ if the limit*

$$\lim_{h \to 0+} G(b + h) - F(b - h) = \alpha \tag{9.28}$$

exists.

Proof. Let $\epsilon > 0$. Since f is (V_s^1)–integrable in $[a, b]$, $\lambda_F(a, b)$ exists and there is a gauge δ_1 on (a, b) and a nondecreasing function τ_1 on $[a, b]$ with $\tau_1(b) - \tau_1(a) < \epsilon/3$ and such that

$$|F(x + h) - F(x - h) - 2f(x)h| \leq \tau_1(x + h) - \tau_1(x - h) \tag{9.29}$$

for all $a < x < b$, $0 < h < \delta_1(x)$. Similarly, since f is (V_s^1)–integrable in $[b, c]$, $\lambda_G(b, c)$ exists and there is a gauge δ_2 on (b, c) and a nondecreasing function τ_2 on $[b, c]$ with $\tau_2(c) - \tau_2(b) < \epsilon/3$ and such that

$$|G(x + h) - G(x - h) - 2f(x)h| \leq \tau_2(x + h) - \tau_2(x - h) \tag{9.30}$$

for all $b < x < c, 0 < h < \delta_2(x)$. Let τ_1 and τ_2 be constant otherwise (i.e. outside these intervals).

Define $\delta(x) = \min\{\delta_1(x), b - x\}$ for $a < x < b$, $\delta(x) = \min\{\delta_2(x), x - b\}$ for $b < x < c$ and $\delta(b)$ so that

$$|G(b + h) - F(b - h) - \alpha| < \epsilon/6 \quad \text{and} \quad |2f(b)h| < \epsilon/6 \qquad (9.31)$$

for $0 < h < \delta(b)$.

For $a < x < b$ write $H(x) = F(x)$ and for $b < x < c$ write $H(x) = G(x) - \alpha$ where α is as in (9.28); take $H(b)$ arbitrary. Construct a nondecreasing function τ_3 constant on $[a, b]$ and on $[b, c]$ with $\tau_3(b+) - \tau_3(b-) = \epsilon/3$. Write $\tau = \tau_1 + \tau_2 + \tau_3$.

Note that $\lambda_H(a, c)$ exists and that τ is a nondecreasing function on $[a, c]$ with $\tau(c) - \tau(a) < \epsilon$. We readily verify, using (9.29), (9.30) and (9.31), that

$$|H(x + h) - H(x - h) - 2f(x)h| \le \tau(x + h) - \tau(x - h) \qquad (9.32)$$

for all $a < x < c, 0 < h < \delta(x)$. By definition then f is (V_s^1)–integrable in $[a, b]$ and the theorem is proved.

Following now are some properties of the (V_s^1)–integral that distinguish it from the (N_s^1)–integral. The first allows us to define an integral of a function that is defined only almost everywhere on an interval.

Theorem 9.14 *Suppose that f is (V_s^1)–integrable on an interval $[a, b]$ and that $f = g$ almost everywhere. Then g is (V_s^1)–integrable on $[a, b]$ and*

$$\int_a^b f(x) \, dx = \int_a^b g(x) \, dx.$$

Proof. Let $\epsilon > 0$. Since f is (V_s^1)–integrable in $[a, b]$ there is an F so that $\lambda_F(a, b)$ exists and there is a gauge δ_1 on (a, b) and a nondecreasing function τ_1 on $[a, b]$ with $\tau_1(b) - \tau_1(a) < \epsilon/2$ and such that

$$|F(x + h) - F(x - h) - 2f(x)h| \le \tau_1(x + h) - \tau_1(x - h) \qquad (9.33)$$

for all $a < x < b, 0 < h < \delta_1(x)$. Choose a nondecreasing function τ_2 on $[a, b]$ so that $\tau_2(b) - \tau_2(a) < \epsilon/2$ and so that $\tau_2'(x) = +\infty$ at every point at which $f(x) \ne g(x)$.

Define $\delta(x) = \delta_1(x)$ at every point x at which $f(x) = g(x)$ and at every point x at which $f(x) \ne g(x)$ choose $\delta(x) < \delta_1(x)$ so that

$$\frac{\tau_2(x + h) - \tau_2(x - h)}{2h} \ge |f(x) - g(x)| \qquad (9.34)$$

for $0 < h < \delta(x)$. Write $\tau = \tau_1 + \tau_2$.

Note that τ is a nondecreasing function on $[a, b]$ and that $\tau(b) - \tau(a) < \epsilon$. We readily verify, using (9.33) and (9.34), that

$$|F(x + h) - F(x - h) - 2g(x)h| \leq \tau(x + h) - \tau(x - h) \qquad (9.35)$$

for all $a < x < c$, $0 < h < \delta(x)$. By definition then (9.35) supplies the fact that g is (V_s^1)–integrable in $[a, b]$ and that the integrals have the same value (i.e. $\lambda_F(a, b)$). This completes the proof.

Theorem 9.15 *Suppose that f is (V_s^1)–integrable on an interval $[a, b]$ and that F is an (V_s^1)–primitive of f. Then F is symmetrically continuous, $F' = f$ almost everywhere in (a, b) and f is measurable.*

Proof. It is clear from Definition 9.11 that any primitive F must be symmetrically continuous. We show that, at almost every point x, SD $F(x) = f(x)$.

Let E_n denote the set of points x in (a, b) for which

$$|F(x + h_k) - F(x - h_k) - 2f(x)h_k| > n^{-1}h_k$$

for some sequence of positive numbers $h_k \to 0$. Every point x in (a, b) at which SD $F(x) = f(x)$ fails lies in some set E_n and so the result we require is proved if we show that each set E_n has measure zero. This we can accomplish using a Vitali argument by constructing a contradiction based on the assumption that, for some n, $|E_n| > c > 0$.

Let $4n\epsilon = c$. There is a gauge δ and a nondecreasing function τ on $[a, b]$ with $\tau(b) - \tau(a) < \epsilon$ and such that

$$|F(x + h) - F(x - h) - 2f(x)h| \leq \tau(x + h) - \tau(x - h) \qquad (9.36)$$

for all $a < x < b$, $0 < h < \delta(x)$. The collection of all intervals $[x - h, x + h]$ with $x \in E_n$, $0 < h < \delta(x)$ and

$$|F(x + h) - F(x - h) - 2f(x)h| > n^{-1}h \qquad (9.37)$$

is a Vitali cover of E_n. Let us choose a finite subcollection of nonoverlapping intervals $[x_i - h_i, x_i + h_i]$ $(i = 1, 2, \ldots p)$ from this cover in such a way that

$$|E_n| < 2 \sum_{i=1}^{p} 2h_i.$$

Then, using (9.36) and (9.37), we obtain

$$c < |E_n| < 2 \sum_{i=1}^{p} 2h_i < 4n \sum_{i=1}^{p} n^{-1}h_i$$

$$< 4n \sum_{i=1}^{p} |F(x_i + h_i) - F(x_i - h_i) - 2f(x_i)h_i|$$

$$\leq 4n \sum_{i=1}^{p} \tau(x_i + h_i) - \tau(x_i - h_i) < 4n\epsilon = c.$$

With this contradiction the proof that SD $F(x) = f(x)$ holds at almost every point x is complete. At almost every such point it follows from the Khintchine Theorem that F' exists and therefore that $F' = f$ almost everywhere. Finally a symmetrically continuous function is measurable (Theorem 2.26) and consequently f, as an a.e. derivative of a measurable function, must be measurable too.

Theorem 9.16 *Suppose that F is symmetrically continuous in (a, b), that*

$$- \infty < \underline{SD}\, F(x) \leq \overline{SD}\, F(x) < +\infty \tag{9.38}$$

for all but countably many x in (a, b) and that $\lambda_F(a, b)$ exists. Then $F'(x) = f(x)$ exists almost everywhere in (a, b), f is (V_s^1)–integrable on $[a, b]$ and

$$\int_a^b f(x)\, dx = \lambda_F(a, b).$$

Proof. By the Khintchine Theorem the condition (9.38) on the symmetric derivates ensures that the derivative $F'(x)$ exists at almost every point of (a, b). Let E denote the set of points where $F'(x)$ exists and let $f(x) = F'(x)$ for $x \in E$ and $f(x) = 0$ otherwise. Let $\epsilon > 0$. Let N denote the set of points $x \in (a, b) \setminus E$ at which

$$-\infty < \underline{SD}\, F(x) \leq \overline{SD}\, F(x) < +\infty$$

and let C denote the set of points x in (a, b) at which $\underline{SD}\, F(x) = -\infty$ or $\overline{SD}\, F(x) = +\infty$. By hypothesis C is countable. Thus $(a, b) = E \cup C \cup N$. Choose a monotonic, increasing function κ on $[a, b]$ that is discontinuous at each point in C and such that $\kappa(b) - \kappa(a) < \epsilon/4$. Choose a monotonic function η so that $\eta(b) - \eta < \epsilon/4$ and with $\eta'(x) = +\infty$ at each $x \in N$. Finally write

$$\tau(x) = \frac{\epsilon(x - a)}{4(b - a)} + \kappa(x) + \eta(x).$$

Then τ is an increasing function on $[a, b]$ with $\tau(b) - \tau(a) < \epsilon$ and for each point $x \in (a, b)$ it is straightforward to verify that there is a $\delta(x) > 0$ such that

$$|F(x + h) - F(x - h) - 2f(x)h| \leq \tau(x + h) - \tau(x - h).$$

For $x \in C$ use κ and the symmetric continuity of F at x; for $x \in N$ use the fact that $\eta'(x) = +\infty$ and bounds on the symmetric derivates of F; finally if $x \in E$ just use the fact that $F'(x) = f(x)$. This verifies the (V_s^1)–integrability of f as required and completes the proof.

Theorem 9.16 has an easy corollary which is proved directly in [111].

Corollary 9.17 (Freiling) *Suppose that F is symmetrically continuous and has finite symmetric derivates nearly everywhere on an interval (a, b). Suppose that $\underline{SD}\,F(x)$ is Denjoy-Perron integrable there. Then there is a scattered set $S \subset (a, b)$ so that*

$$\int_c^d \underline{SD}\,F(x)\,dx = F(d) - F(c) \qquad (c, d \in (a, b) \setminus S).$$

The (V_s^1)–integral shares the usual property of non-absolutely convergent integrals, that it is identical to the Lebesgue integral when applied to nonnegative functions. This is the analogue of Theorem 9.8 for the (N_s^1)–integral. The analogue of Theorem 9.9 will be proved in the next section.

Theorem 9.18 *Suppose that $f \geq 0$, that f is (V_s^1)–integrable on $[a, b]$ and that F is a (V_s^1)–primitive for f. Then f is Lebesgue integrable and F is an indefinite Lebesgue integral for f on (a, b).*

Proof. Since f is nonnegative the primitive function F may be shown to be essentially monotonic. But then, by Theorem 9.15, f is a.e. the derivative of a monotonic function and is accordingly Lebesgue integrable.

It remains, thus, to verify the monotonicity assertion. Let m be any natural number. Since f is (V_s^1)–integrable in $[a, b]$ with primitive F there is a gauge δ_m and a nondecreasing function τ_m on $[a, b]$ with $\tau_m(b) - \tau_m(a) < 1/m$ and such that

$$F(x + h) - F(x - h) \geq 2f(x)h - (\tau_m(x + h) - \tau_m(x - h)) \qquad (9.39)$$

for all $a < x < b$, $0 < h < \delta_m(x)$.

By Lemma 9.1 there is a countable set C so that for every $c, d \in (a, b) \setminus C$, $c < d$ and each m there is a partition

$$c = x_0 < x_1 < \ldots < x_p < x_{p+1} = d$$

with $x_{i+1} - x_i < \delta_m((x_i + x_{i+1})/2)$, $(i = 0, 1, \ldots p)$. From this, (9.39) and the fact that f is nonnegative it follows that

$$F(d) - F(c) \geq -1/m$$

for each m and hence that $F(d) - F(c) \geq 0$, for c and d outside of this countable set. Evidently then F is equivalent to a nondecreasing function and the proof is complete.

9.2.6 Symmetric Absolute Continuity

The Lebesgue integral permits a well-known descriptive characterization: f is Lebesgue integrable on an interval $[a, b]$ with an indefinite integral F if and only if F is absolutely continuous and $F'(x) = f(x)$ almost everywhere in $[a, b]$. The Denjoy-Perron integral permits the same characterization but replaces absolute continuity by the notion ACG$_*$. See [251] for a full account.

This can be done here too with an appropriate symmetric version of absolute continuity. Let us recall first the definition of the symmetric variation of a function. (Section 8.2). Let $E \subset \mathbb{R}$, let δ be a gauge on E and let F be an arbitrary function. Then we write

$$S_\delta(F, E) = \sup \sum_{i=1}^n |F(x_i + h_i) - F(x_i - h_i)|$$

where the supremum is with regard to all sequences $\{[x_i - h_i, x_i + h_i]\}$ of nonoverlapping intervals with centers $x_i \in E$ and with $h_i < \delta(x_i)$. We write then

$$VS_F(E) = \inf S_\delta(F, E)$$

where the infimum is taken over all gauges δ on E. This expression is called the *symmetric variation* of F on E and the set function VS_F is called the *symmetric variational measure* associated with F. We have seen that VS_F is an outer measure on the real line for which all Borel sets are measurable. The properties of VS_F are closely related to the symmetric differentiation structure of F.

Definition 9.19 Let F be defined on an interval (a, b). We say that F is *symmetrically–ACG$_*$* on (a, b) if

(i) $\lambda_F(a, b)$ exists,

(ii) VS_F is σ–finite on (a, b) and

(iii) $VS_F(Z) = 0$ for every set $Z \subset (a, b)$ of Lebesgue measure zero.

A function that is ACG$_*$ on $[a, b]$ can be shown to be symmetrically–ACG$_*$ in (a, b) although this is not immediate from the usual definitions. A function that is symmetrically–ACG$_*$ in (a, b) is easily seen to be symmetrically continuous at every point in that interval. If F is symmetrically–ACG$_*$ on (a, b) then $F'(x)$ exists almost everywhere in (a, b); we state this as a lemma.

Lemma 9.20 *If F is symmetrically–ACG$_*$ on (a, b) then $F'(x)$ exists for almost every point $x \in (a, b)$.*

Proof. See Theorem 8.8.

These concepts allow us to characterize the (V_s^1)–integral in a way analogous to the well-known Lusin characterization of the Denjoy–Perron integral.

Theorem 9.21 *A function f is (V_s^1)–integrable on an interval $[a, b]$ if and only if there exists a function F symmetrically–ACG_* on (a, b) with $F' = f$ almost everywhere there. In that case*

$$\int_a^b f(x)\, dx = \lambda_F(a, b).$$

Proof. Suppose that f is (V_s^1)–integrable on $[a, b]$ with primitive F. For each m let $N_m = \{x \in (a, b) : |f(x)| < m\}$. We show that $VS_F(N_m) < +\infty$ and that $VS_F(Z \cap N_m) = 0$ for each set $Z \subset (a, b)$ of measure zero.

Let $\epsilon > 0$. By definition there is a gauge δ_1 and a nondecreasing function τ on $[a, b]$ with $\tau(b) - \tau(a) < \epsilon$ and such that

$$|F(x + h) - F(x - h) - 2f(x)h| \le \tau(x + h) - \tau(x - h) \qquad (9.40)$$

for all $a < x < b$, $0 < h < \delta_1(x)$.

From this we evidently get for any gauge $\delta \le \delta_1$,

$$VS_F(N_m) \le S_\delta(F, N_m) \le m(b - a) + \epsilon. \qquad (9.41)$$

If Z has measure zero then there is an open set G containing Z with $|G| < \epsilon$. Let δ be defined on Z so that $\delta < \delta_1$ and so that

$$(x - \delta(x), x + \delta(x)) \subset G$$

for each $x \in Z$. Then from (9.40) we obtain

$$VS_F(Z \cap N_m) \le S_\delta(F, Z \cap N_m) \le m\epsilon + \epsilon. \qquad (9.42)$$

We see, from (9.41), that $VS_F(N_m) < +\infty$ for each $m \in Numbers$ and so the measure VS_F must be σ–finite. From (9.42) we see that $VS_F(Z \cap N_m) = 0$ for each set $Z \subset (a, b)$ of measure zero and so $VS_F(Z) = 0$ for each such set. This is enough to establish that the function F is symmetrically–ACG_* on (a, b). By Theorem 9.15, $F' = f$ almost everywhere there and so the necessity of each of the conditions stated in the theorem has been proved.

Let us prove the converse. Suppose F has the properties stated in the theorem, let $\epsilon > 0$ and let E denote the set of points x in (a, b) at which $F'(x) = f(x)$. Choose $\delta_0(x)$ for $x \in E$ so that

$$|F(x + h) - F(x) - f(x)h| < \epsilon h \quad (0 < |h| < \delta_0(x)). \qquad (9.43)$$

For each $m \ge 1$ let $Z_m = \{x \in (a, b) \setminus E : m - 1 \le |f(x)| < m\}$. As each such set has measure zero there is an open set $G_m \supset Z_m$ with $|G_m| < \epsilon m^{-1} 2^{-m}$. By the assumption that F is symmetrically–ACG_* on (a, b) we may choose $\delta_m(x)$ for each $x \in Z_m$ so that

$$(x - \delta_m(x), x + \delta_m(x)) \subset G_m$$

and so that

$$\sum_{i=1}^{n} |F(x_i + h_i) - F(x_i - h_i)| < \epsilon 2^{-m}$$

where the sum is taken over any sequence $\{[x_i - h_i, x_i + h_i]\}$ of nonoverlapping intervals with centers $x_i \in Z_m$ and with $h_i < \delta_m(x_i)$. Note that for such intervals

$$\sum_{i=1}^{n} |F(x_i + h_i) - F(x_i - h_i) - 2f(x_i)h_i| < \epsilon 2^{-m}(1 + mm^{-1}) = \epsilon 2^{-m+1}. \quad (9.44)$$

Let $\tau_m(x)$ be the supremum of such sums in (9.44) for sequences $\{[x_i - h_i, x_i + h_i]\}$ of nonoverlapping subintervals of $[a, x]$ with centers $x_i \in Z_m$ and with $h_i < \delta_m(x_i)$. Note that $\tau_m(a) = 0$ and $\tau_m(b) < \epsilon 2^{-m+1}$. Write

$$\tau(x) = \sum_{m=1}^{\infty} \tau_m(x) + \epsilon(x - a)$$

and write $\delta(x) = \delta_0(x)$ for $x \in E$ and $\delta(x) = \delta_m(x)$ for $x \in Z_m$. This defines δ on (a, b).

Thus we have obtained an increasing function τ on $[a, b]$ with $\tau(b) - \tau(a) < \epsilon(2 + b - a)$ and such that

$$|F(x + h) - F(x - h) - 2f(x)h| \leq \tau(x + h) - \tau(x - h) \quad (9.45)$$

for all $a < x < b, 0 < h < \delta(x)$. To see the inequality (9.45) for $x \in E$ we merely use (9.43) and the fact that $2\epsilon h < \tau(x + h) - \tau(x - h)$. For x in some set Z_m we have

$$|F(x + h) - F(x - h) - 2f(x)h| \leq \tau_m(x + h) - \tau_m(x - h)$$

since τ_m has been defined by using sums that appear in (9.44) and again the inequality (9.45) follows. From (9.45) we see that, by definition, F is a (V_s^1)–primitive of f on $[a, b]$ and the theorem is proved.

This characterization allows us to prove the analogue of Theorem 9.9, that a (V_s^1)–integrable function must be Lebesgue integrable on a dense set of intervals.

Theorem 9.22 *Suppose that f is (V_s^1)–integrable on an interval $[a, b]$. Then there is a set of intervals dense in $[a, b]$ on each of which f is Lebesgue integrable.*

Proof. If F is the (V_s^1)–primitive of f then, by Theorem 9.22, $F' = f$ almost everywhere in (a, b), F is symmetrically continuous and has a σ–finite variational measure VS_F. By Theorem 8.5, the function F has bounded variation on a dense set of intervals and accordingly f, as the a.e. derivative of a function of bounded variation, is Lebesgue integrable on each of these intervals.

The next theorem is useful in applications to trigonometric series. This property, that a periodic function should be integrable over every period if it is integrable over at least one period, seems hardly unexpected. However the integral has been tailored in Definition 9.11 to have this property; it is by no means shared by other symmetric integrals. For example in the case of the SCP–integral of Burkill [35] a 2π–periodic function f might be integrable on intervals $[x, x + 2\pi]$ only for almost every x.

Theorem 9.23 *Suppose that f is 2π–periodic and is (V_s^1)–integrable on some interval $[a, a + 2\pi]$. Then f is (V_s^1)–integrable on every period and to the same value.*

Proof. Let F be a (V_s^1)–primitive for f on a fixed interval $[a, a+2\pi]$. We define G by writing

$$G(x + 2k\pi) = F(x) + k\lambda_F(a, a + 2\pi)$$

for $a < x < a + 2\pi$ and $k \in \mathbb{N}$. Note that G is symmetrically continuous. For points of the form $a + 2k\pi$ this needs to be checked. At all other points it is obvious. For example at the point $a + 2\pi$ we see that

$$\lim_{h \to 0} G(a + 2\pi + h) - G(a + 2\pi - h) =$$
$$\lim_{h \to 0} F(a + h) - F(a + 2\pi - h) + \lambda_F(a, a + 2\pi) = 0.$$

Note also that for any interval $[c, c+2\pi]$ the expression $\lambda_G(c, c+2\pi)$ exists. For example if $a < c < a + 2\pi$ then

$$\lim_{h \to 0} G(c + 2\pi - h) - G(c + h) =$$
$$\lim_{h \to 0} \{-F(c + h) + F(c - h) + \lambda_F(a, a + 2\pi)\} = \lambda_F(a, a + 2\pi).$$

A similar computation works for any such interval.

Let $\epsilon > 0$. There is a gauge δ_1 on $(a, a + 2\pi)$ and a nondecreasing function τ_1 on $[a, a + 2\pi]$ with $\tau_1(a + 2\pi) - \tau_1(a) < \epsilon/3$ and such that

$$|F(x + h) - F(x - h) - 2f(x)h| \leq \tau_1(x + h) - \tau_1(x - h) \tag{9.46}$$

for all $a < x < a + 2\pi$, $0 < h < \delta_1(x)$. Extend τ_1 to a monotonic function τ_2 on the real line so that $\tau_2(x+2k\pi) = \tau_1(x)+k(\tau_1(a+2\pi)-\tau_1(a))$ for $a < x < a+2\pi$.

We show that f is (V_s^1)–integrable on any interval $[c, c+2\pi]$ with primitive G. For example if $a < c < a + 2\pi$ then the computations are as follows. Define $\delta(x) = \delta_1(x - 2\pi)$ if $x - 2\pi \in (a, c)$, and $\delta(x) = \delta_1(x)$ if $x \in (c, a + 2\pi)$ and $\delta(a+2\pi)$ so that $|G(a+2\pi+h) - G(a+2\pi-h)| < \epsilon/6$ and $|2f(a+2\pi)h| < \epsilon/6$ for all $0 < h < \delta(a+2\pi)$.

Construct a nondecreasing function τ_3 constant on $[c, a + 2\pi)$ and on $(a + 2\pi, c + 2\pi]$ with $\tau_3((a + 2\pi)+) - \tau_3((a + 2\pi)-) = \epsilon/3$. Write $\tau = \tau_2 + \tau_3$. Then we obtain $\tau(c + 2\pi) - \tau(c) < \epsilon$ and, from (9.46), that

$$|G(x + h) - G(x - h) - 2f(x)h| \leq \tau(x + h) - \tau(x - h) \qquad (9.47)$$

for all $c < x < c+2\pi, 0 < h < \delta(x)$. From this it follows that f is (V_s^1)–integrable on $[c, c + 2\pi]$ as required.

9.2.7 A Further Application to Trigonometric Series

As in Section 9.2.3 we give a further elementary result showing how a symmetric integral may be used to solve the coefficient problem for trigonometric series when the conventional integrals fail to exist. Here we are placing severe restrictions on the coefficients of the series in order to obtain the existence of the integrals in this sense. In Sections 9.3.3, 9.3.8 and 9.4.5 we show how these conditions may be dropped and a more general symmetric integral used.

Theorem 9.24 *Suppose that the trigonometric series*

$$a_0/2 + \sum_{n=1}^{\infty} (a_n \cos nx + b_n \sin nx) \qquad (9.48)$$

has bounded partial sums at nearly every point x and that the coefficients satisfy

$$\sum_{k=1}^{n} k\sqrt{a_k^2 + b_k^2} = o(n) \qquad (9.49)$$

as $n \to +\infty$. Then the series (9.48) converges almost everywhere to a (V_s^1)– integrable function f and is the (V_s^1)–Fourier series for f.

Proof. Write

$$L(x) = \frac{1}{2}a_0 x + \sum_{n=1}^{\infty} (a_n \sin nx - b_n \cos nx)/n$$

for the Lebesgue function for the series (obtained by a single integration). Under the condition (9.49) the function L exists and is continuous. Also

$$L(x + h) - L(x - h) = O(h) \quad (h \to 0)$$

at every point where the series (9.48) has bounded partial sums. These observations follow from [313, Vol. I, Theorem (2.16), p. 322].

By the Khintchine Theorem the derivative $L'(x) = f(x)$ exists almost everywhere. Applying again [313, Vol. I, Theorem (2.16), p. 322] we conclude that the series (9.48) converges at each such point x to $f(x)$ and so converges almost everywhere to $f(x)$.

Thus we have a continuous function L, $L'(x) = f(x)$ a.e. and $L(x + h) - L(x - h) = O(h)$ at nearly every point. It follows from Theorem 9.16 that f is (V_s^1)–integrable on any interval $[a, b]$ and that

$$\int_a^b f(x)\,dx = \lambda_L(a,b)$$

in this sense. With $[a,b] = [0, 2\pi]$ this gives the usual Fourier formula for the coefficient a_0. The remainder of the proof continues in much the same way using these ideas and the formal multiplication of series techniques of Theorem 9.10.

If the coefficients satisfy instead the stronger condition $a_n, b_n = o(n^{-1})$ as $n \to +\infty$ then the series is known to be a Fourier series in the usual Lebesgue sense; the weaker condition (9.49) does not in general require this. For example the series $\sum n^{-1/2} \cos 2^n x$ satisfies (9.49) but is not a Fourier series (in the Lebesgue sense). Nor, however, does this series satisfy the other conditions of Theorem 9.24 since it converges almost nowhere (cf. [313, Vol. I, p. 203]). We must, however, leave it unanswered whether under *all* of the assumptions of Theorem 9.24 the function f may fail to be Lebesgue integrable.

A further variant on Theorem 9.24 is available by replacing the condition (9.49) by a big "O" growth condition. Lebesgue summability of a series is defined in [313, Vol. I, p. 321].

Theorem 9.25 *Suppose that the series (9.48) has bounded partial sums at nearly every point x and that the coefficients satisfy*

$$\sum_{k=1}^n k\sqrt{a_k^2 + b_k^2} = O(n) \qquad (9.50)$$

as $n \to +\infty$. Then the series (9.48) is Lebesgue summable almost everywhere to a (V_s^1)–integrable function f and is the (V_s^1)–Fourier series for f.

Proof. By using the methods of [313, Vol. I, Theorem (2.16), p. 322] we can show that, under these hypotheses and with the notation of the last proof,

$$\left| \frac{L(x+h) - L(x-h)}{2h} - s_N(x) \right|$$

is bounded uniformly in x as $h \to 0$ where $N = N(h) = [1/h]$ and s_n are the partial sums of (9.48). Then again as before $L'(x) = f(x)$ exists a.e. and the series is the (V_s^1)–Fourier series of f. Here, although we cannot conclude the convergence of the series at points where $L'(x)$ exists, we can conclude the Lebesgue summability of the series at these points as required.

9.2.8 A Symmetric Perron Integral

In this section we present an elementary version of a symmetric Perron type integral that is a direct generalization of the classical Perron integral as defined in Saks [251]. Ponomarev [229] has defined a similar integral but using continuous major and minor functions. The variant given here differs in that, rather than assume continuous major and minor functions H, we assume only that the increment $\lambda_H(a, b)$ exists. The resulting integral is more general and more in the spirit of symmetric integrals.

For a real function f, defined on an interval (a, b), we shall say that a real function H is a *symmetric majorant* of f on $[a, b]$ provided that $\lambda_H(a, b)$ exists and that

$$\underline{\mathrm{SD}}\, H(x) \geq f(x)$$

everywhere in (a, b). Symmetric minorants G are then defined so that $-G$ is a majorant of $-f$ and the Perron integral is defined in more or less the usual manner by taking extremes over all majorants and minorants. If

$$\inf_H \lambda_H(a, b) = \sup_G \lambda_G(a, b) = c$$

then f is said to be (P_s^1)–integrable on $[a, b]$ and we write

$$(\mathrm{P}_s^1) \int_a^b f(x)\, dx = c.$$

Such a symmetric–Perron type integral would be justified by an elementary monotonicity theorem. If H is a symmetric majorant of f and G is a symmetric minorant of f then, with $K = H - G$, $\underline{\mathrm{SD}}\, K(x) \geq 0$ ($a < x < b$) and it follows from Lemma 9.1 that K is nondecreasing off a countable set. As both $\lambda_H(a, b)$ and $\lambda_G(a, b)$ exist so too does $\lambda_K(a, b)$ and this must be nonnegative since K is essentially nondecreasing. It follows that $\lambda_H(a, b) \geq \lambda_G(a, b)$. Thus the integral exists if for every $\epsilon > 0$ there is a pair of major/minor functions H and G with $\lambda_H(a, b) - \lambda_G(a, b) < \epsilon$.

Majorants in the ordinary Perron sense are also majorants in this sense. This is because, for any function F, $\underline{\mathrm{SD}}\, F(x) \geq \underline{\mathrm{D}}\, F(x)$. It follows that such an integral extends the ordinary Perron integral and integrates all exact symmetric derivatives of continuous functions. We shall show that this integral contains the (V_s^1)–integral defined earlier. The integrals seem likely to be distinct but their exact relation is not known.

Theorem 9.26 *If a function f is (V_s^1)–integrable on an interval $[a, b]$ then f is (P_s^1)–integrable on $[a, b]$ and*

$$(\mathrm{V}_s^1) \int_a^b f(x)\, dx = (\mathrm{P}_s^1) \int_a^b f(x)\, dx.$$

Proof. Let $\epsilon > 0$. Let F be a (V_s^1)–primitive of f on (a, b). Then there is a $\delta(x) > 0$ ($x \in (a, b)$) and a nondecreasing function τ on $[a, b]$ with $\tau(b) - \tau(a) < \epsilon/2$ and such that

$$|F(x+h) - F(x-h) - 2f(x)h| \leq \tau(x+h) - \tau(x-h) \qquad (9.51)$$

for all $a < x < b$, $0 < h < \delta(x)$.

Write $H(x) = F(x) + \tau(x)$ and $G(x) = F(x) - \tau(x)$. Note that $\lambda_H(a, b)$ and $\lambda_G(a, b)$ exist and that $0 \leq \lambda_H(a, b) - \lambda_G(a, b) < \epsilon$. From (9.51) one sees that H is a (P^1_s)-major function for f and that G is a (P^1_s)-minor function for f. This proves the (P^1_s)-integrability of f and, from $\lambda_G(a, b) \leq \lambda_F(a, b) \leq \lambda_H(a, b)$, we see that the integral has the value $\lambda_F(a, b)$. This completes the proof.

9.2.9 A Symmetric Riemann Integral

The expression of an integral as a limit of Riemann sums dominated the notion of the integral in the last century. In this century the integral as an aspect of measure theory or functional analysis has taken over. The Riemann sums approach nonetheless retains some utility. It is possible to tailor the limit in the Riemann integral in such a way as to solve specific and special problems. For example the classical Lebesgue integral itself and the integral of Denjoy and Perron can be expressed in this manner. In fact associated with most derivation processes is a generalization of the Riemann integral; the symmetric derivative permits such a generalization.

A Riemann type integral based on the symmetric derivative is alluded to in Henstock [131, p. 222] but not developed; such an integral would not quite fit into the general scheme of integration theory presented there but, with appropriate modifications, many of the details can be carried over. The first account of such an integral appears to be that of Kurzweil and Jarník [168] although, again, they do not give all the details. Later in [169] more complete proofs were supplied.

The symmetric Riemann integral that we introduce in this section is somewhat more general than the integrals of these earlier definitions; it is developed with some extra symmetry so that it more closely connects with the integrals discussed in the preceding sections. There is a number of technical reasons for preferring this definition over other possible definitions.

By a *symmetric partition* of an interval $[a, b]$ we shall mean a set of points

$$a < x_1 < x_2 < \ldots < x_n < b$$

such that $x_1 - a = b - x_n$. Such a partition is said to be *symmetrically finer* than a gauge δ on $[a, b]$ provided

$$x_{i+1} - x_i < \delta \left(\frac{x_i + x_{i+1}}{2} \right) \text{ for } i = 1, 2, \ldots, n - 1$$

and

$$x_1 - a < \delta(a) \quad \text{and} \quad b - x_n < \delta(b).$$

(Note that the only "symmetry" imposed by the definition is the way it treats the endpoints and the interpretation of the fineness with respect to δ.)

Definition 9.27 Let f be defined on $[a, b]$. We say that f is (R_s^1)–integrable on $[a, b]$ if there is a number C such that for every $\epsilon > 0$ there is a gauge δ on $[a, b]$ with the property for any symmetric partition

$$a < x_1 < x_2 < \ldots < x_n < b \quad (x_1 - a = b - x_n)$$

that is symmetrically finer than δ

$$\left| \sum_{i=1}^{n-1} f\left(\frac{x_{i+1} + x_i}{2}\right)(x_{i+1} - x_i) - C \right| < \epsilon.$$

In this case we write

$$(R_s^1) \int_a^b f(x)\, dx = C.$$

Justification for this definition requires that, for any gauge δ, such partitions do in fact exist. This is readily supplied by Lemma 9.1. It is almost immediately clear that a function that is Riemann integrable is integrable in this sense and to the same value.

We shall omit stating and proving the more obvious elementary properties of this integral; any integral expressed as a limit of Riemann sums will have some of the usual obvious properties. Additivity over intervals however fails, just as it has for the (N_s^1), (V_s^1), and (P_s^1)–integrals. We do have the analogue of Theorem 9.4.

Theorem 9.28 If f is (R_s^1)–integrable on $[a, c]$ and on $[a, b]$ $(a < b < c)$ then it is (R_s^1)–integrable on $[b, c]$ and

$$\int_b^c f(x)\, dx = \int_a^c f(x)\, dx - \int_a^b f(x)\, dx \qquad (9.52)$$

in the sense of this integral.

Proof. Let $\epsilon > 0$ and write C_1 and C_2 for the values of the two integrals on the right side of (9.52). Then there is a gauge δ_1 on $[a, c]$ with the property that for any symmetric partition of $[a, c]$

$$a < x_1 < x_2 < \ldots < x_n < c \quad (x_1 - a = c - x_n)$$

that is symmetrically finer than δ_1

$$\left| \sum_{i=1}^{n-1} f\left(\frac{x_{i+1} + x_i}{2}\right)(x_{i+1} - x_i) - C_1 \right| < \epsilon/3.$$

Similarly there is a gauge δ_2 on $[a, b]$ with $\delta_2(x) < \delta_1(x)$ for $a \leq x \leq b$ and possessing the property that for any symmetric partition of $[a, b]$

$$a < x_1 < x_2 < \ldots < x_n < b \quad (x_1 - a = b - x_n)$$

that is symmetrically finer than δ_2

$$\left| \sum_{i=1}^{n-1} f\left(\frac{x_{i+1} + x_i}{2}\right)(x_{i+1}x_i) - C_2 \right| < \epsilon/3.$$

We shall now choose a gauge $\delta < \delta_1$ on $[b, c]$ and a strictly decreasing mapping $h : [b, c] \to [a, b]$ with the following property: for any symmetric partition of $[b, c]$

$$b < x_1 < x_2 < \ldots < x_n < c \quad (x_1 - b = c - x_n)$$

that is symmetrically finer than δ the reflected points

$$a < x_{-n} < x_{-n+1} < \ldots < x_{-1} < b$$

where $x_{-i} = h(x_i)$, $(i = 1, 2, \ldots n)$ form a symmetric partition of $[a, b]$ that is symmetrically finer than δ_2. We can insist, too, that $\delta(b) < \epsilon(6|f(b)| + 1)^{-1}$.

Assuming for the moment that δ and h can be so chosen the theorem can be proved. We notice that

$$a < x_{-n} < x_{-n+1} < \ldots < x_{-1} < x_1 < x_2 < \ldots < x_n < c$$

is a symmetric partition of $[a, c]$ that is symmetrically finer than δ_1 and that

$$a < x_{-n} < x_{-n+1} < \ldots < x_{-1} < b$$

is a symmetric partition of $[a, b]$ that is symmetrically finer than δ_2. This gives the estimate

$$\left| \sum_{i=1}^{n-1} f\left(\frac{x_{i+1} + x_i}{2}\right)(x_{i+1} - x_i) - (C_1 - C_2) \right| \leq$$

$$\left| \sum_{i=-n}^{n-1} f\left(\frac{x_{i+1} + x_i}{2}\right)(x_{i+1} - x_i) - C_1 \right| + 2|f(b)|\delta(b)$$

$$+ \left| \sum_{i=-n}^{-2} f\left(\frac{x_{i+1} + x_i}{2}\right)(x_{i+1} - x_i) - C_2 \right| < \epsilon$$

and the theorem is proved.

There remains only to show how to choose δ and h. If b is the midpoint of $[a, c]$ this is particularly easy. Write $h(x) = 2b - x$ and $\delta(x) = \delta_2(h(x))$ for $b \leq x \leq c$. This guarantees that $b - x_{-1} = x_1 - b$ and that $x_{-n} - a = c - x_n$ so that the partitions are symmetric. In the general case any decreasing continuous map

h of $[b, c]$ onto $[a, b]$ may be chosen as long as $h'(x) = -1$ in a neighbourhood of b and c.

The (R_s^1) integral is at least as general as the integrals we have seen so far. The next theorem establishes the inclusions. The exact relations are unknown.

Theorem 9.29 *The (R_s^1)-integral includes each of the (N_s^1), (V_s^1), and (P_s^1)- integrals.*

Proof. In view of Theorem 9.26, it is enough to show that any function f that is (P_s^1)-integrable on an interval $[a, b]$ is (R_s^1)-integrable there and with the same value.

Let $\epsilon > 0$. Let H be an arbitrary symmetric majorant for f on $[a, b]$. Choose $\delta(x) > 0$ for $a < x < b$ so that

$$H(x + h) - H(x - h) > 2(f(x) - \epsilon)h$$

for $0 < h < \delta(x)$ and choose $\delta(a) = \delta(b) = \delta_0$ so that

$$|H(b - h) - H(a + h) - \lambda_H(a, b)| < \epsilon$$

for $0 < h < \delta_0$. Then for any symmetric partition

$$a < x_1 < x_2 < \ldots < x_n < b \quad (a - x_1 = b - x_n)$$

that is symmetrically finer than δ we obtain

$$\sum_{i=1}^{n-1} f\left(\frac{x_{i+1} + x_i}{2}\right)(x_{i+1} - x_i)$$
$$< H(x_n) - H(x_1) + \epsilon(b - a) < \lambda_H(a, b) + \epsilon(b - a + 1).$$

In the same way we use an arbitrary symmetric minorant G and find a δ so that the corresponding symmetric Riemann sums exceed $\lambda_G(a, b) - \epsilon(b - a + 1)$. It follows now easily that if f is (P_s^1)-integrable on $[a, b]$ it must be (R_s^1)-integrable too and with the same value assigned for the integrals. This completes the proof.

9.2.10 Variational Characterization of the (R_s^1)–Integral

To complete the study of the symmetric Riemann integral we shall require a variational characterization. Note the similarity with the definition of the (V_s^1)- integral. This should be compared, too, with an analogous statement in [169, Theorem 3.9, p. 217].

Lemma 9.30 *In order for a function f to be (R_s^1)-integrable on an interval $[a, b]$ it is necessary and sufficient that there exist a countable set N and a function F defined on $(a, b) \setminus N$ in such a way that*

(i) *for every $\epsilon > 0$ there is a $\delta(x) > 0$ ($x \in (a,b)$) and a nondecreasing function τ on $[a,b]$ with $\tau(b) - \tau(a) < \epsilon$ and such that*

$$|F(x+h) - F(x-h) - 2f(x)h| \le \tau(x+h) - \tau(x-h) \qquad (9.53)$$

for all $a < x < b$, $x + h \notin N$, $x - h \notin N$, $0 < h < \delta(x)$ and

(ii)

$$\lim_{h \to 0,\, a+h \notin N,\, b-h \notin N} F(b-h) - F(a+h) = \int_a^b f(t)\, dt. \qquad (9.54)$$

Proof. We prove first that the conditions are necessary. Assume that f is (R_s^1)-integrable on $[a,b]$ with integral C. Then there is a sequence of positive functions δ_m on $[a,b]$ with the property for any symmetric partition

$$a < x_1 < x_2 < \ldots < x_n < b \quad (a - x_1 = b - x_n)$$

that is symmetrically finer than δ_m

$$\left| \sum_{i=1}^{n-1} f\left(\frac{x_{i+1} + x_i}{2} \right) (x_{i+1} - x_i) - C \right| < 1/m. \qquad (9.55)$$

By Lemma 9.1 there is a countable set $N \subset (a,b)$ so that given any pair c, $d \in (a,b) \setminus N$, $c < d$ and any m there is a partition $c = x_1 < x_2 < \ldots < x_n = d$ that is symmetrically finer than δ_m. Given any two such partitions of $[c,d]$ say $c = x_1 < x_2 < \ldots < x_n = d$ and $c = x_1' < x_2' < \ldots < x_p' = d$ both symmetrically finer than δ_m we obtain the inequality

$$\left| \sum_{i=1}^{n-1} f\left(\frac{x_{i+1} + x_i}{2} \right) (x_{i+1} - x_i) - \sum_{i=1}^{p-1} f\left(\frac{x_{i+1}' + x_i'}{2} \right) (x_{i+1}' - x_i') \right| < \frac{2}{m}$$

by augmenting each partition to a symmetric partition of $[a,b]$ also symmetrically finer than δ_m and using (9.55). This ensures that there is a number $G(c,d)$ associated with each such interval $[c,d]$ with the property that

$$\left| \sum_{i=1}^{n-1} f\left(\frac{x_{i+1} + x_i}{2} \right) (x_{i+1} - x_i) - G(c,d) \right| \le \frac{2}{m}$$

for any partitions of $[c,d]$, $c = x_1 < x_2 < \ldots < x_n = d$ symmetrically finer than δ_m.

By standard methods (see [169] for example) in the theory of Riemann type integrals this argument will produce a function F defined on $(a,b) \setminus N$ with $G(c,d) = F(d) - F(c)$. For any pair c, $d \in (a,b) \setminus N$, $c < d$ and any m and for any partitions of $[c,d]$, $c = x_1 < x_2 < \ldots < x_n = d$ symmetrically finer than δ_m we will obtain

$$\sum_{i=1}^{n-1} \left| f\left(\frac{x_{i+1} + x_i}{2}\right)(x_{i+1} - x_i) - (F(x_{i+1}) - F(x_i))\right| \le \frac{4}{m}. \qquad (9.56)$$

Define then $\tau_m(c, d)$ as the supremum of the sums

$$\sum_{i=1}^{n-1} \left| f\left(\frac{x_{i+1} + x_i}{2}\right)(x_{i+1} - x_i) - (F(x_{i+1}) - F(x_i))\right|$$

for any partitions of $[c, d]$, $c = x_1 < x_2 < \ldots < x_n = d$ symmetrically finer than δ_m.

Because of (9.56) there is a monotone function τ_m defined everywhere on $[a, b]$ so that $\tau_m(b) - \tau_m(a) \le 4/m$ and so that $\tau_m(c, d) = \tau_m(d) - \tau_m(c)$. This has moreover the property that

$$|F(x + h) - F(x - h) - 2f(x)h| \le \tau_m(x + h) - \tau_m(x - h)$$

provided $0 < h < \delta_m(x)$, $x + h \in (a, b) \setminus N$, $x - h \in (a, b) \setminus N$.

The proof of this part is complete if we now prove the final statement (ii) of the lemma. Let $h > 0$ so that $a + h \notin N$, $b - h \notin N$, $0 < h < \delta_m(a)$ and $0 < h < \delta_m(b)$. Then from (9.55) and (9.56) we obtain that there is a partition $a + h = x_1 < \ldots < x_n = b - h$ that will supply the inequality

$$|F(b - h) - F(a + h) - C| \le 5/m.$$

From this we can deduce (9.54) as required.

Let us turn now to the converse. Suppose that such an F and N exist with the limit in (9.54) existing and equal to C. Let $\epsilon > 0$ and choose τ and $\delta(x) > 0$ for $a < x < b$ according to the conditions in the lemma. Because of (9.54) we may choose $\delta(a) = \delta(b)$ so that

$$|F(b - h) - F(a + h) - C| < \epsilon \qquad (9.57)$$

for all $0 < h < \delta(a)$ and $b - h$, $a + h \notin N$.

Let $a < x_1 < x_2 < \ldots < x_n < b$ be a symmetric partition of (a, b) that is symmetrically finer than δ. We will show that

$$\left| \sum_{i=1}^{n-1} f\left(\frac{x_{i+1} + x_i}{2}\right)(x_{i+1} - x_i) - C \right| \le 3\epsilon \qquad (9.58)$$

and the proof of the lemma is complete since then f is integrable on $[a, b]$ and the value of the integral is C.

If, by chance, none of the points x_i in this partition belong to N then this is particularly easy. Suppose, for the moment, that this is so. Then using (9.54) and (9.57) we have

$$\left| \sum_{i=1}^{n} f\left(\frac{x_{i+1} + x_i}{2} \right) (x_{i+1} - x_i) - C \right|$$

$$\leq \left| \sum_{i=1}^{n} f\left(\frac{x_{i+1} + x_i}{2} \right) (x_{i+1} - x_i) - F(x_{i+1}) + F(x_i) \right|$$

$$+ |F(x_n) - F(x_1) - C| \leq \sum_{i=1}^{n} \tau(x_{i+1}) - \tau(x_i) + \epsilon < 2\epsilon.$$

This proves the inequality (9.58) at least in the special case that none of the points of the partition belong to N.

In the general case we have only to note that given any such partition as above we may choose another symmetric partition $a < z_1 < z_2 < \ldots < z_m < b$ with no z_i in N, that is also symmetrically finer than δ and such that

$$\left| \sum_{i=1}^{n} f\left(\frac{x_{i+1} + x_i}{2} \right) (x_{i+1} - x_i) - \sum_{i=1}^{m} f\left(\frac{z_{i+1} + z_i}{2} \right) (z_{i+1} - z_i) \right| < \epsilon. \qquad (9.59)$$

This easily gives the inequality (9.58) (with 3ϵ) that we require. The choice of the partition can be effected by using a simple device that in [107] is called "jiggling" and in [169] is called an "h–modification" of a partition. Just write $z_i = x_i + (-1)^i h$ for $i = 1, 2, \ldots n$ and $h > 0$. If n is even take $m = n$; if n is odd take $m = n + 1$ and $z_m = x_n + h$. For all sufficiently small h, $a < z_1 < z_2 < \ldots < z_m < b$ is a symmetric partition of $[a, b]$ that is also finer than δ. Note that in the even case the centers of the intervals $[x_{i-1}, x_i]$ and $[z_{i-1}, z_i]$ coincide; in the odd case this is so again but there is one extra small interval added which must be argued separately. Now it is easy to arrange, by taking h small, that (9.59) holds. Also h may evidently be chosen in such a way that each point z_i fails to belong to the countable set N. This completes the proof of the lemma.

From the characterization in Lemma 9.30 we can complete some of the properties of the (R_s^1)–integral.

Theorem 9.31 *If f is (R_s^1)–integrable then f is measurable.*

Proof. We sketch the ideas briefly. The variational property in Lemma 9.30 may be employed, exactly as in the proof of Theorem 9.15, to show that $F' = f$ almost everywhere, where here of course F is defined only off some countable set and so the derivative is taken relative to the domain of F. The function F is symmetrically continuous, again relative its domain, and it may be shown that such functions are measurable (in the same way that everywhere defined symmetrically continuous functions are measurable). It follows that f, as the derivative of such a function, must be measurable too.

Theorem 9.32 *If f is (R_s^1)–integrable and f is non-negative then f is Lebesgue integrable.*

Proof. Again we provide only a sketch. As in the preceding proof the variational property in Lemma 9.30 may be employed, exactly as in the proof of Theorem 9.18, to show that f is almost everywhere the derivative of an essentially monotonic function. Here the monotonic function F will be defined only off some countable set and the derivative is relative to the domain of course. It follows that f as the a.e. derivative of such a function must be Lebesgue integrable.

We conclude with a version of the "Cauchy property" for this integral. This, too, can be proved with the help of Lemma 9.30.

Theorem 9.33 *Suppose that f is (R_s^1)–integrable on each interval $[c, d]$ for c, $d \in (a, b) \setminus N$ where N is countable. If*

$$\lim_{h \to 0, \, a+h \notin N, \, b-h \notin N} \int_{a+h}^{b-h} f(t) \, dt = \lambda$$

exists then f is (R_s^1)–integrable on $[a, b]$ and $\int_a^b f(t) \, dt = \lambda$.

9.2.11 Lusin Type Characterization

The characterization of the (V_s^1)–integral in Section 9.2.6 can now be extended to the (R_s^1)–integral with a few modifications. Lemma 9.30 is the key. We shall omit the details.

Let F be a function defined on a subset of the interval (a, b). Then, as before, we write

$$S_\delta(F, E) = \sup \sum_{i=1}^n |F(x_i + h_i) - F(x_i - h_i)|$$

where the supremum is with regard to all sequences $\{[x_i - h_i, x_i + h_i]\}$ of nonoverlapping intervals with centers $x_i \in E$, with $h_i < \delta(x_i)$ and with both endpoints $x_i + h$ and $x_i - h$ *in the domain* of F. Again we write

$$VS_F(E) = \inf S_\delta(F, E)$$

where the infimum is taken over all gauges δ on E. We call this the *symmetric variation* of F on E and we call the set function VS_F the *symmetric variational measure* associated with F. VS_F is an outer measure on the real line.

We now can extend Definition 9.19 to functions whose domain is just a subset of (a, b). Again we say that F is *symmetrically–ACG* on (a, b) if F is defined on a subset of (a, b) and if (*i*) $\lambda_F(a, b)$ exists,

(*ii*) VS_F is σ–finite on (a, b) and

(*iii*) $VS_F(Z) = 0$ for every set $Z \subset (a, b)$ of Lebesgue measure zero.

We state the analogue of Theorem 9.21 for the (R_s^1)–integral. Note that the derivative F' is defined only relative to the domain of the function F. Of course the same applies to the computation of the limit for the expression $\lambda_F(a, b)$.

Theorem 9.34 *A function f is (R_s^1)–integrable on an interval $[a, b]$ if and only if there exists a function F that is defined on all of (a, b) except possibly some countable set and is symmetrically ACG_* on (a, b) with $F' = f$ almost everywhere there. In that case*

$$\int_a^b f(x)\, dx = \lambda_F(a, b).$$

9.3 The Approximate Symmetric Integral

9.3.1 Preliminaries

It had been known for some time that an integral based on the approximate symmetric derivative would have applications in the study of trigonometric series. A first step in constructing such an integral is to obtain a monotonicity theorem for this derivative. Over the years a number of authors had provided incorrect proofs. The first correct proof is that of Freiling and Rinne [112]. Their theorem is the following (quoted from Section 3.6.2 and Section 5.6). Here $\underline{ASD}\, F(x)$ denotes the lower approximate symmetric derivative of F at x defined as

$$\text{ap-}\lim_{h \to 0}\inf \frac{F(x + h) - F(x - h)}{2h}.$$

Theorem 9.35 (Freiling–Rinne) *Let F be a measurable function defined almost everywhere on an interval (a, b) and such that $\underline{ASD}\, F(x) \geq 0$ for every $a < x < b$. Then F is equivalent to a nondecreasing function.*

The measurability of F in the statement of the theorem must be assumed (see Section 6.2.2) and cannot be deduced from the fact that $\underline{ASD}\, F(x)$ is nonnegative for every $a < x < b$.

For the purposes of the integration theory associated with this derivative we require a more geometric version of the Freiling–Rinne theorem. We shall use the following definition.

Definition 9.36 By a *density gauge* on a measurable set $E \subset \mathbb{R}$ we shall mean a measurable set $\Delta \subset E \times (0,1)$ such that, for each $x \in E$,

$$| \{t : t \in (0,h), \quad (x,t) \notin \Delta \} | = o(h)$$

as $h \to 0+$.

Note especially that Δ is taken as measurable with respect to planar Lebesgue measure and that each section $\{t : (x,t) \in \Delta\}$ has inner density 1 on the right at 0. Note that if δ is a gauge on a set E (in the sense of Section 9.2.1) then the set $\Delta = \{(x,t) : x \in E, \ 0 < t < \delta(x)\}$ is a density gauge provided the set is measurable; for example if δ is a measurable function on a measurable set E then this would be the case.

The covering lemma we now state is proved in Section 3.6 and may be considered the basis for the theory of approximate symmetric integrals. It is the correct approximate analogue of Lemma 9.1.

Lemma 9.37 *Let Δ be a density gauge on (a,b). Then there is a set N of measure zero such that for every interval $[c,d]$ with $c, d \in (a,b) \setminus N$ there is a partition*

$$c = x_0 < x_1 < \ldots < x_m < x_{m+1} = d$$

with

$$\left(\frac{x_i + x_{i+1}}{2}, x_{i+1} - x_i \right) \in \Delta$$

$(i = 0, 1, \ldots, m)$.

On the basis of Theorem 9.35 or Lemma 9.37 a theory of approximate symmetric integrals has developed. The first is the Riemann type integral and the Perron type integral in [232]; Lee [180] also contains an account of the Perron version.

The theory that we will now develop is rather more complete and pleasant in detail than is the situation for the ordinary symmetric integrals discussed so far. The applications, too, are deeper.

9.3.2 An Approximate Symmetric Newton Integral

An approximate symmetric Newton integral can be based on the approximate symmetric derivative in the same way that the (N_s^1)–integral was developed in Section 9.2.2. The notion is, of course, rather restrictive but Newton integrals in general are frequently useful in discussions of generalized integrals and derivatives and this integral serves both as an introduction to some more complicated ideas and a reminder of more familiar ones. Moreover it has an immediate, if formal, application to the study of convergent trigonometric series (see Section 9.3.3).

The use of the increment $\alpha_F(a, b)$ in the definition below, is the natural approximate analogue of the notion $\lambda_F(a, b)$ used in Section 9.2.2. The measurability assumption is needed here since without it the primitive is not essentially unique (Section 6.2.2).

Definition 9.38 A finite function f defined everywhere on an interval (a, b) is said to be (N_{as}^1)-*integrable* on $[a, b]$ if there is a measurable function F defined almost everywhere on (a, b) so that ASD $F(x) = f(x)$ everywhere on (a, b) and such that the expression

$$\alpha_F(a, b) = \text{ap-}\lim_{h \to 0} (F(b - h) - F(a + h)) \tag{9.60}$$

exists. We write then

$$\alpha_F(a, b) = (N_{as}^1) \int_a^b f(x) \, dx$$

If the function F is measurable, defined a.e. on (a, b) and ASD $F(x) = f(x)$ everywhere on an interval (a, b) then F is essentially unique (i.e. any two measurable functions with identical approximate symmetric derivatives are equivalent). This follows directly from Theorem 9.35. Thus $\alpha_F(a, b)$ in (9.60), if it exists, does not depend on the choice of F.

We begin with some elementary properties of the (N_{as}^1)-integral that recall the material of Section 9.2.2. The proofs are nearly immediate.

Theorem 9.39 *Let f be (N_s^1)-integrable on an interval $[a, b]$. Then f is (N_{as}^1)-integrable and the values coincide.*

Proof. An (N_s^1)-primitive F for a function f is measurable (since symmetrically differentiable functions are measurable) and clearly it is then an (N_{as}^1)-primitive too. Certainly $\alpha_F(a, b) = \lambda_F(a, b)$ if the latter exists. ∎

Theorem 9.40 *Let f and g be (N_{as}^1)-integrable on an interval $[a, b]$. Then so too is any linear combination $sf + tg$ and*

$$\int_a^b (sf(x) + tg(x)) \, dx = s \int_a^b f(x) \, dx + t \int_a^b g(x) \, dx$$

in the sense of this integral.

This integral is as fragile as the other symmetric integrals and for the same reasons. It does not have the usual interval additivity properties. A product of an integrable function and a linear function need not be integrable. See Section 9.2.2 for a discussion and examples.

The analogue of Theorem 9.4 holds with a similar proof. We give the details to indicate how the change from ordinary limits to approximate limits may be effected.

Theorem 9.41 *If f is (N_{as}^1)-integrable on $[a, b]$ and on $[a, c]$ ($a < b < c$) then it is integrable on $[b, c]$ and*

$$\int_b^c f(x)\,dx = \int_a^c f(x)\,dx - \int_a^b f(x)\,dx \tag{9.61}$$

in the sense of this integral.

Proof. Let F be an (N_{as}^1)-primitive for f on $[a, c]$. By hypothesis $\alpha_F(a, b)$ and $\alpha_F(a, c)$ exist; also F must be approximately symmetrically continuous at b. We can choose measurable sets D_1, D_2 and D_3 each having density 1 on the right at 0 so that each of the limits

$$\lim_{h\to 0,\ h\in D_1} F(c - h) - F(a + h) = \alpha_F(a, c) \tag{9.62}$$

$$\lim_{h\to 0,\ h\in D_2} F(b - h) - F(a + h) = \alpha_F(a, b), \tag{9.63}$$

and

$$\lim_{h\to 0,\ h\in D_3} F(b + h) - F(b - h) = 0 \tag{9.64}$$

exists. Set $D = D_1 \cap D_2 \cap D_3$ so that D also is a measurable set with density 1 on the right at 0. Consider then the identity

$$F(c - h) - F(b + h) = (F(c - h) - F(a + h))$$
$$- (F(b - h) - F(a + h)) - (F(b + h) - F(b - h))).$$

It readily follows from this and from (9.62), (9.63) and (9.64) that

$$\alpha_F(b, c) = \alpha_F(a, c) - \alpha_F(a, b). \tag{9.65}$$

The identity (9.65) supplies (9.61) and the theorem is proved.

We know that (N_s^1) or (N_{as}^1)-integrability on an interval $[a, b]$ does not require integrability on all subintervals. We have following analogue of Theorem 9.5.

Theorem 9.42 *Suppose that f is (N_{as}^1)-integrable on an interval $[a, b]$ with F an (N_{as}^1)-primitive of f there. Then F is approximately continuous almost everywhere in $[a, b]$ and*

$$(N_{as}^1) \int_c^d f(t)\,dt = F(d) - F(c) \tag{9.66}$$

at any points c, $d \in (a, b)$ at which F is approximately continuous.

Proof. Since F is an (N_{as}^1)-primitive of f then, by definition, F must be measurable on (a, b) and so approximately continuous almost everywhere in that interval. If F is approximately continuous at c and d then certainly $\alpha_F(c,d) = F(d) - F(c)$. This supplies the identity (9.66) and the theorem follows since if F is an (N_{as}^1)–primitive on $[a,b]$ it must be as well on $[c,d]$.

We conclude this section with two familiar kinds of theorems that arise in the most studies of nonabsolutely convergent integrals. The first is that the integral is merely the Lebesgue integral for nonnegative functions and the second that an integrable function must be Lebesgue integrable on a dense set of intervals. These are the direct analogues of Theorem 9.8 and Theorem 9.9.

Theorem 9.43 *Suppose that $f \geq 0$ is (N_{as}^1)–integrable. Then f is Lebesgue integrable.*

Proof. Let F be an (N_{as}^1)–primitive for f. By definition, ASD $F(x) = f(x)$ everywhere in (a,b). By Theorem 7.54 this requires that the approximate derivative of F exists and is f a.e.. Also, by Theorem 9.35, F is equivalent to a monotonic function. Since f is a.e. the approximate derivative of a monotonic function it is in fact the ordinary derivative almost everywhere of such a function (see, for example, [18, p. 154]). It follows now from this that f must be Lebesgue integrable.

Theorem 9.44 *Suppose that f is (N_{as}^1)–integrable on an interval $[a,b]$. Then there is a set of intervals dense in $[a,b]$ on each of which f is Lebesgue integrable.*

Proof. It is known that an exact approximate symmetric derivative is Baire 1 (Theorem 7.45) and so the same proof as for Theorem 9.9 holds.

9.3.3 An Application to Trigonometric Series

As a quick application of the (N_{as}^1)–integral, one which serves to motivate its introduction, let us show that it solves, albeit formally, the coefficient problem for trigonometric series. That problem is to determine the coefficients of an everywhere convergent trigonometric series.

Our main tool in this section is the observation (Theorem 1.19) that the approximate symmetric derivative of the integrated series recovers the sum of a trigonometric series at any point at which it converges. As a result of this it is fairly clear that an approximate symmetric Newton integral will solve the coefficient problem for everywhere convergent trigonometric series.

Theorem 9.45 *Suppose that the trigonometric series*

$$f(x) = a_0/2 + \sum_{n=1}^{\infty} (a_n \cos nx + b_n \sin nx)$$

converges everywhere. Then f is (N^1_{as})–integrable on every period and

$$a_n = \frac{1}{\pi}\int_c^{c+2\pi} f(t)\cos nt\, dt\ ,\quad b_n = \frac{1}{\pi}\int_c^{c+2\pi} f(t)\sin nt\, dt$$

in the sense of that integral.

Proof. The formally integrated series

$$F(x) = xa_0/2 + \sum_{k=0}^{\infty}(b_k\cos kx - a_k\sin kx)/k$$

converges on a set B of full measure in every interval and has everywhere $f(x)$ for its approximate symmetric derivative by Theorem 1.19. Consequently, since F is measurable, it is an (N^1_{as})–primitive for f on every interval $[a, b]$ for which $\alpha_F(a, b)$ exists.

As F is approximately symmetrically continuous everywhere and $F(x) - xa_0/2$ is 2π–periodic it is clear that $\alpha_F(a, a + 2\pi)$ must exist for any real a. Simply write

$$F(a + 2\pi - h) - F(a + h) = F(a + 2\pi + h) - F(a + 2\pi - h) - (2\pi a_0/2)$$

and this is evident. Moreover for every c we may integrate through a period $[c, c + 2\pi]$ and obtain the correct formula for a_0, i.e. that $\pi a_0 = \alpha_F(c, c + 2\pi)$.

We obtain a formula for the remaining coefficients by a well known technique, called the formal multiplication of trigonometric series. Multiply the series

$$a_0/2 + \sum_{n=0}^{\infty} a_n\cos nx + b_n\sin nx$$

by $\cos kx$, use elementary trigonometric identities for $\cos nx\cos kx$ and for $\sin nx\cos kx$, and rearrange the series to obtain a trigonometric series that converges everywhere to $f(x)\cos kx$. For example the term corresponding to $a_n\cos kx\cos nx$ is replaced by

$$a_n(\cos(n + k)x + \cos(n - k)x)/2.$$

As before, its formally integrated series converges on a set B of full measure and has everywhere $f(x)\cos kx$ for its approximate symmetric derivative. Consequently (much as before) $f(x)\cos kx$ is integrable and we may integrate it over a period $[c, c + 2\pi]$ for every c. Since the constant term in the series for $f(x)\cos kx$ is $a_k/2$ this gives the required formula for a_k. A similar argument supplies the formula for b_k.

9.3.4 An Approximate Symmetric Variational Integral

The considerations in Section 9.2.5 that led us to define the (V_s^1)–integral lead by the same line of reasoning to an approximate version of a symmetric variational integral. This integral will turn out to be equivalent to a collection of natural "approximate symmetric" integrals.

Definition 9.46 Let f be defined on an interval (a, b). We say that f is (V_{as}^1)–integrable in $[a, b]$ if there is a measurable function F defined almost everywhere on (a, b) such that $\alpha_F(a, b)$ exists and with the property that for every $\epsilon > 0$ there is a density gauge Δ on (a, b) and a nondecreasing function τ on $[a, b]$ with $\tau(b) - \tau(a) < \epsilon$ and such that

$$|F(x + h) - F(x - h) - 2f(x)h| \leq \tau(x + h) - \tau(x - h) \qquad (9.67)$$

for all $(x, h) \in \Delta$. In this case we write

$$(V_{as}^1) \int_a^b f(x)\, dx = \alpha_F(a, b).$$

We can drop the reference to (V_{as}^1) when it is apparent which integral is intended. We shall refer to any function F satisfying the properties here as a (V_{as}^1)–*primitive* for f. It is obvious, in view of the discussion in Section 9.2.5, that this integral extends the (N_{as}^1) integral. While it is clearly the approximate analogue of the (V_s^1)–integral it is not immediate that it is an extension of that integral (because of the measurability assumption in the definition of a density gauge); that it is in fact an extension will emerge later.

The definition requires an argument to justify that the value of the integral does not depend on the choice of function F; this is supplied by the following lemma.

Lemma 9.47 *Suppose that the functions F and G are defined almost everywhere on (a, b) and that for every $\epsilon > 0$ there is a density gauge Δ on (a, b) and a nondecreasing function τ on $[a, b]$ with $\tau(b) - \tau(a) < \epsilon$ and such that*

$$|F(x + h) - F(x - h) - (G(x + h) - G(x - h))| \leq \tau(x + h) - \tau(x - h) \quad (9.68)$$

for all $a < x < b$, $(x, h) \in \Delta$. Then, if both $\alpha_F(a, b)$ and $\alpha_G(a, b)$ exist, $\alpha_F(a, b) = \alpha_G(a, b)$.

Proof. We apply (9.68), with $\epsilon = 1/n$ for each integer n, to obtain a sequence of monotonic functions τ_n with $\tau_n(b) - \tau_n(a) < 1/n$ and a sequence of density gauges Δ_n.

By Lemma 9.37 there is a set N of measure zero so that for every $c, d \in (a, b) \setminus N$ and each n there is a partition

$$c = x_0 < x_1 < \ldots < x_m < x_{m+1} = d$$

with

$$\left(\frac{x_i + x_{i+1}}{2}, x_{i+1} - x_i \right) \in \Delta_n$$

$(i = 0, 1, \ldots m)$. From this it follows that

$$|F(d) - F(c) - (G(d) - G(c))| < 1/n$$

for each n and hence that $F(d) - F(c) = G(d) - G(c)$, for c and d outside of this measure zero set. Thus, if both $\alpha_F(a, b)$ and $\alpha_G(a, b)$ exist, then $\alpha_F(a, b) = \alpha_G(a, b)$ as required.

The (V_{as}^1)–integral again lacks many of the usual additivity or multiplicative properties of integrals, just as does the (N_{as}^1)–integral and the (V_s^1)–integral, but has most of the familiar properties of these symmetric integrals. The analogue of Theorem 9.13 requires only conditions for an "approximate symmetric" join.

Theorem 9.48 *Let $a < b < c$ and suppose that f is (V_{as}^1)–integrable on $[a, b]$ with a primitive F and on $[b, c]$ with a primitive G. Then f is (V_{as}^1)–integrable on $[a, c]$ if the limit*

$$\text{ap-} \lim_{h \to 0+} G(b + h) - F(b - h) = \alpha \tag{9.69}$$

exists.

Proof. Let $\epsilon > 0$. Since f is (V_{as}^1)–integrable in $[a, b]$, $\alpha_F(a, b)$ exists and there is a density gauge Δ_1 on (a, b) and a nondecreasing function τ_1 on $[a, b]$ with $\tau_1(b) - \tau_1(a) < \epsilon/3$ and such that

$$|F(x + h) - F(x - h) - 2f(x)h| \leq \tau_1(x + h) - \tau_1(x - h) \tag{9.70}$$

for all $(x, h) \in \Delta_1$.

Similarly, since f is (V_{as}^1)–integrable in $[b, c]$, $\alpha_G(b, c)$ exists and there is a density gauge Δ_2 on (b, c) and a nondecreasing function τ_2 on $[b, c]$ with $\tau_2(c) - \tau_2(b) < \epsilon/3$ and such that

$$|G(x + h) - G(x - h) - 2f(x)h| \leq \tau_2(x + h) - \tau_2(x - h) \tag{9.71}$$

for all $(x, h) \in \Delta_2$. Let τ_1 and τ_2 be constant otherwise (i.e. outside these intervals).

Define $\Delta = \Delta_1 \cup \Delta_2 \cup W$ where W is the set of all (b, h) such that

$$|G(b + h) - F(b - h) - \alpha| < \epsilon/6 \quad \text{and} \quad |2f(b)h| < \epsilon/6. \tag{9.72}$$

It is straightforward to show that Δ is a density gauge on (a, c).

For $a < x < b$ write $H(x) = F(x)$ and for $b < x < c$ write $H(x) = G(x) - \alpha$ where α is as in (9.69); take $H(b)$ arbitrary. Construct a nondecreasing function τ_3 constant on $[a, b]$ and on $[b, c]$ with $\tau_3(b+) - \tau_3(b-) = \epsilon/3$. Write $\tau = \tau_1 + \tau_2 + \tau_3$.

Note that $\alpha_H(a, c)$ exists and that τ is a nondecreasing function on $[a, c]$ with $\tau(c) - \tau(a) < \epsilon$. We readily verify, using (9.70), (9.71) and (9.72), that

$$|H(x + h) - H(x - h) - 2f(x)h| \le \tau(x + h) - \tau(x - h) \qquad (9.73)$$

for all $(x, h) \in \Delta$. By definition then f is (V_{as}^1)–integrable in $[a, b]$ and the theorem is proved.

The analogue of Theorem 9.42 holds. Here there is little to prove because the measurability of the primitive F is assumed and so, immediately, F is approximately continuous almost everywhere.

Theorem 9.49 *Suppose that f is (V_{as}^1)–integrable on an interval $[a, b]$ and that F is a (V_{as}^1)–primitive of f there. Then F is approximately continuous almost everywhere in $[a, b]$ and*

$$(N_{as}^1) \int_c^d f(t)\, dt = F(d) - F(c)$$

at any points c, $d \in (a, b)$ at which F is approximately continuous.

The following now are some properties of the (V_{as}^1)–integral analogous to those given in Section 9.2.5 for the (V_s^1)–integral. The first, the analogue of Theorem 9.14, allows us to define an integral of a function that is defined only almost everywhere on an interval.

Theorem 9.50 *Suppose that f is (V_{as}^1)–integrable on an interval $[a, b]$ and that $f = g$ almost everywhere. Then g is (V_{as}^1)–integrable on $[a, b]$ and*

$$\int_a^b f(x)\, dx = \int_a^b g(x)\, dx.$$

Proof. Let $\epsilon > 0$. Since f is (V_{as}^1)–integrable in $[a, b]$, there is an F so that $\alpha_F(a, b)$ exists and there is a density gauge Δ_1 on (a, b) and a nondecreasing function τ_1 on $[a, b]$ with $\tau_1(b) - \tau_1(a) < \epsilon/2$ and such that

$$|F(x + h) - F(x - h) - 2f(x)h| \le \tau_1(x + h) - \tau_1(x - h) \qquad (9.74)$$

for all $(x, h) \in \Delta_1$.

Let $Z = \{x \in (a, b) : f(x) \ne g(x)\}$. Since Z has measure zero we may choose a nondecreasing function τ_2 on $[a, b]$ so that $\tau_2(b) - \tau_2(a) < \epsilon/2$ and so that $\tau_2'(x) = +\infty$ at every point $x \in Z$. Define Δ_2 to contain all points (x, h)

with $x \in (a, b) \setminus Z$ and $0 < h < 1$ and to contain all points (x, t) with $x \in Z$ and

$$\frac{\tau_2(x + h) - \tau_2(x - h)}{2h} \geq |f(x) - g(x)| \qquad (9.75)$$

for $0 < h < 1$. Write $\tau = \tau_1 + \tau_2$.

We observe that $\Delta = \Delta_1 \cap \Delta_2$ is a density gauge on (a, b). Note, too, that τ is a nondecreasing function on $[a, b]$ and that $\tau(b) - \tau(a) < \epsilon$. We readily verify, using (9.74) and (9.75), that

$$|F(x + h) - F(x - h) - 2g(x)h| \leq \tau(x + h) - \tau(x - h) \qquad (9.76)$$

for all $(x, h) \in \Delta$. By definition then (9.76) supplies the fact that g is (V_{as}^1)-integrable in $[a, b]$ and that the integrals have the same value (i.e. $\alpha_F(a, b)$). This completes the proof.

Theorem 9.51 *Suppose that f is (V_{as}^1)-integrable on an interval $[a, b]$ and that F is an (V_{as}^1)-primitive of f. Then F is approximately symmetrically continuous, $ADF(x) = f(x)$ almost everywhere in (a, b) and f is measurable.*

Proof. We shall show first that, at almost every point x, ASD $F(x) = f(x)$. By Theorem 7.54 this means that AD $F(x) = f(x)$ almost everywhere in (a, b); finally then f, as the a.e. approximate derivative of a measurable function, is measurable.

Let E_n denote the set of points x in (a, b) for which

$$\left\{ h : |F(x + h) - F(x - h) - 2f(x)h| > n^{-1}h \right\}$$

has positive upper density at 0 on the right. Every point x in (a, b) at which ASD $F(x) = f(x)$ fails lies in some set E_n and so the result we require is proved if we show that each set E_n has measure zero. This we can accomplish by a Vitali argument by constructing a contradiction based on the assumption that, for some n, $|E_n| > c > 0$.

Let $4n\epsilon = c$. There is a density gauge Δ on (a, b) and a nondecreasing function τ on $[a, b]$ with $\tau(b) - \tau(a) < \epsilon$ and such that

$$|F(x + h) - F(x - h) - 2f(x)h| \leq \tau(x + h) - \tau(x - h) \qquad (9.77)$$

for all $(x, h) \in \Delta$. The collection of all intervals $[x - h, x + h]$ with $x \in E_n$, $(x, h) \in \Delta$ and

$$|F(x + h) - F(x - h) - 2f(x)h| > n^{-1}h \qquad (9.78)$$

is a Vitali cover of E_n. Let us choose a finite subcollection of nonoverlapping intervals $[x_i - h_i, x_i + h_i]$ $(i = 1, 2, \ldots p)$ from this cover in such a way that

$$|E_n| < 2 \sum_{i=1}^{p} 2h_i.$$

Then, using the inequalities (9.77) and (9.78), we obtain

$$c < |E_n| < 2 \sum_{i=1}^{p} 2h_i < 4n \sum_{i=1}^{p} n^{-1}h_i$$

$$< 4n \sum_{i=1}^{p} |F(x_i + h_i) - F(x_i - h_i) - 2f(x_i)h_i|$$

$$\leq 4n \sum_{i=1}^{p} \tau_1(x_i + h_i) - \tau_1(x_i - h_i) < 4n\epsilon = c.$$

With this contradiction the proof that $ASD\, F(x) = f(x)$ holds at almost every point x is complete.

We now check the remaining statements in the theorem. It is clear from Definition 9.46 that any primitive F must be approximately symmetrically continuous. By Theorem 7.54 the fact that $ASD\, F(x) = f(x)$ holds at almost every point x means that $AD\, F(x) = f(x)$ almost everywhere in (a, b); finally then f, as the a.e. approximate derivative of a measurable function, is measurable.

Theorem 9.52 *Suppose that F is approximately symmetrically continuous in (a, b), that*

$$-\infty < \underline{ASD}\, F(x) \leq \overline{ASD}\, F(x) < +\infty \qquad (9.79)$$

for all but countably many x in (a, b) and that $\alpha_F(a, b)$ exists. Then

$$AD\, F(x) = f(x)$$

exists almost everywhere in (a, b), f is (V_{as}^1)–integrable on $[a, b]$ and

$$\int_a^b f(x)\, dx = \alpha_F(a, b).$$

Proof. With appropriate changes the proof of the analogous Theorem 9.16 can be made to work here. This can also be made to follow from Theorem 9.58 below. We omit the details.

The (V_{as}^1)–integral shares the usual property of non-absolutely convergent integrals, that it is identical to the Lebesgue integral when applied to nonnegative functions.

Theorem 9.53 *Suppose that $f \geq 0$, that f is (V_{as}^1)–integrable on $[a, b]$ and that F is a (V_{as}^1)–primitive for f. Then f is Lebesgue integrable and F is equivalent to an indefinite Lebesgue integral for f on (a, b).*

Proof. Again, with appropriate changes, the proof of the analogous Theorem 9.18 can be made to work here. One requires an appeal to Lemma 9.37 but otherwise the details are not much different.

We require the next theorem for our applications to trigonometric series. This property, that a periodic function should be integrable over every period if it is integrable over at least one period, has been established for the ordinary symmetric integrals (Theorem 9.23). Many other versions of symmetric integrals would fail to have such a property.

Theorem 9.54 *Suppose that f is 2π-periodic and is (V_{as}^1)-integrable on some interval $[a, a + 2\pi]$. Then f is (V_{as}^1)-integrable on every period and to the same value.*

Proof. Let F be a (V_{as}^1)-primitive for f on a fixed interval $[a, a + 2\pi]$. We define G for almost every real x by writing

$$G(x + 2k\pi) = F(x) + k\alpha_F(a, a + 2\pi)$$

for $a < x < a + 2\pi$. This defines G almost everywhere and G is measurable. Note that G is approximately symmetrically continuous. For points of the form $a + +2k\pi$ this needs to be checked. At all other points it is obvious. For example at the point $a + 2\pi$ we see that

$$\text{ap} \lim_{h \to 0} G(a + 2\pi + h) - G(a + 2\pi - h) =$$
$$\text{ap} \lim_{h \to 0} F(a + h) - F(a + 2\pi - h) + \alpha_F(a, a + 2\pi) = 0.$$

Note also that for any interval $[c, c + 2\pi]$ the expression $\alpha_G(c, c + 2\pi)$ exists. For example if $a < c < a + 2\pi$ then

$$\text{ap} \lim_{h \to 0} G(c + 2\pi - h) - G(c + h)$$
$$= \text{ap} \lim_{h \to 0} -F(c + h) + F(c - h) + \alpha_F(a, a + 2\pi) = \alpha_F(a, a + 2\pi).$$

A similar computation works for any such interval.

Let $\epsilon > 0$. There is a density gauge Δ_1 on $(a, a + 2\pi)$ and a nondecreasing function τ_1 on $[a, a + 2\pi]$ with $\tau_1(a + 2\pi) - \tau_1(a) < \epsilon/3$ and such that

$$|F(x + h) - F(x - h) - 2f(x)h| \le \tau_1(x + h) - \tau_1(x - h)$$

for all $(x, h) \in \Delta_1$. Extend τ_1 to a monotonic function τ_2 on the real line so that $\tau_2(x + 2k\pi) = \tau_1(x) + k(\tau_1(a + 2\pi) - \tau_1(a))$ for $a < x < a + 2\pi$.

We show that f is (V_{as}^1)-integrable on any interval $[c, c + 2\pi]$ with primitive G. For example if $a < c < a + 2\pi$ then the computations are as follows. Define Δ as the collection of all pairs (x, h) with $x - 2\pi \in (a, c)$, $(x, h) \in \Delta_1$, or with $x \in (c, a + 2\pi)$, $(x, h) \in \Delta_1$, and as well all pairs $(a + 2\pi, h)$ with $|G(a + 2\pi + h) - G(a + 2\pi - h)| < \epsilon/6$ and $|2f(a + 2\pi)h| < \epsilon/6$. It can be seen that Δ is a density gauge on $(c, c + 2\pi)$.

Construct a nondecreasing function τ_3 constant on $[c, a + 2\pi)$ and on $(a + 2\pi, c + 2\pi]$ with $\tau_3((a + 2\pi)+) - \tau_3((a + 2\pi)-) = \epsilon/3$. Write $\tau = \tau_2 + \tau_3$. Then we obtain $\tau(c + 2\pi) - \tau(c) < \epsilon$ and

$$|G(x + h) - G(x - h) - 2f(x)h| \leq \tau(x + h) - \tau(x - h)$$

for all $(x, h) \in \Delta$. From this it follows that f is (V_{as}^1)-integrable on $[c, c + 2\pi]$ as required.

9.3.5 Approximate Symmetric Absolute Continuity

The usual connection between the variational properties and the differentiation properties of a function have been extended to a wide variety of settings. We can do much the same for the approximate symmetric derivative in this regard as we did in Section 9.2.6 for the symmetric derivative.

We define an approximate version of the symmetric variation of a function F defined in Chapter 8 and repeated in Section 9.2.6. Let $E \subset \mathbb{R}$, let Δ be a density gauge on $\mathbb{R}$ and suppose that F is a measurable function, defined almost everywhere. Then we write

$$S_\Delta(F, E) = \sup \sum_{i=1}^{n} |F(x_i + h_i) - F(x_i - h_i)|$$

where the supremum is with regard to all sequences $\{[x_i - h_i, x_i + h_i]\}$ of nonoverlapping intervals with centers $x_i \in E$ and with $(x_i, h_i) \in \Delta$. Since F is defined almost everywhere we can take the density gauge so that $F(x + h)$ and $F(x - h)$ are defined for all $(x, h) \in \Delta$. We write then

$$\text{VAS}_F(E) = \inf S_\Delta(F, E)$$

where the infimum is taken over all density gauges Δ.

This expression shall be called the *approximate symmetric variation* of F on E and the set function VAS_F is called the *approximate symmetric variational measure* associated with F. It is possible to verify that VAS_F is an outer measure on the real line for which all Borel sets are measurable. The proof would follow the methods of Chapter 8, complicated somewhat here by measurability assumptions on the density gauges.

Definition 9.55 Let F be a measurable function defined almost everywhere on an interval (a, b). We say that F is *as*-ACG$_*$ on (a, b) if

(*i*) $\alpha_F(a, b)$ exists,

(*ii*) VAS_F is σ-finite on (a, b) and

(*iii*) $\text{VAS}_F(Z) = 0$ for every set $Z \subset (a, b)$ of Lebesgue measure zero.

A function that is ACG$_*$ on $[a, b]$ can be shown to be as–ACG$_*$ in (a, b) although this is not immediate from the usual definitions. Note that a function as–ACG$_*$ on (a, b) must be approximately symmetrically continuous at every point in that interval.

If F is as–ACG$_*$ on (a, b) then AD $F(x)$ exists almost everywhere in (a, b); this follows from the lemmas we now prove. The first lemma carries most of the information needed to relate the measure VAS$_F$ to the differential structure of F. As usual we write $\underline{ASD}\, F(x)$ and $\overline{ASD}\, F(x)$ to denote the upper and lower approximate symmetric derivates of a measurable function F.

Lemma 9.56 *If at each point x of a set E one of the two inequalities*

$$-\alpha > \underline{ASD}\, F(x) \quad or \quad \overline{ASD}\, F(x) > \alpha$$

is true then VAS$_F(E) \geq \alpha|E|$.

Proof. We may assume that E is bounded. Let Δ be any density gauge and let $\mathcal{C}$ denote the class of all intervals $[x, y]$ for which $((x + y)/2, y - x) \in \Delta$ and

$$|F(y) - F(x)| > \alpha(y - x), \quad y - x < \delta((x + y)/2), \quad (x + y)/2 \in E.$$

Since either $-\alpha > \underline{ASD}\, F(x)$ or $\overline{ASD}\, F(x) > \alpha$ is true at each point $x \in E$ it is clear that $\mathcal{C}$ is a Vitali cover of E. For any $\beta < |E|$ choose a nonoverlapping collection $\{[x_i - h_i, x_i + h_i]\} \subset \mathcal{C}$ so that

$$\beta < \sum_{i=1}^{n} 2h_i = \alpha^{-1} \sum_{i=1}^{n} 2\alpha h_i$$

$$\leq \alpha^{-1} \sum_{i=1}^{n} |F(x_i + h_i) - F(x_i - h_i)| \leq \alpha^{-1} S_\Delta(F, E).$$

From this it follows that $\alpha\beta \leq$ VAS$_F(E)$ and finally that $\alpha|E| \leq$ VAS$_F(E)$ as required.

Theorem 9.57 *Let F be a measurable function and let E be a measurable set. If* VAS$_F(E) < +\infty$ *then the approximate derivative* AD $F(x)$ *exists for almost every $x \in E$. If* VAS$_F(E) = 0$ *then* AD $F(x) = 0$ *for almost every $x \in E$.*

Proof. Write

$$A = \left\{ x \in E \ : \ \underline{ASD}\, F(x) = -\infty \quad or \quad \overline{ASD}\, F(x) = +\infty \right\}$$

and

$$B = \{ x \in E \ : \ ASD\, F(x) \neq 0 \}$$

and for any natural number n write

$$A_n = \left\{ x \in E : \underline{\text{ASD}}\, F(x) < -n \text{ or } \overline{\text{ASD}}\, F(x) > n \right\}$$

and

$$B_n = \left\{ x \in E : -n^{-1} > \underline{\text{ASD}}\, F(x) \text{ or } \overline{\text{ASD}}\, F(x) > n^{-1} \right\}.$$

Note that $A = \bigcap_{n=1}^{\infty} A_n$ and $B = \bigcup_{n=1}^{\infty} B_n$. It follows directly from the lemma that $n|A_n| \leq \text{VAS}_F(E)$ and that $|B_n| \leq n\text{VAS}_F(E)$. Thus in the case that $\text{VAS}_F(E) < +\infty$ the set A has measure zero and it follows that the approximate symmetric derivates of F are finite almost everywhere in E.

Once again, by Theorem 7.54, this requires that $\text{AD}\, F(x)$ exists for almost every $x \in E$. In the case that $\text{VAS}_F(E) = 0$, B has measure zero and it follows that the approximate symmetric derivative of F vanishes almost everywhere in E and so $\text{AD}\, F(x) = 0$ for almost every $x \in E$.

These concepts allow us to give a Lusin type characterization of the (V_{as}^1)–integral.

Theorem 9.58 *A function f is (V_{as}^1)–integrable on an interval $[a, b]$ if and only if there exists a function F as–ACG_* on (a, b) with $\text{AD}\, F(x) = f(x)$ almost everywhere there. In that case*

$$\int_a^b f(x)\, dx = \alpha_F(a, b).$$

Proof. Suppose that f is (V_{as}^1)–integrable on $[a, b]$ with primitive F. For each m let $N_m = \{ x \in (a, b) : |f(x)| < m \}$. We show that $\text{VAS}_F(N_m) < +\infty$ and that $\text{VAS}_F(Z \cap N_m) = 0$ for each set $Z \subset (a, b)$ of measure zero.

Let $\epsilon > 0$. By definition there is a density gauge Δ_1 on (a, b) and a nondecreasing function τ on $[a, b]$ with $\tau(b) - \tau(a) < \epsilon$ and such that

$$|F(x + h) - F(x - h) - 2f(x)h| \leq \tau(x + h) - \tau(x - h) \qquad (9.80)$$

for all $(x, h) \in \Delta_1$.

From this we evidently get with $\Delta = \Delta_1$,

$$\text{VAS}_F(N_m) \leq S_\Delta(F, N_m) \leq m(b - a) + \epsilon. \qquad (9.81)$$

If Z has measure zero then there is an open set G containing Z with $|G| < \epsilon$. Let Δ be defined on Z so that if $(x, h) \in \Delta$ and $x \in Z$ then $(x, h) \in \Delta_1$ and $(x - h, x + h) \subset G$. From (9.80) we obtain

$$\text{VAS}_F(Z \cap N_m) \leq S_\Delta(F, Z \cap N_m) \leq m\epsilon + \epsilon. \qquad (9.82)$$

As $\epsilon > 0$ is arbitrary we see, from (9.81), that $\text{VAS}_F(N_m) < +\infty$ and so the measure VAS_F must be σ–finite. From (9.82) we see that $\text{VAS}_F(Z \cap N_m) = 0$

for each set $Z \subset (a, b)$ of measure zero and so $\mathrm{VAS}_F(Z) = 0$ for each such set. This is enough to establish that the function F is as-ACG_* on (a, b). By Theorem 9.51, AD $F(x) = f(x)$ almost everywhere there and so the necessity of each of the conditions stated in the theorem has been proved.

Let us prove the converse now; suppose F has the properties stated in the theorem. Let $\epsilon > 0$. Let E denote the set of points x in (a, b) at which AD $F(x) = f(x)$. Choose a density gauge Δ_0 so that $|F(x + h) - F(x) - f(x)h| < \epsilon h$ for $x \in E$ and $(x, h) \in \Delta_0$.

For each $m \geq 1$ let

$$Z_m = \{x \in (a, b) \setminus E \; : \; m - 1 \leq |f(x)| < m\}.$$

As each set Z_m has measure zero there is an open set $G_m \supset Z_m$ with $|G_m| < \epsilon m^{-1} 2^{-m}$. By the assumption that F is as– ACG_* on (a, b) we may choose Δ_m so that if $(x, h) \in \Delta_0$ and $x \in Z_m$ then $(x - h, x + h) \subset G_m$ and so that

$$\sum_{i=1}^{n} |F(x_i + h_i) - F(x_i - h_i)| < \epsilon 2^{-m}$$

where the sum is taken over any sequence $\{[x_i - h_i, x_i + h_i]\}$ of nonoverlapping intervals with centers $x_i \in Z_m$ and with $(x_i, h_i) \in \Delta_m$. For such intervals

$$\sum_{i=1}^{n} |F(x_i + h_i) - F(x_i - h_i) - 2f(x_i)h_i| < \epsilon 2^{-m}(1 + mm^{-1}) = \epsilon 2^{-m+1}.$$

Let $\tau_m(x)$ be the supremum of such sums for sequences $\{[x_i - h_i, x_i + h_i]\}$ of nonoverlapping subintervals of $[a, x]$ with centers $x_i \in Z_m$ and with $(x_i, h_i) \in \Delta_m$. Write

$$\tau(x) = \sum_{m=1}^{\infty} \tau_m(x) + \epsilon(x - a)$$

and construct a density gauge Δ so that if $(x, h) \in \Delta$ and $x \in E$ then $(x, h) \in \Delta_0$ while if $x \in Z_m$ then $(x, h) \in \Delta_m$.

Thus we have obtained τ on $[a, b]$ with $\tau(b) - \tau(a) < \epsilon(2 + b - a)$ and such that

$$|F(x + h) - F(x - h) - 2f(x)h| \leq \tau(x + h) - \tau(x - h) \qquad (9.83)$$

for all $a < x < b$, $(x, h) \in \Delta$. From (9.83) we see that, by definition, F is a (V_{as}^1)-primitive of f on $[a, b]$ and the theorem is proved.

We conclude this section with a version of the "Cauchy property" for this integral. This follows easily from the Lusin characterization of this integral. This is the approximate analogue of Theorem 9.33.

Theorem 9.59 *Suppose that f is (V_{as}^1)-integrable on each interval $[c, d]$ for c, $d \in (a, b) \setminus N$ where N has measure zero. If*

$$\operatorname{ap} \lim_{h \to 0} \int_{a+h}^{b-h} f(t)\, dt = \lambda$$

exists then f is (V_{as}^1)–integrable on $[a, b]$ and $\int_a^b f(t)\, dt = \lambda$.

9.3.6 An Approximate Symmetric Perron Integral

The fact that a monotonicity theorem is available for the approximate symmetric derivative allows a Perron type integral to be defined based on this derivative in much the same fashion as the original Perron integral. It seems that the first published account of such an integral is Kubota [160] but with restrictive hypotheses on the major and minor functions, too restrictive for applications to trigonometric series; the monotonicity theorem asserted there is correct although the proof is not (see the remarks in Section 5.6). As part of the development of a Riemann type integral based on the approximate symmetric derivative Preiss and Thomson [232] give an approximate symmetric version of the Perron integral that is sufficient to solve the coefficient problem for trigonometric series. Lee [180] contains an account too.

The presentation we now give is partly from [232] but the version there does not exploit the endpoint symmetry that is a feature of our presentation here. For a real function f, defined on (a, b), we shall say that a measurable real function H is a *as–majorant* of f on $[a, b]$ provided that H is defined almost everywhere on (a, b), that $\alpha_H(a, b)$ exists and that

$$\underline{\mathrm{ASD}}\, H(x) \geq f(x)$$

everywhere in (a, b). As–minorants G are then defined so that $-G$ is a majorant of $-f$ and the Perron integral is defined in more or less the usual manner by taking extremes over all majorants and minorants. If

$$\inf_H \alpha_H(a, b) = \sup_G \alpha_G(a, b) = c$$

then f is said to be (P_{as}^1)–integrable on $[a, b]$ and we write

$$(P_{as}^1) \int_a^b f(x)\, dx = c.$$

Such an approximate symmetric–Perron type integral needs to be justified by a monotonicity theorem. If H is an as–majorant of f and G is an as–minorant of f then, with $K = H - G$, $\underline{\mathrm{ASD}}\, K(x) \geq 0$ $(a < x < b)$ and it follows from Theorem 9.35 that K is equivalent to a nondecreasing function. As both $\alpha_H(a, b)$ and $\alpha_G(a, b)$ exist so too does, $\alpha_K(a, b)$ and this must be nonnegative since K is essentially nondecreasing. It follows that $\alpha_H(a, b) \geq \alpha_G(a, b)$. Thus the integral exists if for every $\epsilon > 0$ there is a pair of major/minor functions H and G with $\alpha_H(a, b) - \alpha_G(a, b) < \epsilon$.

Such an integral extends both the ordinary Perron integral and the symmetric Perron integral (from Section 9.2.8); ordinary Perron majorants or symmetric majorants are majorants in this sense. To check that a symmetric majorant in the sense of Section 9.2.8 is measurable one needs to see that it is, in any case, symmetrically semi-continuous and then the material in Section 6.4.1 supplies measurability.

This integral clearly integrates all exact approximate symmetric derivatives of measurable functions and so includes the (N^1_{as})–integral. We shall show that it is equivalent to the (V^1_{as})–integral defined in Section 9.3.4 thus giving us, so far, three characterizations of this integral.

Theorem 9.60 *A function f is (V^1_{as})–integrable on an interval $[a, b]$ if and only if it is (P^1_{as})–integrable on $[a, b]$ and then*

$$(V^1_{as}) \int_a^b f(x)\, dx = (P^1_{as}) \int_a^b f(x)\, dx.$$

Proof. Suppose first that f is (V^1_{as})–integrable on $[a, b]$. Let $\epsilon > 0$ and let F be a (V^1_{as})–primitive of f on (a, b). Then there is a density gauge Δ on (a, b) and a nondecreasing function τ on $[a, b]$ with $\tau(b) - \tau(a) < \epsilon/2$ and such that

$$|F(x + h) - F(x - h) - 2f(x)h| \leq \tau(x + h) - \tau(x - h) \qquad (9.84)$$

for all $a < x < b$, $(x, h) \in \Delta$.

Write $H(x) = F(x) + \tau(x)$ and $G(x) = F(x) - \tau(x)$. Both H and G are a.e. defined and measurable. Moreover $\alpha_H(a, b)$ and $\alpha_G(a, b)$ both exist and $0 \leq \alpha_H(a, b) - \alpha_G(a, b) < \epsilon$. From (9.84) one sees that H is a (P^1_{as})–major function for f and that G is a (P^1_{as})–minor function for f. This proves the (P^1_{as})–integrability of f and, from the inequality $\alpha_G(a, b) \leq \alpha_F(a, b) \leq \alpha_H(a, b)$, we see that the integral has the value $\alpha_F(a, b)$. This completes the proof in one direction.

Now suppose that f is (P^1_{as})–integrable on $[a, b]$. For each integer n we may choose sequences of measurable real functions H_n and G_n so that

$$\underline{\text{ASD}}\, H_n(x) \geq f(x) \geq \overline{\text{ASD}}\, G_n(x)$$

everywhere in (a, b) and so that

$$\alpha_{H_n}(a, b) - \alpha_{G_n}(a, b) < 1/n.$$

Recall that $H_n - G_n$ is essentially nondecreasing on (a, b).

By subtracting appropriate constants from each member of the sequences we may assume that there is a function F with $F(x) = \lim_{n \to \infty} H_n(x) = \lim_{n \to \infty} G_n(x)$ and $H_n(x) \geq F(x) \geq G_n(x)$ almost everywhere in (a, b). Let us suppose these hold on $B \subset (a, b)$. Note that $H_n - F$ and $F - G_n$ are monotonic on B; in particular if $x + h \in B$ and $x - h \in B$ then $H_n(x+h) - F(x+h) \geq H_n(x-h) - F(x-h)$ and $G_n(x+h) - F(x+h) \leq G_n(x-h) - F(x-h)$ and we shall find use for this in the next paragraph.

Let Δ_n denote the set of all pairs (x, h) with $a < x < b$, $0 < h$, $x + h \in B$, $x - h \in B$ and

$$H_n(x + h) - H_n(x - h) > 2(f(x) - n^{-1})h,$$

$$G_n(x + h) - G_n(x - h) < 2(f(x) + n^{-1})h.$$

We obtain, after some computations, that for $(x, h) \in \Delta_n$,

$$|F(x + h) - F(x - h) - 2f(x)h| \leq 2n^{-1}h +$$
$$H_n(x + h) - G_n(x + h) - (H_n(x - h) - G_n(x - h)).$$

Since the H_n and G_n are majorants and minorants of f each Δ_n is a density gauge on (a, b).

If we set $\tau(x) = n^{-1}x + H_n(x) - G_n(x)$ for $x \in B$ we have a monotonic function on B; extend this to a monotonic function on $[a, b]$ and we will have

$$\tau(b) - \tau(a) < 2n^{-1}(b - a) + n^{-1}.$$

Then we have obtained

$$|F(x + h) - F(x - h) - 2f(x)h| \leq \tau(x + h) - \tau(x - h)$$

for $(x, h) \in \Delta_n$ and evidently F is a (V_{as}^1)-primitive for f on $[a, b]$.

9.3.7 An Approximate Symmetric Riemann Integral

The catalogue of integrals on the real line and in Euclidean space that can be characterized as Riemann type integrals is growing. While this may be partly a curiosity it is frequently useful.

The approximate symmetric Riemann integral that we present in this section is equivalent to the other approximate symmetric integrals presented above. It is closely related to the integral first introduced in [232] and is the approximate analogue of the symmetric Riemann integral from Section 9.2.9. The ordinary symmetric Riemann integral may be considered elementary in as much as the covering lemma which justifies its existence is entirely elementary. The approximate symmetric Riemann integral must be considered to lie deeper as its source is Lemma 9.37.

As before, by a *symmetric partition* of an interval $[a, b]$ we shall mean a set of points

$$a < x_1 < x_2 < \ldots < x_n < b$$

such that $x_1 - a = b - x_n$. Such a partition is said to be *symmetrically finer* than a density gauge Δ on $[a, b]$ provided

$$\left(\frac{x_1 + x_{i+1}}{2}, x_{i+1} - x_i \right) \in \Delta \quad \text{for } i = 1, 2, \ldots, n - 1,$$

$$(a, x_1 - a) \in \Delta \quad \text{and} \quad (b, b - x_n) \in \Delta.$$

(Note that the only "symmetry" imposed by the definition is the way it treats the endpoints and the interpretation of the fineness with respect to Δ.)

Definition 9.61 Let f be defined on $[a, b]$. We say that f is (R^1_{as})–integrable on $[a, b]$ if there is a number C such that for every $\epsilon > 0$ there is a density gauge Δ on $[a, b]$ with the property for any symmetric partition

$$a < x_1 < x_2 < \ldots < x_n < b \quad (a - x_1 = b - x_n)$$

that is symmetrically finer than Δ

$$\left| \sum_{i=1}^{n-1} f\left(\frac{x_{i+1} + x_i}{2} \right) (x_{i+1} - x_i) - C \right| < \epsilon.$$

In this case we write

$$(\mathrm{R}^1_{as}) \int_a^b f(x) \, dx = C.$$

Justification for this definition requires that for any density gauge Δ such partitions do in fact exist. This is readily supplied by Lemma 9.37.

It is immediately clear that a function that is Riemann integrable is integrable in this sense and to the same value. It may also appear to be clear that this integral extends the (R^1_s)–integral too; but it is easy to forget that there is a measurability assumption in the definition of a density gauge and there is no measurability assumptions on the gauge for the (R^1_s)–integral.

Our main theorem in this section asserts that this integral is equivalent to the other approximate symmetric integrals so far defined. This, together with the variational characterization in Section 9.3.4, the Lusin type characterization in Section 9.3.5 and the Perron characterization in Section 9.3.6 provides four quite different equivalent expressions for this one integral.

Theorem 9.62 *The (R^1_{as})–integral is equivalent to the (V^1_{as}) and (P^1_{as})–integrals.*

Proof. We recall, by Theorem 9.60, that the (V^1_{as}) and (P^1_{as})–integrals are equivalent. Let us suppose that f that is (P^1_{as})–integrable on an interval $[a, b]$; we show it is (R^1_{as})–integrable there and with the same value.

Let $\epsilon > 0$. Let H be an arbitrary major function for f on $[a, b]$. Choose a density gauge Δ for $[a, b]$ so that

$$H(x + h) - H(x - h) > 2(f(x) - \epsilon)h$$

for $(x, h) \in \Delta$ with $a < x < b$ and so that

$$|H(b - h) - H(a + h) - \alpha_H(a, b)| < \epsilon$$

for (a, h) or $(b, h) \in \Delta$. Then for any symmetric partition

$$a < x_1 < x_2 < \ldots < x_n < b \quad (a - x_1 = b - x_n)$$

that is symmetrically finer than Δ we obtain

$$\sum_{i=1}^{n-1} f\left(\frac{x_{i+1} + x_i}{2}\right)(x_{i+1} - x_i) <$$
$$H(x_n) - H(x_1) + \epsilon(b - a) < \alpha_H(a, b) + \epsilon(b - a + 1).$$

In the same way we use an arbitrary minor function G and find a Δ so that the corresponding symmetric Riemann sums exceed $\alpha_G(a, b) - \epsilon(b - a + 1)$. It follows now easily that if f is (P_{as}^1)-integrable on $[a, b]$ it must be (R_{as}^1)-integrable too and with the same value assigned for the integrals. This completes the proof in one direction.

Now let us suppose that f is (R_{as}^1)-integrable on an interval $[a, b]$ with integral C; we show it is (V_{as}^1)-integrable there. Evidently the first part of the proof will already supply that the values are the same. There is a sequence of density gauges Δ_m on $[a, b]$ with the property for any symmetric partition

$$a < x_1 < x_2 < \ldots < x_n < b \quad (a - x_1 = b - x_n)$$

that is symmetrically finer than Δ_m

$$\left|\sum_{i=1}^{n-1} f\left(\frac{x_{i+1} + x_i}{2}\right)(x_i - x_{i-1}) - C\right| < 1/m. \tag{9.85}$$

By Lemma 9.37 there is a set Z of measure zero, $Z \subset (a, b)$ so that given any pair $c, d \in (a, b) \setminus Z$, $c < d$ and any m there is a partition $c = x_1 < x_2 < \ldots < x_n = d$ that is symmetrically finer than Δ_m. Since $|Z| = 0$ we may assume at this point for all $x \in [a, b]$ and $(x, h) \in \Delta_m$ that $x \pm h \notin Z$.

Given any two such partitions of $[c, d]$ say $c = x_1 < x_2 < \ldots < x_n = d$ and $c = x_1' < x_2' < \ldots < x_p' = d$ both symmetrically finer than Δ_m we obtain the inequality

$$\left|\sum_{i=1}^{n-1} f\left(\frac{x_{i+1} + x_i}{2}\right)(x_i - x_{i-1}) - \sum_{i=1}^{p-1} f\left(\frac{x_{i+1}' + x_i'}{2}\right)(x_i' - x_{i-1}')\right| < \frac{2}{m}$$

by augmenting each partition to a symmetric partition of $[a, b]$ also symmetrically finer than Δ_m and using (9.85). This ensures that there is a number $G(c, d)$ associated with each such interval $[c, d]$ with the property that

$$\left| \sum_{i=1}^{n-1} f\left(\frac{x_{i+1} + x_i}{2}\right)(x_i - x_{i-1}) - G(c,d) \right| \leq \frac{2}{m}$$

for any partitions of $[c,d]$, $c = x_1 < x_2 < \ldots < x_n = d$ symmetrically finer than δ_m.

We now apply our familiar methods (cf. the proof of Lemma 9.30) in the theory of Riemann type integrals to produce a measurable function F defined on $(a,b) \setminus Z$ with $G(c,d) = F(d) - F(c)$. For any pair c, $d \in (a,b) \setminus Z$, $c < d$ and any m and for any partitions of $[c,d]$, $c = x_1 < x_2 < \ldots < x_n = d$ symmetrically finer than Δ_m we will obtain

$$\sum_{i=1}^{n-1}\left| f\left(\frac{x_{i+1} + x_i}{2}\right)(x_i - x_{i-1}) - (F(x_i) - F(x_{i-1})) \right| \leq \frac{4}{m}. \tag{9.86}$$

Define then $\tau_m(c,d)$ as the supremum of the sums

$$\sum_{i=1}^{n-1}\left| f\left(\frac{x_{i+1} + x_i}{2}\right)(x_i - x_{i-1}) - (F(x_i) - F(x_{i-1})) \right|$$

for any partitions of $[c,d]$, $c = x_1 < x_2 < \ldots < x_n = d$ symmetrically finer than Δ_m.

There is a monotone function τ_m defined everywhere on $[a,b]$ so that $\tau_m(b) - \tau_m(a) \leq 4/m$ and so that $\tau_m(c,d) = \tau_m(d) - \tau_m(c)$. This has the property that

$$|F(x+h) - F(x-h) - 2f(x)h| \leq \tau_m(x+h) - \tau_m(x-h)$$

provided $(x,h) \in \Delta_m$. (Recall that $x + h \in (a,b) \setminus Z$ and $x - h \in (a,b) \setminus Z$ whenever $(x,h) \in \Delta_m$.)

We wish now to prove that $\alpha_F(a,b) = C$. Let $t > 0$ so that $a + t \notin Z$, $b - t \notin Z$, $(a, a+t) \in \Delta_m$ and $(b, b-t) \in \Delta_m$. Then from (9.85) and (9.86) we obtain that there is a partition $a + t = x_1 < \ldots < x_n = b - t$ that can be used to prove the inequality

$$|F(b-t) - F(a+t) - C| \leq 5/m.$$

From this we can deduce that $\alpha_F(a,b) = C$ as required.

This completes the proof since we have exhibited each of the details to show that F is a (V_{as}^1)-primitive of f.

9.3.8 Applications to Trigonometric Series

In this section we shall return to the coefficient problem for trigonometric series initiated in Section 9.3.3 above. We shall show that a nearly everywhere convergent trigonometric series is the Fourier series of its sum, if the integral is interpreted in an approximate symmetric sense. This is a considerable extension of Theorem 9.45 in that the integral used here has a number of characterizations. Also a countable exceptional set is allowed here that would not be allowed in Theorem 9.45.

We must begin with a number of tools from the study of trigonometric series. Much of this is early work of Rajchman and Zygmund and all can be found explicitly or implicitly in [313, Chap. IX, §2]. Suppose that we are given a trigonometric series

$$a_0/2 + \sum_{n=1}^{\infty} (a_n \cos nx + b_n \sin nx)$$

with bounded coefficients. We follow the strategy of investigating the integrated series

$$L(x) = a_0 x/2 + \sum_{n=1}^{\infty} (a_n \sin nx - b_n \cos nx)/n.$$

This latter series converges almost everywhere; the function L is called the *Lebesgue function* for the series. The symmetry properties of the Lebesgue function are the key to the approach of Rajchman and Zygmund.

We recall (Theorem 1.19) that the approximate symmetric derivative of L at any point x recovers the value of the series if it converges at x. We need also to study the continuity properties of the integrated series. This theorem is again due to Rajchman and Zygmund and is cited in Zygmund [313, Vol. I, Ex. 9, p. 371]; a proof is not included there but the relevant paper appears in the collected works [314, Vol. I, pp. 81–92]. The full version here, including the approximate continuity at the points where the series converges, follows from the same methods; for convenience we include a proof. It should be pointed out that the function F in (9.88), which will be shown to be approximately symmetrically continuous everywhere, is also L_p-symmetrically continuous ($1 \leq p < +\infty$). In fact the true picture which generalizes both of these results is that F has vanishing mean oscillation (see Ash, Cohen, Freiling and Rinne [7]). This last named concept is due to Sarason [252].

Theorem 9.63 (Rajchman–Zygmund) *Assume that*

$$a_n, b_n = o\left(\tfrac{1}{n}\right) \tag{9.87}$$

as $n \to \infty$. Then the set B where the series

$$F(t) = a_0/2 + \sum_{n=1}^{\infty} (a_n \cos nt + b_n \sin nt) \tag{9.88}$$

converges has full measure, F is everywhere approximately symmetrically continuous and is approximately continuous at each point of B.

*   *If (9.87) is replaced by the weaker assumption that $a_n, b_n = O\left(\tfrac{1}{n}\right)$ then this conclusion is no longer valid.*

Proof. The proof is similar to that used in Zygmund [313, Vol. I, Theorem 2.22, p. 324] for a related theorem. Because of the conditions on the coefficients of the series the sum $L(x)$ of (9.88) is defined almost everywhere. To prove the approximate symmetric continuity of F at every point $t \in \mathbb{R}$ let $N = 1, 2, \ldots,$ and $h \in (2^{-N-1}, 2^{-N})$ be such that $L(t+h)$ and $L(t-h)$ are defined. As usual write $A_n(t) = a_n \cos nt + b_n \sin nt$ and $B_n(t) = a_n \sin nt - b_n \cos nt$. Then

$$(L(t+h) - L(t-h))/2 = \sum_{n=1}^{\infty} -B_n(t) \sin nh. \qquad (9.89)$$

We estimate

$$\left| \sum_{n=1}^{2^N-1} -B_n(t) \sin nh \right|$$

$$\leq h \sum_{n=1}^{2^N-1} n(|a_n| + |b_n|) = o(2^N h) = o(1), \qquad (9.90)$$

and

$$\int_{2^{-N-1}}^{2^{-N}} \left(\sum_{n=2^N}^{\infty} -B_n(t) \sin nh \right)^2 dh$$

$$\leq \int_{-\pi}^{\pi} \left(\sum_{n=2^N}^{\infty} -B_n(t) \sin nh \right)^2 dh$$

$$\leq 2\pi \sum_{n=2^N}^{\infty} (a_n^2 + b_n^2) = o\left(\sum_{n=2^N}^{\infty} 1/n^2 \right) \leq \epsilon_N 2^{-N-1}, \qquad (9.91)$$

where $\epsilon_N \to 0$. Let

$$E = \bigcup_{N=1}^{\infty} \left\{ h \in (2^{-N-1}, 2^{-N}) : \left| \sum_{n=2^N}^{\infty} -B_n(t) \sin nh \right| \geq \sqrt[4]{\epsilon_N} \right\}.$$

Clearly, (9.90) shows that

$$\lim_{h \notin E, h \searrow 0} (L(t+h) - L(t-h))/2 = 0.$$

This proves our statement since (9.91) implies that the measure of the set $E \cap (2^{-N-1}, 2^{-N})$ does not exceed $\sqrt{\epsilon_N} 2^{-N-1}$. Thus E has density 0 at $h = 0$ and so

$$\text{ap} \lim_{h \searrow 0} (L(t+h) - L(t-h))/2 = 0$$

as required.

Now suppose that $t+h, t \in B$ and N is as before; by very much the same methods we show that

$$\text{ap}\lim_{h\to 0} L(t+h) - L(t) = 0. \tag{9.92}$$

A similar expression to (9.89) is

$$L(t+h) - L(t) = \sum_{n=1}^{\infty} -2B_n(t+\frac{h}{2})\sin\frac{nh}{2}$$

and nearly identical arguments will supply the required statement.

As before

$$\left|\sum_{n=1}^{2^N-1} -2B_n(t+\frac{h}{2})\sin\frac{nh}{2}\right| = o(1)$$

and

$$\int_{2^{-N-1}}^{2^{-N}} \left(\sum_{n=2^N}^{\infty} -2B_n(t+\frac{h}{2})\sin\frac{nh}{2}\right)^2 dh$$

$$\leq \int_0^{\pi} \left(\sum_{n=2^N}^{\infty} -2B_n(t+\frac{h}{2})\sin\frac{nh}{2}\right)^2 dh$$

$$= 2\int_0^{\pi/2} \left(\sum_{n=2^N}^{\infty} -2B_n(t+h)\sin nh\right)^2 dh.$$

From here the remainder of the proof will work to prove (9.92).

Finally an example suffices to show that the weaker assumption on the coefficients $a_n, b_n = O\left(\frac{1}{n}\right)$ does not produce an approximately symmetrically continuous function. Consider

$$F(x) = \sum_1^{\infty} \frac{\sin nx}{n}.$$

This converges everywhere but 0 is a point of discontinuity for F in either of these senses.

As a result of Theorem 9.63 and Theorem 1.19 an approximate symmetric integral offers a solution to the coefficient problem. The first account is [232] where it is presented for a Riemann type integral essentially the same as that here.

Theorem 9.64 *Suppose that the trigonometric series*

$$a_0/2 + \sum_{n=1}^{\infty} (a_n \cos nx + b_n \sin nx) \tag{9.93}$$

converges to a function $f(x)$ at nearly every point x. Then f is (R^1_{as})-integrable on any period $[a, a+2\pi]$ and the series is the Fourier series for f in this sense, i.e. for each n

$$\pi a_n = (R_{as}^1) \int_0^{2\pi} f(t) \cos nt\, dt$$

and

$$\pi b_n = (R_{as}^1) \int_0^{2\pi} f(t) \sin nt\, dt.$$

Proof. Since the series converges nearly everywhere the coefficients are tending to zero. If L denotes the Lebesgue function for the series then, by Theorem 1.19, ASD $L(x) = f(x)$ at all but a countable set and, by Theorem 9.63, L is everywhere approximately symmetrically continuous. It follows from Theorem 9.52 that f is integrable on any interval $[a, b]$ for which $\alpha_L(a, b)$ exists. The proof then is nearly identical with that for Theorem 9.45.

A corollary allows one to integrate a trigonometric series.

Corollary 9.65 *Suppose that the trigonometric series (9.93) converges to a function $f(x)$ at nearly every point and let B denote the set of points at which the Lebesgue function for the series exists. Then B has full measure in any interval and f is (R_{as}^1)–integrable on any interval $[a, b]$ $(a, b \in B)$. The integral can be expressed as*

$$(R_{as}^1) \int_a^b f(t)\, dt = a_0(b-a)/2 + \sum_{n=1}^{\infty} \frac{a_n(\sin nb - \sin na) - b_n(\cos nb - \cos na)}{n}.$$

Proof. As before if L denotes the Lebesgue function for the series then ASD $L(x) = f(x)$ at all but a countable set and L is everywhere approximately symmetrically continuous. Also by Theorem 9.63 L is approximately continuous at every point of B and so $\alpha_L(a, b) = L(b) - L(a)$ and the corollary follows from Theorem 9.52.

A further corollary expresses conditions under which a series is a Fourier series in more conventional senses (cf. [313, Vol. I, Theorem (3.18), p. 328]).

Corollary 9.66 *Suppose that the trigonometric series (9.93) converges nearly everywhere to a function f. If $f \geq g$ where g is Lebesgue [Denjoy–Perron] integrable then the series is the Fourier series for f in that sense.*

Proof. If $f \geq g$ where g is Lebesgue [Denjoy–Perron] integrable then f is integrable in the same sense by Theorem 9.53.

As a final corollary we observe that this integral allows a transparent proof that countable sets are sets of uniqueness, a fact first proved by W. H. Young.

Ash [5] points out that the use of these ideas to obtain uniqueness proofs in trigonometric series might suggest techniques in higher dimensions.

Corollary 9.67 (Young) *Suppose that the series (9.93) converges to zero at nearly every point. Then each of the coefficients must vanish.*

9.4 Second Order Symmetric Integrals

Our goal in this section is to introduce an integral which inverts the second order symmetric derivative. Suppose that F is a continuous function for which $SD_2 F(x) = f(x)$ everywhere and f is given. How can we recover F?

Of course, since F is unique only up to the addition of a linear function, this is not quite possible. In fact we actually wish to recover the expression

$$\Delta_{K_p} F = F(a) + F(b) - F(a+p) - F(b-p) \tag{9.94}$$

for all 2–intervals $K_p = ([a, b], [a+p, b-p])$. One easily checks that the expression in (9.94) does not depend on the particular choice of primitive.

For most applications, however, it is more to the point to recover the expression

$$\Lambda_F(a, b) = \lim_{p \to 0+} \frac{F(a) - F(a+p) - F(b-p) + F(b)}{p} \tag{9.95}$$

when this exists. Note that this expression can also be written, in the notation of (9.94), as

$$\Lambda_F(a, b) = \lim_{p \to 0+} \frac{\Delta_{K_p} F}{p}. \tag{9.96}$$

At the simplest level we can invert second order symmetric derivatives in the Newton style. If $SD_2 F(x) = f(x)$ everywhere on an interval (a, b) and F is continuous on $[a, b]$ then we can declare the expression $\Lambda_F(a, b)$, if it exists, as the value of a Newton integral of f on $[a, b]$. We shall do this in Section 9.4.1.

For a Perron or variational attack on the same problem directly based on the second symmetric derivative see the literature cited in the introductory section of this chapter. Our interest here is focussed on the Riemann sums approach to this problem; we take this from Freiling, Rinne and Thomson [116].

The second symmetric derivative itself, without some preliminary modifications, does not seem to allow a Riemann-type integral. Henstock [130] has already pointed this out. An approach is available by observing that the identity $SD_2 F(x) = f(x)$ is equivalent to requiring that

$$\lim_{h \to 0+, 0 \le 3k \le h} \frac{F(x+h) + F(x-h) - F(x+k) - F(x-k)}{h^2 - k^2} = f(x). \tag{9.97}$$

We can rewrite (9.97) as the statement that for every $\epsilon > 0$ there is a gauge δ so that

$$\left|F(x+h) + F(x-h) - F(x+k) - F(x-k) - f(x)(h^2 - k^2)\right| < \epsilon(h^2 - k^2)$$
$$(9.98)$$

for $0 \le 3k \le h < \delta(x)$. Expressed in this form one sees fairly quickly a variational definition of an integral can be given that will solve our problem.

Further thought leads to a Riemann-type integral that will recover (9.94). Here we need to invoke the language and notation of 2–intervals and 2–partitions from Section 3.7. For any sequence $(K_1, K_2, \ldots, K_n)$ forming a 2–partition of the 2–interval K_p and such that each K_i is δ–fine at its center and $\frac{1}{3}$–regular we must have have

$$F(a) + F(b) - F(a+p) - F(b-p) = \Delta_{K_p} F = \sum_{i=1}^{n} \Delta_{K_i} F \qquad (9.99)$$

by additivity. Because of (9.98) the expressions in (9.99) are close to the sum

$$\sum_{i=1}^{n} f(c(K_i)) \|K_i\|. \qquad (9.100)$$

This suggests that (9.100) be interpreted as a type of Riemann sum that when taken to a limit should recover the expressions in (9.99).

This motivates our definition of a Riemann-type integral which inverts second order symmetric derivatives. One extra complication enters into the picture because we cannot hope that such simple partitions as suggested above always exist for any gauge δ. Our covering theorem, Theorem 3.44, does allow such partitions if we introduce a countable exceptional set at which we permit further 2–intervals with outer endpoints in the exceptional set. Since F is always taken as continuous the additions do not affect the total sum by more than a small amount which we can control. The form of the definition may now be anticipated.

9.4.1 A Second Order Symmetric Newton Integral

We present here an account inspired by Jeffrey [142, p. 27] and Mařík [196] who develop a Newton integral based on the second order symmetric derivative as part of an introduction to the P^2–integral of James.

The use of the increment $\Lambda_F(a, b)$ in Definition 9.2 is due to Mařík. It is motivated by a desire to retain a kind of symmetry at the endpoints, to overcome the ambiguity arising from the fact that two different functions may have the same second symmetric derivative and to reduce the integral to some kind of first order process.

Definition 9.68 A finite function f is said to be (N_s^2)–*integrable* on an interval $[a, b]$ if there is a continuous function F on $[a, b]$ so that $SD_2 F(x) = f(x)$ everywhere on (a, b) and such that the expression

$$\Lambda_F(a,b) = \lim_{h \to 0} \frac{F(a) - F(a+p) - F(b-p) + F(b)}{p} \qquad (9.101)$$

exists. We write then

$$\Lambda_F(a,b) = (\mathrm{N}_s^2) \int_a^b f(x)\, dx.$$

If $\mathrm{SD}_2 F(x) = f(x)$ everywhere on an interval (a,b) and F is continuous then F is unique up to the addition of a linear function (Theorem 1.12), and so $\Lambda_F(a,b)$ if it exists is unique. This is enough to justify the definition.

The integral shares some of the properties we have already seen for Newton-type integrals. We shall not develop any further properties except to obtain a relation with the (N_s^1)–integral. In this theorem we should note that a general (N_s^1)–primitive need not be integrable; it is continuous off a scattered set but it may be unbounded.

Theorem 9.69 *Let f be (N_s^1)–integrable on an interval $[a,b]$ with an integrable primitive. Then f is (N_s^2)–integrable on $[a,b]$ and the integrals agree.*

Proof. Let G be an (N_s^1)–primitive of f on $[a,b]$ and write $F(x) = \int_a^x G(t)\, dt$. A direct computation shows that if $\lambda_G(a,b)$ exists then

$$\lambda_G(a,b) = \Lambda_F(a,b). \qquad (9.102)$$

Also it is easy to check that at any point x where $\mathrm{SD}\, G(x)$ exists so too does $\mathrm{SD}_2 F(x)$ and they are equal. Thus F is a (N_s^2)–primitive of f on $[a,b]$ and the agreement of the integrals follows from (9.102).

The main application of the (N_s^2)–integral is to the coefficient problem for trigonometric series. This offers a formal solution that serves to focus the problem. The basic tool is the classical observation of Riemann (Theorem 1.9) that the second symmetric derivative of the twice integrated series recovers the sum of a trigonometric series

$$a_0/2 + \sum_{n=1}^{\infty} (a_n \cos nx + b_n \sin nx) \qquad (9.103)$$

at any point at which this exists. From this and the definition we show that everywhere convergent trigonometric series are (N_s^2)–integrable and may be expressed in Fourier form.

Theorem 9.70 *Suppose that the trigonometric series (9.103) converges everywhere. Then f is (N_s^2)–integrable on every period and*

$$a_n = \frac{1}{\pi} \int_c^{c+2\pi} f(t) \cos nt\, dt \ , \quad b_n = \frac{1}{\pi} \int_c^{c+2\pi} f(t) \sin nt\, dt$$

in the sense of that integral.

Proof. Let G denote the Riemann function of the series (9.103) Then G is smooth and $G(x) - x^2 a_0/4$ is 2π-periodic. Let $g(x) = x^2 a_0/4$. We easily compute that

$$\Lambda_G(c, c + 2\pi) = \Lambda_{G-g}(c, c + 2\pi) + \Lambda_g(c, c + 2\pi) = 0 + \pi a_0$$

for any real c. Since $SD_2 G(x) = f(x)$ everywhere f is (N_s^2)–integrable and we have the correct formula for a_0, i.e. that

$$(N_s^2) \int_c^{c+2\pi} f(t)\,dt = \Lambda_G(c, c + 2\pi) = \pi a_0$$

as required.

In the same way we obtain a formula for the remaining coefficients; multiply the series

$$a_0/2 + \sum_{n=0}^{\infty} a_n \cos nx + b_n \sin nx$$

by $\cos kx$, use elementary trigonometric identities for $\cos nx \cos kx$ and for $\sin nx \cos kx$, and rearrange the series to obtain a trigonometric series that converges everywhere to $f(x) \cos kx$. For example the term corresponding to $a_n \cos kx \cos nx$ is replaced by

$$a_n(\cos(n + k)x + \cos(n - k)x)/2$$

before rearrangement. (This is exactly the argument in [186, p. 41] where they point out that it follows from the Rajchman theory of the formal multiplication of trigonometric series.)

As before, the twice formally integrated series converges uniformly and has everywhere $f(x) \cos kx$ for its second symmetric derivative. Consequently (much as before) $f(x) \cos kx$ is integrable and we may integrate it over a period $[c, c + 2\pi]$ for every c. Since the constant term in the series for $f(x) \cos kx$ is $a_k/2$ this gives the required formula for a_k. A similar argument supplies the formula for b_k.

9.4.2 The Definitions and an Integrability Criterion

We now define the integral that is our main object of study. This is based on the covering theorem of Section 3.7.3. Let δ be a gauge and let

$$K_p = ([a, b], [a + p, b - p]) \quad (0 < p \le (a + b)/2),$$

be a fixed 2–interval. We say that a countable set E is *an associated exceptional set* for δ if the conclusion of Theorem 3.44 holds. Thus for any gauge γ on $E \times \mathbb{N}$ there is a 2–partition of K_p where each element is $\frac{1}{3}$–regular and each is either δ-fine at its center or else has a corner in E and is γ–fine at that corner.

Definition 9.71 Let the function f be defined everywhere on an interval (a, b) and let

$$K_p = ([a, b], [a + p, b - p]) \quad (0 < p \le (a + b)/2),$$

be a 2–interval. Then f is said to be (R_s^2)–integrable on K_p with value $I(f, K_p)$ if for every $\epsilon > 0$ there is a gauge δ so that for any associated exceptional set $E = \{x_1, x_2, x_3, \ldots\}$ there is a gauge η on $E \times \mathbb{N}$ such that for any sequence

$$(K_1, K_2, \ldots, K_n)$$

forming a 2–partition of the 2–interval K_p that is $\frac{1}{3}$–regular, and such that each K_i is either δ–fine at its center or else has an outer corner x_j in E with $|K_i| < \eta(x_j, i)$ we have

$$\left| \sum_{i=1}^{n} {}' f\left(c\left(K_i\right)\right) \|K_i\| - I(f, K) \right| < \epsilon$$

where $\sum'$ indicates that the sum is taken only over the K_i that are δ–fine at the center.

It is not immediately clear that the integral is well defined. The main covering theorem, Theorem 3.44, supplies this.

Lemma 9.72 *Let the function f be defined everywhere on an interval (a, b) and let K be a 2–interval. If f is (R_s^2)–integrable on K with value $I(f, K)$ then $I(f, K)$ is unique.*

Proof. Let $I(f, K) = c_1$ and $I(f, K) = c_2$ and $\epsilon > 0$. By definition there are two gauges δ_1 and δ_2 for which the statement of the definition holds. Let $\delta = \min\{\delta_1, \delta_1\}$. Then δ is a gauge and any associated exceptional set E for δ is also an associated exceptional set for both δ_1 and δ_2. For any gauges η_1 and η_2 on $E \times \mathbb{N}$ let $\eta_1 = \min\{\eta_1, \eta_1\}$. By Theorem 3.44 there exists a sequence

$$(K_1, K_2, \ldots, K_n)$$

forming a 2–partition of K that is $\frac{1}{3}$–regular, and such that each K_i is either δ–fine at its center or else has an outer corner x_j in E with $|K_i| < \eta(x_j, i)$. This sequence can be used with c_1 or c_2 to show that

$$\left| \sum_{i=1}^{n} {}' f\left(c\left(K_i\right)\right) \|K_i\| - c_i \right| < \epsilon \quad (i = 1, 2)$$

and hence that $|c_1 - c_2| < 2\epsilon$. The conclusion now follows.

The "integral" $I(f, K_p)$ taken over the 2–interval K_p is an unfamiliar concept and may not prove too useful. We convert this to an ordinary first order integral by the following definition.

Definition 9.73 A function f defined on an interval (a, b) is said to be (R_s^2)–integrable on $[a, b]$ and we write

$$(R_s^2) \int_a^b f(x)\, dx = \lim_{p \to 0+} \frac{I(f, ([a, b], [a + p, b - p]))}{p}$$

if this limit exists.

Notice that, in order for the integral $(R_s^2) \int_a^b f(x)\, dx$ to be defined, the integral $I(f, K_p)$ must exist for $K_p = ([a, b], [a + p, b - p])$ and all sufficiently small $p > 0$ and the limit in the definition must exist.

We begin with a useful integrability criterion.

Theorem 9.74 *A sufficient condition that f be (R_s^2)–integrable on a 2–interval $K_p = ([a, b], [a + p, b - p])$ is that there is a continuous function G on $[a, b]$ so that for every $\epsilon > 0$ there is a convex function H with $\Delta_{K_p} H < \epsilon$ and there is a gauge δ such that*

$$\Big| G(x + h) + G(x - h) - G(x + k) - G(x - k) - f(x)(h^2 - k^2) \Big|$$

$$< H(x + h) + H(x - h) - H(x + k) - H(x - k)$$

for all $x \in (a, b)$ and $0 \leq 3k \leq h < \delta(x)$. In that case $I(f, K_p) = \Delta_{K_p} G$.

Proof. If the inequality of the theorem holds then, writing K for the 2–interval $([x - h, x + h], [x - k, x + k])$, we have

$$\Big| \Delta_K G - f(x) \|K\| \Big| \leq \Delta_K H \qquad (9.104)$$

for $0 \leq 3k \leq h < \delta(x)$. Let $\epsilon > 0$ and suppose that δ, K_p, G and H are as in the statement of the theorem. Let $E = \{x_1, x_2, x_3, \ldots\}$ be any exceptional set associated with δ. Since G is continuous we may select positive numbers c_{ij} so that

$$|\Delta_J G| < \epsilon 2^{-i-j} \qquad (9.105)$$

whenever J is a 2–interval with an outer endpoint $x_j \in E$ and $|J| < c_{ij}$. Define the gauge $\eta(x_j, i) = c_{ij}$ for each $x_j \in E$ and $i \in \mathbb{N}$.

Now consider any sequence

$$(K_1, K_2, \ldots, K_n)$$

forming a 2–partition of the 2–interval K_p that is $\frac{1}{3}$–regular, and such that each K_i is either δ–fine at its center or else has an outer endpoint x_j in E with $|K_i| < \eta(x_j, i) = c_{ij}$. We use the notation $\sum'$ to indicate a sum taken over the K_i that are δ–fine at the centers and $\sum''$ to indicate the remaining sum where K_i has an outer endpoint x_j in E. Then, using (8.19), (9.104) and (9.105), we have

$$\left| \sum_{i=1}^{n}{}' f\left(c\left(K_i\right)\right) \|K_i\| - \Delta_{K_p} G \right| \le$$

$$\sum_{i=1}^{n}{}' \left| f\left(c\left(K_i\right)\right) \|K_i\| - \Delta_{K_i} G \right| + \sum_{i=1}^{n}{}'' \left| \Delta_{K_i} G \right|$$

$$< \sum_{i=1}^{n}{}' \Delta_{K_i} H + \epsilon \le \Delta_{K_p} H + \epsilon < 2\epsilon.$$

The last inequalities use the convexity of H (i.e. the fact that $\Delta_K H \ge 0$ for every K). By definition f is (R_s^2)–integrable on K_p and the value of the integral is $\Delta_{K_p} G$ as required.

9.4.3 Properties of the Integral

As stated in introducing the (R_s^2)–integral, our main purpose is to invert the second symmetric derivatives of continuous functions. The next theorem and its corollary show that this objective has been reached.

Theorem 9.75 *Suppose that G is a continuous function on $[a, b]$ and that $K_p = ([a, b], [a+p, b-p])$ is a 2–interval. If $SD_2\, G(x) = f(x)$ exists everywhere on (a, b) then f is (R_s^2)–integrable on K_p and*

$$I(f, K_p) = \Delta_{K_p} G$$

Proof. If $SD_2\, G(x) = f(x)$ exists everywhere on (a, b) then for every $\epsilon > 0$ there is a gauge δ with

$$\left| G(x+h) + G(x-h) - 2G(x) - f(x)h^2 \right| < \epsilon h^2$$

for $a < x < b$ and $0 < h < \delta(x)$. It follows then, using the inequality (8.6) and the notation from (8.7), that

$$\left| G(x+h) + G(x-h) - G(x+k) - G(x-k)) - f(x)(h^2 - k^2) \right|$$

$$< \epsilon C_{\frac{1}{3}}(h^2 - k^2)$$

for all $x \in (a, b)$ and $0 \le 3k \le h < \delta(x)$. Accordingly, we may use $H(x) = \epsilon C_{\frac{1}{3}} x^2$ in Theorem 9.74 and integrability follows. $\blacksquare$

As a corollary we see that the (R_s^2)–integral includes the (N_s^2)–integral.

Corollary 9.76 *Suppose that f is (N_s^2)–integrable on $[a, b]$. Then f is (R_s^2)–integrable on $[a, b]$ and*

$$(R_s^2) \int_a^b f(x)\, dx = (N_s^2) \int_a^b f(x)\, dx.$$

Proof. This follows directly from (9.96), Theorem 9.75 and the definition of the integral.

The integral is an extension of the Lebesgue integral; we prove this now. Note that because of Corollary 9.76 it is clearly not identical to the Lebesgue integral as it will invert ordinary derivatives of continuous functions and symmetric derivatives of integrable functions.

Theorem 9.77 *Suppose that f is Lebesgue integrable on an interval $[a, b]$ and that R_0 is a rectangle that projects into $[a, b]$. Then, for any subrectangle R of R_0, the function f is (R_s^2)–integrable on $\pi(R)$,*

$$I(f, \pi(R)) = \int\int_R f(x + y)\, dx\, dy$$

and

$$(R_s^2) \int_a^b f(x)\, dx = \int_a^b f(x)\, dx.$$

Proof. Let $\epsilon > 0$. Write $F(x) = \int_a^x f(t)\, dt$, $G(x) = \int_a^x F(t)\, dt$ and $H(x, y) = G(x + y)$. Since F is continuous

$$\Lambda_G(a, b) = \lambda_F(a, b) = F(b) - F(a). \tag{9.106}$$

For any subrectangle R of R_0 we must have

$$H(R) = \Delta_{\pi(R)} G = \int\int_R f(x + y)\, dx\, dy.$$

To see this let $R = [c_1, c_2] \times [d_1, d_2]$. Then the integral can be computed as

$$\int_{d_1}^{d_2} \int_{c_1}^{c_2} f(x + y)\, dx\, dy = \int_{d_1}^{d_2} (F(c_2 + y) - F(c_1 + y))\, dy$$

$$= G(c_2 + d_2) - G(c_1 + d_2) - (G(c_2 + d_1) - G(c_1 + d_1)) = \Delta_{\pi(R)} G = H(R)$$

where the final identity was given in (8.18).

We fix R and show that $I(f, \pi(R)) = H(R)$. Since $F' = f$ a.e. we have $\mathrm{SD}_2\, G(x) = f(x)$ a.e. also. Let N denote the set of points in (a, b) where $f(x) = \mathrm{SD}_2\, G(x)$ does not hold and write $N_s = \{x \in N : s - 1 \le |f(x)| < s\}$, $s = 1, 2, \ldots$. Each set N_s has measure zero.

Choose a sequence of open sets $O_s \supset N_s \cup \{a\} \cup \{b\}$ with measure sufficiently small so that

$$\sum |H(R_i)| < \epsilon/2^{s+2} \tag{9.107}$$

and

$$\sum \|R_i\| < \epsilon/(s2^{s+2}) \tag{9.108}$$

for any disjoint sequence $\{R_i\}$ of subrectangles of R_0 with each $\pi(R_i) \subset O_s$. The first inequality uses the absolute continuity of the integral and the second inequality just arranges for the set of points in R_0 that project into O_s to have small measure.

Let $K_p = \pi(R)$. Define a gauge δ by setting $\delta(x)$ as the distance from x to $\mathbb{R} \setminus O_s$ if $x \in N_s$ and for $x \notin N$ choose $\delta(x)$ so that

$$\left| \Delta_K G - f(x) \|K\| \right| < \frac{\epsilon \|K\|}{4 \|K_p\|} \tag{9.109}$$

where $x = c(K)$ and K is any $\frac{1}{3}$-regular, 2-interval with $|K| < \delta(x)$. This just uses the fact that $SD_2 G(x) = f(x)$ everywhere in $(a, b) \setminus N$ and the inequality (8.10).

Let $E = \{x_1, x_2, x_3, \ldots\}$ be a countable exceptional set associated with δ. We choose a gauge η on $E \times \mathbb{N}$ so that

$$|H(R')| < \epsilon 2^{-i-j-2} \tag{9.110}$$

whenever R' is a subrectangle of R that has an outer corner that projects onto a point x_j in E for which $|\pi(R')| < \eta(x_j, i)$. This just uses the continuity of the integral.

Now suppose that we have a sequence

$$(K_1, K_2, \ldots, K_n)$$

forming a 2-partition of the 2-interval $\pi(R)$ that is $\frac{1}{3}$-regular, and such that each K_i is either δ-fine at its center or else has an outer endpoint x_j in E with $|K_i| < \eta(x_j, i)$. We can assume that $K_i = \pi(R_i)$ where the $\{R_i\}$ form a partition of R.

Recall that $\Delta_{K_i} G = H(R_i)$, $\|K_i\| = \|R_i\|$ and $H(R) = \sum_{i=1}^{n} H(R_i)$. As before we use the notation $\sum'$ to indicate a sum taken over the K_i that are δ-fine at the centers and $\sum''$ to indicate the remaining sum where K_i has an outer endpoint x_j in E. We have then that

$$\left| \sum_{i=1}^{n}{}' f(c(K_i)) \|K_i\| - H(R) \right|$$

$$\leq \sum_{i=1}^{n}{}' \left| f(c(K_i)) \|K_i\| - H(R_i) \right| + \sum_{i=1}^{n}{}'' |H(R_i)|$$

$$\leq \sum_{c(K_i) \notin N}{}' \left| f(c(K_i)) \|K_i\| - \Delta_{K_i} G \right|$$

$$+ \sum_{s=1}^{\infty} \sum_{c(K_i) \in N_s}{}' \{ s \|K_i\| + |H(R_i)| \} + \sum_{i=1}^{n}{}'' |H(R_i)|.$$

The sum is seen to be smaller than ϵ by using (9.107), (9.108), (9.109) and (9.110). By definition then f is (R_s^2)–integrable and $I(f, \pi(R)) = H(R)$. That the value of the integral $(\mathrm{R}_s^2) \int_a^b f(x)\, dx$ is $F(b) - F(a) = \int_a^b f(x)\, dx$ now follows from (9.106), (9.102) and the definition of the integral.

9.4.4 Integration and Variation

The second order variation developed in Chapter 8 allows us to state some sufficient conditions for integrability that extend those of the preceding sections. For the definitions of the terms see Section 8.3.

Theorem 9.78 *Let f be a continuous function and and suppose that $\Lambda_f(a, b)$ exists. Let g be an arbitrary function. Write*

$$\xi([x - h, x + h], [x - k, x + k])$$
$$= f(x + h) + f(x - h) - f(x - k) - f(x + k) - g(x)(h^2 - k^2)$$

for every 2–interval $K = ([x - h, x + h], [x - k, x + k])$. If $\xi^*((a, b)) = 0$ then g is (R_s^2)–integrable on $[a, b]$ and

$$(\mathrm{R}_s^2) \int_a^b g(x)\, dx = \Lambda_f(a, b).$$

Proof. The proof is similar to that for Theorem 9.74 and is omitted.

The corollaries which now follow can be deduced readily from Theorem 9.78. The first shows that the (R_s^2)–integral inverts almost everywhere second symmetric derivatives provided the exceptional set where that derivative fails to exist has both Lebesgue and variational measure zero.

Corollary 9.79 *Let f be continuous and suppose that $\Lambda_f(a, b)$ exists. If $SD_2 f(x) = g(x)$ almost everywhere and $\mathrm{VS}_f^{(2)}$–almost everywhere in (a, b) then g is (R_s^2)–integrable on $[a, b]$ and*

$$(\mathrm{R}_s^2) \int_a^b g(x)\, dx = \Lambda_f(a, b).$$

As a further corollary we obtain a Lusin type condition that is sufficient for integrability. It is not known if there is a necessary condition of this form.

Corollary 9.80 *Let f be continuous, and suppose that $\Lambda_f(a, b)$ exists. If $\mathrm{VS}_f^{(2)}$ is σ–finite on (a, b) and vanishes on every set of Lebesgue measure zero then*

$SD_2 f(x) = g(x)$ *exists almost everywhere in* (a, b), g *is* (R_s^2)*-integrable on* $[a, b]$ *and*

$$(R_s^2) \int_a^b g(x)\, dx = \Lambda_f(a, b).$$

From properties of the variation itself we can conclude also the last corollary.

Corollary 9.81 *Let* f *be continuous and smooth, suppose that* $\Lambda_f(a, b)$ *exists and that* $-\infty < \underline{SD}_2 f(x) \leq \overline{SD}_2 f(x) < +\infty$ *nearly everywhere in* (a, b). *Then* $SD_2 f(x) = g(x)$ *exists almost everywhere in* (a, b), g *is* (R_s^2)*-integrable on* $[a, b]$ *and*

$$(R_s^2) \int_a^b g(x)\, dx = \Lambda_f(a, b).$$

9.4.5 An Application to Trigonometric Series

Our main tool in this section, as it was in Section 9.3.8, is the classical observation of Riemann (Theorem 1.9) that the second symmetric derivative of the twice integrated series recovers the sum of a trigonometric series at any point at which this exists. From this and Theorem 9.75 one shows easily that everywhere convergent trigonometric series are (R_s^2)-integrable and may be expressed in Fourier form.

Theorem 9.82 *Suppose that the trigonometric series*

$$a_0/2 + \sum_{n=1}^{\infty} (a_n \cos nx + b_n \sin nx) \qquad (9.111)$$

converges nearly everywhere. Then f *is* (R_s^2)*-integrable on every period and*

$$a_n = \frac{1}{\pi} \int_c^{c+2\pi} f(t) \cos nt\, dt\ , \quad b_n = \frac{1}{\pi} \int_c^{c+2\pi} f(t) \sin nt\, dt$$

in the sense of that integral.

Proof. The proof is nearly identical with that for Theorem 9.70 and can be omitted.

This theorem can be expressed directly in terms of our integral over 2–intervals in which case it assumes a less familiar form (cf. [313, Vol. II, pp. 90–91]).

Theorem 9.83 *Suppose that the trigonometric series (9.111) converges nearly everywhere. Then* f *is* (R_s^2)*-integrable on the 2-interval*

$$([-2\pi, 2\pi], [0, 0])$$

and the coefficients may be given by the formulas

$$a_n = \frac{1}{\pi^2} I(f(t) \cos nt, ([-2\pi, 2\pi], [0, 0]))$$

and

$$b_n = \frac{1}{\pi^2} I(f(t) \sin nt, ([-2\pi, 2\pi], [0, 0]))$$

in the sense of that integral.

For a variant we can repeat the arguments of [313, Vol II, pp. 90–1]) to obtain that a trigonometric series whose partial sums are bounded at every point is the (R_s^2)–Fourier series of its Riemann sum. An exceptional countable set is not permitted here unless we assume also that the coefficients tend to zero.

Theorem 9.84 *Suppose that the partial sums of the trigonometric series (9.111) are bounded at every point. Then the series has a finite sum $f(x)$ by the Riemann method of summation at almost every point x, f is (R_s^2)—integrable on any period $[a, a + 2\pi]$ and the series is the Fourier series for f in this sense.*

9.5 Incompatibilities

We conclude our study of the (R_s^2)–integral by showing that it is incompatible with the (R_{as}^1)–integral. Since both solve the coefficient problem for trigonometric series perhaps this may not be anticipated.

It was noted, in passing, by Kubota [160] that an approximate symmetric Perron integral in any fairly general sense would be incompatible with integrals based on the second order symmetric derivative (such as the (SCP)–integral of Burkill, the (T)–integral of Marcinkiewicz-Zygmund, and the (P^2)–integral of James). We present computations from [232] illustrating this fact.

Our first observation is that the (R_{as}^1)–integral is neither contained in nor does it contain the (R_s^2)–integral. This arises merely because of the continuity requirements of these integrals and elementary examples easily show this. Let $F(x) = x^{-2}$ and $f(x) = F'(x) = -2x^{-3}$, its derivative. Because of the symmetry of the function F at the origin and that fact that $F' = f$ everywhere else it is clear that f has both an (N_s^1)–integral and an (N_{as}^1)–integral and that F is a primitive for f in both of these senses on any interval $[a, b]$ with $a < 0 < b$.

The function $G(x) = -x^{-1}$ would represent a second order indefinite integral of f on any interval that does not include the origin. Because it is unbounded at the origin the Riemann sums used in defining

$$(\mathrm{R}_s^2) \int_{-1}^{1} f(x)\,dx$$

cannot converge.

This behavior at a single point is also enough to cause this problem in the other direction. Define an integrable function F that is everywhere differentiable except at the origin and such that

$$\lim_{h \searrow 0} 1/h \left\{ \int_{0}^{h} F(t)\,dt - \int_{-h}^{0} F(t)\,dt \right\} = 0$$

and yet F is not approximately symmetrically continuous at the origin. In this case F' is (R_s^2)–integrable but certainly there is no approximate symmetric Perron integral.

To construct such a function let each interval $[1/(n+1), 1/n]$ be split into the three adjacent intervals I_{1n}, I_{2n} and I_{3n} (in that order) by choosing a small centered interval I_{2n} of length 2^{-n} times the length of $[1/(n+1), 1/n]$. Define $F(x) = (-1)^n$ on each I_{1n}, $F(x) = (-1)^{n+1}$ on each I_{3n}, and choose F on I_{2n} so that $|F(x)| \leq 1$, F is continuously differentiable on $[1/(n+1), 1/n]$ and

$$\int_{1/(n+1)}^{1/n} F(t)\,dt = 0.$$

This defines F on $(0, 1]$; extend F by writing $F(-x) = -F(x)$ and $F(0) = 0$. We evidently have

$$\lim_{h \searrow 0} 1/h \int_{0}^{h} F(t)\,dt = \lim_{h \searrow 0} 1/h \int_{-h}^{0} F(t)\,dt = 0$$

and F' exists except at 0. But the set $\{h > 0 : |F(h) - F(-h)| \geq 1\}$ has density 1 at 0 and so F is far from symmetrically approximately continuous here. Thus, while F serves as an indefinite (R_s^2)–integral of its derivative, F' cannot be integrable in an approximate symmetric sense.

While these trivial differences separate the integrals there is still a compatibility problem: if a function f is integrable in two different symmetric senses then do the integrals computed by the two different procedures yield the same values? The answer is no. We may produce a function f integrable in both the approximate symmetric sense and in the (R_s^2)–sense but with different integrals in the two senses. This is not deep but is just a reflection of the fact that the continuity requirements of the two integrals differ.

To construct such a function, as before, let each interval $[1/(n+1), 1/n]$ be split into the three adjacent intervals I_{1n}, I_{2n} and I_{3n} by choosing a small centered interval I_{2n} of length 2^{-n} times the length of $[1/(n+1), 1/n]$. Let F vanish on each interval I_{1n} and I_{3n} and be defined on I_{2n} so as to be nonnegative and continuously differentiable on $[1/(n+1), 1/n]$ with

$$\int_{1/(n+1)}^{1/n} F(t)\, dt = 1/n - 1/(n+1).$$

We set F_1 to vanish on $[-1,0]$ and to equal F on $(0,1]$ and we define F_2 to have the constant value 1 on $[-1,0]$ and to equal F on $(0,1]$. Note that F_1 has been chosen so as to be approximately continuous at 0 while F_2 has been chosen so that its integral is smooth. Let $f = F_1' = F_2'$. Then F_1 is an indefinite integral for f in the approximate symmetric sense and F_2 is an indefinite integral for f in the (R_s^2)–sense; this exhibits the essential incompatibility of the integrals.

As part of this discussion we might mention that any one of these symmetric integrals that is sufficiently general to solve the coefficient problem for trigonometric series cannot be compatible with the Denjoy-Khintchine integral. The following Theorem of Skljarenko [260] shows this. Further information on this integral can be found in [251].

Theorem 9.85 (Skljarenko) *There is a trigonometric series*

$$f(x) = a_0/2 + \sum_{n=1}^{\infty} (a_n \cos nx + b_n \sin nx)$$

converging everywhere to a function integrable in the Denjoy-Khintchine sense but

$$(DK) - \int_0^{2\pi} f(x)\, dx \neq \pi a_0.$$

Corollary 9.86 *Both the (R_s^2)–integral and the (R_{as}^1)–integral are incompatible with the Denjoy-Khintchine integral.*

Appendix

In this appendix we collect a number of methods and ideas that are used in the main body of the text but whose inclusion there would break the flow, such as it is, of the symmetric ideas.

A.1 Scattered Sets

Countable exceptional sets arise in many studies in analysis. For example, in Chapter 2 there are several special classes of countable sets that have played an important role. In Theorem 2.27 the countable exceptional set has also countable closure. In the Charzyński Theorem of Section 2.3 the countable exceptional set is scattered. We require as well several further refinements.

For a discussion of scattered sets in general topological spaces see, for example, Kuratowski [166, p. 78]. We discuss the nature of scattered sets on the real line and then, by using the natural order on the line, refine this to one-sided versions.

A set is *scattered* (*clairsemé*) if every nonempty subset has an isolated point. One sided versions are useful too: a set is *right [left] scattered* if every nonempty subset has a point isolated on the right [left]. It is such sets that we call *semi-scattered*. A set is *splattered* if it may be expressed as the union of a right scattered set and a left scattered set. Scattered sets may, similarly, be viewed as coming from the intersection of a right scattered set and a left scattered set.

The first explicit use of such ideas is in Cantor [40] where he uses the term *separierte Mengen* for a set that contains no subset dense-in-itself. G. C. Young and W. H. Young, in the first decades of this century, made considerable use of scattered sets including left and right versions but employed no terminology. Denjoy introduced the term *clairsemé* into French language accounts and Hausdorff employed *zerstreute Mengen* in his writing. Hobson [133] in his account of the period used the notions but failed to employ any terminology (even avoiding the French term which, by then, was well known). Viola [302] (a student of Denjoy) studied the one sided versions and introduced the terminology *clairsemé à droit/gauche*. Freiling [111] introduced the language *semi-scattered* as well as the new notion of a splattered set (every nonempty subset contains a point isolated on one side at least) and showed that such sets are unions of semi-scattered sets; in addition his article contains a number of games characterizing all of these concepts. We summarize.

Definition A.1

(i) A set E is said to be *scattered* if each nonempty subset contains an isolated point.

(ii) A set E is said to be *right scattered* (*left scattered*) if each nonempty subset contains a point isolated on the right (left).

(iii) A set E is said to be *semi-scattered* if it is right scattered or left scattered.

(iv) A set E is said to be *splattered* if each nonempty subset contains a point that is isolated on one side at least.

These, together with sets having countable closure, form a hierarchy of countable exceptional sets. Every set having countable closure is scattered, but not conversely; every scattered set is splattered but not conversely. The collection of midpoints to the complementary intervals of the Cantor set form a scattered set which, since its closure contains the entire Cantor set, does not have countable closure. The right hand (left hand) endpoints of these intervals forms a left scattered (right scattered) set that is not scattered. The set of all the endpoints of these intervals is splattered but not scattered on either side.

Let us cite some of the elementary properties of scattered and splattered sets. Any proofs omitted can be found in Freiling [111].

Theorem A.2 *Scattered sets, semi-scattered and splattered sets are necessarily countable.*

Scattered sets can be characterized as those countable sets that are not too complicated as Borel sets. This next theorem is true in more general topological settings (eg. in completely metrizable, separable spaces). It was proved by Davies and Galvin in answer to a query posed in the first issue of the Real Analysis Exchange although it seems likely to have been known since the 1920's at least. Papers of Sierpiński from that period contain similar ideas and may, perhaps, even contain this explicitly.

Theorem A.3 *A set is scattered if and only if it is a countable Borel set of type $\mathcal{G}_\delta$.*

Proof. First observe that a countable $\mathcal{G}_\delta$ set E cannot be dense-in-itself; this is an elementary consequence of the Baire category theorem. For if $\{e_1, e_2, \ldots\}$ is an enumeration of E then one of the sets $\{e_i\}$ must be dense in a portion of E and so e_i must be isolated in E. But every subset of a countable $\mathcal{G}_\delta$ set E is of the same type since each subset is of the form

$$E \setminus C = E \cap (\mathbb{R} \setminus C) = E \cap \bigcap_{c \in C} (\mathbb{R} \setminus \{c\})$$

for a countable set C. Consequently a countable $\mathcal{G}_\delta$ set E can contain no set that is dense-in-itself.

On the other hand suppose that E is scattered. By Theorem A.2 it must be countable. Let E_0 denote the set of points $x \in E$ for which there is a number $\delta(x)$ with $0 < \delta(x) < 1$ and the interval $(x - \delta(x), x + \delta(x))$ contains no other points of E. For any ordinal α let E_α denote the set of points

$$x \in E \setminus \bigcup_{\beta < \alpha} E_\beta$$

for which there is a number $\delta(x)$ with $0 < \delta(x) < 1$ and the interval $(x - \delta(x), x + \delta(x))$ contains no other points of $E \setminus \bigcup_{\beta < \alpha} E_\beta$. If for some α the set $E_\alpha = \emptyset$ then $E \setminus \bigcup_{\beta < \alpha} E_\beta$ must be dense-in-itself which would violate that E is scattered unless this set is empty too.

Thus we have assigned to each $x \in E$ a positive number $\delta(x)$ as well as a set E_α to which it belongs. Define

$$A = \bigcap_{n=1}^{\infty} \bigcup_{x \in E} (x - \delta(x)/n, x + \delta(x)/n).$$

Clearly $A \supset E$; we claim $A = E$ which expresses E as a $\mathcal{G}_\delta$ set and completes the proof.

If $x \in A$ then there is a point $y \in E$ with $|x - y| < \delta(y)/2$; we choose such a point y so that it appears in the set E_α with the least ordinal α, say in E_η. For any $n > 2/\delta(y)$ there is a $z_n \in E$ with $|x - z_n| < \delta(z_n)/n$ and, by the minimality of our choice, z_n cannot belong to any set E_α for $\alpha < \eta$. But

$$|y - z_n| \leq |y - x| + |x - z_n| < \delta(y)/2 + \delta(z_n)/n < \delta(y)/2 + \delta(y)\delta(z_n)/2 < \delta(y)$$

so z_n cannot belong to any set E_α for $\alpha > \eta$. We conclude that $z_n = y$ for all n so that $|x - y| < \delta(y)/n$ for all n which requires $x = y$. Thus $x \in E$ and we have shown that $E = A$ as required.

Theorem A.4 (Freiling) *A set is scattered if and only if it is the intersection of a right-scattered set and a left-scattered set.*

Theorem A.5 (Freiling) *A set is splattered if and only if it is the union of a right-scattered set and a left-scattered set.*

Theorem A.6 *The union of any finite collection of scattered sets (splattered sets) is of the same type.*

The following construction, using ideas of Szpilrajn as presented by Freiling [111], is useful in obtaining examples involving scattered sets.

Lemma A.7 *Let the set E be right scattered [resp. left scattered]. Then there is a nonnegative function g with*

$$E = \{x \,:\, g(x) > 0\}$$

and $g(x - h) \le h^2$ [resp. $g(x + h) \le h^2$] for all x and all sufficiently small $h > 0$.

Proof. Suppose that E is right scattered. Write for each ordinal α the expression $D_r^\alpha(E)$ for the right derived set of E at level α. (That is to say $D_r^0(E) = E$, $D_r^{\alpha+1}(E)$ is the set of points in $D_r^\alpha(E)$ that are not isolated on the right and, for a limit ordinal α, $D_r^\alpha(E) = \bigcap_{\beta < \alpha} D_r^\beta(E)$.) For each $x \in E$ there must be an ordinal $\alpha = \alpha(x)$ and a positive number $\rho(x)$ so that

$$x \in D_r^\alpha(E) \setminus D_r^{\alpha+1}(E)$$

and

$$(x, x + \rho(x)) \cap D_r^\alpha(E) = \emptyset.$$

We define our function $g(x)$ to be zero for every $x \notin E$ and to be $\rho(x)^2$ for $x \in E$.

We now argue that $g(x - h) \le h^2$ for all x and all sufficiently small $h > 0$. Fix an x. Then one of two situations occurs. Either there is a real number c for which the interval $(c, c + \rho(c))$ contains the point x or no such c exists. In the latter case then our inequality is immediate since evidently $g(x - h) = \rho(x - h)^2 \le h^2$ whenever $x - h \in E$ and $g(x - h) = 0 \le h^2$ whenever $x - h \notin E$. Thus it is the former case that needs an argument.

Thus we suppose that there is a real number c for which the interval $(c, c + \rho(c))$ contains the point x; we may choose the point c in such a way that the ordinal associated $\alpha(c)$ is minimal. Then there are no points in E in the interval (c, x) that also belong to $D_r^{\alpha(c)}(E)$; this means that any point $x - h \in E$ that lies in the interval (c, x) must come from an earlier set $D_r^\beta(E)$ with $\beta < alpha(c)$. By the minimality of $\alpha(c)$ then the interval $(x - h, x - h + \rho(x - h))$ cannot contain x. Hence we have again $g(x - h) = \rho(x - h)^2 \le h^2$. Of course, once again too, $g(x - h) = 0 \le h^2$ whenever $x - h \notin E$. Thus in either case the desired inequality is verified and the proof for right scattered sets is complete. The left scattered sets allow a similar proof.

A.2 Scattered Baire Theorem

A convenient method for establishing that certain exceptional sets arising in analysis are scattered has been put forward in [117]. This theorem, which we now state and prove, we shall refer to as "the scattered Baire theorem". Note that it includes a recipe for splattered and semi-scattered exceptional sets as well as scattered ones.

Theorem A.8 (Scattered Baire) *Let $\mathbb{R} = \bigcup_{i=1}^\infty E_i$ where $E_1 \subset E_2 \subset E_3 \ldots$. Then there is a sequence, possibly transfinite, of open sets $G_0 \subset G_1 \subset G_2 \ldots$ such that*

(i) $G_0 = \emptyset$,

(ii) for some countable ordinal ξ, $G_\xi = \mathbb{R}$,

(iii) *if λ is a limit ordinal then $G_\lambda = \bigcup_{\eta < \lambda} G_\eta$,*

(iv) *for each ordinal η if $G_\eta \neq \mathbb{R}$ then $G_{\eta+1} = (a_\eta, b_\eta) \cup G_\eta$ where (a_η, b_η) is an interval not contained in G_η and for some i the set E_i is dense in $(a_\eta, b_\eta) \setminus G_\eta$.*

Let R and L denote, respectively, the set of right and left endpoints of the components of the G_η; then R is left-scattered, L is right-scattered, $R \cap L$ is scattered and $R \cup L$ is splattered.

Proof. This is a direct corollary of the Baire category theorem. Assertions (i) and (iii) define the sequence initially and at limit ordinals. Otherwise for any ordinal η stop the sequence if $\mathbb{R} \setminus G_\eta = \emptyset$ and if $\mathbb{R} \setminus G_\eta \neq \emptyset$ choose an open interval I that meets this set and a set E_i so that E_i is dense in $I \setminus G_\eta$. Set $G_{\eta+1} = G_\eta \cup I$. It is clear that the set of right [left] endpoints of the G_η are left [right] scattered. For example let $\emptyset \neq L_0 \subset L$. Then associated with each point $x \in L_0$ is a first ordinal with $x \in G_\eta$. Choose a point $x_0 \in L_0$ with the least such associated ordinal. This point must be isolated on the right in L_0 since it is a left endpoint of some earlier G_η.

As an illustration of the method let us prove again the scattered part of the Charzyński theorem that played a central role in Section 2.3. This lemma, in one form or other, may be found elsewhere. For example it essentially includes Viola's [302] observation that a function with a right hand derivative must be continuous off a left scattered set. (This is reproduced, as is the scattered Baire theorem, from [117].)

Lemma A.9 *Let g be a nonnegative function and N a countable set. Suppose that*

$$\limsup_{t \to 0+} t^{-1} g(x - t) < \infty$$

for every $x \in \mathbb{R} \setminus N$. Then $\{x : g(x) > 0\}$ is right scattered.

Proof. Let x_1, x_2, x_3, ... be an enumeration of the countable set N and write

$$E_n = \{x_1, x_2, x_3 \ldots x_n\} \cup \{x \notin N : g(x - t) < nt \text{ for } 0 < t < n^{-1}\}.$$

These sets cover the real line and are expanding so the scattered Baire theorem may be applied. Let G_η denote the open sets and let R and L denote the set of right and left endpoints from that application.

We claim that g vanishes off of the right scattered set L. Suppose that $x \notin L$. Then there is a first ordinal so that $x \in G_{\eta+1}$ but $x \notin G_\eta$. Let E_i be dense in $(a_\eta, b_\eta) \setminus G_\eta$. For any $\delta > 0$ the set $(x, x + \delta) \setminus G_\eta$ must be infinite (otherwise x would be a left endpoint of a component of G_η) and so E_i is infinite in $(x, x + \delta)$ too. Choose $x < z < x + \delta$ so that $z \in E_i$ but $z \neq x_j$ for any $j \leq i$. Then

$$g(x) < i(z - x) < i\delta.$$

Hence $g(x) = 0$ for each x not in L as required.

Left scattered, scattered and splattered versions of this lemma can be similarly obtained.

A.3 A Density Computation

We require occasionally some elementary density computations. The following lemma shows how one can guarantee the selection of a point in common to a number of sets provided some common density condition is met. We are interested, for example, in assuring that an intersection

$$A \cap (B - x) \cap (0, x)$$

is nonempty when 0 is a point of density of A and B and x is sufficiently small. By using 0 as the common density point we save ourselves some unimportant notational and computational clumsiness.

Lemma A.10 *If 0 is a density point for each of the sets* $A_1, A_2, \ldots A_k$, *all but one of which are measurable, then there is a positive number* δ *so that whenever* $0 < \tau < \delta$ *and* $\max\{|\alpha_1|, |\alpha_2|, \ldots |\alpha_k|\} \leq \tau$ *then the sets*

$$(0, \tau) \cap \bigcap_{i=1}^{k} (\alpha_i + A_i)$$

and

$$(-\tau, 0) \cap \bigcap_{i=1}^{k} (\alpha_i + A_i)$$

have positive outer measure.

Proof. Suppose that A is measurable and has 0 as a point of density. Write γ for the characteristic function of A and $\Gamma(t) = \int_0^t \gamma(s) \, ds$. By the density requirements on A we have $\Gamma(t)/t \to 1$ as $t \to 0$. The set $(\alpha + A) \cap (0, t)$ with $|\alpha| \leq t$ has measure

$$\int_0^t \gamma(s - \alpha) \, ds = \Gamma(t - \alpha) - \Gamma(-\alpha).$$

which is again asymptotically equal to t for small t. Thus

$$(0, \tau) \cap \bigcap_{i=2}^{k} (\alpha_i + A_i)$$

also has measure asymptotically equal to t for small t if $A_2, A_3, \ldots A_k$ are measurable. This proves the lemma if all sets are measurable.

Even if A_1 is not measurable the set

$$(0, \tau) \cap (\alpha_1 + A_1)$$

has outer measure exceeding $\tau/2$ for small τ and thus the set

$$(0, \tau) \cap \bigcap_{i=1}^{k} (\alpha_i + A_i)$$

must have positive outer measure if τ is small enough.

Other variants on this are useful. Occasionally one needs points further away from the density point, say in the interval $(\theta\tau, \tau)$ rather than merely in $(0, \tau)$ as in the lemma. For instances see [313, Vol. II, p. 78] or [186, p. 12] or [214, pp. 207–208].

A.4 Density Points

Many of our proofs require a more careful handling of density arguments than is afforded by just the one result of the preceding section.

We define the collection of density points of a set E of real numbers as

$$d(E) = \left\{ x : \lim_{h \to 0} \frac{|E \cap (x - h, x + h)|}{2h} = 1 \right\}.$$

For the dual category notion we take the notion of the "second category points" of E given as

$$c(E) = \mathbb{R} \setminus \text{Closure} \left(\bigcup \{(a, b) : (a, b) \cap E \text{ is first category}\} \right).$$

(Note that this set is somewhat smaller than the analogous notion used in Kuratowski [166, p. 83]).

The following properties (taken directly from Uher [294]) are readily established for the density notion and are all that are needed in constructing the geometric arguments used in all of the proofs involving density.

Theorem A.11 *Let $d(E)$ denote the set of exterior density points of an arbitrary set E defined as*

$$d(E) = \left\{ x : \lim_{h \to 0} \frac{|E \cap (x - h, x + h)|}{2h} = 1 \right\}.$$

Then the following properties for all sets of real numbers.

(1) *If $A \subset B$ then $d(A) \subset d(B)$.*

(2) *If $\bigcup A_n = A$ then $d(A) \setminus \bigcup d(A_n)$ has measure zero.*

(3) *If $A \setminus B$ and $B \setminus A$ both have measure zero then $d(A) = d(B)$.*

(4) *$d(d(A)) = d(A)$ for any set A.*

(5) *$A \setminus d(A)$ has measure zero for any set A.*

(6) *$d(\alpha + \beta A) = \alpha + \beta d(A)$ for any set A and real numbers α and β.*

(7) *A is measurable if and only if $|d(A) \setminus A| = 0$.*

(8) *If A is measurable then $d(A \cap B) = d(A) \cap d(B)$ for any set B.*

(9) *If 0 is a density point for each of the sets $A_1, A_2, \ldots A_k$ all but one of which are measurable then for every $p > 0$ there is a positive number δ so that whenever $0 < \tau < \delta$, $(a, b) \subset (-\tau, +\tau)$ with $b - a > \tau/p$ and $\max\{|\alpha_1|, |\alpha_2|, \ldots |\alpha_k|\} \leq \tau$ then the set*

$$(a, b) \cap \bigcap_{i=1}^{k} (\alpha_i + A_i)$$

has positive measure.

A.5 Category Density Points

Recall that for the dual category notion we have defined the "second category points" of E given as

$$c(E) = \mathbb{R} \setminus \text{Closure}\left(\bigcup \{(a,b) : (a,b) \cap E \text{ is first category}\}\right).$$

Subject to the correct translation, the dual properties for the category notion are identical to those we have just seen for measure-theoretic density points. For "measure zero" read "first category", for "positive measure" read "second category", for "measurable set" read "set with the Baire property" and, of course, replace $d(A)$ by $c(A)$. The above properties now translate directly into the following. By displaying them as identical formally with the density properties it is easy to see how a proof for the density case translates quickly to a proof for the category case. Again this is taken from Uher [294].

Theorem A.12 *Let $c(E)$ denote the set of second category points of an arbitrary set E defined as*

$$c(E) = \mathbb{R} \setminus \text{Closure}\left(\bigcup \{(a,b) : (a,b) \cap E \text{ is first category}\}\right).$$

Then the following properties hold for all sets of real numbers.

(1) *If $A \subset B$ then $c(A) \subset c(B)$.*

(2) *If $\bigcup A_n = A$ then $c(A) \setminus \bigcup c(A_n)$ is first category.*

(3) *If $A \setminus B$ and $B \setminus A$ are both first category then $c(A) = c(B)$.*

(4) *$c(c(A)) = c(A)$ for any set A.*

(5) *$A \setminus c(A)$ is first category for any set A.*

(6) *$c(\alpha + \beta A) = \alpha + \beta c(A)$ for any set A and real numbers α and β.*

(7) *A has the Baire property if and only if $c(A) \setminus A$ is first category.*

(8) *If A has the Baire property then $c(A \cap B) = c(A) \cap c(B)$ for any set B.*

(9) *If 0 belongs to each of $c(A_1)$, $c(A_2)$, $\ldots c(A_k)$ where all but one of the sets A_1, A_2, $\ldots A_k$ have the Baire property then for every $p > 0$ there is a positive number δ so that whenever $0 < \tau < \delta$, $(a,b) \subset (-\tau, +\tau)$ with $b - a > \tau/p$ and $\max\{|\alpha_1|, |\alpha_2|, \ldots |\alpha_k|\} \leq \tau$ then the set*

$$(a,b) \cap \bigcap_{i=1}^{k} (\alpha_i + A_i)$$

is second category.

A.6 Hamel Bases

A *Hamel basis* H for the real numbers is a vector space basis for the real numbers considered as a vector space over the rationals. Thus H must be linearly independent in the sense that if the sum

$$\sum_{i=1}^{n} r_i h_i = 0$$

for rational $\{r_i\}$ and distinct elements $h_i \in H$ then necessarily

$$r_1 = r_2 = \ldots = r_n = 0,$$

and H must span $\mathbb{R}$ in the sense that every $x \in \mathbb{R}$ has a representation

$$x = \sum_{i=1}^{n} r_i h_i$$

again for appropriate rationals $\{r_i\}$ and elements $h_i \in H$. The existence of such a set was established by Hamel [127] in 1905 as part of a study of the functional equation

$$f(x + y) = f(x) + f(y). \tag{A.1}$$

Every Hamel basis has the power of the continuum ([161, Theorem 3, p. 82]). It is known that Hamel bases cannot be analytic sets ([161, Theorem 3, p. 258]), Hamel bases have inner measure zero ([161, Theorem 1, p. 255]) and Hamel bases can contain no second category subset with the Baire property ([161, Theorem 1, p. 257]).

Hamel bases may be chosen quite large; there is a Hamel basis that intersects every uncountable Borel set ([161, Theorem 3, p. 259]). They may equally be chosen quite small; there is a Hamel basis of measure zero ([161, Corollary 3, p. 261]). If C is the usual Cantor ternary set there is a Hamel basis H so that $H \setminus C$ is dense and countable; H has measure zero and is first category. Thus a Hamel basis may be measurable and have the Baire property although, as observed above, it must then be measure zero and first category ([44] use this fact to illustrate that there is a measurable function ξ_H with the Baire property that is nowhere symmetrically continuous.)

It is possible to select a Hamel basis that contains a perfect set; see [161, Theorem 2, p. 270] or the discussion of rationally independent sets in [201, p. 205 and p. 220].

From the fact that Hamel bases exist one easily produces nonmeasurable functions which satisfy the functional equation (A.1) and hence are linear. Let f map H onto $\mathbb{R}$ and extend the domain of f from H to all of $\mathbb{R}$ by linearity. It is clear that $f(x + y) = f(x) + f(y)$ no matter how f has been defined on H. For our purposes we note that any such function must also satisfy

$$f(x + t) + f(x - t) - 2f(x) = 0$$

identically and so is in a sense extremely symmetric but otherwise bizarre.

Weil [304] (cf. also Jones [145]) shows how it is possible to arrange this construction so that f has the Darboux property. Split $H = \bigcup_{n=1}^{\infty} H_n$ where the sets H_n

each have the power of the continuum and the collection is pairwise disjoint. We can adjust each element of H_n by multiplication by a rational number in such a way that $H_n \subset (1 - 1/n, 1 + 1/n)$. Let f map each H_n onto $\mathbb{R}$ and extend the domain of f from H to all of $\mathbb{R}$ by linearity. It is clear once again that $f(x + y) = f(x) + f(y)$. We show that f attains every real value in every interval. Let $(a, b) \subset \mathbb{R}$, $r \in (a, b)$ and $y \in \mathbb{R}$ be given. We find a point in (a, b) at which f attains the value y. Choose n so that $rx \in (a, b)$ for every point x in $(1 - 1/n, 1 + 1/n)$ and choose $h \in H_n$ so that $f(h) = y/r$. Then $rh \in (a, b)$ and $f(rh) = y$ as we wished. In particular this function has a dense set of zeros; this observation will be useful too.

On the other hand we can easily arrange for f to have a countable range so that f is not Darboux.

A.7 Weak Quasi–Continuity

We shall require a number of weak continuity conditions in our study. The following definition is a version of two forms of continuity known in the literature as qualitative-continuity and quasi-continuity. (For these notions, due to S. Marcus and S. Kempisty respectively, see [282, p. 21 and p. 26].) Recall that $c(A)$ denotes the set of category density points (as defined in Section A.5 of this Appendix). This material is mostly reproduced from [284].

Definition A.13 A function f is said to be *quasi-qualitatively continuous* (or simply qq–continuous) at a point x if

$$\{y : |f(x) - f(y)| < \epsilon\} \cap A \neq \emptyset$$

for every positive number ϵ and for every set A with $x \in c(A)$.

In order to better understand the notion recall that in order for f to be continuous at a point x then for every $\epsilon > 0$ the set

$$\{y : |f(x) - f(y)| < \epsilon\}$$

contains an open interval about x. In order for f to be quasi-continuous there that set must contain an open subinterval of every neighborhood of x. In order for f to be qualitatively continuous it must be residual in an open interval about x. Finally then in order for f to be qq–continuous there it must be residual in some open subinterval of every neighborhood of x. Our only applications of this notion will use the observation that if f is qq–continuous at a point x and the point x belongs to $c(A)$ then there is a sequence of points $a_n \in A$ with $a_n \to x$ and $f(a_n) \to f(x)$.

The notion of qq–continuity while formally apparently very weak is equivalent to the property of Baire. We prefer to use the "weaker" version in the statements of the theorems that require some kind of continuity, but it is well to keep in mind that we have not obtained stronger results.

Theorem A.14 *A function f has the property of Baire if and only if it is qq–continuous at the points of a dense set.*

Proof. For any function f and any set H let $\omega(f, H, x)$ denote the oscillation of the restriction of the function f to the set H taken at the point x. If f is qq-continuous at the points of a residual set (or even just a dense set) then we shall show that f has the Baire property. It is evidently enough to show that for every integer n there is a residual set A_n such that $\omega(f, A_n, x) < 1/n$ for every point $x \in A_n$. In this case f is continuous relative to the residual set $\bigcap_{n=1}^{\infty} A_n$ and so has the Baire property.

Let n be fixed. Let $\mathcal{H}$ denote the set of pairs (I, B) where I is an open interval, $B \subset I$, B is residual in I and $\omega(f, B, x) < 1/n$ for every $x \in B$. Then for every interval J there is a pair $(I, B) \in \mathcal{H}$ such that $I \subset J$. Indeed if $x \in J$ is a point of qq-continuity of f then the set

$$C = \{y : |f(x) - f(y)| < 1/(2n)\}$$

must be residual in some open subinterval I of J and then $(I, I \cap C) \in \mathcal{H}$. This implies that there is a sequence $(I_k, B_k) \in \mathcal{H}$ such that $I_k \cap I_m = \emptyset$ for $k \neq m$ and $\bigcup_{k=1}^{\infty} I_k$ is everywhere dense in $\mathbb{R}$. Thus the set $A_n = \bigcup_{k=1}^{\infty} B_k$ is residual and $\omega(f, A_n, x) < 1/n$ holds for every $x \in A_n$. This completes the proof.

A.8 Weak Approximate Continuity

The notion of qq–continuity suggests that we should require also a similar weakening of the notion of approximate continuity. Again this is reproduced from [284].

Definition A.15 A function f is said to be *quasi-approximately continuous* (or simply qa–continuous) at x if
$$\{y : |f(x) - f(y)| < \epsilon\} \cap A \neq \emptyset$$
for every positive number ϵ and for every set A with $x \in d(A)$.

We shall use this terminology since it allows a parallel development between the measure-theoretic and category-theoretic versions of certain theorems. However, again, it should be noted that this formally weaker requirement than approximate continuity is almost everywhere the same as the stronger requirement. A function is measurable if and only if it is almost everywhere approximately continuous; the same is true for qa–continuity.

Theorem A.16 *A function f is measurable if and only if it is qa–continuous at the points of a set of full outer measure.*

Proof. It is evident that a measurable function, since it is almost everywhere approximately continuous, must be almost everywhere qa–continuous. Thus it is enough to show the converse. Suppose that f is qa–continuous at the points of a set of full outer measure. It is enough to show that for every $a < b$ the sets

$$U = \{x : f(x) < a\}$$

and

$$V = \{x : f(x) > b\}$$

can be separated by a measurable set. If not then by Lemma A.20 there is a measurable set P of positive measure so that U and V are of full outer measure in P. Since f is almost everywhere qa–continuous almost every point x in U has the property that $x \notin d(V)$ since that would require $U \cap V \neq \emptyset$; for the same reason almost every point x in V has the property that $x \notin d(U)$. But this is impossible for then any point of density p of P, which is then also a point of density for both U and V since these sets are full in P, can belong to neither U nor V. But almost every point of P is a point of density which provides our contradiction.

Thus U and V can be separated by a measurable set for all $a < b$ and so f is measurable. This completes the proof.

A.9 Baire Class

An extended real-valued function f is said to be in the first Baire class if it may expressed as a pointwise limit

$$f(x) = \lim_{n \to \infty} f_n(x)$$

of a sequence of continuous functions $\{f_n\}$. Equivalently f is in this Baire class if the associated sets

$$\{x : f(x) > a\} \text{ and } \{x : f(x) < b\}$$

are Borel sets of type F_σ for all real a and b.

It is often convenient to obtain Baire classification results by appealing to an observation of Denjoy asserting that such a property may be characterized as a separation property. We cite the following result from Preiss [230, Theorem 1, p. 376] where the property of Denjoy was independently rediscovered. The proof given there applies to any completely metrizable topological space.

Theorem A.17 (Denjoy) *A function f belongs to the first Baire class if and only if there is no perfect set P and numbers $\alpha < \beta$ such that both sets*

$$\{x \in P : f(x) \leq \alpha\}$$

and

$$\{x \in P : f(x) \geq \beta\}$$

are dense in P.

A.10 Goffman Theorem

The upper and lower densities of a set E (measurable or not) at a point x_0 are defined as the upper and lower limits of the ratio

$$|E \cap [x_0 - h, x_0 + k]|/(h + k)$$

as $h \to 0+$, $k \to 0+$ with $h + k > 0$. If the limit exists then the density is said to exist.

.The following observation from Goffman [123] will prove useful for a counterexample in Section 7.2.1.

Theorem A.18 (Goffman) *Let Z be an arbitrary bounded set of measure zero. Then there is a measurable set E whose metric density fails to exist at every point of Z.*

In particular then

$$f(x) = \int_{-\infty}^{x} \chi_E(t) \, dt$$

is a Lipschitz function (all of its derivates, symmetric or ordinary, are bounded in absolute value by 1) and $f'(x)$ cannot exist at any point of the set Z. This puts a limitation on what can be said in the Khintchine Theorem. Even if we know a function has bounded symmetric derivates we cannot say more about the set of points of nondifferentiability than that is has measure zero. Here Z can be second category or non σ–porous or have any Hausdorff dimension for example.

A.11 Measurability

We shall obtain a number of our measurability results in Chapter 6 as separation properties and for that we shall require the lemmas of this section. Recall that sets A and B are said to be *separated* by a set M if $A \subset M$ and $B \cap M = \emptyset$. The first lemma is just the elementary and well known observation that measurability may be characterized as a separation property.

Theorem A.19 *A function f is measurable if and only if for every $a < b$ the sets*

$$U = \{x : f(x) < a\}$$

and

$$V = \{x : f(x) > b\}$$

can be separated by a measurable set.

Proof. Certainly the condition is necessary. To show it is sufficient we show, under its assumption, that for any real number c the set of points

$$E_c = \{x : f(x) \leq c\}$$

is measurable. For each natural number n choose a measurable set M_n so that

$$M_n \supset \{x : f(x) \leq c\}$$

and

$$M_n \cap \{x : f(x) > c + 1/n\} = \emptyset.$$

It is an easy matter to verify that

$$\bigcap_{n=1}^{\infty} M_n = \{x : f(x) \le c\}$$

which exhibits the set E_c as measurable.

The next lemma and its proof are reproduced from Preiss and Thomson [232]. This gives a useful criterion of measurability in light of Lemma A.19.

Theorem A.20 *Suppose that the subsets U and V of the reals cannot be separated by a measurable set and that h is a positive function defined on $U \cup V$. Then there are a positive number ε and a nonempty compact subset P of $\mathbb{R}$ with the following properties.*

(1) *The intersection $I \cap P$ has positive measure whenever I is an open interval meeting P.*

(2) *The sets $\{x \in U \cap P; h(x) > \epsilon\}$ and $\{x \in V \cap P; h(x) > \epsilon\}$ are both of full outer measure in P.*

Proof. If for each $k = 1, 2, \dots$ the sets

$$\{x \in U; h(x) > 1/k\} \quad \text{and} \quad \{x \in V; h(x) > 1/k\}$$

could be separated by measurable sets, say, M_k, then U and V would be separated by $\bigcup_{j=1}^{\infty} \bigcap_{k=j}^{\infty} M_k$. Hence there is a positive number ϵ such that the sets

$$\{x \in U; h(x) > \epsilon\} \quad \text{and} \quad \{x \in V; h(x) > \epsilon\}$$

cannot be separated by a measurable set.

Next we observe that whenever $E_1, E_2, \dots$ are measurable sets such that the sets $A \cap E_i$ and $B \cap E_i$ are separated by measurable sets, say, M_i, then the sets

$$A \cap \bigcup E_i \quad \text{and} \quad B \cap \bigcup E_i$$

are separated by $\bigcup(E_i \cap M_i)$. From this we easily deduce that there is a measurable set $A \subset \mathbb{R}$ such that the sets

$$\{x \in U \cap A; h(x) > \epsilon\} \quad \text{and} \quad \{x \in V \cap A; h(x) > \epsilon\}$$

can be separated by a measurable set and that the set $\mathbb{R} \setminus A$ contains no measurable set B of positive measure for which the sets

$$\{x \in U \cap B; h(x) > \epsilon\} \quad \text{and} \quad \{x \in V \cap B; h(x) > \epsilon\}$$

would be separated by a measurable set.

Since the set $\mathbb{R} \setminus A$ has positive measure, it contains a nonempty compact set P for which A.20(1) holds. Clearly, P is the required set, since the assumption that, for example, the set

$$P \setminus \{x \in U \cap P; h(x) > \epsilon\}$$

contains a measurable set B of positive measure immediately implies that the sets

$$\{x \in U \cap B; h(x) > \epsilon\} \quad \text{and} \quad \{x \in V \cap B; h(x) > \epsilon\}$$

are separated by B.

A.12 The Baire Property

A set E is said to have the Baire property (see [166, p. 87]) if it is of the form

$$E = (G \setminus P) \cup R$$

where G is open and P and R are first category sets. Equivalently (see [166, p. 88]) E has this property if it is the union of a G_δ set and a set of the first category.

A function f is said to have the Baire property if the set $f^{-1}(G)$ has the Baire property for every open set G. All Baire functions have this property. A more useful characterization is provided by the following theorem (see [166, p. 400] for a proof) .

Theorem A.21 *A necessary and sufficient condition for a function f to have the Baire property is that there exist a set Z residual in $\mathbb{R}$ so that f is continuous relative to Z.*

Analogous to the measurability results proved in Section A.11, it is possible to characterize functions with the Baire property in terms of separation properties and for that we shall require some preliminary lemmas. Recall that sets A and B are said to be separated by a set M if $A \subset M$ and $B \cap M = \emptyset$.

Lemma A.22 *A function f has the Baire property if and only if for every $a < b$ the sets*

$$U = \{x : f(x) < a\}$$

and

$$V = \{x : f(x) > b\}$$

can be separated by a set which has the Baire property .

Proof. Certainly the condition is necessary. To show it is sufficient we show, under its assumption, that for any real number c the set of points

$$E_c = \{x : f(x) \leq c\}$$

has the Baire property. For each natural number n choose a set M_n with the Baire property so that

$$M_n \supset \{x : f(x) \leq c\}$$

and

$$M_n \cap \{x : f(x) > c + 1/n\} = \emptyset.$$

It is an easy matter to verify that

$$\bigcap_{n=1}^{\infty} M_n = \{x : f(x) \leq c\}$$

which exhibits the set E_c as having the Baire property as we require.

Lemma A.23 *Suppose that the subsets U and V of the reals cannot be separated by a set with the Baire property and that h is a positive function defined on $U \cup V$. Then there are a positive number ϵ and a nonempty G_δ set P of $\mathbb{R}$ with the following properties.*
(a) P is residual in some interval I.
(b) The sets

$$P \setminus \{x \in U \cap P; h(x) > \epsilon\} \quad and \quad P \setminus \{x \in V \cap P; h(x) > \epsilon\}$$

contain no second category set with the Baire property.

Proof. Just for this proof let us say that sets A and B are simply separated if they can be separated by a set with the Baire property. If for each $k = 1, 2, \ldots$ the sets

$$\{x \in U; h(x) > 1/k\} \quad and \quad \{x \in V; h(x) > 1/k\}$$

could be separated by a set, say M_k, with the Baire property then U and V would be separated by $\bigcup_{j=1}^{\infty} \bigcap_{k=j}^{\infty} M_k$. Hence there is a positive number ϵ such that the sets

$$U_1 = \{x \in U; h(x) > \epsilon\} \quad and \quad V_1 = \{x \in V; h(x) > \epsilon\}$$

cannot be separated.

Next we observe that whenever $E_1, E_2, \ldots$ are pairwise disjoint sets with the Baire property such that the sets $A \cap E_i$ and $B \cap E_i$ are separated, say by M_i, then the sets

$$A \cap \bigcup E_i \quad and \quad B \cap \bigcup E_i$$

are separated by $\bigcup (E_i \cap M_i)$. From this we can deduce that there is a set $A \subset \mathbb{R}$ such that A has the Baire property, the sets $U_1 \cap A$ and $V_1 \cap A$ can be separated and that the set $\mathbb{R} \setminus A$ contains no second category set B with the Baire property for which the sets $U_1 \cap B$ and $V_1 \cap B$ can be separated.

To see this choose, if possible, a second category set E_1 with the Baire property so that $U_1 \cap E_1$ and $V_1 \cap E_1$ can be separated. Continue inductively choosing for any ordinal β a second category set

$$E_\beta \subset \mathbb{R} \setminus \bigcup_{\alpha < \beta} E_\alpha$$

with the Baire property so that $U_1 \cap E_\beta$ and $V_1 \cap E_\beta$ can be separated. While the process may continue transfinitely we know that any collection of disjoint second category sets with the Baire property is countable (eg. [166, p. 256]) so that by the observation just made in the preceding paragraph the sets

$$U_1 \cap \bigcup E_\alpha \quad and \quad V_1 \cap \bigcup E_\alpha$$

can be separated.

The set $\mathbb{R} \setminus \bigcup E_\alpha$ has the Baire property and must itself be of second category otherwise U_1 and V_1 can be separated. Thus if we take $A = \bigcup E_\alpha$ then, by its construction, the set $\mathbb{R} \setminus A$ contains no second category set B with the Baire property for which the sets $U_1 \cap B$ and $V_1 \cap B$ can be separated.

We select a set $P \subset A$ so that the assertion A.23(a) holds. Clearly P is the required set since the assumption that, for example, the set $P \setminus \{x \in U \cap P; \ h(x) > \epsilon\}$ contains a second category set B with the Baire property immediately implies that the sets

$$\{x \in U \cap B; \ h(x) > \epsilon\} \quad \text{and} \quad \{x \in V \cap B; \ h(x) > \epsilon\}$$

are separated by B.

Problems

1. Characterize the set of points of discontinuity of everywhere symmetrically continuous functions.

2. Characterize the set of points of discontinuity of measurable, everywhere symmetric functions.

3. Is there a difference in the classes of sets of points of discontinuity of everywhere symmetrically continuous functions and those for measurable, everywhere symmetric functions? (One set is a subset of the other. Is it proper?)

4. Can any N–set appear as the set of points of discontinuity of a measurable, everywhere symmetric functions? (cf. Theorem 2.33.)

5. Let $\{\xi_i\}$ be a sequence $(0 < \xi_i < \frac{1}{2})$. For what symmetric perfect sets $C(\{\xi_i\})$ can there exist a symmetrically continuous [symmetric] function that is discontinuous at every point of $C(\{\xi_i\})$. (The text includes only the case ξ_i constant.)

6. (Evans [84]) Is the set of discontinuities of a measurable, symmetric function necessarily σ–porous? (The odd analogue has a negative answer; see Section 2.9.)

7. Characterize the set of points of discontinuity of everywhere symmetrically differentiable functions (allowing infinite derivatives).

8. Characterize the set of points of discontinuity of functions f satisfying, at each point x,
$$f(x+t) - f(x-t) = O(t^\alpha)$$
for all $0 < \alpha < 1$.

9. Characterize the set of points of discontinuity of measurable functions f satisfying, at each point x,

$$f(x+t) + f(x-t) - 2f(x) = O(t^\alpha)$$

for all $0 < \alpha < 1$.

10. If a measurable function f satisfies

$$f(x+t) + f(x-t) - 2f(x) = O(t^\alpha)$$

for some $0 < \alpha < 1$ at each point x then it is continuous on a dense open set. Given a dense open set G can we find a function f that satisfies this and is continuous at every point in G?

11. (Evans [84]) Characterize the set of points at which a measurable, L_p-symmetric function is L_p-discontinuous.

12. (Evans [84]) Characterize the set of points at which a measurable, L_p-smooth function is L_p-discontinuous.

13. (Larson [173]) Characterize the set of points at which a symmetrically differentiable function (allowing infinite derivative) is discontinuous. (The set is σ-porous; can it be uncountable?)

14. (S. Marcus) Characterize the set of points at which a function can be symmetrically continuous. (Recall that there is no upperbound on the topological complexity of such sets. See Theorem 2.46.)

15. Characterize the set of points at which a measurable function can be symmetric.

16. Characterize the measurable functions f satisfying

$$\limsup_{t \to 0} \left| \frac{f(x+t) + f(x-t) - 2f(x)}{t} \right| \leq C$$

at every point. (The odd version, Charzynski–Wolibner Theorem, is just that the function is in substantial agreement with a Lipschitz function. The *uniform* even version of this gives a condition on the modulus of continuity of f.)

17. Does there exist a measurable function f satisfying

$$f(x+t) + f(x-t) - 2f(x) = O(h^\alpha)$$

for every $0 < \alpha < 1$ and having an uncountable set of points of discontinuity? (It would have to be nowhere dense of course.)

18. In the space of bounded, symmetric functions with supremum norm does the typical function have a c-dense set of discontinuities?

19. In the space of bounded, symmetrically continuous functions with supremum norm does the typical function have a c-dense set of discontinuities?

20. In the space of uniformly quasi-smooth functions (Zygmund's space Λ) is nowhere differentiability typical?

21. In the space of uniformly smooth functions (Zygmund's space λ) is almost nowhere differentiability typical?

22. In the other Zygmund spaces $(\Lambda_\alpha, \lambda_\alpha)$ what other properties are typical?

23. Characterize the property of local midpoint convexity for measurable functions, i.e. that
$$f(x+h) + f(x-h) - 2f(x) \geq 0$$
just locally at each x. (We know f is convex on each component of an open dense set.)

24. Characterize the properties of measurable, semi–symmetric functions, that is
$$\liminf_{h \to 0} f(x+h) + f(x-h) - 2f(x) \geq 0$$
at each x.

25. (Ciesielski and Larson [44]) Is there a bounded function that is nowhere weakly symmetrically continuous?

26. (C. M. Lee) Characterize the class of sets E that have the property that F measurable and approximately symmetrically continuous and ASD $F(x) = 0$ for all $x \notin E$ implies that F is equivalent to a constant function.

27. Characterize the class of sets E that have the property that F continuous and smooth and $SD_2 F(x) = 0$ for all $x \notin E$ implies that F is linear.

28. (C. Freiling [111]) Suppose that $\underline{SD}\, F(x) > -\infty$ everywhere. Then F is upper semicontinuous on a substantial set. In fact it is continuous except on a set of measure zero and first category. Is F in fact upper semicontinuous except on a countable set? Except on a scattered set?

29. Let f be continuous and suppose that VS_f is σ–finite on (a,b) and absolutely continuous with respect to Lebesgue measure (see Definition 8.1). Is f necessarily ACG?

30. Let f be continuous and suppose that $VS_f^{(2)}$ is σ–finite on (a,b) and absolutely continuous with respect to Lebesgue measure (see Section 8.3.5). Is f necessarily ACG?

31. Does there exist a measurable, symmetric function that is everywhere approximately differentiable and nowhere differentiable? (cf. Theorem 4.65.)

32. Characterize the set of points at which a continuous, smooth function fails to have a derivative. (The set of points where a continuous, smooth function has a derivative is necessarily c–dense but may have measure zero.)

33. (Evans [84]) Is a measurable, approximately smooth function necessarily in Baire∗1?

34. Let E be a measure zero set of type F_σ. Then there is a continuous, uniformly smooth function f so that $f'(x) = +\infty$ for every $x \in E$. Find a real variable proof.

35. A complex-valued, uniformly smooth function can be nowhere differentiable. Find a real variable proof.

36. A continuous, uniformly smooth function is differentiable on a set of Hausdorff dimension one. Find a real variable proof.

37. Let f be a continuous, periodic, uniformly quasi–smooth function. Then f fulfils Banach's condition T_2. Find a direct proof (i. e. more direct than [1]).

38. Let f be a continuous, quasi–smooth function. Does f fulfil Banach's condition T_2? (This is the pointwise version of the preceding problem.)

39. Let f be a continuous quasi–smooth function. Must the graph of f have σ–finite length. (This is a stronger condition than T_2; for uniformly quasi–smooth functions this is proved in [1]).

40. Let f be a continuous, periodic, uniformly smooth function. Then f fulfils Banach's condition T_1. Find a direct proof (i. e. more direct than [1]).

41. Let f be a continuous, smooth function. Does f fulfil Banach's condition T_1. (This is the pointwise version of the preceding problem.)

42. Characterize the set of points at which a symmetrically differentiable function may fail to be differentiable.

43. Characterize the support set of a symmetric devil (see Section 7.3.7).

44. Characterize the set of points at which a measurable, approximately symmetrically differentiable function may fail to be approximately differentiable.

45. A necesary and sufficient condition for a continuous periodic function f to satisfy a Lip(α) condition for $0 < \alpha < 1$ is that

$$f(x + h) + f(x - h) - 2f(x) = O(h^\alpha)$$

uniformly in x. Is there an elementary proof? (The one in [312, Lemma 4, p. 63] uses best approximation theory.)

46. (Kostryko et al. [158]) Characterize the functions which are uniform limits of locally symmetric functions.

47. If a function f is (R_s^2)–integrable on an interval must it have a primitive there (i.e. is there a continuous function F with SD $F(x) = f(x)$ a.e. there)?

48. Does there exist a trigonometric series $a_0/2 + \sum_{n=1}^\infty (a_n \cos nx + b_n \sin nx)$ with bounded partial sums at every point x and so that the coefficients satisfy

$$\sum_{k=1}^{n} k\sqrt{a_k^2 + b_k^2} = o(n)$$

as $n \to +\infty$, but the series is not a Lebesgue–Fourier series. (cf. Theorems 9.24 and 9.25).

49. (A. Kechris) Classify the complexity of determining the coefficients from the sum function of an everywhere convergent trigonometric series. (See Dougherty and Kechris [71] for the language within which this problem is properly expressed.)

References

[1] J. M. Anderson and L. D. Pitt. Probabilistic behaviour of functions in the Zygmund spaces Λ^* and λ^*. *Proc. London Math. Soc. (3)*, 59: 558–592, 1989.

[2] T. Anghelutza. Sur une propriété des polynômes. *Bull. Sci. Math. (2)*, 63:239–246, 1939.

[3] J. Ash, J. Cohen, C. Freiling, L. Gluck, E. Rieders, and G. Wang. Characterizations and generalizations of continuity. *Proc. Amer. Math. Soc.*, (to appear) 1993.

[4] J. M. Ash. Generalizations of the Riemann derivative. *Trans. American Math. Soc.*, 126:181–191, 1967.

[5] J. M. Ash. A new proof of uniqueness for [multiple] trigonometric series. *Proc. American Math. Soc.*, 107:409–410, 1989.

[6] J. M. Ash. A new, harder proof that continuous functions with Schwarz derivative 0 are lines. *Contemporary Aspects of Fourier Analysis, Bray and Stanojevi'vc eds.*, 1993.

[7] J. M. Ash, J. Cohen, C. Freiling, and D. Rinne. Generalizations of the wave equation. *Trans. Amer. Math. Soc.*, 337: 1993.

[8] J. M. Ash, C. Freiling, and D. Rinne. Uniqueness of rectangularly convergent trigonometric series. *Annals of Math.*, 137:145–166, 1993.

[9] H. Auerbach. Sur les dérivées generalisées. *Fund. Math.*, 8:49–55, 1926.

[10] C. E. Aull. The first symmetric derivative. *American Math. Monthly*, 74:708–711, 1967.

[11] N. Bary. *A treatise on trigonometric series, Vol. I and II*. Pergamon Press, New York, 1964.

[12] C. L. Belna. Symmetric continuity of real functions. *Proc. American Math. Soc.*, 87:99–102, 1983.

[13] C. L. Belna, M. J. Evans, and P. Humke. Symmetric and ordinary differentiation. *Proc. American Math. Soc.*, 72:261–267, 1978.

[14] C. L. Belna, M. J. Evans, and P. Humke. Symmetric and strong differentiation. *American Math. Monthly*, 86:121–123, 1979.

[15] F. Berstein and G. Doetsch. Zur Theorie der konvexen Funktionen. *Math. Ann.*, 76:514–526, 1915.

[16] H. Blumberg. On convex functions. *Trans. American Math. Soc.*, 29:40–44, 1919.

[17] M. W. Botsko. A unified treatment of various theorems in elementary analysis. *Amer. Math. Monthly*, 94:450–452, 1987.

[18] A. M. Bruckner. *Differentiation of Real Functions*. Lecture Notes in Mathematics 659. Springer-Verlag, Berlin, 1978.

[19] A. M. Bruckner, J. Ceder, and M. Weiss. On uniform limits of Darboux functions. *Colloq. Math.*, 45:65–77, 1966.

[20] Yu. A. Brudnyi. On the maximum modulus of a quasi-smooth function. *Uspehi Mat. Nauk (N.S.)*, 12:273–275, 1957.

[21] Z. Buczolich. Convexity and symmetric derivates of measurable functions. *Real Anal. Exchange*, 16:187–196, 1990/91.

[22] Z. Buczolich, M. J. Evans, and P. D. Humke. Approximate high order smoothness. Acta Math. Hung., 61 (3-4), 369-388 (1993).

[23] Z. Buczolich and M. Laczkovich. Concentrated Borel measures. *Acta. Math. Hungar.*, 57:349–362, 1991.

[24] P. S. Bullen. Construction of primitives of generalized derivatives with applications to trigonometric series. *Canad. J. Math.*, 13:48–58, 1961.

[25] P. S. Bullen. Nonabsolute integrals: a survey. *Real Anal. Exchange*, 5:195–259, 1979.

[26] P. S. Bullen. Denjoy's index and porosity. *Real Anal. Exchange*, 10:85–144, 1984.

[27] P. S. Bullen and C. M. Lee. On the integrals of Perron type. *Trans. American Math. Soc.*, 182:481–501, 1973.

[28] P. S. Bullen and C. M. Lee. The SC_nP–integral and the P^{n+1}–integral. *Canad. J. Math.*, 25:1274–1284, 1973.

[29] P. S. Bullen and S. N. Mukhopadhyay. Integration by parts formulae for some trigonometric integrals. *Proc. London Math. Soc.*, (3) 29:159–173, 1974.

[30] P. S. Bullen and S. N. Mukhopadhyay. Generalized continuity of higher order symmetric derivatives. *Southeast Asian Bull. Math.*, 13:127–137, 1989.

[31] P. S. Bullen and S. N. Mukhopadhyay. The integrability of Riemann summable trigonometric series. *Canad. Math. Bull.*, 33:273–281, 1990.

[32] H. Burkill. A note on trigonometric series. *J. Math. Anal. Appl.*, 40:39–44, 1972.

[33] H. Burkill. Fourier series of SCP–integrable functions. *J. Math. Anal. Appl.*, 57:587–609, 1977.

[34] J. C. Burkill. The expresssion of trigonometric series in Fourier form. *J. London Math. Soc.*, 11:43–48, 1936.

[35] J. C. Burkill. Integrals and trigonometric series. *Proc. London Math. Soc.*, (3) 1:46–57, 1951.

[36] J. C. Burkill. Corrigendum: Integrals and trigonometric series. *Proc. London Math. Soc.*, (3) 47:192, 1983.

[37] P. Butzer. Bezeihungen zwischen den Riemannscen, Taylorscen und gewohnlichen reelwertiger Funktionen. *Math. Ann.*, 144:275–298, 1961.

[38] P. Butzer and W. Kozakiewicz. On the Riemann derivatives for integrable functions. *Canad. J. Math.*, 6:572–581, 1954.

[39] A. Calderon and A. Zygmund. Local properties of solutions of elliptic partial differential equations. *Studia Math.*, 20:171–225, 1961.

[40] G. Cantor. Über unendliche, lineare Punktmannigfaltigkeiten. *Math. Ann.*, 23:453–488, 1884.

[41] G. Cantor. *Gessamelte Abhandlungen*. Georg Olms, Hilldesheim, 1962.

[42] Z. Charzyński. Sur les fonctions dont la derivée symetrique est partout finie. *Fund. Math.*, 21:214–225, 1931.

[43] M. Chlebík. On symmetrically continuous functions. *Real Anal. Exchange*, 13:34, 1987/88.

[44] K. Ciesielski and L. Larson. Uniformly antisymmetric functions. *Real Anal. Exchange*, 19(1), 1993/94.

[45] E. Corominas. Dérivation de Riemann-Schwarz. *C. R. Acad. Sci. Paris Sér. I Math.*, 224:176–177, 1947.

[46] P. Cousin. Sur les fonctions de n variables complexes. *Acta Math.*, 19:1–61, 1895.

[47] G. E. Cross. The expression of trigonometric series in Fourier form. *Canad. J. Math.*, 12:694–698, 1960.

[48] G. E. Cross. The relation between two definite integrals. *Proc. American Math. Soc.*, 11:578–579, 1960.

[49] G. E. Cross. The relation between two symmetric integrals. *Proc. American Math. Soc.*, 14:185–190, 1963.

[50] G. E. Cross. On the generality of the AP–integral. *Canad. J. Math.*, 23:557–561, 1971.

[51] G. E. Cross. The P^n–integral. *Canad. Math. Bull.*, 18:493–497, 1975.

[52] G. E. Cross. Additivity of the P^n–integral. *Canad. J. Math.*, 30:783–796, 1978.

[53] G. E. Cross. The representation of (C,k)–summable trigonometric series. *Canad. Math. Bull.*, 21:149–158, 1978.

[54] G. E. Cross. The $SC_{k+1}P$–integral and trigonometric series. *Proc. American Math. Soc.*, 69:297–302, 1978.

[55] G. E. Cross. Higher order Riemann complete integrals. *Real Anal. Exchange*, 11:347–364, 1985/86.

[56] G. E. Cross and B. S. Thomson. Symmetric integrals and trigonometric series. *Dissertationes Math. (Rozprawy Mat.)*, Vol. 319, 1992.

[57] U. B. Darji. Symmetric behavior in functions. *Proc. American Math. Soc.*, November 1993.

[58] U. B. Darji. Symmetric functions, Lebesgue measurability and the Baire property. *Proc. American Math. Soc.*, 118:1151–1158, 1993.

[59] U. B. Darji. Coanalytic sets and symmetric behavior in functions. *J. Math. Anal. Appl.*, (to appear).

[60] R. O. Davies. Symmetric sets are measurable. *Real Anal. Exchange*, 4:87–89, 1978/79.

[61] R. O. Davies and F. Galvin. Solution to Query 5. *Real Anal. Exchange*, 2:74–75, 1976.

[62] S. De Sarkar and A. G. Das. Riemann derivatives and general integrals. *Bull. Austral. Math. Soc.*, 35:187–211, 1987.

[63] M. Dehn Über die Zerlegung von Rechtecken in Rechtecke. *Math. Ann.*, 57:314–332, 1903.

[64] A. Denjoy. Calcul de la primitive de la fonction dérivée la plus générale. *C. R. Acad. Sci. Paris Sér. I Math.*, 154:1075–1078, 1912.

[65] A. Denjoy. Sur l'absolue convergence des séries trigonométrique. *C. R. Acad. Sci. Paris Sér. I Math.*, 156:135–136, 1912.

[66] A. Denjoy. Une extension de l'intégrale de M. Lebesgue. *C. R. Acad. Sci. Paris Sér. I Math.*, 154:859–862, 1912.

[67] A. Denjoy. Calcul des coefficients d'une séries trigonométrique partout convergente. *C. R. Acad. Sci. Paris Sér. I Math.*, 172:1218–1221, 1921.

[68] A. Denjoy. *Leçons sur le calcul des coefficients d'une série trigonométrique.* Hermann, Paris, 1941–49.

[69] A. Denjoy. Totalization des derivées premieres generalizés, I. *C. R. Acad. Sci. Paris Sér. I Math.*, 241:617–620, 1955.

[70] A. Denjoy. Totalization des derivées premieres generalizés, II. *C. R. Acad. Sci. Paris Sér. I Math.*, 241:829–832, 1955.

[71] R. Dougherty and A. S. Kechris. The complexity of antidifferentiation. *Advances in Math.*, 88:145–169, 1991.

[72] P. L. Dure, H. S. Shapiro, and A. L. Shields. Singular measures and domains not of Smirnov type. *Duke Math. J.*, 33:247–254, 1966.

[73] W. L. Duren. Mathematical induction in sets. *Amer. Math. Monthly*, 64:19–21, 1957.

[74] T. K. Dutta. Generalized smooth functions. *Acta. Math. Hungar.*, 40:29–37, 1982.

[75] T. K. Dutta. On uniform generalized symmetric derivatives. *Indian J. Math.*, 33:63–74, 1991.

[76] T. K. Dutta and S. N. Mukhopadhyay. Generalized smooth functions. II. *Acta. Math. Hungar.*, 55:47–56, 1990.

[77] A. F. Dzyadyk, V. K.Timan. On best approximation of quasi-smooth functions by ordinary polynomials. *Doklady Akad. Nauk-SSSR (N.S.)*, 75:499–501, 1950.

[78] The Editors. Query 37. *Real Anal. Exchange*, 3:107, 1977/78.

[79] H. W. Ellis. On the relation between the P^2-integral and the Cesàro-Perron scale of integrals. *Trans. Roy. Soc. Canada (4)*, III (3) 46:29–32, 1952.

[80] P. Erdös. Some remarks on subgroups of real numbers. *Colloq. Math.*, 42:119–120, 1979.

[81] M. Esser and O. Shisha. A modified differentiation. *American Math. Monthly*, 71:904–906, 1964.

[82] M. J. Evans. On continuous functions and approximate symmetric derivatives. *Colloq. Math.*, 31:129–136, 1974.

[83] M. J. Evans. A symmetric condition for monotonicity. *Bull. Math. Inst. Acad. Sinica*, 6:85–91, 1978.

[84] M. J. Evans. Symmetric and smooth functions: a few questions and fewer answers. *Real Anal. Exchange*, 9:381–385, 1983/84.

[85] M. J. Evans. Peano differentiation and high order smoothness in L_p. *Bull. Inst. Math. Acad. Sinica*, 13:197–209, 1985.

[86] M. J. Evans. Approximate smoothness of continuous functions. *Colloq. Math.*, 54:307–313, 1987/88.

[87] M. J. Evans. Points of approximate continuity, approximate symmetry and L-points. *Real Anal. Exchange*, 13:25–27, 1987/88.

[88] M. J. Evans. A note on symmetric and ordinary differentiation. *Real Anal. Exchange*, 17:820–826, 1991.

[89] M. J. Evans. Some theorems whose σ–porous exceptional sets are not σ–symmetrically porous. *Real Anal. Exchange*, 17:809–814, 1991.

[90] M. J. Evans and S. Fu. Mean value properties for symmetrically differentiable functions. *Real Anal. Exchange*, 17:657–667, 1991.

[91] M. J. Evans and P. Humke. Approximate continuity points and L–points of integrable functions. *Real Anal. Exchange*, 11:390–410, 1985/86.

[92] M. J. Evans, P. Humke, and K. Saxe. A symmetric porosity conjecture of Zajícek. *Real Anal. Exchange*, 17:258–271, 1985/86.

[93] M. J. Evans and P. D. Humke. On equality of unilateral derivates. *Proc. American Math. Soc.*, 79:609–613, 1980.

[94] M.-J. Evans and P.-D. Humke. A pathological approximately smooth function. *Acta Math. Hungar.*, 46:211–215, 1985.

[95] M. J. Evans and L. Larson. Qualitative differentiation. *Trans. American Math. Soc.*, 280:303–320, 1983.

[96] M. J. Evans and L. Larson. The continuity of symmetric and smooth functions. *Acta. Math. Hungar.*, 43:251–257, 1984.

[97] M. J. Evans and R. W. Vallin Qualititative symmetric differentiation. *Real Anal. Exchange* 18:575–585 1992/93.

[98] M. J. Evans and C. E. Weil. Peano derivatives: a survey. *Real Anal. Exchange*, 7:5–23, 1981/82.

[99] C. C. Ezzell and J. E. Nymann. An analogue of the Denjoy theorem for the symmetric derivative. *Rev. Roumaine Math. Pures Appl.*, 17:237 – 241, 1972.

[100] H. Fejzić. Convex functions and Schwarz derivates. *Proc. Amer. Math. Soc.* (to appear).

[101] H. Fejzić and C. E. Weil. A repair of the proof of Ash–Marcinkiewicz–Zygmund theorem. *Real Anal. Exchange* (to appear).

[102] F. M. Filipczak. Sur la structure de l'ensemble des points oú une fonction n'admet pas de dérivée symétrique. *Dissertationes Math. (Rozprawy Mat.)*, 130:1–49, 1975.

[103] F. M. Filipczak. Sur les dérivées symétriques des fonctions approximativement continues. *Colloq. Math.*, 34:249–256, 1976.

[104] L. Filipczak. Exemple d'une fonction continue privé de dérivée symétrique partout. *Colloq. Math.*, 20:249–253, 1969.

[105] J. Foran. Some relations of subgroups of the real numbers to Hausdorff measures. *J. London Math. Soc.*, (2) 7:651–661, 1974.

[106] J. Foran. The symmetric and ordinary derivative. *Real Anal. Exchange*, 2:105–108, 1977.

[107] J. Foran and L. Larson. The symmetric derivation basis, approximate symmetric derivatives and montonicity. *Real Anal. Exchange*, 13:278–284, 1987/88.

[108] M. Foran. Symmetric functions. *Real Anal. Exch.*, 1:38–40, 1976.

[109] L. R. Ford. Interval-additive propositions. *Amer. Math. Monthly*, 64:106–142, 1957.

[110] C. Freiling. A converse to a theorem of Sierpiński on almost symmetric sets. *Real Anal. Exchange*, 15:760–767, 1989/90.

[111] C. Freiling. Symmetric derivates, scattered and semi-scattered sets. *Trans. American Math. Soc.*, 318:705–720, 1990.

[112] C. Freiling and D. Rinne. A symmetric density property, monotonicity and the approximate symmetric derivative. *Proc. American Math. Soc.*, 104:1098–1102, 1988.

[113] C. Freiling and D. Rinne. A symmetric density property for measurable sets. *Real Anal. Exchange*, 14:203–209, 1988–89.

[114] C. Freiling and D. Rinne. Symmetric derivates of non-measurable functions. *Real Anal. Exchange*, 14:517–522, 1988–89.

[115] C. Freiling and D. Rinne. An approximate analog of a theorem of Khintchine. *Fund. Math.*, 135:55–59, 1990.

[116] C. Freiling, D. Rinne, and B. S. Thomson. A Riemann-type integral based on the second symmetric derivative. *(manuscript)*, 1993.

[117] C. Freiling and B. S. Thomson. Scattered sets, chains and the Baire cateegory theorem. *(manuscript)*, 1993.

[118] E. Freund. S-Null functions. *Real Anal. Exchange*, 16:319–321, 1990/91.

[119] H. Fried. Über die symmetrische Steigkeit von Funktionen. *Fund. Math.*, 29:134–137, 1937.

[120] W. H. Gage and R. D. James. A generalised integral. *Proc. Roy. Soc. Canada (4)*, 40:25–36, 1946.

[121] K. Garg. A new notion of derivative. *Real Analysis Exch.*, 75:65–84, 1981/82.

[122] N. Giovannelli. On a generalization of the k–pseudosymmetric derivative. *Atti Accad. Sci. Lett. Arti Palermo Ser. (5)*, 38:219 –229, 1978.

[123] C. Goffman. On Lebesgue's density theorem. *Proc. Amer. Math. Soc.*, 1:384–388, 1950.

[124] C. Goffman and C. J. Neugebauer. On approximate derivatives. *Proc. American Math. Soc.*, 11:962–966, 1960.

[125] E. Gorlich and R. Nessel. Uber Peano- und Riemann-Ableitungen in deer Norm. *Arch. Math. (Basel)*, 18:399–410, 1967.

[126] E. Goursat. Sur la définition générale des fonctions analytiques d'après cauchy. *Trans. Amer. Math. Soc.*, 1:14–16, 1900.

[127] G. Hamel. Eine Basis aller Zahlen und die unstetige Losungen der funktionalgleichung $f(x + y) = f(x) + f(y)$. *Math. Ann.*, 60:459–462, 1905.

[128] L. Hammill. On sets of symmetric points. *AMS Abstracts*, 13:46, 1992.

[129] F. Hausdorff. Problème No. 62. *Fund. Math.*, 25:578, 1935.

[130] R. Henstock. N–variation and N–variational integrals of set functions. *Proc. London Math. Soc.*, (3) 11:109–133, 1961.

[131] R. Henstock. *Linear Analysis*. Butterworths, London, 1967.

[132] E. Hewitt and K. R. Stromberg. Some examples of nonmeasurable sets. *J. Austrail. MAth. Soc.*, 18:236–238, 1974.

[133] E. W. Hobson. *The theory of functions of a real variable and the theory of Fourier series I, II (2nd Ed.)*. Cambridge, Cambridge, 1926.

[134] P. D. Humke and M. Laczkovich. Monotonicity theorems for generalized Riemann derivatives. *Rend. Circ. Mat. Palermo (2)*, 38:437–454, 1989.

[135] P. D. Humke and M. Laczkovich. An elementary proof of Freiling's symmetric covering lemma. *Real Anal. Exchange*, 16:549–551, 1990/91.

[136] P. D. Humke and M. Laczkovich. Parametric semicontinuity implies continuity. *Real Anal. Exchange*, 17:668–680, 1991/92.

[137] P. D. Humke and T. Salát. Remarks on strong and symmetric differentiability of real functions. *Acta Math. univ. Comenian.*, 52/53:235–241, 1987.

[138] R. D. James. A generalised integral (II). *Canad. J. Math.*, 2:297–306, 1950.

[139] R. D. James. Integrals and summable trigonometric series. *Bull. American Math. Soc.*, 61:1–15, 1955.

[140] J. Jaskula and B. Szkopińska. On the set of points of symmetric continuity. *An. Univ. Bucureşti Mat.*, 37:29–35, 1988.

[141] J. M. Jedrzewski. Approximately smooth functions. *Zeszyty Nauk. Uniw. Lodzk. Nauki Mat. Przyrod. Ser. II*, 52:7–14, 1973.

[142] R. L. Jeffrey. Trigonometric series. *Canad. Math. Congress Lecture Series (#2)*, 1956.

[143] J. L. W. V. Jensen. Om konvekse funktioner og uligheder imellem middel-vaerdier. *Nyt. Tidsskrift for Mathematik*, 16 B:49–69, 1905.

[144] F. John and L. Nirenberg. On functions of bounded mean oscillation. *Comm. Pure Appl. Math.*, 14:415–426, 1961.

[145] F. B. Jones. Measure and other properties of a Hamel basis. *Bull. Amer. Math. Soc.* 48:472–481, 1942.

[146] G. Jungck. Interval induction. *Amer. Math. Monthly*, 73:295–297, 1966.

[147] J.-P. Kahane. Trois notes sur les ensembles parfaits linéaires. *Enseignemant Math.*, 15:187–192, 1969.

[148] I. P. Kārkliņa. On quasi-smooth functions. *Latvijas Valsts Univ. Zinatn. Raksti*, 41:43–46, 1961.

[149] C. Kassimatis. Functions which have generalized Riemann derivatives. *Canad. J. Math.*, 10:413–420, 1958.

[150] J. H. B. Kemperman. A general functional equation. *Trans. American Math. Soc.*, 86:28–56, 1957.

[151] A. Khintchine. Recherches sur la structure des fonctions mesurables. *Fund. Math.*, 9:212–279, 1927.

[152] P. Komjáth and S. Shelah. On uniformly antisymmetric functions. *Real Anal. Exchange*, 19(1), 1993/94.

[153] P. Kostyrko. On the symmetric derivative. *Colloq. Math.*, 25:265 – 267, 1972.

[154] P. Kostyrko. On a local form of Jensen's functional equation. *Aequationes Math.*, 30:65–69, 1986.

[155] P. Kostyrko. The symmetric derivative and the Darboux property. *Real Anal. Exchange*, 13:305–309, 1987/88.

[156] P. Kostyrko. Approximate symmetric derivatives are uniformly closed. *Real Anal. Exchange*, 17:652–656, 1991.

[157] P. Kostyrko. On locally symmetric and locally Jensen functions. *Real Anal. Exchange*, 17:405–408, 1991.

[158] P. Kostyrko, T. Neubrunn, Šalát T., and J. Smital. On locally symmetric and symmerically continuous functions. *Real Anal. Exchange*, 6:67–76, 1980.

[159] B. D. Kotljar. On quasi-smooth functions of two variables. *Ukrain. Mat. Z.*, 16:383–385, 1964.

[160] Y. Kubota. An integral with basis and its application to trigonometric series. *Bull. Fac. Sci. Ibaraki Univ. Ser. A*, 5:1–8, 1973.

[161] M. Kuczma. *An introduction to the theory of functional equations and inequalities.* Uniwersytet Ślaski, Warsaw, 1985.

[162] M. Kulbacka. Sur les ensembles stationaires et déterminants pour certaines classes de dérivées symétriques. *Colloq. Math.*, 19:255–259, 1968.

[163] N. K. Kundu. On some conditions of symmetric derivatives implying monotonicity of functions. *Rev. Roumaine Math. Pures Appl.*, 15:561–568, 1970.

[164] N. K. Kundu. On approximate symmetric derivative. *Colloq. Math.*, 28:275–285, 1973.

[165] N. K. Kundu. On some properties of symmetric derivatives. *Ann. Polon. Math.*, 30:9–18, 1974.

[166] K. Kuratowski. *Topology.* Academic Press, London, 1966.

[167] S. Kurepa. Convex functions. *Glasnik Mat. Fiz. Astronom.*, (2) 11:89–93, 1956.

[168] J. Kurzweil and J. Jarník. On some extensions of the Perron integral on one-dimensional intervals. An approach by integral sums fulfilling a symmetry condition. *Funct. Approx. Comment. Math.*, 17:49–55, 1987.

[169] J. Kurzweil and J. Jarník. On a generalization of the Perron integral on one-dimensional intervals. *Ann. Polon. Math.*, 51:205–218, 1990.

[170] M. Laczkovich. A Baire two function with non-Borel upper symmetric derivative. *Real Anal. Exchange*, 13:258–264, 1987/88.

[171] L. Larson. The Baire class of approximate symmetric derivates. *Proc. American Math. Soc.*, 87:125–130, 1983.

[172] L. Larson. The symmetric derivative. *Trans. American Math. Soc.*, 277:589–599, 1983.

[173] L. Larson. Symmetric real analysis: a survey. *Real Anal. Exchange*, 9:154–178, 1983/84.

[174] L. Larson. The generalized Zahorski class structure of symmetric derivatives. *Trans. American Math. Soc.*, 282:45–58, 1984.

[175] L. Larson. A method for showing generalized derivatives are in Baire class one. *Classical real analysis, Series: Amer. Math. Soc., Providence, R.I.*, Contemp. Math., 42.:87–95, 1985.

[176] L. Larson. Approximate smoothness and Baire∗1. *Acta. Math. Hungar.*, 57:187–200, 1991.

[177] C. M. Lee. On the integrals of Perron type. *Thesis*, (U. B. C.), 1972.

[178] C. M. Lee. On integrals and summable trigonometric series. *RAE*, 4:66–68, 1978.

[179] C. M. Lee. Generalizations of Cesàro continuous functions and integrals of Perron type. *Trans. Amer. Math. Soc.*, 266:461–481, 1981.

[180] C. M. Lee. A symmetric approximate Perron integral for the coefficient problem of convergent trigonometric series. *Real Anal. Exchange*, 16:329–339, 1990.

[181] N. Lusin. Sur l'absolue convergence des séries trigonométrique. *C. R. Acad. Sci. Paris Sér. I Math.*, 155:580–582, 1912.

[182] N. Lusin. Sur une théorème fondamental du calcul intégral. *Mat. Sbornik*, 28:266–294, 1912.

[183] W. A. J. Luxemburg. Problem 68. *Niew Archief voor Wiskund*, p. 121, 1965.

[184] N. G. Makarov. On the radial behaviour of Bloch functions. *Soviet Math. Dokl.*, 40:505–508, 1990.

[185] J. Marcinkiewicz. Sur les séries de Fourier. *Fund. Math.*, 27:38–69, 1936.

[186] J. Marcinkiewicz and A. Zygmund. On the differentiability of functions and the summability of trigonometrical series. *Fund. Math.*, 26:1–43, 1936.

[187] S. Marcus. Mulţimile F_σ şi continuitatea simetrică. *Bul. Şti. Acad. Rep. Pop. Rom. Seria Matem. Fiz.*, 7(4):871–886, 1955.

[188] S. Marcus. Sur un problème de F. Hausdorff concernant les fonctions symmétriques continues. *Bull. Acad. Pol. Sci. (classe III)*, 4(4):201–205, 1956.

[189] S. Marcus. Sur un probleme de la theorie de la mesure de H. Steinhaus et S. Ruziewicz. *Bull. Acad. Plon. des Sciences*, 4:197–199, 1956.

[190] S. Marcus. On anticonvex sets. *Rev. Roumaine Math. Pures Appl.*, 13(9):1399–1401, 1968.

[191] S. Marcus. Symmetry in one dimension and symmetric continuity. *Noesis*, 12:98–103, 1986.

[192] S. Marcus. Symmetry in the simplest case: the real line. *Comput. Math. Appl.*, 17:103–115, 1989.

[193] Solomon Marcus. Fonctions convexes et fonctions internes. *Bull.-Sci.-Math.*, 81:66–70, 1957.

[194] J. Matoušek. A typical property of the symmetric differential quotient. *Coll. Math.*, 57:339–343, 11989.

[195] J. Matoušek. Approximate symmetric derivative and monotonicity. *Comment. Math. Univ. Carolin.*, 27:83–86, 1986.

[196] J. Mařík. *Generalized integrals and trigonometric series.* (unpublished manuscript).

[197] S. Mazurkiewicz. On the first generalized derivative (in Polish). *Prace Matematyczno-Fizyczne*, 28:79–85, 1917.

[198] S. Mazurkiewicz. On the relation between the existence of the second generalized derivative and the continuity of a function (in Polish). *Prace Matematyczno-Fizyczne*, 30:225–242, 1919.

[199] J. McGrotty. A theorem on complete sets. *J. London Math. Soc.*, 37:338–340, 1962.

[200] M. R. Mehdi. On convex functions. *JLMS*, 39:321–326, 1964.

[201] J. C. Morgan. *Point set theory.* Marcel Dekker, New York, 1990.

[202] M. I. Morozov. On the question of the approximation of periodic quasi-smooth functions and functions satisfying a Lipschitz condition. *Aviacion. Inst. Sergo Ord. Trudy Inst.*, 61:41–47, 1956.

[203] R. M. F. Moss and G. T. Roberts. A creeping lemma. *Amer. Math. Monthly*, 75:649–652, 1968.

[204] S. N. Mukhopadhyay. On Schwartz differentiability, II. *Rev. Roumaine Math. Pures Appl.*, 9:859–869, 1964.

[205] S. N. Mukhopadhyay. Some results on Schwartz derivative. *Indian J. Mech. Math.*, 3:1–6, 1965.

[206] S. N. Mukhopadhyay. On approximate Schwartz differentiability, IV. *Monatsh. Math.*, 70:454–460, 1966.

[207] S. N. Mukhopadhyay. On Schwartz differentiability, I. *Proc. Indian Nat. Sci. Acad. Part A*, 36:525–533, 1966.

[208] S. N. Mukhopadhyay. On differentiability. *Bull. Calcutta Math. Soc.*, 59:181–183, 1967.

[209] S. N. Mukhopadhyay. On the regularity of the P^n-integral and its application to summable trigonometric series. *Pacific J. Math.*, 55:233–247, 1974.

[210] I. Natanson. *Theory of functions of a real variable.* Ungar Publishing Co., New York, 1964.

[211] C. J. Neugebauer. A theorem on derivates. *Acta Sci. Math. (Szeged)*, 23:79–81, 1962.

[212] C. J. Neugebauer. Smoothness and differentiability in l_p. *Studia Math.*, 25:81–91, 1964.

[213] C. J. Neugebauer. Symmetric, continuous and smooth functions. *Duke Math. J.*, 31:23–32, 1964.

[214] C. J. Neugebauer. Differentiability almost everywhere. *Proc. American Math. Soc.*, 1205–1210, 1965.

[215] R. J. O'Malley. Approximate maxima. *Fund. Math.*, 94:77–83, 1976.

[216] R. J. O'Malley. Baire* 1 functions. *Proc. American Math. Soc.*, 60:187–192, 1976.

[217] D. Oppegaard Generalizations of continuity, symmetry and symmetric continuity. *Thesis, North Carol. State Univ.*, 1993.

[218] A. Ostrowski. Zur Theorie der konvexen Funktionen. *Comment. Math. Helv.*, 1:157–159, 1929.

[219] O. Perron. Ueber den Integralbegriff. *Sitzber. Heidelberg, Akad. Wiss. Abt. A.*, 16:1–16, 1914.

[220] I. N. Pesin. On the measurability of symmetrically continuous functions. *Teor. Funkcii Funkcional. Anal. i Prilozen*, 5:99–101, 1967.

[221] G. Piranian. Two monotonic, singular, uniformmly almost smooth functions. *Duke Math. J.*, 33:255–262, 1966.

[222] I. Pokorny. Symmetric selective derivatives. *Acta Univ. Carolin. Math. Phys.*, 99–100, 1987.

[223] V. G. Pokrovskii, Slicings of n-dimensional parallelopipeds. *Mat. Zametki*, 33(2):273–280, 1983.

[224] S. P. Ponomarev. On the measurability of symmetrically continuous functions. *Uspehi Mat. Nauk.*, 20:149–150, 1965.

[225] S. P. Ponomarev. O simetričeski něpreryvnych funkcijach. *Akademiya Nauk Soyuza SSR. Matematicheskie Zametki*, 1:385–390, 1967.

[226] S. P. Ponomarev. On the monogeneity of symmetrically differentiable functions. *Teoret. Prikl. Mat. Vip.*, 2:11–14, 1967.

[227] S. P. Ponomarev. On a problem of Hausdorff. *Akademiya Nauk Soyuza SSR. Matematicheskie Zametki*, 14:197–200, 1973.

[228] S. P. Ponomarev. Symmetrically differentiable functions with a perfect set of points of nondifferentiability. *Akademiya Nauk Soyuza SSR. Matematicheskie Zametki*, 38:80–88, 1985.

[229] S. P. Ponomarev. Functions with a D_*-integrable symmetric derivative. *Akademiya Nauk Soyuza SSR. Matematicheskie Zametki*, 39:221–227, 1986.

[230] D. Preiss. A note on symmetrically continuous functions. *Časopis Pěst. Mat.*, 96:262–264, 1971.

[231] D. Preiss and B. Thomson. A symmetric covering theorem. *Real Anal. Exchange*, 14:253–254, 1988/89.

[232] D. Preiss and B. Thomson. The approximate symmetric integral. *Canad. J. Math.*, 41:508–555, 1989.

[233] H. H. Pu and H. W. Pu. Derivates for symmetric functions. *Colloq. Math.*, 46:293–301, 1982.

[234] H.-H. Pu and H.-W. Pu. A generalization of Khintchine's theorem. *Tamkang J. Math.*, 17:57–61, 1986.

[235] H. H. Pu and H. W. Pu. On level sets of approximate derivative and approximate symmetric derivative. *Colloq. Math.*, 54:139–142, 1989.

[236] H. H. Pu and H. W. Pu. Pseudo-symmetric differention. *Colloq. Math.*, 57:137–143, 1989.

[237] H.-H. Pu, H.-W. Pu, and T.-H. Teng. Symmetric and approximate symmetric derivates for symmetric functions. *Soochow J. Math.*, 5:1–7, 1979.

[238] H. W. Pu. Derivates for symmetric functions. *Real Anal. Exchange*, 4(2):164–166, 1978/79.

[239] H. W. Pu and H. H. Pu. Dini's type theorem for symmetric derivatives. *Rev. Roumaine Math. Pures Appl.*, 18:1087–1090, 1973.

[240] H. W. Pu and H. H. Pu. Measurability of real functions having symmetric derivatives everywhere. *Real Anal. Exchange*, 10:214–219, 1984.

[241] H. W. Pu and H. H. Pu. A problem of Denjoy-Young-Saks theorem for symmetric derivates. *Rev. Roumaine Math. Pures Appl.*, 33:517–518, 1988.

[242] A. Rajchman. Séries trigonométrique sommables par le procédé de Poisson (in Polish). *Prace Mat.-Fiz.*, 30:19–88, 1919.

[243] M. Repický. An example which discerns porosity and symmetric porosity. *Real Anal. Exchange*, 17:416–420, 1991.

[244] M. Riesz. Uber summierbare trigonometrische Reihen. *Math. Annalen*, 71:54–75, 1911.

[245] A. W. Roberts and D. E. Varberg. *Convex functions*. Academic Press, New York, 1973.

[246] C. A. Rogers. *Hausdorff Measures*. Cambridge University Press, Cambridge, 1970.

[247] G. Russo. Sui numeri derivati approssimati simmetrici di une funzioune. *Atti Accad. Sci. Lett. Arti Palermo Ser. (5)*, 4:185–196, 1984.

[248] G. Russo and S. Valenti. On the k-pseudo-symmetrical differentiability. *Fund. Math.*, 94:79–83, 1981.

[249] S. Ruziewicz. Contribution à l'étude des ensembles de distances de points. *Fund. Math.*, 7:141–143, 1925.

[250] I. Z. Ruzsa. Locally symmetric functions. *Real Anal. Exchange*, 4:84–86, 1978/79.

[251] S. Saks. *Theory of the Integral*. Monografie Matematyczne 7, Warsaw, 1937.

[252] D. Sarason. Functions of vanishing mean oscillation. *Trans. American Math. Soc.*, 207:391–405, 1975.

[253] A.-I. Savotin. Some cases of the nondifferentiability of continuous functions that can be represented by series. *Applications of functional analysis in approximation theory. Kalinin. Gos. Univ.*, 90–97, 1987.

[254] P. Shanahan. A unified proof of several basic theorems of real analysis. *Amer. Math. Monthly*, 79:890–891, 1972.

[255] H. S. Shapiro. Monotonic singular functions of high smoothness. *Michigan Math. J.*, 15:265–275, 1968.

[256] W. Sierpinski. Sur les fonctions convexes mesurables. *Fund. Math.*, 1:125–128, 1920.

[257] W. Sierpinski. Sur une hypothese de S. Mazurkiewicz. *Fund. Math.*, 11:148–150, 1928.

[258] W. Sierpinski. Sur une fonction non mesurable partout presque symétrique. *Acta. Litt. Scient (Szeged)*, 8:1–6, 1936.

[259] V. A. Skljarenko. Certain properties of the P^2-primitive. *Mat. Zametki*, 12:693–700, 1972.

[260] V. A. Skljarenko. On Denjoy integrable sums of everywhere convergent trigonometric series. *Soviet Math. Dokl.*, 210:771–775, 1973.

[261] V. A. Skljarenko. Integration by parts in the SCP Burkill integral. *Mat. Sb. (N.S.)*, 112 (154), 4(8):630–646, 1980.

[262] V. A. Skvorcov. Interrelation between general Denjoy integrals and totalization $(T_{2S})_0$. *Mat. Sb. (N.S.)*, 52 (94):551–578, 1960.

[263] V. A. Skvorcov. Interelation between Denjoy's general integral and the totalization T_{2S}). *Dokl. Akad. Nauk SSR*, 127:975–976, 1962.

[264] V. A. Skvorcov. Interrelation between the general denjoy integral and the totalization $t_{2S})_0$. *Mat. Sb. (N.S.)*, 52:551–578, 1962.

[265] V. A. Skvorcov. On the relation between the D integral and the totalization T_{2S}). *Vestnik Moskov. Univ. Ser. I Mat. Mekh.*, no. 6:20–25, 1962.

[266] V. A. Skvorcov. On integrating the exact Schwarzian derivative. *Mat. Sb. (N.S.)*, 63 (105):329–340, 1964.

[267] V. A. Skvorcov. Concerning definitions of P^2- and SCP-integrals. *Vestnik Moskov. Univ. Ser. I Mat. Mekh.*, 21:12–19, 1966.

[268] V. A. Skvorcov. The mutual relationship between the AP-integral of Taylor and the P^2-integral of James. *Mat. Sb. (N.S.)*, 70 (112):380–393, 1966.

[269] V. A. Skvorcov. The connection between certain integrals. *Vestnik Moskov. Univ. Ser. I Mat. Mekh.*, 22:68–72, 1967.

[270] V. A. Skvorcov. The Marcinkiewicz–Zygmund integral and its relation to Burkill's SCP-integral. *Vestnik Moskov. Univ. Ser. I Mat. Mekh.*, 27:78–82, 1972.

[271] V. A. Starcev. The smoothness of functions with respect to a set. *Mat. Zametki*, 15:431–436, 1974.

[272] E. M. Stein. *Singular Integrals and Differentiability Properties of Functions*. Princeton University Press, Princeton, New Jersey, 1970.

[273] E. M. Stein. The development of square functions in the work of A. Zygmund. *Conference on Harmonic Analysis in honor of Antoni Zygmund (Wadsworth)*, 1–30, 1983.

[274] E. M. Stein and A. Zygmund. On the differentiability of functions. *Studia Math.*, 23:295–307, 1960.

[275] E. M. Stein and A. Zygmund. Smoothness and differentiability of functions. *Ann. Univ. Sci. Budapest. Eötvös Sect. Math.*, III–IV:295–307, 1964.

[276] H. Steinhaus. Problème. *Fund. Math.*, 1: 1920.

[277] H. Steinhaus. Problème #23. *Fund. Math.*, p. 368, 1923.

[278] K. R. Stromberg. Universally nonmeasurable subgroups of ℝ. *Amer. Math. Monthly*, March:253–255, 1992.

[279] E. Szpilrajn. Remarque sur la dérivée symmétrique. *Fund. Math.*, 21:226–228, 1933.

[280] S. J. Taylor. An integral of Perron's type defined with the help of trigonometric series. *Quart. J. Math. Oxford Ser. (2)*, (2) 6:255–274, 1955.

[281] B. S. Thomson. On full covering properties. *Real Anal. Exchange*, 6:77–93, 1980/81.

[282] B. S. Thomson. *Real Functions*. Lecture Notes in Mathematics 1170. Springer-Verlag, Berlin, 1985.

[283] B. S. Thomson. Derivation bases on the real line. I. II. *Real Anal. Exchange*, 8:67–207, 278–442, 1986.

[284] B. S. Thomson. Some symmetric covering lemmas. *Real Anal. Exchange*, 15:346–383, 1989.

[285] B. S. Thomson. An analogue of Charzyński's theorem. *Real Anal. Exchange*, 15:743–753, 1989/90.

[286] B. S. Thomson. Derivates of interval functions. *Mem. Amer. Math. Soc.*, 93, no. 452, 96 pp., 1991.

[287] B. S. Thomson. Symmetric variation. *Real Anal. Exchange*, 17:409–415, 1991/92.

[288] B. S. Thomson. The range of symmetric derivatives. *Real Anal. Exchange*, (to appear) 1993.

[289] A. F. Timan. Quasi-smooth functions. *Uspehi Matem. Nauk (N.S.)*, 5:128–130, 1950.

[290] A. F. Timan. On quasi-smooth functions. *Izvestiya Akad. Nauk-SSSR Ser. Mat.*, 15:243–254, 1951.

[291] L. Tonelli. Su una proposizione fondamentale dell'analisi matematica. *Rend R. Accad. die Lincei*, 23:161–165, 1936.

[292] T. C. Tran. Symmetric functions whose set of points of discontinuity is uncountable. *Real Anal. Exchange*, 12:496–508, 1986/87.

[293] J. Uher. Symmetrically differentiable functions are differentiable almost everywhere. *Real Anal. Exchange*, 8:253–260, 1982/83.

[294] J. Uher. Symmetric continuity implies continuity. *Trans. American Math. Soc.*, 293:421–429, 1986.

[295] J. Uher. Symmetric continuity implies continuity. *Real Anal. Exchange*, 13:35–38, 1987/88.

[296] D. C. Ullrich. Differentiation of Zygmund functions. *Proc. Amer. Math. Soc.*, 117:195–197, 1993.

[297] S. Valenti. Sur la derivation k-pseudo-symetrique des fonctions numeriques. *Fund. Math.*, 74:147–152, 1972.

[298] R. W. Vallin. Shell porous, porous, totally porous, hyperporous sets. *Real Anal. Exchange*, 18:294–320, 1992/93.

[299] S. Verblunsky. The generalized fourth derivative. *J. London Math. Soc.*, 6:82–84, 1931.

[300] S. Verblunsky. On the theory of trigonometrical series (VI). *Proc. London Math. Soc.*, 38:284–326, 1934.

[301] S. Verblunsky. On the theory of trigonometrical series (VII). *Fund. Math.*, 23:192–236, 1934.

[302] T. Viola. Sur la détermination d'une fonction discontinue par sa dérivée unilatérale. *Ann. École Norm.*, (3) 50:72–125, 1933.

[303] E. Wajch. On symmetric derivatives of functions of the first class of Baire. *Demonstratio Math.*, 19:189–195, 1986.

[304] C. E. Weil. Monotonicity, convexity and symmetric derivatives. *Trans. American Math. Soc.*, 222:225–237, 1976.

[305] C. E. Weil. The Peano derivative: what's known and what isn't. *Real Anal. Exchange*, 9:354–365, 1983/84.

[306] M. Weiss. On symmetric derivatives in L^p. *Studia Math.*, 24:89–100, 1964.

[307] M. Weiss and A. Zygmund. A note on smooth functions. *Konin. Nederlandse Akad. van Wetenschappen Proc.*, 62:52–58, 1959.

[308] W. H. Young and G. C. Young. On the reduction of sets of intervals. *Proc. London Math. Soc. (2)*, 14:111–130, 1915.

[309] Z. Zahorski. Sur la première derivée. *Trans. American Math. Soc.*, 69:1–54, 1950.

[310] L. Zajíček. Porosity and σ-porosity. *Real Anal. Exchange*, 13:314–350, 1987/88.

[311] L. Zajíček. A note on the symmetric and ordinary derivative. *(manuscript)*, 1991.

[312] A. Zygmund. Smooth functions. *Duke Math. J.*, 12:47–76, 1945.

[313] A. Zygmund. *Trigonometric series.* Cambridge University Press, London, 1968.

[314] A. Zygmund. Selected Papers of Antoni Zygmund, edited by A. Hulanicki, P. Wojtaszczyk, and W. Zelazko. *Kluwer Academic Publ. (London)*, 1989.

Index

Milton Keynes UK
Ingram Content Group UK Ltd.
UKHW021902071024
449327UK00021B/1604